CHIRALITY IN DRUG DESIGN AND DEVELOPMENT

CHIRALITY IN DRUG DESIGN AND DEVELOPMENT

EDITED BY

INDRA K. REDDY
Texas A&M University School of Pharmacy
Kingsville, Texas, U.S.A.

REZA MEHVAR
Texas Tech University Health Sciences Center
Amarillo, Texas, U.S.A.

MARCEL DEKKER, INC.

NEW YORK · BASEL

FIRST INDIAN REPRINT, 2009

Although great care has been taken to provide accurate and current information, neither the author(s) nor the publisher, nor anyone else associated with this publication, shall be liable for any loss, damage, or liability directly or indirectly caused or alleged to be caused by this book. The material contained herein is not intended to provide specific advice or recommendations for any specific situation.

Trademark notice: Product or corporate names may be trademarks or registered trademarks and are used only for identification and explanation without intent to infringe.

Library of Congress Cataloging-in-Publication Data
A catalog record for this book is available from the Library of Congress.

ISBN: 0-8247-5062-4

Headquarters
Marcel Dekker, Inc., 270 Madison Avenue, New York, NY 10016, U.S.A.
tel: 212-696-9000; fax: 212-685-4540

Distribution and Customer Service
Marcel Dekker, Inc., Cimarron Road, Monticello, New York 12701, U.S.A.
tel: 800-228-1160; fax: 845-796-1772

Eastern Hemisphere Distribution
Marcel Dekker AG, Hutgasse 4, Postfach 812, CH-4001 Basel, Switzerland
tel: 41-61-260-6300; fax: 41-61-260-6333

World Wide Web
http://www.dekker.com

The publisher offers discounts on this book when ordered in bulk quantities. For more information, write to Special Sales/Professional Marketing at the headquarters address above.

Foreword

The importance of stereochemistry in drug research and development as well as pharmacotherapy is well acknowledged. This is mainly due to a collective understanding of scientists involved in the area in the 1980s and 1990s. The main emphasis of the activities was on the potential therapeutic benefits of these new findings. Additionally, the industry was also exploring the intellectual properties involved. The potential benefit of developing stereochemically pure drugs from available racemates occupied many minds. This was thought to have two main advantages: first, the stereochemically pure drug was intuitively presumed to be superior to the racemate, particularly when beneficial effects were mainly attributable to one of the enantiomers; second, the development of the single enantiomer was to provide an opportunity to expand market exclusivity. In both instances, it was thought that the availability of the vast amount of data already generated from the use of racemate would facilitate the development of the enantiomer. In this context, the use of "bridging" data was an essential part of all discussions. The interest in the development of stereochemically pure drugs and the belief that they may provide safer and more efficacious alternatives to racemates resulted in the introduction of guidance by many regulatory agencies. In some of these guidance documents, sponsors of racemic drugs were required to provide convincing data as to why the racemate and not the enantiomer was being developed. This clearly reflected the belief of the time that the single enantiomer was intuitively superior to the racemate, a notion that, with a few exceptions, has not been supported with many robust clinical data. While the belief of superiority of single enantiomers was the motive behind a great deal of activity, the potential complication involved in stereochemically pure drugs was, for some time, a deterrent in choosing chiral molecules for development of new medicinal products. The term "stereophobia" has been used to reflect the feeling of the time.[1] Now that the dust has settled, we can say that the single enantiomer does not necessarily provide a better alternative to the racemate. Indeed, except for a few examples, the effort put into development of single enantiomers to switch from racemic drugs has been found to be fruitless for many reasons including lack of clinically significant therapeutic advantages and overestimation of the value of the bridging data. Thus, the interest in developing single enantiomers of racemic drugs has subsided over the years. It follows that the opportunity to generate robust clinical data comparing enantiomers with racemate no longer exists since new chiral molecules are likely to be developed as stereochemically pure drugs.

Whether the emergence of regulatory guidance and the resultant stereophobia prevented development of useful racemic drugs is hard to discern. Nevertheless, presently the stereochemical considerations in drug action and disposition are well recognized in all steps of research and development. In addition, the emergence of more sophisticated chiral separation and synthesis techniques has helped to eliminate the stereophobia of the 1980s and 1990s.

Stereochemistry of the molecule must be carefully considered in all fields of pharmaceutical sciences ranging from discovery to the bedside. An understanding of the role of chirality in the properties of the molecule of interest is undoubtedly a main key to the rational process of developing as well as clinical use of chiral drugs. Keeping in mind the issue of "racemate versus single enantiomer" during various steps of drug discovery and development may save money and time.

Physicochemical properties of a racemate may be drastically different from the enantiomers depending on the nature of the former.[2] Chirality may also influence drug delivery because a single enantiomer or a non-racemic blend may have improved solubility, dissolution, and stability (see Chapter 1). In addition, many available pharmaceutical excipients (e.g., cellulose and its derivatives) either naturally occur as single enantiomers or are derivatives of the latter chiral molecules. These stereochemically pure molecules may interact with other chiral molecules (i.e., the active ingredient) and form stereoisomers. The latter will have physicochemical properties different from the original chiral molecule (see Chapter 2). For example, the presence of heptakis(2,6-di-O-ethyl)-beta-cyclodextrin results in stereoselective dissolution of tiaprofenic acid.[3] While this stereoselective release did not result in stereoselective bioavailability, it highlights the potential implication of the effect of chirality on physicochemical properties of drugs.

Similar to the solid dosage forms containing chiral excipients, biological membranes may provide chiral environments (see Chapter 3). Most drugs cross the gastrointestinal membrane through simple passive diffusion; thus, no stereoselectivity in the process is expected. It appears, however, that stratum corneum possesses chiral discrimination properties. Indeed, in vitro data indicate significant stereoselectivity in skin penetration of ketoprofen and propranolol. In addition, preferential transport and bioavailability of racemic drugs has been achieved by administration of prodrugs (see Chapter 4). Although the therapeutic significance of these findings is not established, the available data point to their potential application.

The two main sources of stereoselectivity in drug disposition are the circulatory proteins and enzymes in both the gastrointestinal tract and the liver. Both binding of drugs to proteins and metabolism by various isozymes are, therefore, often stereoselective. Many examples of stereoselective systemic clearance and presystemic metabolism exist (see Chapters 6 and 7). For a few classes of drugs, metabolism may include chiral inversion (see Chapter 8). This, if unidirectional, adds to the overall stereoselectivity in disposition of drugs. Bidirectional bioinversion (see Chapter 8), on the other hand, similar to chemical racemization (thalidomide, see Chapter 5), may diminish stereoselectivity.

Enantiomers may interact with one another at both pharmacokinetic and pharmacodynamic levels.[4] One of the earlier and more interesting examples of such interactions was that of propoxyphene enantiomers.[5] The analgesic activity of propoxyphene is due mainly to the *d* enantiomer. However, at least in the rat, the racemate is more potent than an equimolar dose of *d*-propoxyphene due to a reduced clearance of the latter caused by the presence of *l* enantiomer in the racemate.

Use of stereospecific data in determination of bioequivalence between products of the same racemic drug has been the topic of many discussions (see Chapter 9). The issue, however, is mainly focused on relatively older drugs for which the approval of the brand product has been based on nonstereospecific data. More recent new drug applications, however, either deal with single enantiomers or contain stereospecific data. In the former case (single enantiomers), and in the absence of chiral conversion, stereospecific approaches become nonissues. However, if the approval of the brand name is based on stereospecific data, stereochemistry in bioequivalence should be considered. The regulatory agencies in various countries have provided guidance in developing safer and efficacious products of streochemically pure drugs or racemates (see Chapter 10).

Stereochemical aspects of drug action have been known for many decades. It is, however, only within the last couple of decades that emphasis has been placed on stereochemistry of drug disposition. The area that still needs far more attention is the effect of chirality on the physicochemical properties of drugs. Nevertheless, the present state of knowledge of the area should assist us to provide better therapeutic interpretations and develop better drugs.

Fakhreddin Jamali, Ph.D.
University of Alberta
Edmonton, Alberta, Canada

REFERENCES

1. Ariens, E.J.; The acceptable face of international competition. Stereophobia: an economic afterthought. Trends Pharmacol. Sci. **1988**, *9*, 317.
2. Mitchell, A. G. Racemic Drugs: Racemic Mixture, Racemic Compound, or Pseudoracemate? J. Pharm. Pharm. Sci. **1998**, *1*, 8–12.
3. Vakily, M., Pasutto, F.M.; Daneshtalab, M.; Jamali, F. Inclusion complexation of heptakis(2,6-di-O-ethyl)-beta-cyclodextrin with tiaprofenic acid: pharmaco-kinetic consequences of a pH-dependent release and stereoselective dissolution. J. Pharm. Sci. **1995**, *84*, 1014–1019.
4. Berry, B.W.; Jamali, F. Enantiomeric interaction of flurbiprofen in the rat. J. Pharm. Sci. **1989**, *78*, 632–634.
5. Murphy, P.J.; Nickander, R.C.; Bellamy, G.M.; Kurtz, W.L. Effect of l-propoxyphene on plasma levels and analgesic activity of *d*-propoxyphene in the rat. J. Pharmacol. Exp. Ther. **1976**, *199*, 415–422.

Preface

The stereoisomeric composition of drug substances has become a critical issue in the development, approval, and clinical use of drugs. Over the past three decades, stereoselectivity in drug action and disposition has become a well-recognized consideration in clinical pharmacology and product development of chiral drugs. Although the initial focus of many investigations in this field was on implications of stereoselectivity for pharmacologic action and, later on, the pharmacokinetics of chiral drugs, more recently other aspects of chirality such as physicochemical properties and formulation issues have also been explored. Additionally, after much debate, the regulatory agencies have provided guidelines for the development and characterization of chiral drugs. However, the information currently available on many relevant issues of chirality, including the recent developments, though extensive, is still fragmented into various disciplines, making meaningful dissemination of knowledge difficult. This is especially important because today more than 50% of marketed drugs are chiral. Therefore, we took a multidisciplinary approach to create this volume so that it may be used as a reference book for the research community and pharmaceutical industries interested in this area or as an educational tool for medical and health-related practitioners.

The choice of developing a drug as a racemate or an individual enantiomer must be based on a sound and critical evaluation of chiral characteristics with respect to their pharmacodynamic, pharmacokinetic, and toxicological considerations. This book entails the relevant topics in chiral drug product development, including an overview of crystal structure and physical properties of chiral drugs, the interaction of chiral drugs with chiral excipients in formulation, pharmaceutical considerations in transdermal delivery, stereoselective drug delivery through a prodrug approach, pharmacokinetics and dynamics, and regulatory issues of chiral drugs with regard to product development. Recent developments in these areas such as the use of nonconventional ratios of isomers are also discussed.

This book serves as a comprehensive review of chirality with respect to biopharmaceutic, pharmacologic, pharmacokinetic, bioequivalence, and regulatory issues. Experts from academia, industry, and/or regulatory agencies have written all chapters of the book. Each chapter presents a detailed discussion of the implications of chirality in various interdisciplinary areas of drug development and integrates the knowledge from various applied areas. The last chapter presents a concise discussion of the "enantiomer versus racemate" debate and provides some regulatory perspectives and guidance on drug product development of chiral molecules. Further, a thorough and thoughtful foreword, from Dr. Fakhreddin (Mo) Jamali, a pioneering researcher in this area, sets the tone for this volume.

This book is targeted toward pharmaceutical scientists in academia and industry, pharmacologists, pharmacokineticists, toxicologists, and graduate students in various research-intensive programs who are dealing with any aspect of drug development for chiral drugs. It may also be adopted as a resource supplement in a graduate-level course.

We gratefully acknowledge our thanks to Dr. Fakhreddin Jamali, Dr. Philip Breen, and Dr. E. Kim Fifer for their helpful suggestions and editorial assistance. We also thank our families for their inspiration, support, and forbearance, which made this undertaking a pleasure.

Indra K. Reddy
Reza Mehvar

Contents

Contributors

Moji Christianah Adeyeye Duquesne University, Pittsburgh, Pennsylvania, U.S.A.

Mohsen I. Afouna University of Arkansas for Medical Sciences, Little Rock, Arkansas, U.S.A.

Dion R. Brocks University of Alberta, Edmonton, Alberta, Canada

Chyung Cook Baxter Healthcare, Round Lake, Illinois, U.S.A.

Neal M. Davies Washington State University, Pullman, Washington, U.S.A.

Biana Godin The Hebrew University of Jerusalem, Jerusalem, Israel

David J. W. Grant University of Minnesota, Minneapolis, Minnesota, U.S.A.

Chong-Hui Gu* University of Minnesota, Minneapolis, Minnesota, U.S.A.

Andrew J. Hutt King's College London, London, England

Teruko Imai Kumamoto University, Kumamoto, Japan

Aziz Karim Takeda Pharmaceuticals North America, Inc., Lincolnshire, Illinois, U.S.A.

Current affiliation: Bristol-Myers Squibb Company, New Brunswick, New Jersey, U.S.A.

Thirumala R. Kommuru   Pfizer Global Research and Development, Ann Arbor, Michigan, U.S.A.

Cherukury Madhu   Pfizer, Inc., Ann Arbor, Michigan, U.S.A.

Reza Mehvar   Texas Tech University Health Sciences Center, Amarillo, Texas, U.S.A.

Masaki Otagiri   Kumamoto University, Kumamoto, Japan

Bhavesh K. Patel   King's College London, London, England

Indra K. Reddy   Texas A&M University School of Pharmacy, Kingsville, Texas, U.S.A.

Chandra Sahajwalla   Food and Drug Administration, Rockville, Maryland, U.S.A.

Elka Touitou   The Hebrew University of Jerusalem, Jerusalem, Israel

Majid Vakily   TAP Pharmaceutical Products, Inc., Lake Forest, Illinois, U.S.A.

CHIRALITY IN DRUG DESIGN AND DEVELOPMENT

1

Effects of Crystal Structure and Physical Properties on the Release of Chiral Drugs

Chong-Hui Gu* and David J. W. Grant
University of Minnesota, Minneapolis, Minnesota, U.S.A.

1. INTRODUCTION

Chiral drugs are a subgroup of drug substances that contain one or more chiral centers. More than one-half of marketed drugs are chiral [1]. It is well established that the opposite enantiomer of a chiral drug often differs significantly in its pharmacological [2], toxicological [3], pharmacodynamic, and pharmacokinetic [4,5] properties. Therefore from the points of view of safety and efficacy, the pure enantiomer is preferred over the racemate in many marketed dosage forms. However, the chiral drug is often synthesized in the racemic form, and it is frequently costly to resolve the racemic mixture into the pure enantiomers. Currently, then, most chiral drugs, including some "blockbuster" drugs, such as fluoxetine hydrochloride (Prozac®) and omeprazole (Losec®), are still marketed as racemates. However, the recent trend is toward marketing more single-enantiomer drugs [6]. In addition, a "racemic switch," which involves the development of a pure enantiomer of a drug that is already marketed as a racemate, is actively pursued by many companies to improve its therapeutic efficacy and to extend patent protection [7]. The decision whether to market the racemate or the enantiomer of a chiral drug is mainly based on pharmacology, toxicology,

---

*Current affiliation: Bristol-Myers Squibb Company, New Brunswick, New Jersey, U.S.A.

and economics. From a pharmaceutical perspective, the physical properties of both the racemate and the enantiomer should be characterized in detail in order to develop a safe, efficacious, and reliable formulation, no matter whether the racemate or the enantiomer is chosen as the marketed form. Furthermore, the chirality of a drug will also influence the efficiency of delivery, which has not been well recognized in the pharmaceutical field. Many physical properties of a crystalline solid, such as density, solubility, dissolution behavior, stability, and mechanical properties, are governed by the crystal structure [8]. Knowledge of the relationship between the crystal structure and the physical properties, and their influence on drug release, may therefore provide a fundamental understanding of the property–delivery relationship. This chapter provides an overview of the physical characterization of chiral drugs with an emphasis on the influence of physical properties on the rate of drug release.

2. ENANTIOMERS, RACEMIC SPECIES (OR RACEMATES), AND DIASTEREOMERS

Molecular *chirality* is a concept that was derived historically from the distinction between the configurational isomers of asymmetric molecules, which was discovered by Pasteur and reported in 1894 [9]. Configurational isomers are compounds with the same molecular formula and the same substituent groups but with different configurations. The asymmetric center in configurational isomers is called the *chiral* center [10]. Configurational isomers, which are a subset of stereoisomers and also termed optical or chiral isomers, may be classified as enantiomers, racemic species (or racemates), and diastereomers. *Enantiomers* are pairs of configurational isomers that are mirror images of each other and yet are not superimposable. Each enantiomer is *homochiral*, meaning that all the molecules have exactly the same configuration. *Diastereomers* are pairs of compounds that contain more than one chiral center, not all of which are superimposable. Enantiomers behave differently only in a chiral medium, such as when exposed to polarized light or when participating in a chemical reaction catalyzed by a chiral catalyst, particularly an enzyme in the body. Diastereomers generally exhibit different physical properties, even in an achiral environment. An equimolar mixture of opposite enantiomers is termed a *racemic species* or *racemate*, and is *heterochiral*, meaning that the molecules have different chiralities. The word *racemate* has at least two different interpretations. Chemists reserve the term racemate for a racemic compound, discussed in Sec. 4, whereas pharmaceutical scientists generalize the term racemate, as in this chapter, to include all types of racemic species. Three types of racemic species, or racemates, have been identified, as discussed in Sec. 4.1.

3. MOLECULAR INTERACTIONS OF CHIRAL MOLECULES

The differences in properties between the enantiomer and the racemic compound arise from the differences in interactions between the same homochiral molecules and those between the heterochiral molecules, and from the different packing arrangements in the crystal structures. In the enantiomer, the molecular interactions are homochiral interactions $(R \cdots R)$, which are defined as the intermolecular nonbonded attractions or repulsions in assemblies of molecules of the same chirality. However, the heterochiral interactions $(R \cdots S)$ in the racemate are those between molecules of opposite chirality [10,11]. These two interactions are unlikely to be identical because they are diastereomerically related. However, the difference is small enough to be neglected in the gaseous or liquid state, or in an achiral solvent [10]. In the solid state or in a chiral environment, the difference is significant enough to result in different physical properties between the racemate and the corresponding pure enantiomer; this difference is termed *enantiomeric discrimination* [10]. Enantiomeric discrimination is observed when comparing the crystal structure of the racemic compound with that of the corresponding enantiomer. The difference between the homochiral and heterochiral interactions leads to different structures, which in turn lead to differences in physical properties and/or biological activities. A natural extension of the concept of enantiomeric discrimination is *diastereomeric discrimination*, which is a result of the difference between the interactions of diastereomeric pairs, such as $R_I \cdots R_{II}$ and $R_I \cdots S_{II}$, where I and II represents different molecules [12]. Diastereomeric discrimination is the basis of chromatographic separation of enantiomers by a chiral stationary phase. The selective release of chiral drugs in a formulation containing chiral excipients is also an example of diastereomeric discrimination (Sec. 5.2).

4. SOLID-STATE PROPERTIES OF CRYSTALLINE ENANTIOMERS AND RACEMATES

4.1. Nature of Racemates

A crystalline racemate may exist as a conglomerate, a racemic compound, or a pseudoracemate (Fig. 1). A racemic conglomerate is an equimolar eutectic mixture consisting of crystals of the two opposite enantiomers,

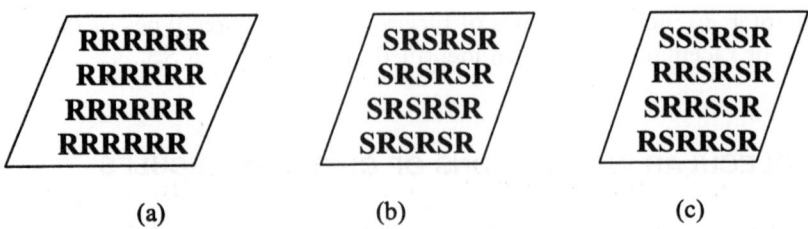

(a)   (b)   (c)

Figure 1   Schematic representation of the molecular arrangements in three types of racemates: (a) enantiomeric or homochiral crystal (same chirality); (b) racemic or heterochiral crystal (paired enantiomers); (c) pseudoracemate or solid solution (randomly arranged enantiomers). (Adapted from Ref. 11.)

which in principle can be separated manually. In a conglomerate, each crystal contains homochiral molecules. Approximately 5–10% of crystalline racemic species are of this type [13]. A racemic compound is a cocrystal containing equal numbers of molecules of the opposite enantiomer, usually paired up. A racemic compound is a heterochiral crystal, because the unit cell of each crystal contains enantiomeric molecules with opposite chirality. A racemic compound possesses physical properties different from those of the enantiomers. Racemic compounds are the most common racemic species, representing 90–95% of them. A rare type of racemic species is a pseudoracemate, which is a solid solution containing equal numbers of molecules of the opposite enantiomers in a more or less random arrangement. A pseudoracemate is also a heterochiral crystal. Less than 1% of racemic species are pseudoracemates. Under given conditions, only one form of racemate is thermodynamically stable. However, any of the three forms may be observed under nonequilibrium conditions. This phenomenon is discussed in Sec. 4.3.

The different types of racemic species can be distinguished by their phase diagrams (Fig. 2). X-ray diffractometry (Fig. 3), and spectroscopic techniques (Figs. 4 and 5) may also distinguish these different types. The molecular basis of the different types of racemic species is the different intermolecular interactions in their crystals, as shown in Fig. 1.

The melting phase diagram, which is often determined by differential scanning calorimetry (DSC) or thermal (hot-stage) microscopy, can be used to characterize the type of racemate, provided that the compound is thermally stable. The phase diagram of the conglomerate shows typical eutectic behavior (Fig. 2a). The liquidus line, which relates the liquid composition to the melting temperature, in the diagram can be calculated by

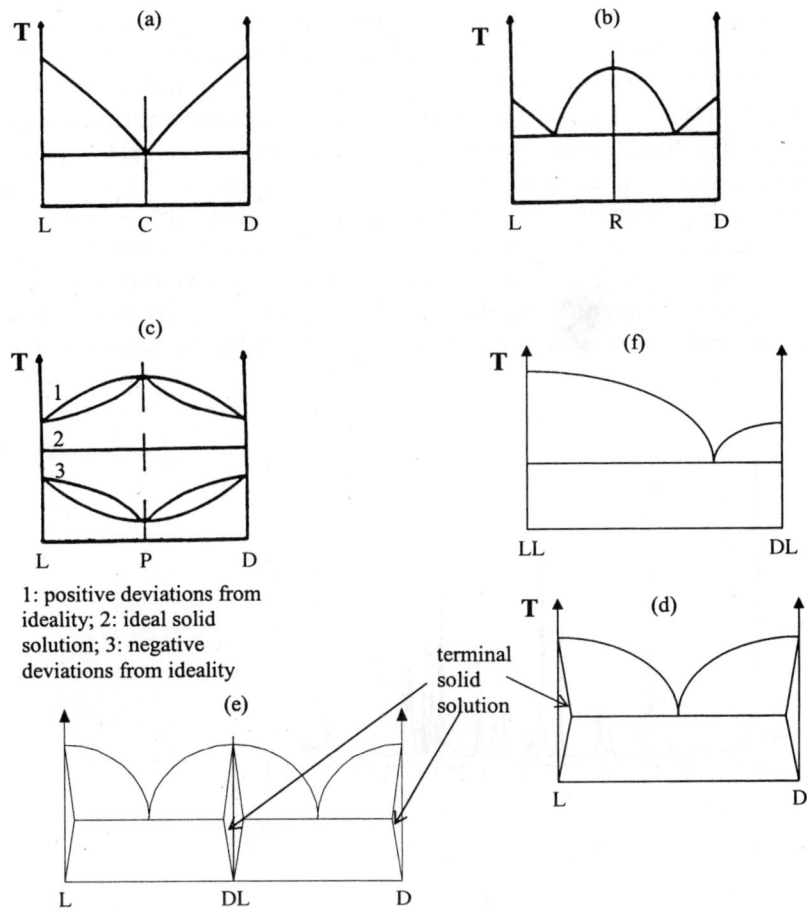

Figure 2 Typical phase diagrams of melting point against composition for (a–c) three types of racemic species, (a) racemic conglomerate, C, (b) racemic compound, R, (c) pseudoracemates, P; (d–e) terminal solid solutions involving opposite enantiomers, (d) solid solution rich in an enantiomer, (e) solid solution rich in an enantiomer or rich in the racemic compound; (f) a pair of diastereomers. (From Ref. 13. Reproduced by permission of John Wiley and Sons.)

the Schröder–van Laar equation (Eq. 1):

$$\ln x = \frac{\Delta H_A^f}{R}\left(\frac{1}{T_A^f} - \frac{1}{T^f}\right)$$

(1)

where x is the mole fraction of the more abundant enantiomer, ΔH_A^f is the molar enthalpy of fusion of the pure enantiomer, R is the gas constant, and T_A^f and T^f are the melting points of the pure enantiomer and the mixture, respectively [13]. Equation (1) assumes that the enantiomers form an ideal mixture in the liquid state and neglects the differences in heat capacity between the solid and the liquid states of the enantiomers.

The phase diagram of a racemic compound (Fig. 2b) is that of a molecular complex, but it is symmetrical about the central vertical line. The racemic compound is a new crystalline species that usually forms eutectic mixtures with the individual enantiomers. The liquidus line between the pure enantiomer and the eutectic point can be calculated by the Schröder–Van

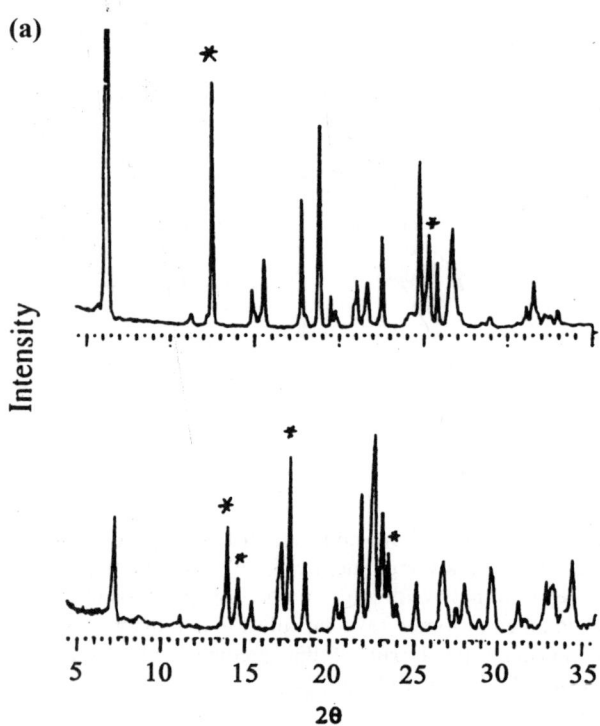

Figure 3 Powder x-ray diffraction patterns of (a) the racemic compound (upper trace) and an enantiomer (lower traces) of norephedrine hydrochloride, and (b) the pseudoracemate (+/−) and an enantiomer (−) of atenolol. Shift of peak position is indicated by an asterisk and is characteristic of a solid solution. (From Ref. 17. Reproduced by permission of the American Pharmaceutical Association.)

(b)

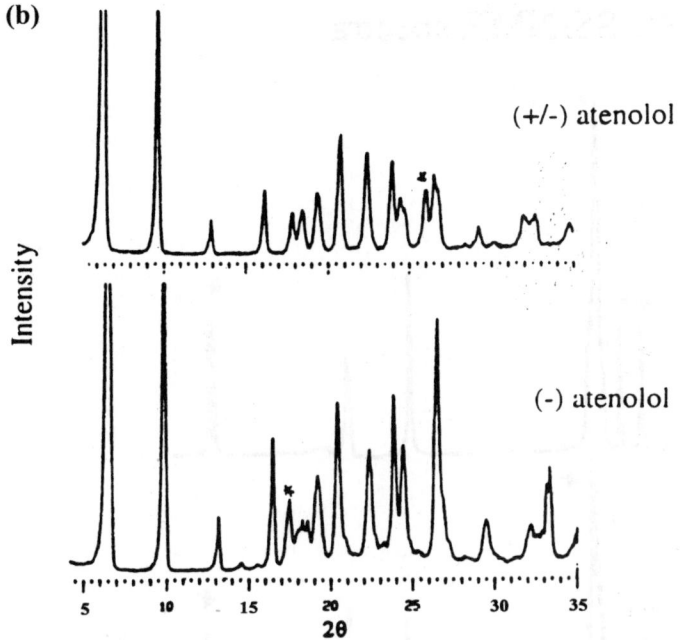

Figure 3 Continued.

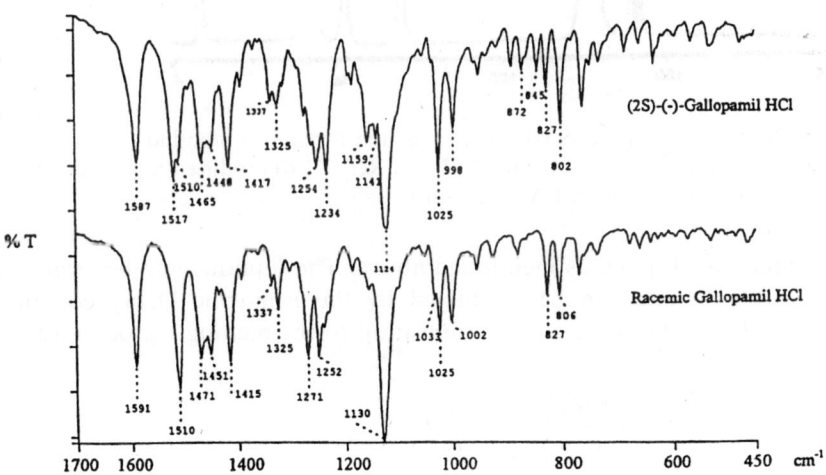

Figure 4 Infrared spectra of the racemic compound and enantiomer of gallopamil hydrochloride. (From Ref. 105. Reproduced by permission of Elsevier.)

^{13}C SS-NMR spectra

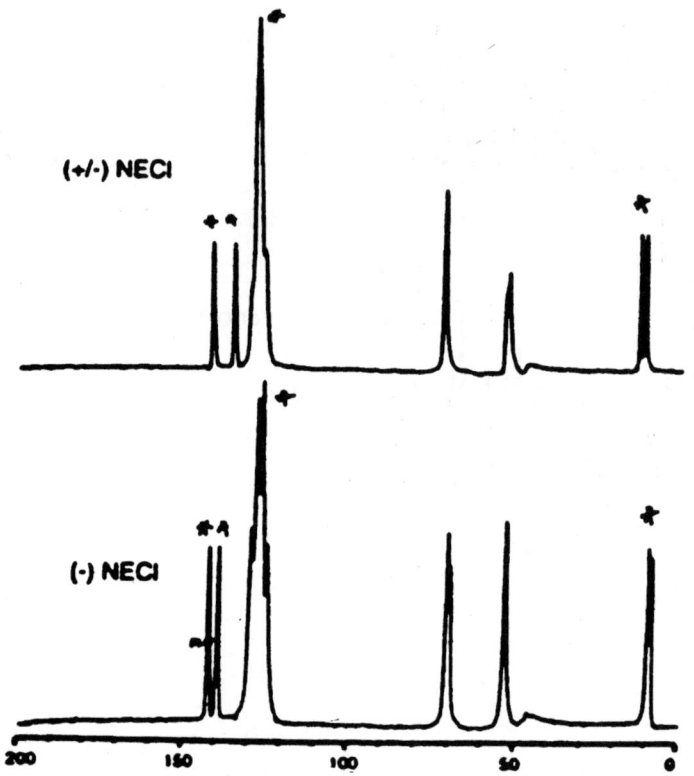

Figure 5 ^{13}C solid-state NMR spectra of the racemic compound and an enantiomer of norephedrine hydrochloride. (From Ref. 17. Reproduced by permission of the American Pharmaceutical Association.)

Laar equation (Eq. 1), as mentioned above. The liquidus line between the two eutectic points can be calculated by the Prigogine–Defay equation (Eq. 2), which relates the solid composition to the melting temperature:

$$\ln 4x(1-x) = \frac{2\Delta H_R^f}{R}\left(\frac{1}{T_R^f} - \frac{1}{T^f}\right) \tag{2}$$

where x is the mole fraction of the more abundant enantiomer and ΔH_R^f is the enthalpy of fusion of the racemic compound, and T_R^f and T^f are the melting points of the racemic compound and of the solid mixture, respectively [13]. Equation (2) assumes that the enantiomer and the racemic compound form an ideal mixture in the liquid state and neglects the differences in heat capacity between the solid and the liquid states of the individual species.

Less than 1% of racemic species are pseudoracemates, which show typical phase diagrams of continuous solid solutions. Figure 2c shows the three types of melting phase diagrams of pseudoracemates, which comprise ideal solid solutions, solid solutions with positive deviations from ideality, and solid solutions with negative deviations from ideality, respectively [13]. In real systems, the enantiomers and the racemic compound may display a small mutual solubility, even if they show eutectic behavior, which corresponds to terminal solid solutions for which the phase diagrams are shown in Figs. 2d and 2e.

Diastereomers may also form eutectic mixtures, molecular complexes, or solid solutions. The most common phase diagram for diastereomers is that of a eutectic system (Fig. 2f) [13]. Because diastereomers have different physical properties, the phase diagram of diastereomers (Fig. 2e) does not exhibit the mirror symmetry shown by enantiomers (Fig. 2a).

From the melting data determined by DSC, chiral purity may be estimated either by measuring the area of the eutectic peak or by applying the Schröder–Van Laar or the Prigogine–Defay equation [13]. When the eutectic peak is too small to be detected, the peak asymmetry factor for melting of the major component may be used to estimate the purity [14].

The enantiomer and the racemic compound possess different crystal structures, which correspond to different intermolecular interactions, as mentioned in Sec. 3. Therefore the enantiomer and the racemic compound exhibit different powder x-ray diffraction (PXRD) patterns, different infrared (IR) and Raman spectra, and different ^{13}C solid-state nuclear magnetic resonance (SSNMR) spectra. However, the opposite enantiomers give identical PXRD patterns, and identical IR, Raman, and SSNMR spectra. Consequently, the PXRD patterns and the above spectra of a conglomerate, which is a physical mixture of opposite enantiomers, are identical to that of the pure enantiomers. In contrast, the diffraction pattern and the various corresponding spectra of the racemic compound usually differ significantly from those of the pure enantiomers. Therefore the type of racemate can be easily determined by comparing the diffraction patterns or the various spectra of the racemic species with that of one of the pure enantiomers (Figs. 3–5). The enantiomeric composition in a racemic mixture may be determined by PXRD, or by IR or SSNMR spectroscopy. Quantitative PXRD has been applied to determine the relative

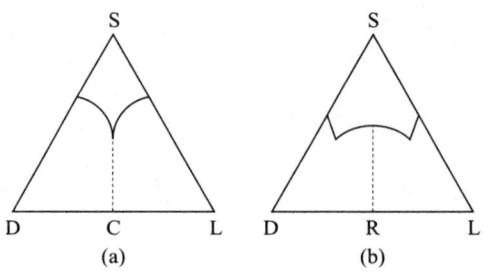

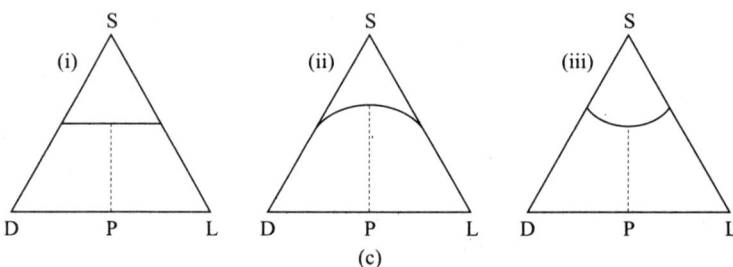

Figure 6 Ternary phase diagram showing the solubility of the racemic species: (a) conglomerate, C, (b) racemic compound, R, (c) pseudoracemate, P; i, ideal; ii, positive deviations; iii, negative deviations. D and L represent the enantiomers, S represents the solvent, at constant temperature. (From Ref. 10. Reproduced by permission of John Wiley and Sons.)

amount of enantiomeric ibuprofen in racemic ibuprofen, which is a racemic compound [15,16].

In addition to melting point phase diagrams, ternary solubility phase diagrams, in which the third compound is a liquid solvent, may also be applied to classify racemates (Fig. 6) at constant temperature.

4.2. Thermodynamic Stability of the Racemic Compound vs. the Conglomerate

The formation of a racemic compound may be represented by the equation

$$D + L = R \tag{3}$$

where the left side represents the conglomerate, which is an equimolar mixture of opposite enantiomers, and the right side represents the racemic compound. Using a thermodynamic cycle [13,17], the free energy of formation of the racemic compound at a given temperature is given by

$$\Delta G_T^o = -\Delta S_R^f (T_R^f - T_A^f) - T_A^f R \ln 2 + (\Delta S_R^f - \Delta S_A^f + R \ln 2)(T_A^f - T)$$

$$+ (C^l - C_R^s)\left[T_R^f - T_A^f - T \ln \frac{T_R^f}{T_A^f}\right]$$

$$+ (C_A^s - C_R^s)\left[T_A^f - T - T \ln \frac{T_A^f}{T}\right] \qquad \text{when } T_A^f < T_R^f \qquad (4)$$

$$\Delta G_T^o = -\Delta S_A^f (T_R^f - T_A^f) - T_R^f R \ln 2 + (\Delta S_A^f - \Delta S_R^f + R \ln 2)(T_R^f - T)$$

$$+ (C^l - C_A^s)\left[T_R^f - T_A^f - T \ln \frac{T_R^f}{T_A^f}\right]$$

$$+ (C_A^s - C_R^s)\left[T_R^f - T - T \ln \frac{T_R^f}{T}\right] \qquad \text{when } T_A^f > T_R^f \qquad (5)$$

where ΔS^f is the entropy of fusion, T^f is the melting temperature, T is the temperature of interest, C is the heat capacity at constant pressure, the subscripts A and R refer to enantiomer and racemic compound, respectively, as mentioned above, and the superscripts l and s refer to liquid and solid states, respectively [13,17].

When the free energy of formation is negative at a given temperature, the racemic compound is more stable than the conglomerate at that temperature. Equations (4) and (5) suggest that the more the melting temperature of the racemic compound exceeds that of its enantiomer, the more stable the racemic compound. A study on 23 chiral drugs supports this conclusion [17].

4.3. Polymorphism and Pseudopolymorphism of Chiral Drugs

Many pharmaceutical compounds exist as distinct polymorphs, which are defined as crystalline solids with the same chemical composition but with different arrangements and/or conformations of the molecules in the crystal lattice. A compound may also form solvates (sometimes known as pseudopolymorphs), which are crystals that contain the solvent of crystallization as part of the crystal structure [18]. Different polymorphs exhibit significantly different physicochemical properties, owing to differences in crystal packing, thermodynamics, kinetics, surface, mechanics, and spectroscopic transitions. Examples include solubility, dissolution rate, processing properties, and stability [19]. The rank order of the solubility and the dissolution rate among polymorphs is reflected in their relative stability, with the most stable form having the lowest solubility and intrinsic dissolution rate [20].

Polymorphism and pseudopolymorphism are known among chiral drugs. Whereas the enantiomers of chiral drugs may display ordinary polymorphism, the racemic species show a special type of polymorphism, namely conversion between the different types of racemate. For example, racemic sodium ibuprofen exhibits three polymorphs, the stable polymorph at room temperature being the conglomerate while the other two polymorphs, which are racemic compounds, are obtained by rapidly cooling the melt of the conglomerate [21].

Based on the thermodynamic stability relationships between the polymorphs, polymorphism may be categorized into two types, namely enantiotropy and monotropy (Fig. 7) [22,23]. In enantiotropy, the polymorphs have a transition temperature below the melting points; on crossing the transition temperature, the relative stability of the polymorphs is inverted (Fig. 7a). At the transition temperature, the free energy difference between the two polymorphs is zero. The transition temperature may be estimated from plots of solubility, dissolution rate, or vapor pressure against temperature [24] or from melting data [25], including the melting of suitable eutectics [26]. In addition, the transition temperature may be estimated from heats of solution and solubility data determined at any one temperature [27].

In monotropy, one polymorph is always more stable than the other at all temperatures below their melting points (Fig. 7b). This definition is based on the assumption that the pressure remains constant. An alternative definition is that, if the pressure–temperature phase diagram does not allow a polymorph to be in equilibrium with its vapor phase below the critical point, it is the unstable monotrope, otherwise it is an enantiotrope. This definition recognizes that some monotropes may be thermodynamically stable at elevated pressures and temperatures, e.g., diamond, which is the metastable polymorph of carbon under ambient conditions.

Table 1 lists theoretically possible polymorphs of chiral drugs and practical examples [13,21,28–35]. The most common are polymorphic enantiomers, polymorphic racemic compounds, and the existence of both a racemic conglomerate and a racemic compound of a chiral drug. Polymorphs may be discovered by crystallization from supersaturated solutions in various solvents, by solvent-mediated polymorphic transformation, by crystallization of amorphous solids under different conditions,

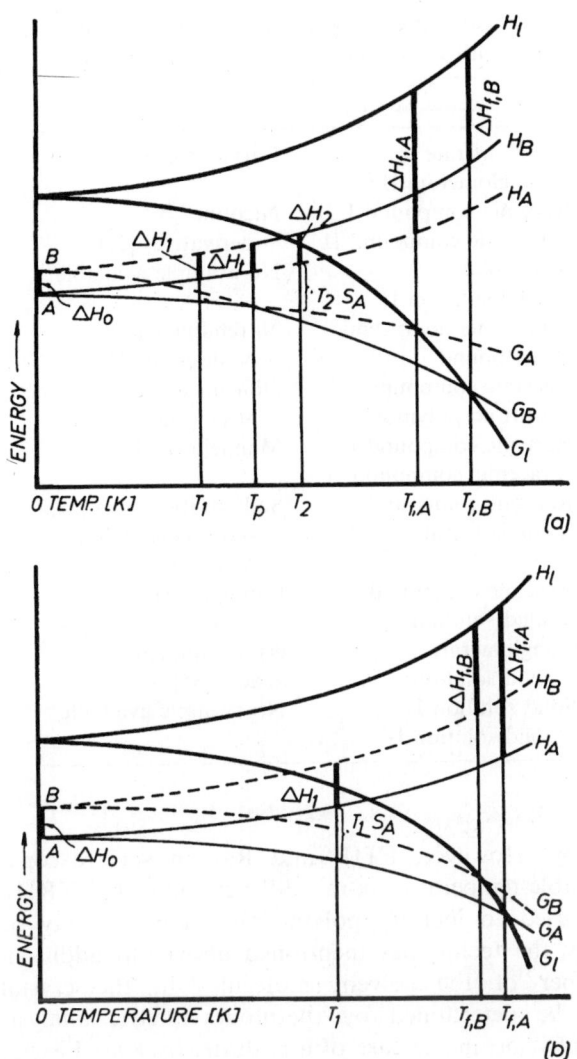

Figure 7 Energy/temperature diagrams of polymorphic systems: (a) an enantio-tropic system, (b) a monotropic system. (From Ref. 22. Reproduced by permission of Springer, Vienna.)

or by polymorphic transformations induced by stress, such as temperature or mechanical pressure [36]. The most compelling evidence for polymorphism is a clear difference in the crystal structure determined by the single-crystal x-ray diffractometry. Valuable but less definitive are PXRD and

Table 1 Several Theoretically Possible Polymorphic Systems of Enantiomers and of the Corresponding Racemates, with Examples.

Enantiomers	Racemic species	Examples
Individual enantiomers exhibit polymorphism	Conglomerate I, conglomerate II	Carvoxime [13]
	Racemic compound I, racemic compound II	Nicotine derivatives [28]
	Solid solution I, solid solution II	No example available
	Conglomerate, racemic compound	Nitrendipine [29] Nimodipine [30]
	Racemic compound with no polymorph	Difficult to prove absence of polymorphism
Individual enantiomers have no polymorphs	Racemic compound I, racemic compound II	Mandelic acid [31]
	Racemic compound, conglomerate	Sodium ibuprofen [21,32] α-Bromoçamphor binaphthyl [33]
	Racemic compound, solid solution	Camphoroxime [34]
	Conglomerate, solid solution	cis-π-Camphanic acid [35]
	Solid solution I, solid solution II	No example available

^{13}C SSNMR spectroscopy. However, FTIR and Raman spectroscopy, thermal microscopy, variable-temperature x-ray diffractometry (VTXRD), and DSC may also be used to identify polymorphs. Solvates may be similarly characterized by the techniques mentioned above. In addition, the stoichiometric number of the solvent molecules in the crystal lattice of solvates may be determined by thermogravimetric analysis (TGA), gas chromatography, or, in the case of a hydrate, by Karl Fischer titrimetry [37].

For chiral drugs, it is usually necessary to apply several techniques to characterize both the enantiomer and the racemic species in order to interpret the origin of the polymorphism among the racemic species. For example, three monotropically related polymorphs of (RS)-nitrendipine were found [29]. Based on the melting phase diagram, IR spectra, and PXRD patterns, the thermodynamically stable form (I) of (RS)-nitrendipine was shown to be a racemic compound, while the other polymorphs, forms II and III, were both found to be conglomerates. Study of one of the pure enantiomers also revealed three polymorphs, among which the enantiomeric form I corresponds to the metastable racemic conglomerate form II, the enantiomeric form II corresponds to the racemic conglomerate form III, while the enantiomeric form III is not related to any of the racemic modifications. Hence both the enantiomer and the racemate display polymorphism in this system. When the designations of the polymorphs of the enantiomer and of the conglomerate do not correspond, it is necessary to pay careful attention to the nomenclature.

Certain chiral drugs may form solvates. When the incorporated solvent is water, the solvates are known as hydrates, which are the most abundant solvates. A hydrate may be prepared by exposing the solid drug to water vapor or by suspending it in an organic solvent + water mixture with the appropriate water activity [38]. The incorporated solvent may be present either in stoichiometric amounts, e.g., histidine monohydrate [34], or in nonstoichiometric amounts, e.g., cromolyn sodium hydrates [39]. The relative stability of the unsolvated crystal (or anhydrate) and the solvate (or hydrate) is determined by the water activity, a_{water}, of the environment and the temperature, T, as expressed by the equation

$$a_{water} = \frac{p_t}{p_s} = \exp\left(\frac{\Delta H_{tr}}{nRT_c} - \frac{\Delta H_{tr}}{nRT}\right) \qquad (6)$$

where p_t is the critical water vapor pressure for equilibrium between the hydrate and the anhydrate, and p_s is the saturated water vapor pressure, all at temperature T. ΔH_{tr} is the enthalpy of transition from the hydrate to the anhydrate and is equal to the difference in enthalpy of aqueous solution between the hydrate and the anhydrate, and n is the number of moles of water in one mole of hydrate [40,41]. At a defined absolute temperature, T, the hydrate is more stable than the anhydrate when the water activity of the medium exceeds a_{water} in Eq. 6.

In chiral systems, the racemic compound and the enantiomer usually undergo different degrees of solvation under given conditions. There are many examples of a change of the degree of solvation as a result of a change in the nature of the racemate [13]. For example, enantiomeric histidine hydrochloride forms a monohydrate when crystallized from water, whereas racemic histidine hydrochloride forms a dihydrate. The racemic dihydrate transforms to the conglomerate monohydrate above 45°C [34].

Diastereomeric pairs also form solvates with different degrees of solvation. The solubility of a solvate with a higher degree of solvation is usually lower than that with a lower degree of solvation in the corresponding solvent at a lower temperature. At a higher temperature, the order tends to be inverted. On the other hand, in a solvent that is miscible with the solvating solvent, the solubility of the more highly solvated solid is usually higher than that of the solid with a lower solvation [13]. These rank orders are based on the free energy relationships, the more stable form having the lower free energy at the solvent activity and temperature of interest. Similarly, the solubility ranking of a diastereomeric pair may be changed when the solvent is altered. For example, in the resolution of DL-leucine by forming salts with (S)-(−)-phenylethanesulfonic acid (PES), the salt, L-leucine-(−)-PES (LS), is less soluble in acetonitrile + methanol mixture than is the diastereomeric salt, D-leucine-(−)-PES (DS). However, in acetonitrile + water mixture, DS monohydrate is formed and is less soluble than LS, which does not form a hydrate [42]. Because of the solubility difference among solvates and non-solvates, the formation of solvates may change the rate of drug release.

Polymorphic and pseudopolymorphic transitions may occur during pharmaceutical processing, such as grinding, milling, and wet granulation. Figure 8 illustrates the possible phase transformation induced by pharmaceutical processing [43]. It was found that physical mixtures of the enantiomers of malic acid, tartaric acid, and serine transformed to racemic compounds after grinding or during storage at 53% relative humidity (RH)

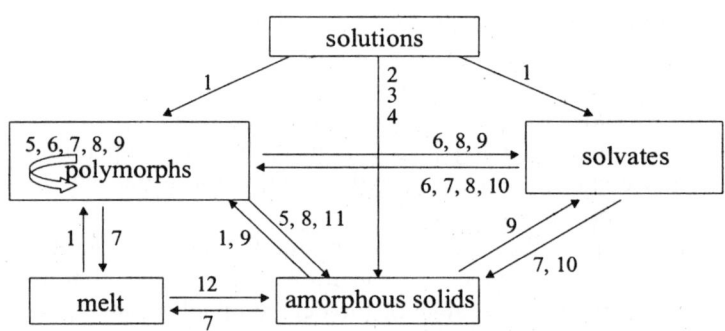

1 crystallization; 2 freeze drying; 3 spray drying; 4 precipitation; 5 milling; 6 slurry conversion; 7 heating; 8 wet granulation; 9 expose to solvent vapor; 10 drying; 11 solid dispersion; 12 quench cooling.

Figure 8 Possible phase transformations induced by pharmaceutical processing. (Adapted from Ref. 43.)

and at 40°C [44]. Both mechanical forces, especially shear, and the environmental RH are believed to exert influences. The applied mechanical energy may be transformed to structural energy in the solid phase(s), facilitating the transition, while the water content of a drug may increase under higher RH, significantly increasing the molecular mobility or providing a medium in which transformation may occur [45]. Dehydration and hydration may also occur when the critical RH of a drug is higher or lower than the environment RH, respectively. These polymorphic and pseudo-polymorphic transformations may lead to changes in release rate of the drug and will then need to be controlled.

4.4. Crystal Structures of Chiral Drugs

Solids may exist in states of various degrees of atomic or molecular order between two extremes, crystalline solids and amorphous solids. A crystalline solid consists of a three-dimensional periodic arrangement of atoms or molecules. Most marketed pharmaceutical products contain crystalline drugs, which are usually molecular crystals. Amorphous solids, like liquids, lack long range order. The amorphous solid is thermodynamically unstable in comparison with its crystalline counterpart. The physical properties of crystalline chiral drugs are manifestations of the intermolecular interactions in the crystal structure. To understand the physical properties at the molecular level, it is necessary to relate the physical properties to the crystal structures. Analysis of crystal structures facilitates enormously our understanding of the factors that determine the various physical and chemical properties, such as the thermodynamic stability of different types of racemates, differences in properties among polymorphs, and resolution efficiency during crystallization.

The crystallographic space group of an enantiomeric crystal is chiral, which implies that the structure must lack elements of improper rotation that correspond to a proper rotation combined with either an inversion center or a reflection plane. Because their crystal structures are dissymmetric, enantiomers can crystallize in only 11 of the 32 enantiomorphous crystal classes, which make up only 66 of the 230 space groups [13]. The racemic compound, however, may crystallize in any of the 230 space groups, including chiral ones. In a survey of 430 enantiomeric crystal structures by Bel'skii and Zorkii [46], the enantiomers crystallized in certain preferred space groups: 67% in the $P2_12_12_1$ space group, 27% in the $P2_1$, 1% in the $C2$, and 5% in the other 63 space groups. A similar survey of 792 crystal structures of racemic compounds revealed 67% in the $P2_1/c$ space group, 15% in the $C2/c$, and 13% in the P-1 [46].

4.5. Crystal Properties of Enantiomers and Racemic Compounds

Statistically, 90–95% of racemic species exist as racemic compounds while only 5–10% of racemic species are conglomerates. These relative frequencies imply that the racemic compound is more stable than the conglomerate. The nature of the racemic species is governed by the relative stability of enantiomeric (homochiral) crystals and racemic (heterochiral) crystals. When the formation of the enantiomeric crystal is favored, the conglomerate is more stable than the racemic compound, and vice versa. On comparing the density of eight pairs of chiral compounds at room temperature, Wallach in 1895 [47] formulated a rule that crystals of the racemic compound tend to be denser than those of the corresponding enantiomeric crystals. A more recent statistical study of 129 compounds in the Cambridge Structural Database [48] led to the conclusion that Wallach's rule is valid only for resolvable enantiomers, i.e., for enantiomers that are chemically stable and that are not in rapid exchange between one another. The greater density of the racemic (heterochiral) compound indicates that the molecules in the racemic crystals are more tightly packed than those in the corresponding enantiomeric (homochiral) crystals. Because the more tightly packed form is generally more stable at a lower temperature, the tighter packing of the racemic compound may partially explain the general tendency for racemic compounds to be more stable than the corresponding enantiomers at ambient temperatures. Sometimes, the enantiomer does not crystallize, whereas the racemic species readily forms stable crystals. For example, racemic ephedrinium salicylate readily crystallizes, while the corresponding enantiomeric species failed to crystallize despite the use of various procedures and solvents [49]. This inability has been attributed to the high energy of the conformations and/or to the poor packing of homochiral ephedrinium salicylate molecules in the crystal lattice. In marked contrast, the diastereomer, pseudoephedrinium salicylate, readily crystallizes, and the racemic species is a conglomerate.

As discussed above, most racemic compounds crystallize in three space groups, namely $P2_1/c$, $C2/c$, and P-1, which possess elements of inverse symmetry. On the other hand, enantiomers can crystallize only into dissymmetric space groups, which are devoid of inverse symmetry elements. The inverse symmetry in the racemic compound may lead to close packing and ultimately contributes to the greater stability of the racemic compound.

The fundamental factor that controls the packing mode in the crystals is the intermolecular interactions. The intermolecular interactions, including electrostatic interactions, van der Waals interactions, and hydrogen bonding interactions, differ significantly between the racemic compound and the enantiomer. These differences are the origin of the differences in packing and symmetry. Recent developments in computational chemistry make it possible to estimate the strength of the intermolecular interactions either qualitatively or quantitatively (Sec. 7).

5. INFLUENCE OF CHIRALITY IN DRUG RELEASE

5.1. Solubility of Chiral Drugs

When the rate of drug absorption is controlled by the dissolution rate, the bioavailability of a drug increases with an increase of its dissolution rate. The dissolution rate is proportional to the solubility, regardless of dissolution mechanism. The solubilities of the two enantiomers are identical in an achiral solvent. The theoretical ternary solubility phase diagrams of racemates are represented by Fig. 6. The solubility phase diagram of a conglomerate (Fig. 6a) shows eutectic behavior.

The solubility of the racemic compound is different from that of the enantiomers (Fig. 6b) and is often less than that of the pure enantiomers (Table 2) [50–53]. The greater solubility of the enantiomer will lead to a greater dissolution rate and possibly to a greater bioavailability, which is an advantage in administering the pure enantiomer. It was observed that the enantiomeric drug in a tropical formulation, being more soluble, exhibits a significantly greater rate of skin permeation than the corresponding racemic drug [51]. When the solubility of the enantiomer is much greater than that of the racemic compound, the presence of small amounts of opposite enantiomer may significantly reduce the apparent solubility of the enantiomer, because the racemic compound will form in the solution and may precipitate from the solution. For example, the solubility of (+)-dexclamol hydrochloride is five times that of (±)-dexclamol hydrochloride. 16 mg of pure (+)-dexclamol hydrochloride form a clear solution in 1.0 mL

Table 2 Solubility of Some Racemic Compounds and the Corresponding Enantiomers

	Solubility (mg/mL)	
	Racemic compound	Enantiomer
Dexclamol hydrochloride [50]	3.3	16.4
1,2-Di(4-piperazine-2,6-dione)propane [51]	3	15
SCH-39304 [52]	3.6	39.2
Ketoprofen [53]	1.42	2.32

water, whereas only 11 mg of (+)-dexclamol hydrochloride containing 1.5% of (−)-dexclamol hydrochloride completely dissolve in 1.0 mL water [49].

Pseudoracemates may exhibit three types of ternary phase diagram, as mentioned previously (Fig. 6c). The change of solubility with composition is relatively small. Therefore we may expect similar solubilities and intrinsic dissolution rates for the enantiomer and the pseudoracemate. For the same reason, resolution of a pseudoracemate by crystallization is practically impossible [10].

Unlike enantiomers, diastereomers have different solubilities in an achiral environment. Many chiral counterions, such as arginine and lysine, are commonly used to prepare salts of an ionic drug in order to improve its pharmaceutically relevant properties. However, when the drug is chiral, a chiral counterion should be used with caution. If the drug or the counterion is racemic, diastereomeric salts will be formed, constituting a heterogeneous system. Because the solubilities of diastereomers differ, a diastereomeric mixture of chiral salts is undesirable in a formulation.

5.2. Stereoselective Release of Chiral Drugs

Many pharmaceutical excipients consist of chiral molecules. The common chiral excipients are listed in Table 3 [54]. When a chiral excipient is used in a formulation, the dissolution rate of the opposite enantiomers may differ, because the interactions between the excipient and each enantiomer are diastereomeric and may therefore differ. In the field of chromatography, the different strengths of the diastereomeric interactions form the basis of enantiomeric separation on a chiral column. Because the opposite enantiomer may exhibit different pharmacological, toxicological, or pharmacokinetic properties, it is of practical value to compare the release rate of the opposite enantiomers from a formulation containing a chiral excipient.

The stereoselectivity in the dissolution of opposite enantiomers from a formulation consisting of a racemic drug and a chiral excipient varies among drug–excipient pairs. Dosage forms, such as propranolol hydrochloride in hydroxypropyl methylcellulose (HPMC) matrices [55], verapamil hydrochloride in matrix tablets containing pectin, galactomannan, or scleroglucan [56], and ketoprofen or ricobendazole in matrix tablets with HPMC [57], show stereoselective release of the enantiomers. However, no stereoselective dissolution was observed in the following dosage forms: propranolol hydrochloride in matrix tablets with β-cyclodextrin (β-CD) [55]; verapamil hydrochloride in matrix tablets coated with HPMC, β-CD, hydroxypropyl-β-CD, or cross-linked amylose [56]; ibuprofen in a solid solution with various chiral excipients [58]; and salbutamol sulfate in HPMC matrices [59]. The differences observed among different systems may be attributed

Table 3 List of Common Chiral Excipients and Their Applications in Pharmaceutics

Excipient	Application
Alginic acid	disintegrant, tablet binder
Carboxymethylcellulose calcium	disintegrant
Carboxymethylcellulose sodium	disintegrant
Microcrystalline cellulose	binder, diluent, disintegrant
Dextrin	adhesive, stiffening agent
Dextrose	sweetening agent, binder
Ethylcellulose	binder, film former
Guar gum	binder
Hydroxyethyl cellulose	binder, film former
Hydroxypropyl cellulose	granulating agent, binder
Hydroxypropyl methylcellulose	film former, in sustained release formulations
Hydroxypropyl methylcellulose phthalate	binder, in preparation of granules
Lactose	diluent, filler
Mannitol	carrier, lubricant
Methylcellulose	binder
Ascorbic acid	antioxidant
Starch	diluent
Sucrose	sweetener
Cyclodextrins and modified cyclodextrins	dissolution enhancers, complexing agents

to the complex nature of the process of drug release from a matrix tablet. The process of drug release may have the following steps: (1) diffusion of water, a nonstereoselective process; (2) dissolution of the enantiomer, which is presumably stereoselective when the enantiomeric molecules of the drug form transition complexes with the chiral excipient; (3) diffusion of the dissolved enantiomer through the hydrated matrix, presumably a stereoselective process; (4) erosion of the hydrated matrix, a nonstereoselective process [55,59]. Drug release is likely to be stereoselective only when it is controlled by one or more stereoselective processes, in particular, dissolution and/or diffusion of the chiral drug. The release rate of propranolol hydrochloride from β-CD tablets, mentioned above, is controlled by the dissolution rate of β-CD, because the enantiomer is completely included in the cavity of β-CD. Therefore in this case the release rate of the drug is not stereoselective. In addition, when the release rate of the drug is high, e.g., when the drug has a high solubility, it is difficult to detect a small difference between the release rates of the enantiomers.

In general, the differences between the release rates of opposite enantiomers is small ($< 10\%$). Furthermore, whether any small but statistically significant difference in release rate, observed in vitro, can lead to significantly different effects in vivo needs to be verified [56].

The dissolution rate is likely to influence the bioavailability only for drugs that are poorly soluble. For highly soluble drugs with poor permeabilities, the bioavailability of the drug is often related to the permeability of the drug. Because the cell membrane is chiral, the permeability of opposite enantiomers may differ, which will lead to different bioavailabilities of the enantiomers. For example, the transport rates of the opposite enantiomers of propranolol across rat skin differ significantly [60]. Furthermore, a chiral enhancer may cause the stereoselective permeability, which may be attributed to the two types of diastereomeric interactions between the chiral enhancer and the two opposite enantiomers. The permeation enhancing effect of *l*-menthol on *S*-propranolol and *R*-metoprolol was found to be greater than that on the corresponding opposite enantiomers of these two drugs [61,62]. However, a stereoselective effect of a chiral enhancer has not been demonstrated in every system. For example, another chiral enhancer, linalool, does not exert a stereoselective effect on the permeability of the enantiomers of metoprolol [62]. More detailed discussion of the influence of chirality on drug permeability can be found in other chapters of this book.

6. INFLUENCE OF STEREOISOMERIC IMPURITIES IN THE CRYSTALLIZATION MEDIUM ON THE PHYSICAL PROPERTIES OF CHIRAL DRUGS

Crystallization is widely used to purify chiral drugs. Inevitably, trace amounts of stereoisomeric impurities are present in the crystallization medium and may modify the nucleation and/or growth of the crystals. Hence the impurities may significantly change the properties of the crystallized chiral drugs, such as morphology [63], polymorphic form [64], and content of defects [65]. Furthermore, the stereoisomeric impurities may become incorporated into the crystal lattice of the chiral drug and thereby modify its crystal properties.

In systems displaying polymorphism, embryonic nuclei with structures resembling each of the mature phases may be present. The appearance of polymorphs is then kinetically controlled by the relative rates of nucleation and crystal growth of the different polymorphs [64]. The presence of impurities may lead to the formation of a particular polymorph by either promoting the formation of this polymorph or by preventing the formation of another polymorph. For example, the less stable conglomerate of histidine hydrochloride is formed in the presence of an enantiomeric polymer (1% of poly(p-(acrylamido)phenylalanine), because it prevents the formation of more stable racemic compounds [66].

Dissolved impurities exert different inhibitory effects on the growth rate of different surfaces, which can modify the crystal morphology. This phenomenon has been observed in numerous crystalline systems and is widely used as a technique for morphological engineering [63]. The changes in crystal habit, including variations in the size, the development of certain faces, and the nature and number of the faces expressed, may lead to changes in particle size distribution, in specific surface area, and in true density [67,68]. These changes can further alter the mechanical properties [69] and the dissolution behavior of the host crystals [70,71].

As mentioned above, stereoisomeric impurities may often become incorporated into the lattice of the host chiral crystals during crystallization. Indeed, commercially available chiral compounds often contain trace amounts of their chiral isomers [72]. In addition to its possible toxicological effects, the impurity may change the physical properties of the host crystals and cause batch-to-batch variations [66–68]. Among the ephedrine derivatives, the incorporation of an enantiomeric or diastereomeric impurity decreases both the enthalpy and the entropy of fusion of the host crystals, indicating increases in both lattice energy and disorder [73–76]. As a result of the replacement of the host molecules by the incorporated impurity molecules, the intermolecular hydrogen bonds in the host crystals are often disrupted, and the repetition in the crystal lattice is usually disturbed. Pseudoephedrinium salicylate [49] and pseudoephedrine hydrochloride [76] provide such examples. Thus the increases in both lattice energy and disorder have a structural basis.

While the enthalpy and entropy of the host crystals are found to increase when chiral impurities are incorporated, the intrinsic dissolution rate of the host may be increased or decreased depending on the system, host + impurity [73–76]. The dissolution rate of a solid is related to the solubility, which is controlled by the free energy of the solid. Although the incorporation of impurity usually increases both the enthalpy and the entropy of the solid, the change in the free energy depends on the relative magnitude of the change in enthalpy (ΔH), and the change in entropy (ΔS) times the temperature T, according to the classical equation,

$$\Delta G = \Delta H - T^* \Delta S \qquad (7)$$

The free energy of the impure crystals is higher than that of the pure crystals, when the amount of incorporated impurity exceeds its solubility limit in the host crystals. In this case, the solid solution thus formed is metastable, so the solubility, and consequently the intrinsic dissolution rate, of the impure crystals are higher than those of the pure crystals. Conversely, when the amount of incorporated impurity is below the solid solubility limit, a stable solid solution is formed, and the impure crystals will have a lower solubility and dissolution rate than the pure crystals. Furthermore, the impurity may enter the dissolution medium and exert an inhibitory effect on the dissolution rate of the host crystals [77,78].

The extent of impurity incorporation is controlled by both the solid solubility of the impurity in the host crystals and the crystallization kinetics, such as the crystal growth rate and degree of supersaturation [79]. The affinity of the impurity for the host crystals is characterized by the segregation coefficient, which is defined as the ratio of the concentration of the impurity in the host crystals to that in the solution, both with respect to the host molecules. In most systems, the segregation coefficient is less than unity, which means that the impurity tends to be rejected by the growing crystals. For these systems, the extent of impurity incorporation is greater when the crystal growth is faster and the degree of supersaturation is greater. In a single batch, the larger crystals are found to incorporate more impurities than smaller crystals for those systems for which the segregation coefficient is less than unity [80]. Considering the significant effects of incorporated impurities on the properties of host crystals, the crystallization conditions must be carefully monitored to control the extent of impurity incorporation.

7. APPLICATION OF MOLECULAR MODELING IN PREDICTING THE PHYSICAL PROPERTIES OF CHIRAL DRUGS

Molecular modeling uses computational methods to simulate the properties of chemical and biological systems. Many molecular modeling tools are commercially available to predict the physical properties of solids with reasonable accuracy. These tools may be applied to the chiral drugs to explain the experimental observations at the molecular level and to predict the properties based on the crystal structure or the molecular structure.

Attempts have been made to rationalize the formation of different racemates by visualizing the crystal structures and calculating their lattice energies. Kinbara et al. [81] determined the crystal structures of a series of salts formed by chiral primary amines with achiral carboxylic acids.

These authors found that those salts that form conglomerates contain a characteristic columnar hydrogen bond network, in which the ammonium cations and the carboxylate anions are aligned around a 2-fold screw axis, constituting a 2_1 column (Fig. 9) [81].

To probe quantitatively the differences in the intermolecular interactions between the conglomerate and the racemic compound, the lattice energies of the racemic compound and the enantiomer are compared. The lattice energy is the sum of the intermolecular interaction energies in the unit cell, which is equivalent to the enthalpy of sublimation. Several commercial software packages, such as Cerius2 and Sable, are available to calculate the lattice energy of a given crystal structure using the appropriate

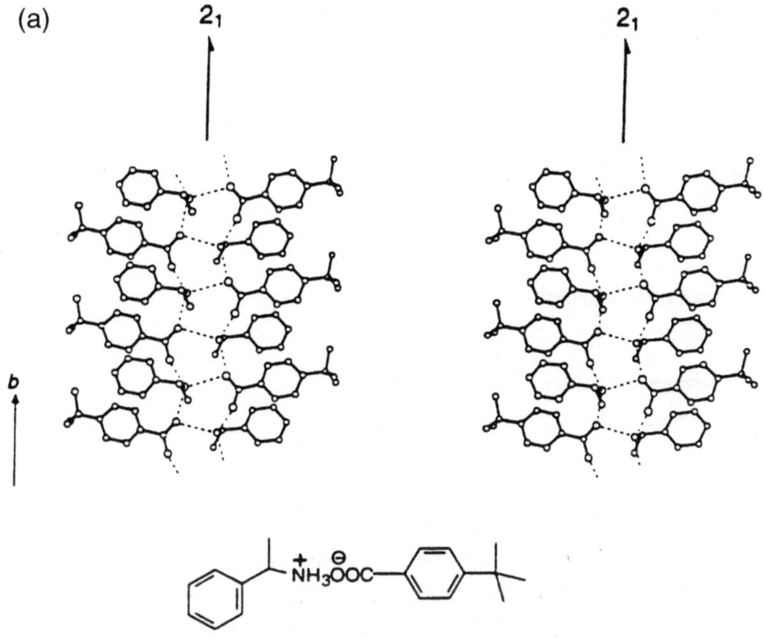

Salt of 1-phenylethylamine with p-butyl benzoic acid

Figure 9 (a) Hydrogen-bond network in the crystal of salt 1, represented by the dotted lines. Two ammonium cation and carboxylate anion pairs form a unit through the hydrogen bonds between the ammonium hydrogens and the carboxylate oxygens. This unit forms an infinite columnar structure around a twofold screw axis along the b axis (2_1-column). (b) Schematic representation of a 2_1-column formed in conglomerate salts of chiral primary monoamines with achiral monocarboxylic acids. (From Ref. 81. Reproduced by permission of the American Chemical Society.)

(b)

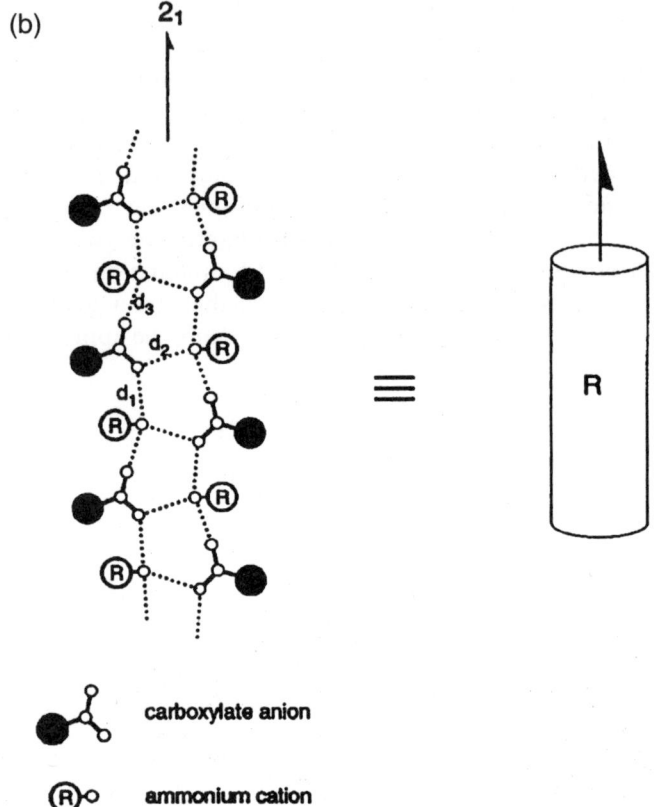

carboxylate anion

(R)○ ammonium cation

Figure 9 Continued.

force field. The stability of the crystal increases with a decrease in lattice energy (more negative), assuming a negligible contribution of the entropy term to the free energy. The lattice energy of seven pairs of enantiomers and racemic compounds was calculated by Li et al. [82] using the Dreiding force field. The lattice energy of the racemic compound, if it can exist, is algebraically less (more negative) than that of the corresponding enantiomer, which explains the preferred formation of the racemic compound. The van der Waals forces most significantly contribute to the difference in the lattice energy. In salts, the contribution of electrostatic energy toward the total lattice energy increases, which offsets the contribution of the van der Waals forces. This result may explain why conglomerates are more frequently formed in chiral salts.

The efficiency of racemate resolution by crystallization of diastereomers may also be predicted by molecular modeling. The resolution efficiency is generally greater when the difference in the solubility between the diastereomers is greater. The difference in solubility may be deduced from the crystal structure by assuming that the solubility decreases with increasing stability of the crystal. Kinbara et al. [83,84] studied the dependence of resolution efficiency by visually comparing the crystal structures of a series of diastereomeric pairs. High-resolution efficiency was achieved with those pairs, for which one crystal is stabilized by both hydrogen bonding and van der Waals interactions. Leusen [85] also found that the resolution efficiency can be explained by comparing the intermolecular bonding patterns and packing modes. However, the lattice energy, calculated using the software available at that time, was not accurate enough to correlate with the resolution efficiency. Similar approaches may be used to estimate the rank order of the solubility and the dissolution rate of chiral isomers.

In addition to the above applications, crystal structure analysis and interaction energy calculation may yield additional information, such as morphology, surface impurity binding energy, and mechanical properties of a crystalline drug. The crystal morphology will significantly affect dissolution behavior and the processing properties of a drug. The morphology grown in vacuum may be predicted by Hartman–Perdok theory [86,87]. This theory identifies planes in crystals containing periodic bond chains and enables the attachment energy of these planes to be calculated. By assuming that the surface area of each plane is inversely proportional to the attachment energy, the morphology of the crystal can be calculated. This approach has been applied to predict the morphologies of two diastereomeric salts of ephedrine with cyclic phosphoric acid [88]. To predict the morphology of crystals grown from solution, the interaction between the solvent and the crystal surface needs to be considered. Cerius2 software, version 4.2 (Accelrys, Inc., San Diego, CA, and Cambridge, UK) includes the Surface-Docker Module to calculate the solvation energy, which may be integrated into the attachment energy for the prediction of morphology [89].

It is sometimes desirable to market a metastable phase, such as a metastable polymorph or an amorphous solid, to exploit its superior pharmaceutical properties. It is then necessary to prevent phase transformation during processing and storage. To inhibit phase transformation, additives may be introduced. The greater the binding energy of the impurity to the crystal surface, the greater the inhibitory effect of the additive on the crystallization process. Additives may therefore be selected based on the calculated binding energy to stabilize a metastable phase for marketing [80,90].

When drugs are exposed to mechanical forces, such as milling and grinding, new surfaces are created. The surface properties of a drug often affect pharmaceutically relevant properties, such as wettability, dissolution rate for a surface-controlled dissolution process, and compactability. The surface created by shear forces may be predicted from the attachment energy, because the plane with the lower attachment energy is more likely to be the fracture plane. After identifying the fracture plane, the surface properties of the plane may be estimated by visualizing the molecular arrangements on the plane. Molecular modeling has been applied to characterize the surface properties of racemic propranolol after milling, which are in agreement with the results obtained by inverse gas chromatography [91].

Although crystal structures are routinely determined by single crystal x-ray diffractometry, it is often difficult and time-consuming to grow single crystals of suitable quality for crystal structure determination. Using molecular modeling, it is now possible to derive the crystal structure from the powder x-ray diffraction pattern [92]. This process consists of the following three steps: (1) Determination of the lattice parameters and assignment of a crystal space group from the position of the diffraction peaks, i.e., d-spacing [93]. (2) Structure solution, in which the atomic positions in the asymmetric unit are derived from the intensity of each diffraction peak, using either the direct method [94], the Patterson method [95], the entropy maximization method [96], or the Monte Carlo simulated annealing method [97]. (3) Structure refinement by the Rietveld method [92,98]. In this last step, the powder diffraction pattern, calculated from the structure model determined in the structure solution step, is compared point by point with the experimental powder diffraction pattern. The structure model is then adjusted to give the best fit to the experimental powder diffraction pattern.

The primary problem in structure solution from the powder x-ray diffraction pattern is the overlap of peaks in the pattern, which leads to severe ambiguities in determining the intensity of individual peaks and consequently the position of the atoms in the structure. High resolution x-ray diffraction patterns, which can be obtained using synchrotron x-radiation, are necessary for accurate structure determinations [92]. In addition to the x-ray diffraction patterns, neuron diffraction data have also been used for structure determination [99].

The ultimate goal is to determine the crystal structures in which a molecule may crystallize and then to predict the physical properties without recourse to experiment. Attempts have been made to calculate the crystal structure directly from the molecular structure of the drug [100,101]. The basic assumptions are that real crystal structures have relatively low lattice energies and that these energies can be reliably calculated by force field methods. Based on these assumptions, thousands of possible crystal structures are generated by the Monte Carlo Simulated Annealing approach. The structure with the lowest lattice energy represents the possible actual structure. Impressive results of crystal structure prediction have been obtained for hydrocarbons [102]. An extension of crystal structure solution from molecular structure is polymorph prediction. The lattice energy is again used as a criterion to screen the possible crystal structures generated by Monte Carlo simulation. All structures with low lattice energy are potential polymorphs. Cerius2 software, version 4.2 (Accelrys, Inc., San Diego, CA, and Cambridge, UK) is commercially available with a function for polymorph prediction. Polymorphs of aspirin [103], primidone and progesterone [104] have been predicted, but not all the predicted polymorphs have been prepared.

Crystal structure determinations either from the powder x-ray diffraction pattern or from the molecular structure are not yet routine procedures. Currently, crystal structure prediction from the molecular structure is feasible only for relatively rigid and nonionic compounds. More sophisticated criteria need to be developed to select the likely polymorphs from the predicted crystal structures. In addition, in order to estimate accurately the stability of a crystal structure, the lattice entropy needs to be taken into account as well as the lattice energy. However, no accurate method is yet available to estimate the lattice entropy. These limitations may be overcome by advances in computational chemistry and computing power.

Many of the applications of molecular modeling, mentioned above, have not yet been applied to chiral drugs. However, we can expect to see more studies on chiral drugs in this field, which will open a new era of understanding of the relation among the physical properties, the crystal structures, and the release rate of chiral drugs.

8. CONCLUDING REMARKS

Chiral drugs constitute an important subgroup of drug substances. As for any achiral drug, the drug release profile of a chiral drug is influenced by its physical properties, such as solubility and morphology, which are governed by the internal structure and the molecular interactions both inside the crystal and with the external environment. When an enantiomer is selected as the marketed form, the drug may be treated as any achiral compound in terms of its physical characterization. However, special attention is needed when a chiral excipient is used in a formulation, or when the enantiomer is reacted with a chiral counter ion to prepare a salt, or when racemization occurs during processing. When the marketed chiral drug is a racemic species (racemate), the nature of the racemate as well as any change in the nature of the racemate, such as polymorphic transformation, must be characterized to ensure a robust formulation with a desirable drug release profile. Classical characterization tools, such as thermodynamics and spectroscopy, along with crystal structure analysis, provide an in-depth understanding of the physical properties. The advancement in molecular modeling makes it possible to predict some physical properties from either the crystal structure or directly from the molecular structure.

REFERENCES

1. Millership, J.; Fitzpatrick, A. Commonly used chiral drugs: a survey. Chirality **1993**, *5*, 573–576.
2. Islam, M.M.; Mahdi, J.G.; Bowen, I.D. Pharmacological importance of stereochemical resolution of enantiomeric drugs. Drug Safety **1997**, *17*, 149–165.
3. Wainer, I. *Drug Stereochemistry: Analytical Methods and Pharmacology*, 2nd Ed.; Marcel Dekker: New York, 1993.
4. Drayer, D. Pharmacodynamic and pharmacokinetic differences between drug enantiomers in humans: an overview. Clin. Pharmacol. Ther. **1986**, *40*, 125–133.
5. Midha, K.M.; McKay, G. Rawson, M.J.; Hubbard, J.W. The impact of stereoisomerism in bioequivalence studies. J. Pharm. Sci. **1998**, *87*, 797–802.
6. Stinson, S. Counting on chiral drugs. Chem. Eng. News **1998**, *76*, 83–103.
7. DiCicco, R. The future of S-(+)-ibuprofen and other racemic switches. In *Chiral' 93 USA*; Spring Innovations: Stockport, UK, 1993.
8. Leusen, F.; Engel, G. Computational approaches to crystal structure and polymorph prediction. J. Pharm. Pharmacol. **1999**, *51* (S), 1.
9. Mislow, K. Molecular chirality. In *Topics in Stereochemistry*; Denmark, S.C. Ed.; John Wiley: New York, 1999; 1–82.
10. Eliel, E.L.; Wilen, S.H.; Mander, L.N. *Stereochemistry of Organic Compounds*; John Wiley: New York, 1994; 153–295.
11. Li, Z., Grant, D.J.W. Relationship between physical properties and crystal structures of chiral drugs. J. Pharm. Sci. **1997**, *86*, 1073–1078.
12. Stewart, M.; Arnett, E. Chiral monolayers at the air–water interface. Topics in Stereochem. **1982**, *13*, 195.
13. Jacques, J.C.; Collet, A.; Wilen, S.H. *Enantiomers, Racemates, and Resolutions*; John Wiley: New York, 1981.
14. Elsabee, M.; Prankerd, R.J. Solid-state properties of drugs. III. Differential scanning calorimetry of chiral drug mixtures existing as racemic solid solutions, racemic mixtures and racemic compounds. Int. J. Pharm. **1992**, *86*, 221–230.

15. Stahly, G.; McKenzie, A. Andres, M.; Russel, C.; Byrn, S.R.; Johnson, P. Determination of the optical purity of ibuprofen using x-ray powder diffraction. J. Pharm. Sci. **1997**, *86*, 970–971.

16. Phadnis, N.; Suryanarayanan, R. Simultaneous quantification of an enantiomer and the racemic compound of ibuprofen by x-ray powder diffractometry. Pharm. Res. **1997**, *14*, 1176–1180.

17. Li, Z.; Zell, M.T.; Munson, E.J.; Grant, D.J.W. Characterization of racemic species of chiral drugs using thermal analysis, thermodynamic calculation, and structural studies. J. Pharm. Sci. **1999**, *88*, 337–346.

18. Byrn, S.R.; Pfeiffer, R.; Stowell, J. *Solid-State Chemistry of Drugs*, 2nd. Ed.; SSCI: West Lafayette, IN, 1999.

19. Grant, D.J.W. Theory and origin of polymorphism, In *Polymorphism in Pharmaceutical Solids*; Brittain, H.G. Ed.; Marcel Dekker: New York, 1999; 1–33.

20. Brittain, H.G.; Grant, D.J.W. Effects of polymorphism and solid-state solvation on solubility and dissolution rate. In *Polymorphism in Pharmaceutical Solids*; Brittain, H.G. Ed.; Marcel Dekker: New York, 1999.

21. Zhang, G.; Paspal, S.; Suryanarayanan, R.; Grant, D.J.W.; Racemic species of sodium ibuprofen: Characterization and polymorphic relationships. J. Pharm. Sci. **2003**, *92*, 1356–1366.

22. Burger, A.; Ramberger, R. On the polymorphism of pharmaceuticals and other molecular crystals. I. Theory of thermodynamic rules. II. Applicability of thermodynamic rules. Mikrochim. Acta. **1979**, *2*, 259–271.

23. Giron, D. Thermal analysis and calorimetric methods in the characterization of polymorphs and solvates. Thermochim. Acta. **1995**, *248*, 1–59.

24. Grant, D.J.W.; Higuchi, T. *Solubility Behavior of Organic Compounds*; John Wiley: New York, 1990; 22–36.

25. Yu, L. Inferring thermodynamic stability relationship of polymorphs from melting data. J. Pharm. Sci. **1995**, *84* (8), 966–974.

26. Yu, L.; Stephenson, G.A.; Mitchell, C.A.; Bunnell, C.A.; Snorek, S.V.; Bowyer, J.J.; Borchardt, T.B.; Stowell, J.G.; Byrn, S.R. Thermochemistry and conformational polymorphism of a hexamorphic crystal structure. J. Am. Chem. Soc. **2000**, *122*, 585–591.

27. Gu, C.; Grant, D.J.W. Estimating the relative stability of polymorphs and hydrates from heats of solution and solubility data. J. Pharm. Sci. **2001**, *90*, 1277–1287.

28. Langhammer, L. Binary systems of enantiomeric nicotine derivatives. Arch. Pharm. **1975**, *308*, 933–939.

29. Burger, A.; Rollinger, J.M.; Brüggeller, P. Binary system of (R)- and (S)-nitrendipine—polymorphism and structure. J. Pharm. Sci. **1997**, *86*, 674–679.

30. Grunenberg, A.; Keil, B.; Henck, J.-O. Polymorphism in binary mixtures, as exemplified by nimodipine. Int. J. Pharm. **1995**, *118*, 11–21.

31. Kuhnert-Brandstätter, M.; Ulmer, R. Beitrag zur thermischen Analyse optischer Antipoden: Mandelsäure. Mikrochim. Acta. **1974**, 927–935.

32. Zhang, G.Z. Influence of Solvents on Properties, Structures, and Crystallization of Pharmaceutical Solids. Ph.D. thesis, Department of Pharmaceutics, University of Minnesota, Minneapolis, 1998, pp. 70–122.

33. Collet, A.; Brienne, M.; Jacques, J. Dédoublements spontanes et conglomerats d'énantiomères. Bull. Soc. Chim. Fr. **1972**, 127–142.

34. Jacques, J.; and Gabard, J. Étude des mélanges d'antipodes optiques. III. Diagrammes de solubilité pour les divers types de racemiques. Bull. Soc. Chim. Fr. **1972**, 342–350.

35. Brienne, M.J.; Jacques, J. Une γ-lactone d'un type rare: l'acide trans π-camphanique. Tetrahedron **1970**, *26*, 5087–5100.

36. Gullory, J.K. Generation of polymorphs, hydrates, solvates, and amorphous solids. In *Polymorphism in Pharmaceutical Solids*; Brittain, H.G., Ed.; Marcel Dekker: New York, 1999.

37. Brittain, H.G., Ed. *Physical Characterization of Pharmaceutical Solids*; Marcel Decker: New York, 1995.

38. Zhu, H.; Yuen, C.; Grant, D.J.W. Influence of water activity in organic solvent plus water mixtures on the nature of the crystallizing drug phase. 1. Theophylline. Int. J. Pharm. **1996**, *135*, 151–160.

39. Chen, L.; Young, V.G., Jr.; Lechuga-Ballesteros, D.; Grant, D.J.W. Solid state behavior of cromolyn sodium hydrates. J. Pharm. Sci. **1999**, *88*, 1191–1200.

40. Chen, L.; Grant, D.J.W. Extension of Clausius–Clapeyron equation to predict hydrate stability at different temperatures. Pharm. Dev. Technol. **1998**, *3*, 487–494.

41. Han, J. Influence of Environmental Conditions on the Dehydration of Pharmaceutical Hydrates. Ph.D. thesis, Department of Pharmaceutics, University of Minnesota, Minneapolis, 1998, pp. 162–180.

42. Yoshioka, R.; Okamura, K.; Yamada, S.; Aoe, K.; Date, T. The role of water-solvation in the optical resolution of DL-leucine with (S)-(−)-2-phenyl-ethanesulfonic acid-characterization and x-ray crystal structures of their diastereomeric salts. Bull. Chem. Soc. Jpn. **1998**, *71*, 1109–1116.

43. Yu, L.; Reutzel, S.; Stephenson, G. Physical characterization of polymorphic drugs—an integrated characterization strategy. Pharm. Sci. Tech. Today **1998**, *1*, 118–127.

44. Piyarom, S.; Yonemochi, E.; Oguchi, T.; Yamamota, K. Effects of grinding and humidification on the transformation of conglomerate to racemic compound in optical active drugs. J. Pharm. Pharmacol. **1997**, *49*, 384–389.

45. Zografi, G. States of water associated with solids. Drug Dev. Ind. Pharm. **1988**, *14*, 1905–1926.

46. Bel'skii, V.K.; Zorkii, P.M. Distribution of molecular crystals by structural classes. Soviet. Phys. Crystallogr. (English translation) **1971**, *15*, 607–610.

47. Wallach, O. Zur Kenntniss der Terpene und der ätherischen Oele. Liebigs Ann. Chem. **1895**, *286*, 90–143.

48. Brock, C.; Schweizer, W.B.; Dunitz, J.D. On the validity of Wallach's rule: on the density and stability of racemic crystals compared with their chiral counterparts. J. Am. Chem. Soc. **1991**, *113*, 9811–9820.

49. Duddu, S.P.; Grant, D.J.W. Effects of configuration around the chiral carbon atoms on the crystal properties of ephedrinium and pseudoephedrinium salicylates. Pharm. Res. **1994**, *11*, 1549–1556.

50. Liu, S.T.; Hurwitz, A. Effect of enantiomeric purity on solubility determination of dexclamol hydrochloride. J. Pharm. Sci. **1978**, *67*, 636–638.

51. Repta, A.J.; Baltezor, M.J.; Bansal, P.C. Utilization of an enantiomer as a solution to a pharmaceutical problem: application to solubilization of 1,2-di(4-piperazine-2,6-dione)propane. J. Pharm. Sci. **1976**, *65*, 238–242.

52. Wearley, L.; Antonacci, B.; Cacciapuoti, A.; Assenza, S.; Chaudry, I.; Eckhart, C.; Levine, N.; Loebenberg, D.; Norris, C.; Parmegiani, R.; Sequeira, J.; Yarosh-Tomaine, T. Relationship among physicochemical properties, skin permeability, and topical activity of the racemic compound and pure enantiomers of a new antifungal. Pharm. Res. **1993**, *10*, 136–140.

53. Kommuru, T.R.; Khan, M.A.; Reddy, I.K. Racemate and enantiomers of ketoprofen: phase diagram, thermodynamic studies, skin permeability, and use of chiral permeation enhancers. J. Pharm. Sci. **1998**, *87*, 833–840.

54. Duddu, S.P. Implications of Chirality of Drugs and Excipients in Physical Pharmacy. Ph.D. thesis, Department of Pharmaceutics, University of Minnesota, Minneapolis, 1993, pp. 191–208.

55. Duddu, S.P.; Vakilynejad, M.; Jamali, F.; Grant, D.J.W. Stereoselective dissolution of propranolol hydrochloride from hydroxypropyl methylcellulose matrices. Pharm. Res. **1993**, *10*, 1648–1653.

56. Maggi, L.; Massolini, G.; De Lorenzi, E.; Conte, U.; Caccialanza, G. Evaluation of stereoselective dissolution of verapamil hydrochloride from matrix tablets press-coated with chiral excipients. Int. J. Pharm. **1996**, *136*, 43–52.

57. Alvarez, C.; Torrado, J.J.; Cadorniga, R. Stereoselective drug release from ketoprofen and ricobendazole matrix tablets. Chirality **1999**, *11*, 611–615.

58. Janjikhel, R.K.; Adeyeye, C.M. Dissolution of ibuprofen enantiomers from coprecipitates and suspensions containing chiral excipients. Pharm. Dev. Tech. **1999**, *4*, 9–17.

59. Solinis, M.A.; Lugara, S.; Calvo, B.; Hernandez, R.M.; Gascon, A.R.; Pedraz, J.L. Release of salbutamol sulfate enantiomers from hydroxypropylmethyl-cellulose matrices. Int. J. Pharm. **1998**, *161*, 37–43.

60. Miyazaki, K.; Kaiho, F.; Inagaki, A.; Dohi, M.; Hazemoto, N.; Haga, M.; Hara, H.; Kato, Y. Enantiomeric difference in percutaneous penetration of propranolol through rat excised skin. Chem. Pharm. Bull. **1992**, *40*, 1075–1076.

61. Zahir, A.; Kunta, J.R.; Khan, M.A.; Reddy, I.K. Effect of menthol on permeability of optically active and racemic propranolol across guinea pig skin. Drug Dev. Ind. Pharm. **1998**, *24*, 77–80.

62. Kommuru, T.; Khan, M.A.; Reddy, I.K. Effect of chiral enhancers on the permeability of optically active and racemic metoprolol across hairless mouse skin. Chirality **1999**, *11*, 536–540.

63. Wright, J.D. *Molecular Crystals*; Cambridge University Press: Cambridge, UK, 1995; 42–65.

64. Weissbuch, I.; Porovitz-Biro, R.; Lahav, M.; Leiserowitz, L. Understanding and control of nucleation, growth, habit, dissolution and structure of two- and three-dimensional crystals using 'tailor-made' auxiliaries. Acta. Crystallogr. **1995**, *B51*, 115–148.

65. Mullin, J.W. *Crystallization*, 3rd Ed.; Butterworth-Heinemann: London, 1993.

66. Weissbuch, I.; Zbaida, D.; Addadi, L.; Lahav, M.; Leiserowitz, L. Design of polymeric inhibitors for the control of crystal polymorphism: induced enantiomeric resolution of racemic histidine by crystallization at 25°C. J. Am. Chem. Soc. **1987**, *109*, 1869–1871.

67. Chow, K.Y.; Go, J.; Mehdizadeh, M.; Grant, D.J.W. Modification of adipic acid crystals: influence of growth in the presence of fatty acid additives on crystal properties. Int. J. Pharm. **1984**, *20*, 3–24.

68. Chow, A.H.L.; Chow, P.K.K.; Wang, Z.; Grant, D.J.W. Modification of acetaminophen crystals: influence of growth in aqueous solutions containing *p*-acetoxyacetanilide on crystal properties. Int. J. Pharm. **1985**, *24*, 239–258.

69. Law, D.; Grant, D.J.W. Influence of traces of *n*-octanoic acid in adipic acid crystals on Hiestand's indices of tableting performance of the crystals. In *Agglomeration and Size Enlargement: A Symposium in the First International Particle Technology Forum of the AIChE*. American Institute of Chemical Engineers: Denver, CO, 1994.

70. Burt, H.M.; Mitchell, A.G. Crystal defects and dissolution. Int. J. Pharm. **1981**, *9*, 137–152.

71. Chow, A.H.L.; Gordon, J.D.; Szeitz, A.; Young, J.W.M. Modification of phenytoin crystals. III. Influence of 3-butanoyloxymethyl-5,5-diphenylhydantoin on solution-phase crystallization and related crystal properties. Int. J. Pharm. **1995**, *126*, 11–19.

72. Duddu, S.P.; Mehvar, R.; Grant, D.J.W. Liquid chromatographic analysis of the enantiomeric impurities in various (+)-pseudoephedrine samples. Pharm. Res. **1991**, *8*, 1430–1433.

73. Duddu, S.P.; Fung, F.K.Y.; Grant, D.J.W. Effect of the opposite enantiomer on the physicochemical properties of (-)-ephedrinium 2-naphthalenesulfonate crystals. Int. J. Pharm. **1993**, *94*, 171–179.

74. Duddu, S.P.; Fung, F.K.Y.; Grant, D.J.W. Effects of crystallization in the presence of the opposite enantiomers on the crystal properties of (SS)-(+)-pseudoephedrinium salicylate. Int. J. Pharm. **1996**, *127*, 53–63.

75. Li, Z.J.; Grant, D.J.W. Effects of excess enantiomer on the crystal properties of a racemic compound: ephedrinium 2-naphthalenesulfonate. Int. J. Pharm. **1996**, *137*, 21–31.

76. Gu, C.H.; Grant, D.J.W. Effects of a diastereomeric impurity on the crystal properties of a chiral drug, (+)-pseudoephedrine hydrochloride. Enantiomer **2000**, *5*, 271–280.

77. Gilman, J.J.; Johnston, W.G.; Sears, G.W. Dislocation etch pit formation in lithium fluoride. J. Appl. Phys. **1958**, *29*, 747–754.

78. Bundgaard, H. Influence of an acetylsalicylic anhydride impurity on the rate of dissolution of acetylsalicylic acid. J. Pharm. Pharmacol. **1974**, *26*, 535–540.
79. Klug, D.L. The influence of impurities and solvents on crystallization. In *Handbook of Industrial Crystallization*; Myerson, A.S. Ed.; Butterworth-Heinemann: Boston, 1993; 65–87.
80. Gu, C. Influence of Solvents and Impurities on the Crystallization Process and Properties of Crystallized Products. Ph.D. thesis, Department of Pharmaceutics, University of Minnesota, Minneapolis, 2001, pp. 123–160.
81. Kinbara, K.; Hashimoto, Y.; Sukegawa, M.; Nohira, H.; Saigo, K. Crystal structures of the salts of chiral primary amines with achiral carboxylic acids: recognition of the common-occurring supramolecular assemblies of hydrogen-bond networks and their role in the formation of conglomerate. J. Am. Chem. Soc. **1996**, *118*, 3441–3449.
82. Li, Z.; Ojala, W. Grant, D.J.W. Molecular modeling study of chiral drug crystals: lattice energy calculations. J. Pharm. Sci. **2001**, *90*, 1523–1539.
83. Kinbara, K.; Saiko, K.; Hashimoto, Y.; Nohira, H.; Saigo, K. Chiral discrimination upon crystallization of the diastereomeric salt of 1-aryethyl-amines with mandelic acid or *p*-methoxymandelic acid: interpretation of the resolution efficiencies on the basis of the crystal structures. J. Chem. Soc. Perkin Trans. **1996**, *2*, 2615–2622.
84. Kinbara, K.; Kobayashi, Y.; Saigo, K. Chiral discrimination of 2-arylalkanoic acids by (1S, 2R)-1-aminoindan-2-ol through the formation of a consistent columnar supramolecular hydrogen-bond network. J. Chem. Soc. Perkin Tran. **2000**, *2*, 111–119.
85. Leusen, F.J.J. Rationalization of Racemate Resolution—A Molecular Modeling Study. Ph.D. thesis, Katholieke University, Nijmegen, The Netherlands, 1993.
86. Hartman, J.; Perdok, W.G. On the relationships between structure and morphology of crystals. Acta Crystallogr. **1955**, *8*, 49–52.
87. Grimbergen, R.F.P.; Reedijk, M.F.; Meekes, H.; Bennema, P. Growth behavior of crystal faces containing symmetry-related connected nets: a case study of naphthalene and anthacene. J. Phys. Chem. B **1998**, *102*, 2646–2653.
88. Strom, C.S.; Leusen, F.J.J.; Geertman, R.M.; Ariaans, G.J.A. Morphology of the diastereomeric salt of the alkaloid ephedrine and a chlorine substituted cyclic phosphoric acid. J. Cryst. Growth **1997**, *171*, 236–249.
89. Molecular Simulation Inc., <http://www.msi.com/user/consortia/pdc/surface_docker/surface_docker.htm2000>.
90. Myerson, A.S.; Jang, S.M. A comparison of binding energy and metastable zone width for adipic acid with various additives. J. Cryst. Growth **1995**, *156*, 459–466.
91. York, P.; Ticehurst, M.D.; Osborn, J.C.; Roberts, R.J.; Rowe, R.C. Characterisation of the surface energetics of milled *dl*-propranolol hydro-chloride using inverse gas chromatography and molecular modeling. Int. J. Pharm. **1998**, *174*, 179–186.

92. Harris, K.D.M.; Tremayne, M. Crystal structure determination from powder diffraction data. Chem. Mater. **1996**, *8*, 2554–2570.
93. Shirley, R. Data accuracy for powder index. *Natl. Bur. Stand. (US) Spec. Publ. No. 567*, **1980**.
94. Altomare, A.; Cascarano, G.; Giacovazzo, C.; Guagliardi, A.; Burla, M.C.; Polidori, G.; Camalli, M. Early finding of preferred orientation: a new method. J. Appl. Crystallogr. **1994**, *27*, 1045–1050.
95. Wilson, C.C. Determination of crystal structures from poor-quality data using Patterson methods. Acta Crystallogr. **1989**, *A45*, 833–839.
96. Shankland, K.; Gilmore, C.J.; Bricogne, G.; Hashizume, H. A multisolution method for phase determination by combined maximization of entropy and likelihood. VI. Automatic likelihood analysis via the student *t* test, with an application to the powder structure of magnesium boron nitride, Mg_3BN_3. Acta Crystallogr. **1993**, *A49*, 493–501.
97. Tremayne, M.; Kariuki, B.M.; Harris, K.D.M. Solution of an organic crystal structure from x-ray powder diffraction data by a generalized rigid-body Monte Carlo method: crystal structure determination of 1-methylfluorene. J. Mater. Chem. **1996**, *6*, 1601–1604.
98. Rietveld, H. A profile refinement method for nuclear and magnetic structures. J. Appl. Crystallogr. **1969**, *2*, 65–71.
99. Vogt, T.; Fitch, A.N.; Cockroft, J.K. A powder neutron diffraction investigation of the solid phases of iodine heptafluoride (IF_7). J. Solid State Chem. **1993**, *103*, 275–279.
100. Karfunkel, H.R.; Gdanitz, R.J. *Ab initio* prediction of possible crystal structures for general organic molecules. J. Comput. Chem. **1992**, *13*, 1171–1183.
101. Verwer, P.; Leusen, F. Computer simulation to predict possible crystal polymorphs. In *Reviews in Computational Chemistry*; Lipkowitz, K., Boyd, D. Eds.; Wiley-VCH: New York, 1998; 327–365.
102. Gdanitz, R.J. Prediction of molecular crystal structures by Monte Carlo simulated annealing without reference to diffraction data. Chem. Phys. Lett. **1992**, *190*, 391–396.
103. Payne, R.S.; Rowe, R.C.; Roberts, R.J.; Charlton, M.H.; Docherty, R. Potential polymorphs of aspirin. J. Comput. Chem. **1999**, *20*, 262–273.
104. Payne, R.S.; Roberts, R.J.; Rowe, R.C.; Docherty, R. Examples of successful crystal structure prediction: polymorphs of primidone and progesterone. Int. J. Pharm. **1999**, *177*, 231–245.
105. Rustichelli, C.; Gamberini, M.C.; Ferioli, V.; Gamberini, G. Properties of the racemic species of verapamil hydrochloride and gallopamil hydrochloride. Int. J. Pharm. **1999**, *178*, 111–120.

2
Use of Chiral Excipients in Formulations Containing Chiral Drugs

Moji Christianah Adeyeye
Duquesne University, Pittsburgh, Pennsylvania, U.S.A.

1. INTRODUCTION

A majority of the chiral excipients employed in pharmaceutical dosage forms are from natural sources and therefore exist in the optically pure form. Chiral excipients have been traditionally used, though unintentionally, in many pharmaceutical dosage forms [1–3] and as stationary phases for gas and liquid chromatography in qualitative and/or quantitative analysis of enantiomeric drugs [4–7]. In the latter case, the interaction between chiral excipients and drugs forms the basis for the separation of enantiomers. Many advances have been made in the use of chiral polymeric materials as stationary phases in liquid and gas chromatography, and in capillary electrophoresis. Fortuitously, some of the investigations that have been documented regarding chiral excipients in pharmaceutical formulations are based on the knowledge of chiral stationary phases. With the advent of stereospecific analytical methods in recent years, more attention has be drawn to the influence of chiral excipients on the modification of in vitro release and in vivo disposition of chiral drugs [1,2,8,9]. However, the significance of drug delivery modulation as a result of chiral drug–excipient interactions has been controversial [1–3,10]. This article presents a concise review on the use of selected chiral excipients in pharmaceutical dosage forms containing chiral therapeutic agents.

2. CHALLENGES AND REGULATORY PERSPECTIVES IN THE DEVELOPMENT OF FORMULATIONS CONTAINING CHIRAL DRUGS

Chiral excipients have been widely used in pharmaceutical dosage forms. A selected list of commonly used excipients along with their major applications is shown in Table 1. The interactions of chiral excipients with drugs and their influence on the therapeutic outcomes have not been thoroughly investigated. This may be due to the misconception that excipients are

Table 1 Selected Chiral Excipients and Their Pharmaceutical Applications

Excipients	Major Applications
A. Celluloses	
Carboxymethylcellulose calcium	Disintegrant, suspending agent, stabilizer
Carboxymethylcellulose sodium	Disintegrant, suspension stabilizer, coating agent
Ethylcellulose	Binder, coating material, viscosity builder
Hydroxyethyl cellulose	Binder, film former, dissolution modifier, viscosity builder
Hydroxypropyl cellulose	Viscosity builder, binder, stabilizer
Hydroxypropyl methylcellulose	Dissolution modifier, suspension stabilizer coating agent, film former, viscosity builder
Hydroxypropyl methylcellulose phthalate	Enteric coating agent, taste masking
Microcrystalline cellulose	Binder, diluent, disintegrant, stabilizer
Cellulose acetate	Dissolution modifier, film coating agent
Cellulose acetate butyrate	Dissolution modifier, film former
Cellulose acetate phthalate	Dissolution modifier, film former
B. Starch, Sugars and Derivatives	
Alginic acid	Binder, disintegrant, viscosity builder
Carrageenan	Sustained release matrix, suspending agent
Dextrose	Diluent, sweetener
Fructose	Diluent, sweetener, dissolution enhancer
Guar gum or Galactomannan	Binder, suspending agent, viscosity builder, disintegrant

(continues)

Table 1   Continued

Excipients	Major Applications
Lactose	Diluent, sweetener
Mannitol	Diluent, sweetner, bulking agent for lyophilized products
Maltose	Sweetener, diluent
Pectin	Dissolution modifier, dispersant
Scleroglucan	Dissolution modifier
Sorbitol	Diluent, humectant, sweetner
Starch	Binder, diluent
Sucrose	Sweetener
Xanthan gum	Viscosity builder, suspension stabilizer
C. Cyclodextrins	
Beta-cyclodextrin	Complexing agent, dissolution enhancer, stabilizer
Hydroxypropyl β-cyclodextrin	Complexing agent, dissolution enhancer, stabilizer
Heptakis-(2,6-di-O-ethyl)-β-cyclodextrin	Complexing agent, dissolution enhancer, stabilizer
D. Acids	
Ascorbic acid	Antioxidant
Lactic acid	Acidifying agent, acidulant
Malic acid	Acidulant, antioxidant, flavor, buffering agent
Tartaric acid	Effervescent agent, diluent
E. Amino Acids, Peptides and Derivatives	
Arginine	Stabilizer
Aspartame	Sweetener
Bovine serum albumin	Solubility enhancer (for biomolecules)
Human serum albumin	Solubility enhancer (for biomolecules)
Lysine	Stabilizer
Protamine	Stabilizer
F. Fats, Oils, Essentials Oils	
Medium chain triglycerides	Emulsifying agent, suspending agent in emulsion systems, absorption enchancer, dissolution modifier
Carvacrol	Flavor, Permeation enhancer
Carvone	Flavor, Permeation enhancer
1,8, cineole	Flavor, Permeation enhancer
± Linalool	Flavor, Permeation enhancer
D-limonene	Flavor, Permeation enhancer
Menthone	Flavor, Permeation enhancer
L-menthol	Flavor, Permeation enhancer
α-tocopherol	Antioxidant

"inactive" materials and are not expected to affect the performance of dosage forms. Another reason may be a lack of proper understanding regarding the stereochemistry of these molecules and their implications on the performance of the dosage form, especially in the presence of chiral drugs. The interplay between some chiral excipients and drugs in pharmaceutical formulations may be advantageous or unwarranted, depending on the characteristics of the drugs and the excipients. On one hand, interactions of chiral drugs with chiral excipients may enhance physicochemical properties including solubility and stability, which can be exploited to meet the objectives of the drug product development. On the other hand, chiral recognition of drugs and excipients may lead to undesirable interactions affecting the physical, chemical, organoleptic, and biopharmaceutical properties of the drug product or the enantiomer/ diastereomer in the drug product. Such properties may include solubility, thermal characteristics (such as melting point and glass transition temperature), crystallinity or crystal packing, intermolecular bonding, taste, odor, dissolution, and bioavailability.

The potential problems or opportunities associated with the interactions of chiral excipients with chiral drugs have not been adequately addressed. For example, there are a number of reports in the literature suggesting the possibility of chiral interactions between drugs and excipients in modified-release dosage forms [2,8]. Nonetheless, the biopharmaceutical consequences of such interactions, if any, have not been investigated. Similarly, many studies have suggested enantioselective differences in the dissolution of chiral drugs, but the significance of these differences in terms of therapeutic outcomes has been controversial [1–3,8]. Although the significance has been noteworthy in some studies, more research in this area is anticipated that may be critical to the development of safe and effective dosage forms and delivery systems for chiral drugs. The challenge, therefore, lies more in understanding the stereochemical implications of the materials used as pharmaceutical excipients on the release and the overall therapeutic outcomes of chiral therapeutic agents. Careful consideration must also be given to the safety of such excipients, since the regulatory requirement of "generally regarded as safe (GRAS)" must be met by all pharmaceutical excipients.

In light of the current trend and rationale for the development of optically pure enantiomeric drugs, one can also make a case for the development of single isomers for chiral excipients when one stereoform of an excipient confers superior attributes to the dosage form over its antipode, which may be translated as unique benefits for the drug product. As an example, only the L-L isomer of aspartame (N-aspartylalanine methyl ester) is marketed as artificial sweetener or sugar substitute with potential applications as a sweetening agent in pharmaceutical formulations, while its L-D diastereomer is bitter [11].

The Food and Drug Administration (FDA) has developed the policy for manufacturers that make stereoisomeric drugs. These guidelines are discussed in Chap. 10. Although the FDA does not have a stated policy for the development of chiral excipients, some of the requirements for chiral drugs may be extended to these excipients. These may include (1) design of GRAS chiral excipients, (2) determination of isomeric composition and purity of chiral excipients, (3) assessment of chiral conversion, racemization, or mutarotation and their consequences on product performance, (4) development of enantiospecific assays for chiral excipients, (5) investigation of pharmacokinetic and pharmacodynamic implications of the drug product due to drug–excipient interaction(s), and (6) determination of the safety of excipients or formed diastereomers.

Some relevant questions that need further investigation with regard to the use of chiral excipients include (1) Is the interaction between chiral excipient and drug so significant as to affect physical/chemical properties, in vitro dissolution or drug release, and consequently bioavailability and efficacy? (2) Are the reported incompatibilities that are associated with many chiral excipients direct or indirect consequences of isomeric drug–excipient interaction? (3) Could knowledge of the utilization of chiral stationary phases in chromatography be applied to the development of novel stereoisomeric delivery systems?

3. PROPERTIES AND CHARACTERIZATION OF CHIRAL EXCIPIENTS

Synthesis of a chiral molecule from achiral precursors always leads to a racemic modification, resulting in the formation of one of the three possible entities: (1) a racemic mixture (or conglomerate), (2) a racemic compound (racemate), and (3) a pseudoracemate (or mixed crystal) [12]. These three racemic modifications emerge from differences in the packing forces in the crystal lattice. In a racemic mixture, each enantiomer has a greater affinity for molecules of its own kind than for those of the other enantiomer. As a result, the two enantiomers crystallize in separate phases. In a racemic compound, each enantiomer has a greater affinity for molecules of the opposite configuration than for its own kind. Consequently, the unit cell of the crystal contains an equal number of molecules of each enantiomer and the product is regarded as a true addition compound. In cases where there is little difference in the affinity between enantiomers of like or opposite type, the two enantiomers exist in an unordered manner in the crystal.

In such cases, the racemic modification shows nearly ideal mixing and forms a racemic solid solution (pseudoracemate or mixed crystal) [13,14].

Many chiral excipients are naturally derived and are therefore isomerically pure. Additionally, the technological advances over the years have made the synthesis of single isomers on a large scale possible. However, storage conditions and handling could cause different batches of the same substance to vary in isomeric purity. The properties of isomers should therefore be studied and contrasted to their corresponding racemates prior to use in pharmaceutical formulations. Some of these properties include optical activity, organoleptic properties, particle/crystal shape, density, thermal properties (such as melting point and heat of mixing), solubility, vapor pressure, infrared spectra, nuclear magnetic resonance (NMR) spectra, x-ray spectra, electrophoretic mobility, and liquid chromatography.

Optical Activity.   Optical activity is the most characteristic index of optical purity in cases where the chiral excipients are suspected to be enantiomerically impure. Chiral excipients with no observable optical activity are assumed to be in a 1 : 1 ratio of enantiomers (racemates). However, the measurement may be accidental due to the storage conditions (such as temperature and medium) in which the determination was done, which could lead to changes in optical activity [15]. In such cases, the sample is not considered racemic but is said to be *cryptochiral* [16].

Organoleptic Properties.   Stereochemical differences of enantiomeric excipients may influence perception by sensory organs. Piutti [17] reported as early as 1886 that the interaction of stereoisomer with chiral receptors led to chiral discrimination as a consequence of the formation of diastereomers. He observed that the dextrorotatory asparagine has a sweet taste whereas the levorotatory form is tasteless. Greenstein and Winitz [18] and Solms et al. [19] reported such differences for many amino acids. Shallenberger et al. [20] reported that for some monosaccharides, both isomers have similar sweetness. In contrast, aspartame (*N*-aspartylalanine methyl ester) is marketed as the L,L isomer because it is more than 100 times as sweet as sucrose. However, the L,D diastereomer of aspartame is bitter [11]. It should be noted that the individual differences of perception of these properties could vary.

Terpenes are widely used as pharmaceutical excipients. The odor of some terpenes such as carvone, limonene, and menthol are stereospecifically different [21–23]. For example, S-(+) carvone has the odor of caraway, while the R enantiomer has a spearmint odor [20]. In the case of limonene, the R isomer has an orange odor, whereas its antipode has a lemon odor.

Emberger and Hopp [24] reported that (−)-menthol exhibits a cooling effect, but its optical antipode is devoid of such property. The differences are thought to be due to stereoselective recognition by the receptor proteins, i.e., the chiral binding site of the receptor is preferentially occupied by one isomer as a result of stronger binding interaction [25].

Density. Racemates can differ in density compared to the pure isomers, and this difference could be as large as 5%. While density differences have been widely reported for chiral drugs [26], they have not been reported for excipients. Ordinarily, the difference is based on the enthalpy change associated with the formation of the racemate. The density differences between the racemate and its isomers could affect the melting point and other properties such as solubility and dissolution rate. Differences in densities could be observed in some cases where the isomers are stable and easily resolvable [27]. Density-based differences are not expected to be significant where mutarotation occurs.

Thermodynamic Characteristics. Thermal events could be very important in formulations containing isomeric/racemic drugs or excipients when there are differences in the melting points of the enantiomers and the racemate. The thermal behavior can be measured using the differential scanning calorimetry (DSC) or differential thermal analysis (DTA). DSC curves can reveal isomeric purity in a mixture. DSC measures the enthalpy absorbed (of an endotherm) by a sample as a function of temperature and is used to determine the isomeric composition based on the area above the endotherm. For example, DSC curves of alpha-methylbenzylammonium cinnamate are shown in Fig. 1. The melting point of the pure enantiomer is given by T_A^f, while T_E represents that of the racemate. The area of the first endotherm (E) is proportional to the heat of fusion ($20.66 \, J \, g^{-1}$) or the energy necessary to melt the racemate at 144.26°C. For the (+) enantiomer, the heat of fusion observed at 160.39°C was $112.6 \, J \, g^{-1}$. DSC studies reveal the absorption of energy during melting of the racemate and the enantiomer. In case of a conglomerate, it is expected that the racemate melts below the corresponding enantiomers [16]. In a racemic compound, the racemate melts at a higher temperature, and addition of an enantiomer lowers the melting temperature.

Piyarom et al. [28] investigated the thermal events of racemic malic acid consequent to grinding and reported a change in a portion of the conglomerate to a racemic compound. Using the physical mixture of the enantiomers, they observed the formation of a eutectic mixture upon melting during the DSC scan, as depicted by an endotherm at 106°C, followed by crystallization of the racemic compound (shown as exotherm)

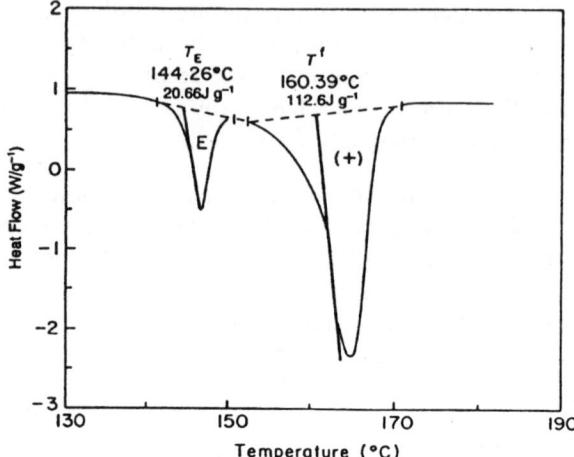

Figure 1 Differential scanning calorimetry curves of melting of an enantiomeric mixture of α-methylbenzylammonium cinnamate. The left peak shows the enthalpy of melting of the eutectic mixture (racemate). The right peak shows the enthalpy of melting of the (+) enantiomer in the mixture. (From Ref. 16.)

at 110°C, and the melting of the racemic compound at 133°C. However, upon grinding, the eutectic melting and recrystallization peaks decreased with grinding time (Fig. 2). There was a gradual conversion of the enantiomers (conglomerate) into racemic compound in up to 30 s. After grinding for 3 min, only the second endotherm peak, i.e., that of the racemic compound, remained. The authors observed similar thermal behavior for tartaric acid.

Attempts have been made to determine enantiospecific differences between enantiomers and racemates in solution using microcalorimetry [29]. The microcalorimetric method can be used to understand the magnitude of stereoselective interactions that could result from the mixing of solutions of enantiomers of a chiral excipient. The heat evolved or heat of solution (ΔH^{soln}) is measured for the racemate as well as the enantiomers and could be indicative of enantioselective discrimination. However, Horeau and Guette [30] reported that enantioselective interaction in an aqueous medium measured using microcalorimetry may be inconsistent and flawed because of the insufficient purity of the optically active samples used and the insensitivity of the measurement relative to small differences in magnitude of the observed effects.

Many authors have reported differences in the DSC melting endotherms of racemates and/or the enantiomers [3,9,31]. In a study that

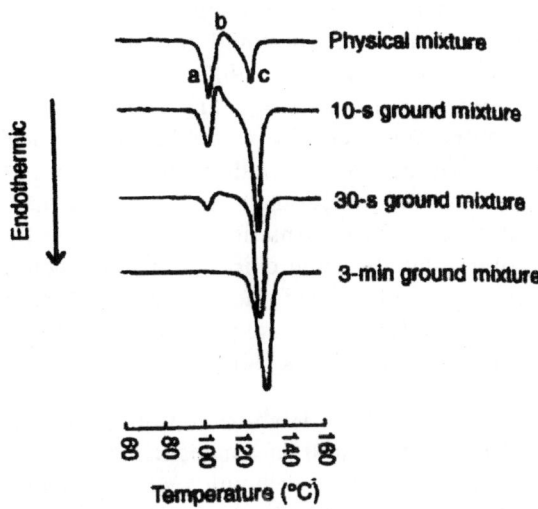

Figure 2 Changes in the DSC curve of physical mixture of malic acid upon grinding. (a) melting of eutectic mixture, (b) crystallization of racemic compound, and (c) melting of racemic compound. (From Ref. 28.)

involved the use of chiral excipients such as tartaric acid, sucrose, hydroxy-propyl methylcellulose (HPMC), hydroxypropyl β-cyclodextrin (HPCD), and ibuprofen in solid and liquid formulations, Janjikhel and Adeyeye [9] used DSC melting endotherms to screen for interactions and reported that there was little or no stereoselective release of the drug, suggesting no chiral interaction.

Vapor Pressure. Differences in vapor pressure (i.e., in the heat of sublimation [ΔH_{sublim}]) have been observed for enantiomers and racemates. It has been suggested that heterochiral interactions between solid enantiomers (as in a racemic compound) can be either stronger or weaker than the homochiral interactions, i.e., between the crystals of the 1 : 1 mechanical mixture [32]. These interactions have not been widely investigated in pharmaceutical formulations containing racemates or enantiomers. The difference in ΔH_{sublim} can also be used to separate enantiomers from a mixture, and it is thought to be a better method of separation, at least in some cases, than recrystallization.

Infrared. Infrared (IR) spectra can reveal whether a racemate is a conglomerate or a racemic compound. If a mixture is a conglomerate, no difference in IR of the enantiomers or corresponding racemate should be observed [16]. In a racemic compound, IR could reveal differences between the racemate and enantiomers. Observed differences in spectra could reveal that certain molecular interactions are modified [16]. For example, using hydrogen phthalate ester as a model carboxylic acid, Eliel and Kofron [33] showed that hydrogen bonding in the racemate is more important than in the individual enantiomers. In carboxylic acids, hydrogen bonds are responsible for dimer formation, and IR revealed that the dimer is a racemic compound. In some chiral excipients such as hydrogen-bonded carboxylic acids, mandelic acid, lactic acid, and tartaric acid and carbohydrates (e.g., arabinose), IR can readily reveal differences in the racemates and the corresponding enantiomers [34].

Nuclear Magnetic Resonance (NMR). High-resolution solid state NMR spectra can be obtained for isomers and racemates using a combination of cross-polarization and magic-angle spinning. Measurement is based on differences in chemical shifts of the NMR spectra for isomerically pure and racemic samples. Differences in solid state ^{13}C NMR of tartaric acid enantiomers and racemate have been reported [35]. Enantiomer discrimination in a solution may reveal little or no difference with the solid-state analysis. Kim and Park [36] used ^{1}H NMR to study stereoselective interaction between terbutaline and β-cyclodextrin (β-CD). Similarly, Kano et al. [37] evaluated enantioselectivity between anionic heptakis (per CO_2-β-CD) and methyl esters of some dipeptides. The binding constant obtained from the NMR titrations suggested that van der Waals forces contributed to the complexation between the CD and the peptides.

X-Ray Diffraction. Crystallography can reveal the symmetry of the constituent molecules in a sample. Isomerically pure samples crystallize in noncentrosymmetric space groups (i.e., absence of center of symmetry). In essence, this can give information on whether the compound exists as a conglomerate. Racemic compounds crystallize mostly in centrosymmetric space groups. Jacques et al. [26] and Brock et al. [27] reported that racemic compounds crystallize in chiral space groups, while conglomerates crystallize in noncentrosymmetric space groups.

Powder x-ray diffraction spectra could show distinct bands for enantiomers and corresponding racemates if the latter is a racemic compound. The nonracemic character was reported for cryptochiral samples of triglycerides and l-lauro dipalmitin [38,39]. Natural triglycerides had little or no optical activity (cryptochiral) while the synthetic racemic triglycerides showed optical activity. This property could be further characterized by inducing piezoelectricity (creation of charges on opposite faces of crystals via mechanical means such as grinding, compression, and torsion). Piyarom et al. [28] also observed changes in chiral molecular state of tartaric acid, serine, and malic acid from conglomerate to racemic compound upon grinding, which were depicted by powder x-ray diffractions. The synthetic triglycerides exhibited no piezoelectricity, while the natural ones did; thus the natural glycerides were nonracemic.

Chromatography. Chromatography (HPLC, GC, and TLC) is commonly used in characterization of chiral drugs and has consequently been used in the evaluation of chiral excipient–drug interactions. Many authors have used HPLC to study stereoselectivity between chiral excipients and drugs [3,8,9,40–42]. The columns used in these studies range form octadesylsilane to chiral selectors such as Chiral AG and molecularly imprinted stationary phases.

4. STEREOSELECTIVE INTERACTIONS OF CHIRAL EXCIPIENTS WITH STEREOISOMERIC DRUGS

Understanding of the stereochemistry of chiral excipients and its implications on product performance can be beneficial in developing formulations of chiral drugs, and it may answer some of the questions raised in previous sections. A relevant discussion on selected chiral excipients and examples of stereoselective interactions between the excipients and chiral drugs is presented here.

4.1. Celluloses

Cellulose is a polysaccharide consisting of D-glucopyranose residues connected by β-1,4 glycosidic linkages (Fig. 3). It consists of a parallel bundle of fibers held together by hydrogen bonding. The D-glucose unit is made up of five chiral centers and three hydroxyl groups with all the ring substituents positioned in an equatorial fashion. The polysaccharide can interact with highly polar drugs with multiple sites for hydrogen bonding. Selected examples of cellulose derivatives and their pharmaceutical applications are described below.

Carbamates. Many examples of cellulose derivatives as chiral stationary phases (CSPs) in liquid chromatography and capillary electrophoresis have been reported in the literature [43,44]. Chankvetadze et al. [43] used cellulose chlorophenyl carbamates as liquid chromatography stationary phases to resolve enantiomers of several chiral drugs including sedatives

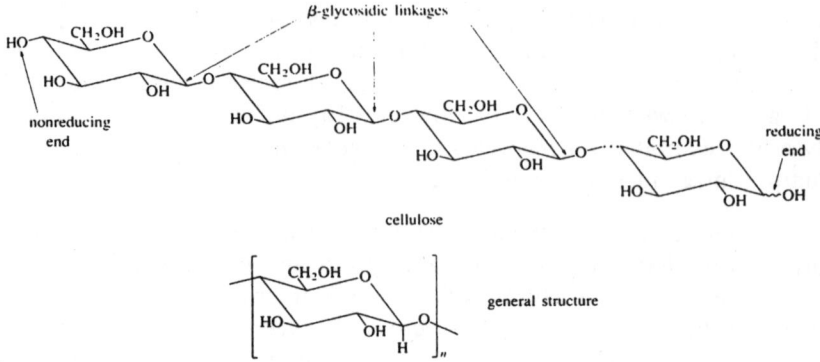

Figure 3 Structure of cellulose.

(hexobarbital, pentobarbital, lorazepam, and camazepam), Ca^{2+} channel blockers (nitredipine and nicardipine), β-blockers (pindolol and acebutolol) and antitussives (clofedanol). Another cellulose derivative [cellulose tris(3,5-dimethyl cabamate)] was used by Krause et al. [44] along with a polyacrylamide-type polymer as chiral stationary phases to resolve chiral drugs such as β-blockers, diuretics, and benzodiazepines. These CSPs could possibly be developed into chiral excipients and marketed as enabling excipients.

Heard and Suedee [40] reported the use of cellulose carbamates in stereoselective adsorption and membrane permeation of racemate and individual enantiomers of propranolol. In this study, three carbamates were used including cellulose tris N-(phenyl carbamate) (R_1); cellulose tris N-(3,5-dichlorophenyl carbamate) (R_2); and cellulose tris N-(3,5-dimethyl-phenyl carbamate) (R_3) (Fig. 4). The S-enantiomer of propranolol was found to be preferentially bound with all the three carbamates used in this study. Further, incubation at 32°C (pH 7.4) increased binding compared to room temperature. The dimethylphenyl carbamate derivative (R_3) had the highest stereroselective binding compared to the dichlorophenyl (R_2) and phenyl (R_1) forms. The ratio R to S enantiomer flux values for R_1, R_2, and R_3 were 2.65, 2.31, and 1.35, respectively. In the control experiment with no carbamates, however, no stereoselective adsorption was observed.

The R:S flux ratios of the enantiomers of propranolol through a silastic membrane of the three cellulose derivatives were equivalent, and the flux rate of the R enantiomer was about twice that of the S enantiomer when pure enantiomers were used. The data for two carbamate derivatives (R_2 or 3,5-Cl and R_3 or 3,5-CH$_3$) are shown in Fig. 5. Differences in flux values of

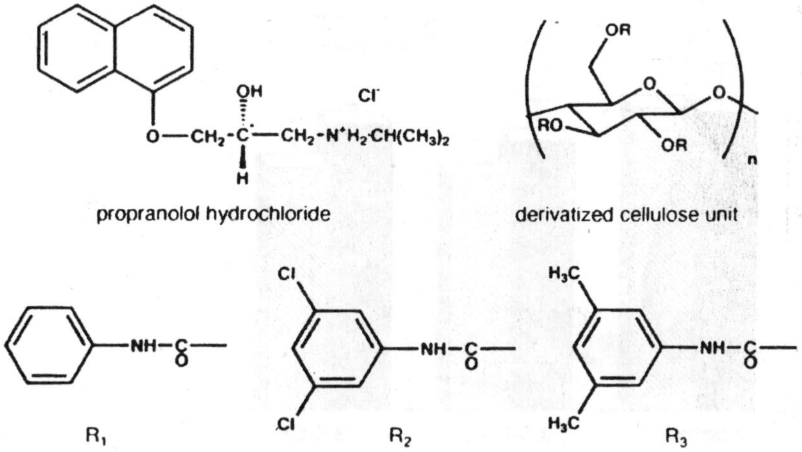

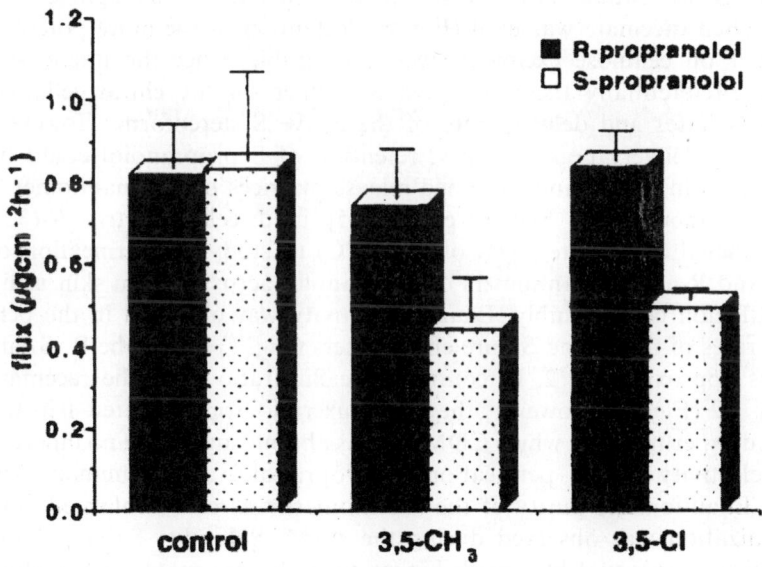

Figure 4 Structure of propranolol hydrochloride and derivatized cellulose, where R1 = N-phenyl carbamate, R2 = N-3,5-dichlorophenyl carbamate, and R3 = N-3,5-dimethylphenylcarbamate. (From Ref. 40.)

Figure 5 Histogram of steady-state flux of optically pure R- and S-propranolol hydrochloride across Silastic (pH 7.4, 32°C). (From Ref. 40.)

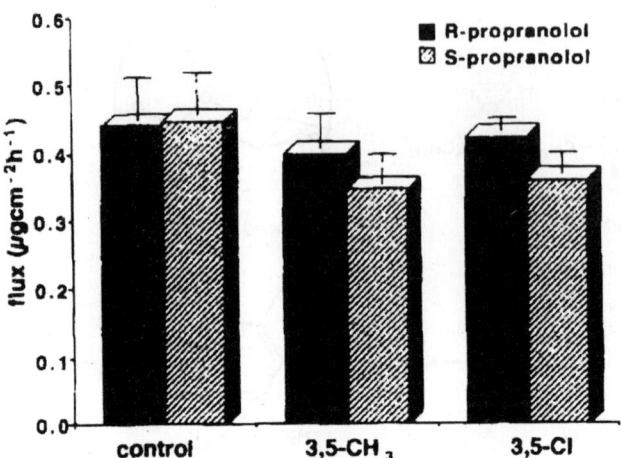

Figure 6 Histogram of steady-state flux of R- and S-propranolol hydrochloride across Silastic (pH 7.4, 32°C) from racemate. (From Ref. 40.)

the two isomers for both carbamates clearly suggest that the S form was more strongly adsorbed on the cellulose than the R isomer, resulting in greater R : S flux ratios. In contrast, the R : S flux ratio was significantly reduced when racemate was used (Fig. 6). Retention of the more potent S enantiomer on cellulose excipients was undesirable, since the intent was to retain preferentially the less active R isomer on the chiral cellulose stationary phases and deliver more of the active S stereoform. However, from a drug delivery point of view, retention of S propranolol could be advantageous in developing sustained release matrices of the enantiomer.

In another study, Suedee et al. [45] used cellulose tris N-(3,5-dimethylphenyl carbamate) –(R₃ or CDMPC) to study the permeation of racemic and R and S enantiomers of propranolol across human skin using Franz diffusion cell assembly. Enantioselectivity was observed in the permeation rates of the R and S pure enantiomers (Fig. 7) where the R : S flux ratio was approximately 2. In contrast, the flux ratio from the racemate was only 1.2 (Fig. 8). However, the R : S flux ratio approximated 1 in the controlled experiments in which CDMPC was absent, suggesting no inherent enantioselectivity in the permeation of propranolol across human skin (Fig. 9). Enantiomeric purity of the enantiomers was also confirmed, and no racemization was observed during the study. Solubility values of the enantiomers were equal but much higher than the racemate, and melting points of the enantiomers were accordingly lower than the racemate. The authors postulated that the observed differences in the permeation of

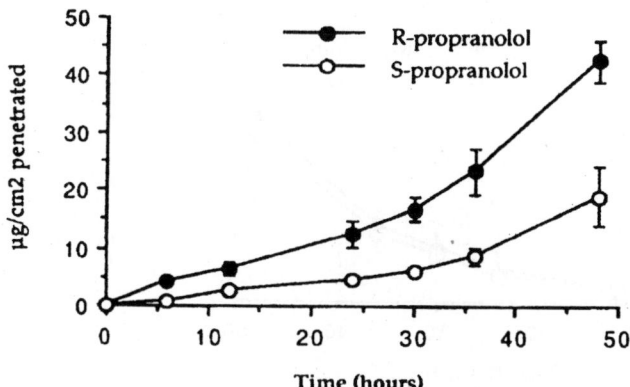

Figure 7 Permeation of enantiomerically pure R- and S-propranolol in the presence of CDMPC across excised human skin from phosphate buffer pH 7.4 ($n = 6 + SE$). (From Ref. 45.)

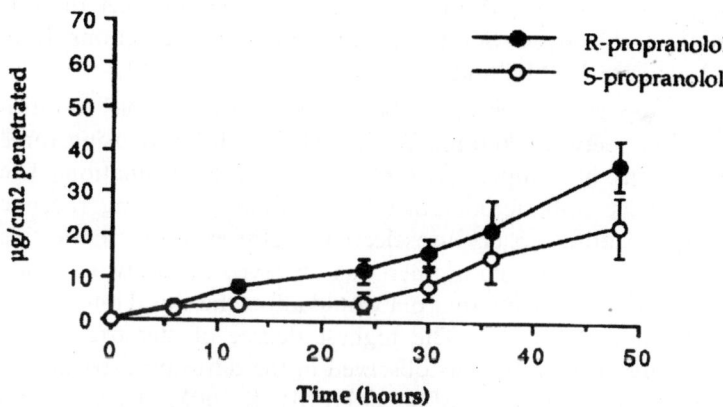

Figure 8 Permeation profiles of R- and S-propranolol from racemate in the presence of CDMPC across excised human skin from phosphate buffer pH 7.4 ($n = 6 + SE$). (From Ref. 45.)

the enantiomers were due to a formulation effect resulting from the imbalance in the donor phase concentrations and the chemical potential of each enantiomer resulting from the enantioselective sorbent [45].

Hydroxypropyl Methylcellulose (HPMC). HPMC is partly *O*-methylated and *O*-2-hydroxypropylated cellulose. It is available in several

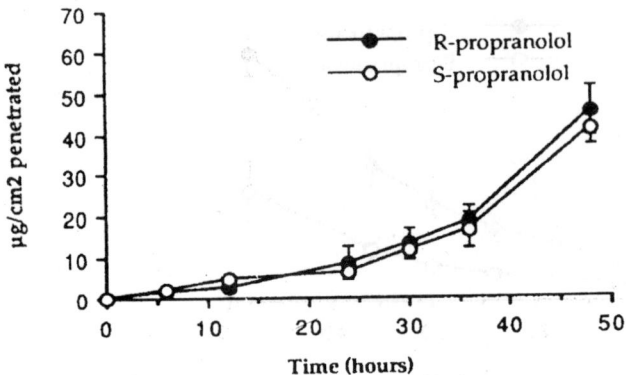

Figure 9 Permeation profiles for propranolol enantiomers in the absence of CDMPC across excised human skin from phosphate buffer pH 7.4 ($n = 6 + SE$). (From Ref. 45.)

grades, varying in degrees of substitution and viscosities to meet the different formulation objectives. Several enantioselective interactions have been described between HPMC and chiral drugs [3,8,9].

In a recent report, Solínis et al. [41] studied the possible enantio-selective interactions between ketoprofen and HPMC K100M using three different formulations of compressed tablets and a gel formulation. The R/S flux ratio for the unbuffered tablet formulation containing racemic drug was unity, indicating no enantioselectivity. However, stereoselective dissolution in favor of the R enantiomer was observed in the two studied buffered formulations, suggesting that pH would affect the chiral interaction between ketoprofen and HPMC. The highest degree of stereoselectivity in release (R/S ratios: 1.07–1.29) was observed in the diffusion experiments with the gel formulation consisting of 0.2% HPMC K 100M. These studies suggest that the chiral interactions between drugs and excipients may be more significant when the mechanism of release is by swelling and diffusion (gel) instead of erosion (tablet) (Fig. 10). Therefore when designing dosage forms with the intent to have chiral recognition between the excipient and the drug, the release mechanism should mainly be diffusion controlled. Nevertheless, in the case of ketoprofen, the stereoselective release even for the gel formulation is expected to be of minor clinical significance.

The release of two chiral drugs, ketoprofen and ricobendazole, from matrix tablets that are produced with two different excipients (a chiral cellulose polymer, HPMC, and an achiral acrylic polymer, Eudragit RL) was investigated by Alvarez et al. [31]. In the case of ketoprofen from

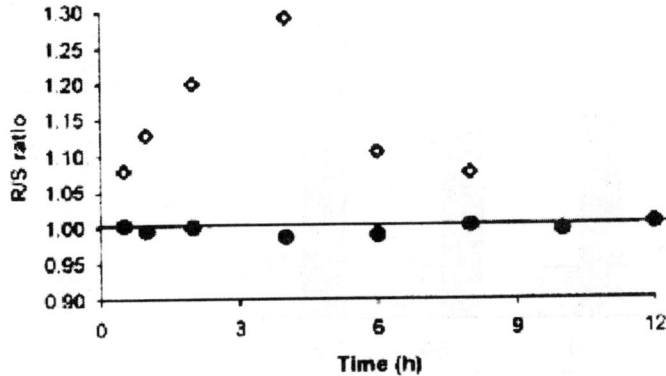

Figure 10 R/S ratio of ketoprofen enantiomers released versus time for unbuffered tablet formulation (●) and gel formulation (◇). (From Ref. 41.)

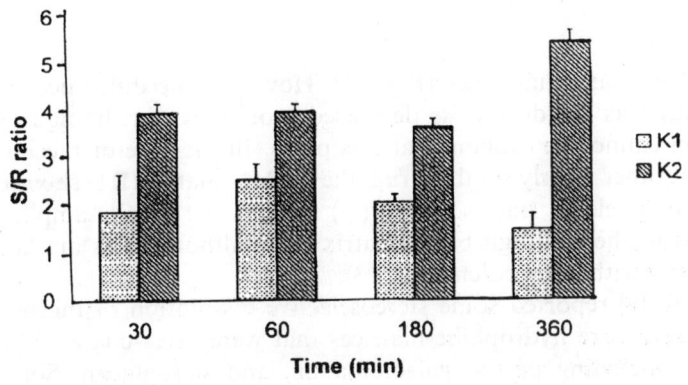

Figure 11 S/R ratio of ketoprofen released at different times for two formulations containing racemic (K1) drug and the physical mixture of the enantiomers (K2). (From Ref. 31.)

HPMC matrix tablets, a greater enantioselectivity was observed with a physical mixture of enantiomers (K2) compared to formulations made with racemate (K1). While from both formulations the S enantiomer was released faster than the R enantiomer, the S/R ratio was significantly higher (S/R ratio >5) in the case of the physical mixture compared to formulations made with racemate (S/R ratio ~1.5) (Fig. 11). When the release of enantiomers of ketoprofen presented as physical mixtures was compared from HPMC (K2) and Eudragit RL (K5), matrices, only the chiral HPMC matrix has shown stereoselectivity with the S/R ratios ranging between

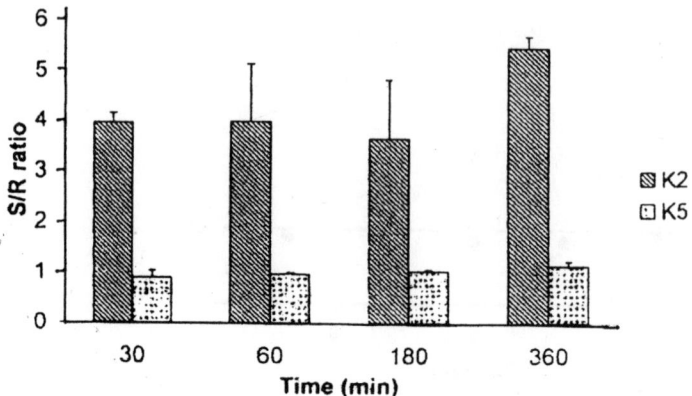

Figure 12   S/R ratio of ketoprofen released at different times for formulations containing physical mixture of the enantiomers in HPMC (K2) or Eudragit RL (K5) matrices. (From Ref. 31.)

3 and 5.5 at all the sampling times (Fig. 12). However, the differences in release disappeared as the drug dose decreased (not shown in the figure). This observation cannot be explained at this point. In the case of racemic ricobendazole, another poorly soluble drug, the HPMC-matrix (R1) showed stereoselective drug release (ratios of (−)/(+) isomers > 1 at all sampling times) compared to the Eudragit-based matrix (R2), although less marked than that obtained with ketoprofen (Fig. 13).

Maggi et al. [8] reported some stereoselective dissolution of racemic (±) verapamil from core hydrophilic matrices that were press-coated with chiral excipients including pectin, galactomanan, and steroglucan. Some excipients such as β-cyclodextrin, hydroxypropyl methylcellulose, and cross-linked amylose did not result in any stereoselectivity in the dissolution of verapamil. Pectin showed the fastest delivery rate, followed by galacto-manan and steroglucan. In the central phase of the release profile, a higher amount of the R enantiomer was always observed with S/R ratios ranging from 0.8 to 0.9. In the same study, stereoselectivity was also observed for a commercial sustained release product of racemic verapamil (Calan®) in which the S/R ratios were below unity for about 14 hours. The authors concluded that stereoselective drug dissolution could occur in formulations if there is enough contact time with chiral excipients such as is the case in sustained release formulations. The dissolution medium penetrates into the core over a lag period only when a matrix is completely hydrated. The drug will then start diffusing outward, and this process of fluid penetration and diffusion of drug could have contributed to possible

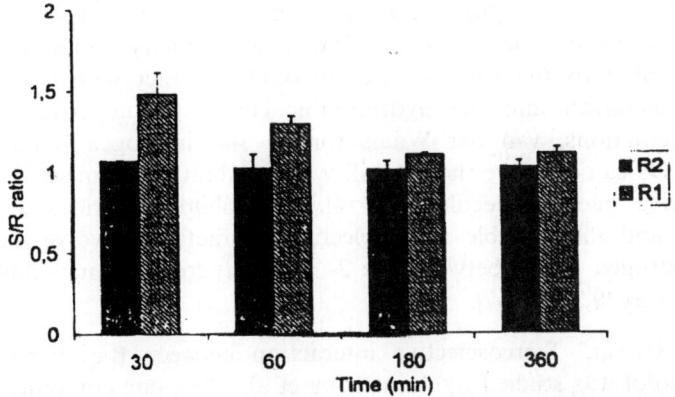

Figure 13 S/R ratio of ricobendazole released at different times for two formulations containing racemic drug in HPMC (R1) or Eudragit RL (R2) matrix. (From Ref. 31.)

interaction between the chiral excipient and the drug [8]. The stereoselectivity in drug release was in agreement with a previous report by Aubry and Wainer [2], where a flow-through-cell device operated at a very low flow rate was utilized.

Cellulose Triacetate.   Cellulose triacetate (triethylcellulose) can interact stereoselectively with enantiomers of chiral drugs [48–50]. It has regions of crystallinity, which allow for enantioselective inclusion of drugs (solutes), especially those having substituent-free phenyl groups. However, it loses its enantioselectivity when solubilized and reprecipitated owing to the breakdown of the crystalline structure.

As previously stated, in order for stereoselective interaction between different celluloses and chiral drugs to occur, a few criteria have to be met. These criteria include that there must be enough contact time between the chiral drugs and cellulose excipients, the polymer must be swellable, and the release mechanism must be mainly by diffusion. The celluloses appear to be useful in developing chiral formulations, but reports are limited in the literature about this potential. Therefore the challenge is to investigate chiral interactions between various celluloses and chiral drugs.

4.2. Cyclodextrins

Cyclodextrins (CDs) are oligosaccharides made up of six, seven, or eight glucopyranose units linked with α-1,4 linkages to form a ring, designated α, β, and γ, respectively. The ring takes a torus or cone shape with hydroxyl groups on the wider (secondary) and the narrower (primary) sides that are usually presented to the solvent medium. CDs interact with many compounds by inclusion into the hydrophobic cavity via noncovalent hydrophobic interactions, van der Waals forces, and hydrogen bonds. Consequently, CDs can improve the solubility and stability, and mask the taste of respective guest molecules [51–54]. The oligosaccharides are optically active, and the possible stereoselective interactions involve the formation of hydrogen bonds between the 2- and 3-hydroxyl groups and the respective isomer [9,35,55–57].

Beta-Cyclodextrin. Stereoselective interaction between β-cyclodextrin and propranolol was studied by Armstrong et al. [55] using computer-assisted molecular modeling based on crystallographic data. The aim was to understand the factors that are responsible for enantioselective inclusion complex formation between cyclodextrin and the enantiomers. They observed that interaction between S-(+)-propranolol and cyclodextrin was greater than that for R-(−) propranolol. Hydrogen bonds between S-(+)-propranolol and the 2- and 3-hydroxyl groups of immobilized β-CD were postulated to be shorter and hence stronger than the bonds between the cyclodextrin and the R(−) isomer.

In another study on the interaction between fenoprofen and β-CD [58], it was reported that a 1:1 complexation occurred. However, in the crystal, each host–guest pair existed as a dimer in a head-to-head alignment in the case of the S-(+) enantiomer, while for the R-(−) antipode, the alignment was tail to tail.

The interaction of racemic cis-ketoconazole with β-CD in the presence of (+); L-tartaric acid, another chiral excipient, was reported by Redenti et al. [57]. Using ^{1}H NMR, ^{13}C NMR, and molecular modeling for the analysis, the authors observed that the inclusion of the dichlorophenyl moiety of ketoconazole into the β-CD cavity occurs from its wider diameter side where the secondary hydroxyl groups reside. The presence of tartaric acid aided the recognition process by establishing electrostatic interaction between the imidazoline ring of the drug and hydrogen bonds with the 2- or 3-hydroxyl groups of the β-CD. Another interaction site was between the phenoxy fragment, which involved van der Waals and electrostatic interaction with the wider rim of β-CD with possible hydrogen bond formation.

The interaction between cyclodextrin and a chiral drug depends on the type of cyclodextrin. Enantioselective interactions, a basis for chiral recognition, may lead to complex supramolecular aggregates. This information can be gathered from NMR studies. For example, the interaction between enantiomers of glutethimide and cyclodextrin are reported by Scriba [4]. In these studies, β-CD and γ-cyclodextrin (γ-CD) displayed an opposite chiral recognition toward enantiomers of the drug glutethimide. For β-CD, inclusion of the drug was via the p-aminophenyl moiety into the cavity from the wider secondary side of CD. In contrast, complexation with γ-CD, albeit via the same p-aminophenyl moiety, was from the narrower primary side of the γ-CD.

Gamma-Cyclodextrin. Srichana and Suedee [59] recently reported a significant chiral interaction between γ-cyclodextrin (γ-CD) and salbutamol as observed from stereoselective drug release in favor of the eutomer (R-salbutamol) from matrix tablets. Other chiral excipients used in the study were sulfobutylether-β-cyclodextrin (SBE-β-CD), heptakis (2,6-di-O-methyl)-β-cyclodextrin (DMCD), HPMC, and egg albumin. The CD tablets were coated while the HPMC and egg albumin tablets were uncoated. In addition to the γ-CD tablets, which exhibited significant stereoselectivity, only the DMCD tablets showed a slight stereoselectivity in release as confirmed by ^{1}H NMR; no stereoselective release was observed for the SBE-β-CD, HPMC, or egg albumin formulations. Scriba [4] also observed chiral recognition between glutethimide and γ-CD.

Hydroxypropyl-β-cyclodextrin (HP-β-CD). Enantioselective inclusion of terbutaline enantiomers into the hydrophobic cavity of hydroxypropyl-β-cyclodextrin was reported by Kim and Park [36]. The inclusion complex was prepared by lyophilization and characterized using DSC and ^{1}H NMR spectroscopy. NMR studies showed that the aromatic moiety of terbutaline was included deep inside the cavity of HP-β-CD, and α(1→4) linkage oxygen of HP-β-CD carried electron densities to the aromatic moiety of the drug. Possible enantioselective binding of each enantiomer with HP-β-CD was also deduced, with the S enantiomer complex showing larger chemical shift than the R enantiomer. This suggests that the methine and methylene protons play a role in chiral discrimination of the terbutaline-HP-β-CD system. The authors concluded that the complexation could enhance the stability of the drug in aqueous media and can be used to develop controlled release terbutaline. In the study by Janjikhel and Adeyeye [9], in which the coprecipitate formulation of ibuprofen and HPCD was developed, no stereoselectivity was reported. This was very likely due to the very fast release manner of the coprecipitate.

Heptakis-(2,6-di-O-ethyl)-β-cyclodextrin (DECD). Anionic heptakis [6-carboxymethylthio-6-deoxy-β-cyclodexrin] (per-CO_2-β-CD) was reported by Kano et al. [37] to interact stereoselectively with methyl esters of some dipeptides such as alanylanine (Ala-Ala-O-Methyl), alanylleucine

(Ala-Leu-*O*-Methyl), trytophylalanine (Try-Ala-*O*-Methyl), and alanyl-tryptophan (Ala-Try-*O*-Methyl). Using ^{1}H NMR and molecular modeling, they observed that van der Waals forces are responsible for the host–guest complexation and that Coulombic interactions between protonated dipeptide and the anionic host also contribute to the interaction. Per-CO_2-β-CD also interacted more strongly with the (*R,R*) isomers more than the (*S,S*) isomers. In complexation of drugs with CD, the most important factor for forming a stable inclusion complex is matching the shape and size between the host and guest molecules. Therefore it is postulated that in solution the asymmetrically twisted native CD recognizes the bended dipeptide methyl ester structure better than the less bended methyl esters of the amino acid monomers. The usual symmetric cyclic cavity of native CD in crystal is stabilized by intermolecular hydrogen bonding between the hydroxyl groups at the 2 and 3 positions of the adjacent glucopyranose units. However, in aqueous solution, symmetry appears to change to a twisted form because of the intermolecular hydrogen bonding between the hydroxyl groups of CD and water molecules. These dipeptides may be useful either as active components or as additives in biopharmaceutical formulations, and the propensity for chiral interaction with excipients such as anionic CDs should be thoroughly investigated.

Heptakis (2,6-di-O-methyl)-β-cyclodextrin (DMCD). Solinís et al. [60] used hydrophilic DMCD along with HPMC (in six of nine tablet formulations), with the presumption that the aromatic ring of the drug could increase the possibility of stereoselective interaction between salbutamol enantiomers and DMCD. However, stereoselective release or interaction was not observed. In the capillary electrophoresis, in which DMCD was added to the buffer, resolution of the enantiomers was noted. In these formulations, cyclodextrins tend to dissolve quickly or disintegrate the matrix too fast to allow for diffusion or enantioselective drug release to take place.

Vakily et al. [61] reported stereoselective release in favor of the R enantiomer for tiaprofenic acid from sustained release formulations containing DMCD when dissolution was carried out at pH 3.0; the S : R ratio of cumulative percent release at 24 h was 0.88 ± 0.04. However, stereoselective release was not observed in the dissolution study carried out at pH 7.4. Nevertheless, the stereoselectivity in the pharmacokinetics of tiaprofenic acid was not different when the powder or the DMCD formulation was administered orally [61], suggesting that the in vitro stereoeselectivity in the release of the drug at pH 3 from the DMCD formulation is not important in vivo.

Tetrakis and Hexakis-β-Cyclodextrins. Other derivatives of β-CD that have been developed and used as chiral stationary phases include tetrakis (endo-exo-6-*O*-norborn-2-ene-5-ylmethoxylsilyl)-β-CD and hexakis (endo-exo-6-*O*-norborn-2-ene-5-ylmethoxysilyl)-β-CD. These CD derivatives have been used to separate the enantiomers of some β-blockers and amino acids [62].

4.3. Molecularly Imprinted Polymers

Molecular imprinting technology facilitates the creation of specific molecular recognition sites in synthetic polymers by the use of appropriate templates. Monomers of specific functional group(s) are allowed to form a complex with a given drug molecule through covalent or noncovalent interactions and subsequently polymerized with cross-linking monomers in which the specific functional group(s) are held in specific positions. The template is removed by solvent extraction or other chemical means, and as a result, the polymer has binding sites specific to the target molecule.

From a drug delivery point of view, molecular imprinting polymer (MIP) beads can be fabricated into matrices and incorporated into different formulations to result in stereoselective drug delivery. The MIP beads selective to S-ibuprofen, S-ketoprofen, and R-propranolol were prepared using multistep swelling and thermal polymerization. MIP beads were also formulated with racemate. The influence of drug/polymer ratio and medium pH on the selective enantiomeric in vitro release of MIP granules was also explored. The release profiles of both S-ibuprofen MIP and R-propranolol MIP granules exhibited enantioselectivity. The enantioselective release of S-ibuprofen MIP and R-propranolol MIP granules appeared to depend on drug/polymer ratio and pH of the dissolution medium. The controlled delivery granules, based on MIPs, demonstrated significant enantioselective release for the chiral drugs tested and thus offer promise as potential excipients in the development of a single enantiomer formulation [63].

The stereoselective release behaviors of "low-swelling" molecularly imprinted polymer bead matrices in pressed-coat tablets were studied using either R- or S-propranolol selective MIPs. The in vitro release profiles of the low-swelling matrices showed a difference in the release of enantiomers, in that the nontemplate isomer was released faster than the template isomer. However, in the last phase of dissolution this difference was reduced and later reversed [64]. Stereoselectivity of release profiles for propranolol enantiomers were identified in MIP synthetic membranes from tablet formulations with significant differences between enantiomers [65]. Release of the enantiomer used as the print was always faster than the release of the nonprint enantiomer. Differences in the release rates of the two propranolol enantiomers were still apparent as the pH was varied between 3 and 7.4 and when the temperature was decreased from 37 to 25°C [65]. In another study, the diffusion of carbobenzyloxy (CBZ) L-tyrosine and CBZ-D-tyrosine through MIP was enantioselective, and diffusion of the template enantiomer (CBZ-L-tyrosine) was faster than that of its optical antipode [66].

4.4. Miscellaneous

Tartaric Acid. D-(−)-tartaric acid occurs in nature and can be obtained from the racemate, but, it is not the form used in pharmaceutical preparations. It is the L-(+)-2,3-dihydroxybutanedioic acid that is used as acidifying agent, flavor enhancer, and sequestering agent. The specific rotation is +12.0° to +12.8° [67]. In a ketoconazole formulation containing cyclodextrin, the presence of tartaric acid provided a basis for chiral recognition between the 2- or 3-hydroxyl group of CD and the imidazoline ring of the drug [57]. Tartaric acid was also incorporated into a coprecipitate formulation of ibuprofen (at drug : tartaric acid ratios of 1 : 1, 1 : 5, and 1 : 10) as reported by Janjikhel and Adeyeye [9], and no stereoselective interaction was observed. This could be due to the faster dissolution of the matrix. A decrease in dissolution was noted as a result of the decrease in pH as the dissolution proceeded.

Dipeptides and Polymeric Di- and Tripeptides. Aspartame is a polymeric dipeptide with a specific rotation of +14.5° to +16.5° [68]. Stereoselective differences in organoleptic properties of aspartame have been documented. Methyl esters of some dipeptides such as alanylalanine and tryptophylalanine have been reported to interact stereoslectively with β-CD. Since these peptides can serve both as active ingredients and as excipients, possible chiral excipient–drug or excipient–excipient interaction should be investigated.

Terpenes. Permeability of the S enantiomer (eutomer) of propranolol was compared to that of racemic propranolol with or without terpene enhancers across guinea pig abdominal skin [69]. Stereoselective permeation of S-propranolol was observed in the presence of menthol, a permeation enhancer, where an increase in membrane permeability coefficient of the eutomer was observed, compared to that of the racemate. The enhancement factors were 2.12 and 0.85 for S enantiomer and racemate, respectively. Lag times were also reduced considerably compared to the control (without enhancer) [70,71].

5. CONCLUSIONS

Several excipients employed in pharmaceutical dosage forms, including celluloses, cyclodextrins, many sugars and sugar derivatives, and terpenes, are chiral in nature. Despite the ubiquitous use of these excipients in pharmaceutical formulations, possible stereoselective interactions that could help or hurt in vitro and in vivo performance of the dosage forms have not been widely investigated. Although a few recent reports have addressed these issues, detailed investigations should be undertaken to enable the pharmaceutical formulation scientists to make proper selection of chiral excipients for improved therapeutic outcomes. A thorough understanding of the implications of interactions between chiral excipients and chiral drug molecules is critical and is the first step in the development of effective dosage forms for chiral drugs. Eventually, multiple "enabling" chiral excipients would be available for enantioselectively designed pharmaceutical delivery systems.

REFERENCES

1. Carr, R.A.; Pasutto, F.R.; Longstreth, J.A.; Foster, R.T. Stereospecific determination of the *in vitro* dissolution of modified release formulations of ± verapamil in serum using achiral-chiral high performance liquid chromatography. Chirality **1993**, *5*, 443–447.
2. Aubry, F.; Wainer, I.W. An *in vitro* study of the stereoselective dissolution of (rac)-verapamil from two sustained release formulations. Chirality **1993**, *5*, 84–90.
3. Duddu, S.P.; Vakilynejad, M.; Jamali, F.; Grant, D.J. Stereoselective dissolution of propranolol hydrochloride from hydroxypropyl methylcellulose matrices. Pharm. Res. **1993**, *10* (11), 1648–1653.
4. Scriba, G.K. Selected fundamental aspects of chiral electromigration techniques and their application to pharmaceutical and biomedical analysis. J. Pharm., Biomed. Anal. **2002**, *27* (3–4), 373–399.
5. De Lorenzi, E.; Massolini, G. Riboflavin binding proteins as chiral selectors in HPLC and CE. Pharmaceutical Science & Technology Today **1999**, *2* (9), 352–364.
6. Gotmar, G.; Fornstedt, T.; Andersson, M.; Guiochon, G. Influence of the solute hydrophobicity on the enantioselective adsorption of beta-blockers on a cellulose protein used as the chiral selector. J. Chromatography A **2001**, *905* (1–2), 3–17.
7. Feibush, B.; Richardson, M.F.; Sievers, R.E.; Springer, C.S., Jr. Complexes of nucleophiles with rare earth chelates. I. Gas chromatographic studies of lanthanide nuclear magnetic resonance shift reagents. Journal of the American Chemical Society **1972**, *94* (19), 6717–6724.

8.   Maggi, L.; Massolini, G.; De Lorenzi, E.; Conte, U.; Caccialanza, G. Evaluation of stereoselective dissolution of verapamil hydrochloride from matrix tablets press-coated with chiral excipients. Int. J. Pharm. **1996**, *136* (1–2), 43–52.

9.   Janjikhel, R.; Adeyeye, C.M. Dissolution of ibuprofen enantiomers from coprecipitates and suspensions containing chiral excipients. Pharm. Develop. and Technol. **1999**, *4* (1), 9–17.

10. Heard, C.M.; Watkinson, A.C.; Brain, K.R.; Hadgraft, J. *In vitro* skin permeation of propranolol enantiomers. Int. J. Pharm. **1993**, *90* (3), R5–R8.

11. Mazur, R.H.; Schlatter, J.M.; Goldkamp, A.H. Structure-taste relationships of some dipeptides. J. Am. Chem. Soc. **1969**, *91* (10), 2684.

12. Eliel, E.T.; Baslo, F. In *Elements of Stereochemistry*; John Wiley: New York, 1969; 16.

13. Li, Z.J.; David Grant, J.W. Relationship between physical properties and crystal structures of chiral drugs. J. Pharm. Sci. **1997**, *86* (10), 1073–1078.

14. Mitchell, A.G. Racemic drugs: racemic mixture, racemic compound, or pseudo-racemate? J. Pharm. Pharmaceut. Sci. **1998**, *1* (1), 8–12.

15. Eliel, E.L.; Wilen, S.H.; Mander, L.S. Chiroptical properties. In *Stereochemistry of Organic Compounds*; John Wiley: New York, 1994; Chap. 13.

16. Eliel, E.L.; Wilen, S.H.; Mander, L.S. Properties of stereoisomers. Stereoisomer discrimination. In *Stereochemistry of Organic Compounds*; John Wiley: New York, 1994; Chap. 6.

17. Piutti, C.R. Sur une nouvelle espece d'asparagine. Hebd. Seances Acad. Sci. **1886**, *103*, 134–141.

18. Greenstein, J.P.; Winitz, M. Chemistry of the Amino Acids, 1 & 2; John Wiley: New York, 1961.

19. Solms, L.; Vuataz, L.; Egli, R.H. Taste of L and D amino acids. Egli, Experientia **1965**, *21*, 692.

20. Shallenberger, R.S.; Acree, T.E.; Lee, C.Y. Sweet taste of D and L-sugars and amino-acids and the steric nature of their chemo-receptor site. Nature (London) **1969**, *221*, 555.

21. Russell, G.F.; Hills, J.I. Odor differences between enantiomeric isomers. Science **1971**, *172*, 1043.

22. Friedman, L.; Miller, J.G. Odor incongruity and chirality. Science **1971**, *172*, 1044.

23. Lietereg, T.J.; Guadagni, D.G.; Harris, J.; Mon, T.R.; Teranishi, R. Evidence for the difference between the odors of the optical isomers (+)- and (−)-carvone. Nature (London) **1971**, *230*, 455.

24. Emberger, R.; Hopp, R. Synthesis and sensory characterization of menthol enantiomers and their derivatives for the use in nature identical peppermint oils. Spec. Chem. **1987**, *7*, 193.

25. Allenmark, S.G. Chromatographic methods for optical purity determination of drugs. Acta Pharm. Nord. **1990**, *2*, 161–170.

26. Jacques, J.; Collet, A.; Willen, S.H. *Enantiomers, Racemates and Resolutions*; John Wiley: New York, 1981.

27. Brock, C.P.; Schweizer, W.B.; Dunitz, J.D. On the validity of Wallach's rule: on the density and stability of racemic crystals compared with their chiral counterparts. J. Am. Chem. Soc. **1991**, *113*, 9811.

28. Piyarom, S.; Yonemochi, E.; Oguchi, T.; Yamamoto, K. Effects of grinding and humidification on the transformation of conglomerate to racemic compound in optically active drugs. J. Pharm. Pharmacol. **1996**, *49*, 384–389.

29. Matsumoto, M.; Amaya, K. The heats of solution of optically active compounds and the corresponding racemic compounds. Bull. Chem. Soc. Jpn. **1980**, *60*, 4139.

30. Horeau, A.; Guetté, J.P. Diastereoisomeric interactions of antipodes in the liquid phase. Tetrahedron **1974**, *30*, 1923.

31. Alvarez, C.; Torrado, J.J.; Cadórniga, R. Stereoselective drug release from ketoprofen and ricobendazole matrix tablets. Chirality **1999**, *11*, 611–615.

32. Chickos, J.S.; Garin, D.L.; Hitt, M.; Schilling, G. Some solid state properties of enantiomers and their racemates.Tetrahedron **1981**, *37*, 2255.

33. Eliel, E.L.; Kofron, J.T. Resolution of *p*-ethylphenylmethylcarbinol. Infrared spectra of enantiomorphs and racemates. J. Am. Chem. Soc. **1953**, *75*, 4585.

34. Wirzing, G. Infrared spectrometric studies of optical antipodes, racemic forms, and inactive mixtures. Z. Anal. Chem. **1973**, *267*, 1.

35. Hill, H.; D.W.; Zens, A.P.; Jacobus, J. Solid-state NMR spectroscopy. Distinction of diastereomers and determination of optical purity. J. Am. Chem. Soc. **1979**, *101*, 7090.

36. Kim, K.H.; Park, Y.H. Enantioselective inclusion between terbutaline enantiomers and hydroxypropyl-β-cyclodextrin. Int. J. Pharm. **1998**, *175* (1), 247–253.

37. Kano, K.; Hasegawa, H.; Miyamura, M. Chiral recognition of dipeptide methyl esters by anionic β-cyclodextrin. Chirality **2001**, *13*, 474–482.

38. Schlenk, W., Jr. New results of configurational investigations. Angew. Chem. Int. Ed. Engl. **1965**, *4*, 139.

39. Schlenk, W., Jr. Synthesis and analysis of optically active triglycerides. J. Am. Oil Chem. Soc. **1965**, *42*, 945.

40. Heard, C.M.; Suedee, R. Tereoselective adsorption and trans-membrane transfer of propranolol enantiomers using cellulose derivatives. Int. J. Pharm. **1996**, *139* (1–2), 15–23.

41. Solínis, M.A.; de la Cruz, Y.; Hernández, R.M.; Gascón, A.R.; Calvo, B.; Pedraz, J.L. Release of ketoprofen enantiomers from HPMC K100M matrices in diffusion studies. Int. J. Pharm. **2002**, *239* (1–2), 61–68.

42. Wulff, G.; Minarik, M. Tailor-made sorbents: a molecular approach to chiral separations. In *Chromatographic Chiral Separations*; Zief, M., Crane, L.J., Eds.; Marcel Dekker: New York, 1988; Chap. 2.

43. Chankvetadze, B.; Chankvetadze, L.; Sidamonidze, S.; Yashima, E.; Okamoto, Y. High performance liquid chromatography enantioseparation of chiral pharmaceuticals using tris(chloro-methylphenylcarbamate)s of cellulose. J. Pharm. Biomed. Anal. **1996**, *14* (8–10), 1295–1303.

44. Krause, K.; Girod, M.; Chankvetadze, B.; Blaschke, G. Enantioseparations in normal- and reversed-phase nano-high-performance liquid chromatography and capillary electrochromatography using polyacrylamide and polysaccharide derivatives as chiral stationary phases. J. Chromatography A **1999**, *837* (1–2), 51–63.
45. Suedee, R.; Brain, K.R.; Heard, C.M. Differential permeation of propranolol enantiomers across human skin *in vitro* from formulations containing an enantioselective excipient. Chirality **1999**, *11* (9), 680–683.
46. Peppas, N.A. Analysis of Fickian and non-Fickian drug release from polymers. Pharm. Acta. Helv. **1985**, *60*, 110–111.
47. Higuchi, T. Mechanism of sustained-action medication. Theoretical analysis of rate of release of solid drugs dispersed in solid matrices. J. Pharm. Sci. **1963**, *52*, 1145–1149.
48. Shibata, T.; Okamoto, I.; Ishii, K. Chromatographic optical resolution on polysaccharides and their derivatives. J. Liq. Chromatogr. **1986**, *9*, 313.
49. Blaschke, G. New analytical methods. 17. Chromatographic racemate separation. Angew. Chem. Int. Ed. Engl. **1980**, *92* (1), 14–25.
50. Ichida, A.; Shibata, T. Cellulose derivatives as stationary phases. In *Chromatographic Chiral Separations*; Zief, M., Crane, L.J., Eds.; Marcel Dekker: New York, 1988; Chap. 9.
51. Bekers, O.; Uijtendaal, E.V.; Beijnen, J.H.; Bult, A.; Underberg, W.J.M. Cyclodextrins in the pharmaceutical field. Drug. Dev. Ind. Pharm. **1991**, *17*, 1503–1509.
52. Ghorab, M.K.; Adeyeye, C.M. Elucidation of solution state complexation in wet-granulated oven-dried ibuprofen and β-cyclodextrin: FT-IR and 1H-NMR studies. Pharm. Develop. and Technology **2001**, *6* (3), 315–324.
53. Badawy, S.F.; Ghorab, M.M.; Adeyeye, C.M. Characterization and Bioavailability of Danazol-hydroxypropyl-?-cyclodextrin coprecipitates. Int. J. Pharm. **1996**, *128*, 45–54.
54. Ghorab, M.K.; Adeyeye, M.C. Enhancement of ibuprofen dissolution via wet granulation with ?-cyclodextrin. Pharm. Develop. and Technology **2001**, *6* (3), 303–312.
55. Armstrong, D.W.; Ward, T.J.; Armstrong, R.D.; Beesley, T.E. Separation of drug stereoisomers by the formation of beta-cyclodextrin inclusion complexes. Science **1986**, *232*, 1132.
56. Ghorab, M.K. Effect of wet granulation of ibuprofen and β-cyclodextrin on complexation, dissolution and bioavailability. Ph.D. diss., Graduate School of Pharmaceutical Sciences, Duquesne University, 2001, p. 7.
57. Redenti, E.; Ventura, P.; Frionza, G.; Selva, A.; Riviara, S.; Plazzi, P.V.; Mor, M. Experimental and theoretical analysis of the interaction of ± cis-ketoconazole with b-cyclodextrin in the presence of (+)-l-tartaric acid. J. Pharm. Sci. **1999**, *88* (6), 599–607.
58. Hamilton, J.A.; Chen, L. Crystal structure of an inclusion complex of β-cyclodextrin with racemic fenoprofen: direct evidence for chiral recognition. J. Am. Chem. Soc. **1988**, *110*, 4379.

59. Srichana, T.; Suedee, R.; Evaluation of stereoselective dissolution of racemic salbutamol matrices prepared with commonly used excipients and 1H-NMR study. Drug Dev. and Ind. Pharm. **2001**, *27* (5), 457–464.

60. Solinís, M.A.; Lugará, S.; Calvo, B.; Hernández, R.M.; Gascón, A.R.; Pedraz, J.L. Release of salbutamol sulfate enantiomers from hydroxypropylmethylcellulose matrices. Int. J. Pharm. **1998**, *161* (1), 37–43.

61. Vakily, M.; Pasutto, F.M.; Daneshtalab, M.; Jamali, F. Inclusion complexation of heptakis(2,6-di-*O*-ethyl)-beta-cyclodextrin with tiaprofenic acid: pharmacokinetic consequences of a pH-dependent release and stereoselective dissolution. J. Pharm. Sci. **1995**, *84* (8), 1014–1019.

62. Mayr, B.; Sinner, F.; Buchmeiser, M.R. Chiral beta-cyclodextrin-based polymer supports prepared via ring-opening metathesis graft-polymerization. Chromatography A **2001**, *907* (1–2), 47–56.

63. Suedee, R.; Srichana, T.; Rattananont, T. Enantioselective release of controlled delivery granules based on molecularly imprinted polymers. Drug Deliv **2002**, *9* (1), 19–30.

64. Suedee, R.; Srichana, T.; Chotivatesin, R.; Martin, G.P. Stereoselective release behaviors of imprinted bead matrices. Drug Dev. Ind. Pharm. **2002**, *28* (5), 545–554.

65. Suedee, R.; Srichana, T.; Martin, G.P. Evaluation of matrices containing molecularly imprinted polymers in the enantioselective-controlled delivery of beta-blockers. J Control Release **2000**, *66* (2–3), 135–147.

66. Dzgoev, A.; Haupt, K. Enantioselective molecularly Imprinted polymer membranes. Chirality **1999**, *11*, 465–469.

67. Vaughan, K.D. Tartaric acid. In *Handbook of Pharmaceutical Excipients*, 3rd Ed.; Kibbe, A.H., Ed.; American Pharmaceutical Association: Washington D.C., 2000; 558–559.

68. Russell, G.; Thurgood, D.M. Aspartame. In *Handbook of Pharmaceutical Excipients*, 3rd Ed.; Kibbe, A.H., Ed.; American Pharmaceutical Association: Washington D.C., 2000, 27–29.

69. Zahir, A.; Kunta, J.R.; Khan, M.A.; Reddy, I.K. Effect of menthol on permeability of an optically active and racemic propranolol across guinea pig skin. Drug Dev. and Ind. Pharm. **1998**, *24* (9), 875–878.

70. Kommuru, T.R.; Khan, M.A.; Reddy, I.K. Effect of chiral enhancers on the permeability of optically active and racemic metoprolol across hairless mouse skin. Chirality **1999**, *11* (7), 536–540.

71. Kommuru, T.R.; Khan, M.A.; Reddy, I.K. Racemate and enantiomers of ketoprofen: phase diagram, thermodynamic studies, skin permeability, and use of chiral permeation enhancers. J. Pharm. Sci. **1998**, *87* (7), 833–840.

3
Transport of Chiral Molecules Across the Skin

Elka Touitou and Biana Godin
The Hebrew University of Jerusalem, Jerusalem, Israel

Thirumala R. Kommuru
Pfizer Global Research and Development, Ann Arbor, Michigan, U.S.A.

Mohsen I. Afouna
University of Arkansas for Medical Sciences, Little Rock, Arkansas, U.S.A.

Indra K. Reddy
Texas A&M University School of Pharmacy, Kingsville, Texas, U.S.A.

1. INTRODUCTION

In the last decade, a number of investigations describing the percutaneous permeation of different stereoforms of chiral compounds have been reported. The rational development of products for transdermal drug delivery requires a thorough understanding of both the biological composition of the skin barrier and the physicochemical properties of drug molecules. For chiral drugs with the biological activity associated with one enantiomer, enantioselective permeation would affect the pharmacodynamic profile of the racemate. In such cases, development of a transdermal delivery system based on a single enantiomer, rather than racemate, could reduce the optimal therapeutic dose required to achieve the desired pharmacological effect, which in turn can minimize or eliminate the adverse effects associated with chiral drugs. This chapter provides an overview of research carried out on the skin permeation of enantiomers as well as racemates of chiral molecules, particularly focusing on the stereoselective skin binding, metabolism, and pharmacodynamic effects in the skin.

67

Further, a mathematical model to predict and correlate the flux characteristics of enantiomer(s) and racemate of chiral drugs is described. Finally, the use of eutectics/chiral terpenes to enhance the percutaneous permeation of selected chiral drugs is also discussed. An outline of the skin structure and its barrier function precedes the discussion on enantioselective skin permeation of chiral drugs in the presence and absence of chiral terpene enhancers.

2. SKIN STRUCTURE AND BARRIER FUNCTION

The skin is the largest organ of the body (weighing more than 10% by mass), its primary function being a permeability barrier to the surrounding milieu [1–3]. It is beyond the scope of this chapter to present a comprehensive review on the skin structure and its barrier function, for which the readers are referred to a book authored by Schaefer and Redelmeier [1].

The skin appendages, which include eccrine and apocrine sweat glands, hair follicles, and sebaceous glands (also referred to as pilosebaceous glands), provide important physiological functions while representing only about 0.1% of the skin surface. They may also play a significant role in the transport of molecules across the skin [1–3]. The appendageal route may serve as an important route of entry through the skin under both transient and steady-state conditions [4].

The skin is a multilayered dynamic structure that is covered by a thin, irregular, discontinuous layer of sebum mixed with sweat, bacteria, and dead cells. The outermost layer of the skin, the stratum corneum or *horny layer*, is about 20 μm thick and has an essential role as a barrier against the transport of water and xenobiotics. A complete turnover of the stratum corneum takes place approximately every two weeks in a normal individual.

The main barrier to percutaneous absorption lies within the outermost layer of the skin, the stratum corneum [1–3]. The stratum corneum provides the main resistance to diffusion, allowing no molecules to pass readily. For electrolytes and large molecules with low diffusion coefficients, shunt diffusion through the appendages may be significant [4]. Histologically, it consists of hexagonal cells, called corneocytes, which are continuously replenished by the slow upward migration of cells produced by the basal layer of the epidermis. Corneocytes are made of insoluble keratins enveloped by cross-linked proteins. They are physiologically and metabolically inactive, have no nuclei, and are embedded in the intercellular lipid domain. The stratum corneum is composed of about 70% protein,

15% lipid, and 15% water [5,6]. The main components of the intercellular lipids are ceramides, free fatty acids, cholesterol, and cholesteryl
esters [5–7]. The lipid composition of different skin layers is shown in
Fig. 1. It is noteworthy that no phospholipids are present in the stratum
corneum, a characteristic feature that distinguishes it from other biological
membranes [6].

The intercellular lipids present the only single continuous region in
the stratum corneum. The stratum corneum can be described as a brick
(corneocytes) and mortar (intercellular lipids) structure [1,2]. The epidermal
cells are held together primarily by highly convoluted interlocking bridges of
desmosomes, tonofibrils (intercellular anchors) and interstitial lipids, which
are responsible for the unique integrity of the skin.

Underlying the stratum corneum is the viable epidermis, which has
a thickness of about 50 to 100 μm. The viable epidermis, which makes a
flat interface with the dead, horny layer, consists of, from top to bottom, the
translucent layer (stratum lucidum), the granular layer (stratum granulosum),
the spinous or prickle layer (stratum spinosum), and the basal layer (stratum
germinativum) [1,2]. The lipid compositions of the stratum granulosum and
stratum germinativum are shown in Fig. 1.

The main cells of the viable epidermis are keratinocytes. They get their
growth factors and nutrients by passive diffusion via the interstitial fluid,
which is estimated to represent about 15% of the total volume of the
epidermis and drains into the lymphatic system. The viable epidermis also
contains melanocytes, Langerhans cells, migrant macrophages, and lymphocytes. The top two layers of the viable epidermis, the stratum lucidum and
the stratum granulosum, are physiologically very important. Removal of
these three epidermal layers results in water loss and an enhancement of
skin permeability [8].

Histochemical tests indicate that in the region of the granular layer
of normal human skin there is a high-energy system responsible for the
synthesis of keratin from polypeptides in the cytoplasm of epidermal cells.
Sulfhydryl groups, phospholipids, and glycogen are concentrated in the
granular layer, but all are absent in the immediately overlying keratin.
At this site the polypeptide chains are presumably unfolded and broken
down and resynthesized into keratin molecules [9].

The dermis follows the epidermis and has a thickness of 1 to 2 mm.
It is a complex structure and consists of cells including fibroblasts, mast
cells, endothelial cells, blood cells, and nerve cells. Fluorescence microscopy
suggests that the dermis is filled with a gel containing oriented tropocollagen
(polypeptide) macromolecules. The network or gel structure is responsible
for the elastic properties of the skin. An approximate composition of the
dermis includes collagen (75%), elastin (4%), reticulin (0.4%), and ground

Touitou et al.

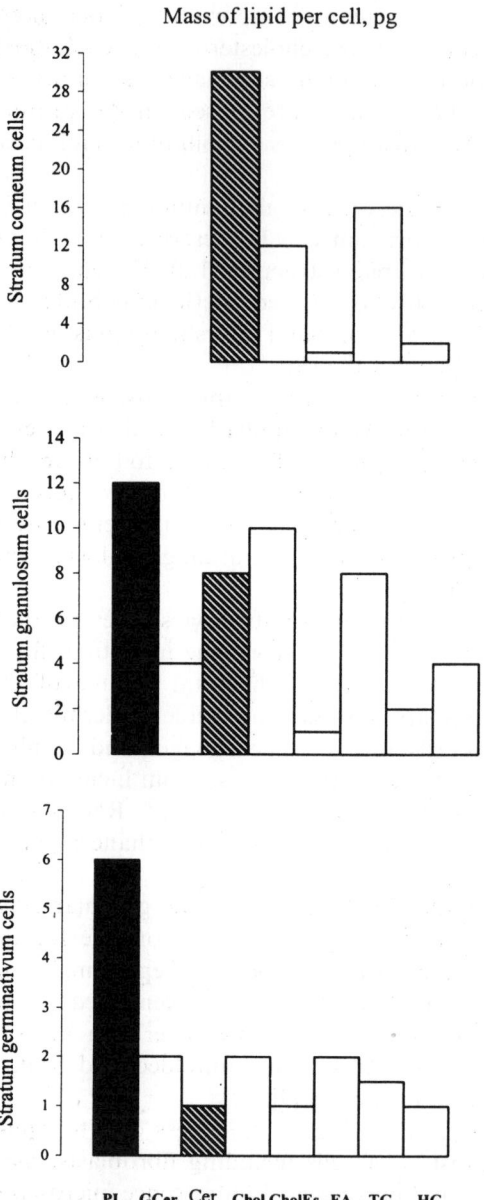

Figure 1   Lipid composition of the stratum corneum and the layers of epidermis. *Key*: PL, phospholipid; Gcer, glucosylceramide; Cer, ceramide; Chol, cholesterol; CholEs, cholesteryl ester; FA, fatty acid; TG, triacylglycerol; HC, hydrocarbon. (Modified from data presented in Ref. 7.)

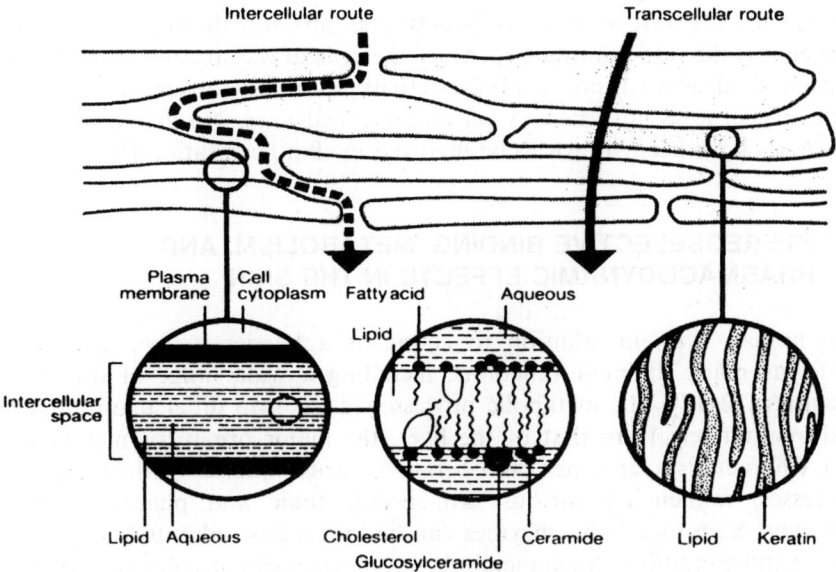

Figure 2 Main routes of penetration for topically applied drugs into the skin.

substance (20%) [1,10,11]. The dermis is rich in blood capillaries, a network of sensory nerves, and a lymphatic network. Therefore, when a topically applied drug molecule reaches the vascularized dermal layer, it becomes available for absorption into the general circulation [2]. Beneath the dermis is the subcutaneous tissue, which is a sheet of fat-containing areolar tissue, known as the superficial fascia, which attaches the dermis to the underlying structures.

The diffusant molecule from a topically applied formulation has three potential routes of entry to the subepidermal tissue; (1) the transappendageal route, (2) the transcellular route and (3) the intercellular route (Fig. 2) [1]. Percutaneous absorption refers to the overall process of mass transport of substances applied topically and includes their transport across each layer of the skin and finally their uptake by the microcirculation of the skin. The process of percutaneous absorption can be described by a series of individual transport events occurring in sequence. First, deposition of a penetrant molecule onto the stratum corneum, then the diffusion through it and through the viable epidermis, the passage through the upper part of the papillary dermis, and finally uptake into the microcirculation for subsequent systemic distribution [1,3,4]. The viable tissue layers and the capillaries are relatively permeable, and the peripheral circulation is sufficiently rapid, so that for the vast majority of penetrants, diffusion through the stratum corneum is the rate-limiting step. Since the cells of the stratum corneum are considered dead with no metabolic activity, diffusion through this layer is a passive process governed by physicochemical laws. No active transport processes have been shown to be involved in skin permeation [1].

3. STEREOSELECTIVE BINDING, METABOLISM, AND PHARMACODYNAMIC EFFECTS IN THE SKIN

The human skin, in addition to acting as a barrier in two directions, performs many different functions, including a wide range of metabolic activities [12–17]. The metabolic processes in the skin differ quantitatively and qualitatively from that of the liver, the major organ for metabolism [12]. Characteristic enzymes can operate in various phases of the metabolic processes, influencing various skin-specific toxic and pharmacological reactions. Skin metabolic enzymes can be induced or inhibited by various agents and conditions. Such metabolizing enzymes can also convert inactive molecules into active pharmacological and toxicological moieties [13,14,17]. Cutaneous metabolism may influence the pharmacokinetic/pharmaco-dynamic parameters of drugs, which could have important implications in the design of transdermal delivery systems.

The role of the skin as a portal of entry for chiral molecules is becoming an increasingly active and exciting field of research. To date, however, only a few investigations have been focused on the effect of enantiomeric differences of chiral drugs with respect to binding or metabolism in the skin on their transdermal delivery. There may be several sources for any observed enantioselectivity in terms of skin permeation of chiral drugs. For example, the components of the stratum corneum such as keratin and ceramides can serve as sources for chiral discrimination that could result in differential diffusion rates, dependent upon the stereochemistry of the solute.

As mentioned previously, phospholipids are absent in the stratum corneum, which is a unique feature that distinguishes the stratum corneum from other biological membranes [6]. The enzyme sphingomyelinase, which is present in the skin, hydrolyzes sphingomyelin to ceramide (an amide formed by linking a fatty acid to sphingosine) and phosphorylcholine. This enzyme degrades all cellular phospholipids before the cells move from the granular layer of the epidermis into the stratum corneum. The phosphorous that is generated is thought to be reabsorbed and reutilized by the epidermis. This may explain the absence of phospholipids in the stratum corneum.

Ceramides, which contain a long-chain or sphingoid base linked to a fatty acid via an amide bond, are formed as the key intermediates in the biosynthesis of all the complex sphingolipids. Relatively high levels of distinctive ceramides are present in the stratum corneum, which include *O*-acylceramides, together with free fatty acids and cholesterol. Some of these skin ceramides have distinctive structures (Fig. 3), which contain the normal range of longer-chain fatty acids (some with hydroxyl groups in position 2), linked both to di-hydroxy bases with trans double bonds in position 4 or to tri-hydroxy bases (structures 1 and 3 in Fig. 3). Further, there are *O*-acylceramides in which the long-chain fatty acid component has a terminal hydroxyl group, which may be in the free form or esterified with either linoleic acid or a 2-hydroxy acid; the sphingoid base can be either a di- or a tri-hydroxy base (structures 2 and 4 in Fig. 3). In addition, the *omega*-hydroxyceramides may be covalently bound to proteins in certain skin cells. The composition of ceramides can vary from one skin layer to the other.

Head groups of ceramides contain multiple hydroxyl groups with a well-defined stereochemistry, so that stereospecific interactions between ceramides and enantiomers of a chiral drug may cause enantioselective permeation. In addition, differential binding of enantiomers of chiral drugs

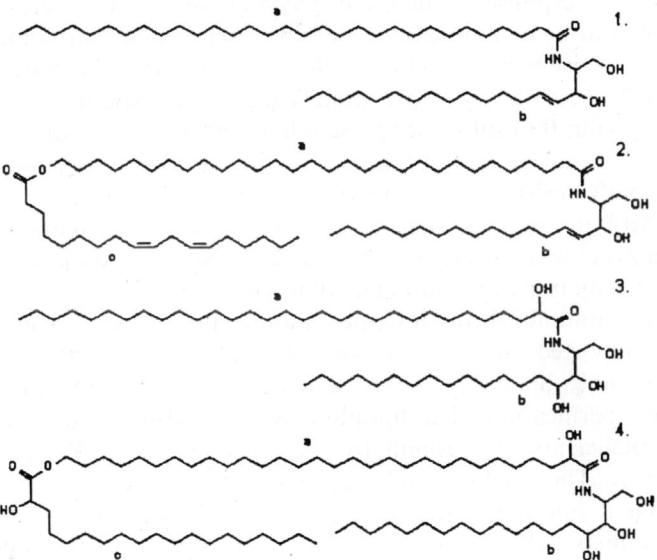

Figure 3 Chemical structures of selected ceramides of the skin.

to keratin or interactions with ceramides may give rise to differences in the permeation profiles of optical antipodes [18,19]. For example, although the percentage of binding of R- and S-isomers of propranolol to solubilized epidermal keratin was not found to be stereoselective $(7.9 \pm 1.7$ vs. $8.3 \pm 2.0\%$, respectively) [19], studies with ceramide monolayers produced qualitative evidence of dose-dependent stereoselective interaction when the pure diastereomers of ephedrine were present in the aqueous phase. This suggested that differences in diffusion rates of ephedrine stereoforms might occur in skin. However, the differences in in vitro permeation rates for these diastereomers through human skin for 12 hours were not statistically significant [119.1 $\pm 2.6\,\mu g/cm^2$ for $(+)$-(1S,2R)-ephedrine versus $107.0 \pm 3.9\,\mu g/cm^2$ for $(-)$-(1R,2S)-ephedrine] [18]. Since evidence for intrinsic stereoselectivity in skin permeation was absent, the authors speculated that the permeant concentration could be high enough to overshadow the ability of the skin to differentiate between stereoisomers.

Unique side-chain conformation responsible for chirality and azimuthal orientation in the molecular alignment of skin collagen was reported. This alignment may account for the axial periodicity and chiral appearance of skin collagen fibrils. However, its implications in transdermal permeability are not yet known [20].

Stereoselective processes were observed within the viable epidermis in contact dermatitis [21,22]. Allergic contact dermatitis (an immunological reaction of the skin in response to the direct physical contact with certain allergens) was shown to be enantiospecific: sensitization to one enantiomer generally does not give rise to any reaction to the nonsuperimposable mirror image allergens [22]. Accordingly, when guinea pigs were experimentally sensitized to $(+)$-tulipalin B (a substance present in tulip bulbs), they did not show any reaction to its antipode, $(-)$-tulipalin B. Similar observations were reported for $(+)$- and $(-)$-β-hydroxy-γ-butyrolactones [21,22].

Stereoselectivity in the metabolism and percutaneous permeation, related to skin enzymatic activity, was reported for several compounds [23–28]. Stereoselectivity in permeation and cutaneous hydrolysis of several ester prodrugs of propranolol through hairless mouse skin was investigated [23]. The authors reported the stereoselective hydrolysis of propranolol prodrugs that is notably biased towards the R-isomer, which resulted in the enantioselective permeation. The lipophilicity of prodrugs, expressed as the partition coefficients, was found to affect the apparent skin permeability coefficients. The more lipophilic prodrugs readily entered into the stratum corneum, but their clearance into hydrophilic deeper strata (epidermis and dermis), where drug hydrolysis takes place, was much less effective. Unlike S-isomers, the R-isomers of propranolol esters were entirely hydrolyzed in epidermis and freely crossed the dermis strata, resulting in enantioselective permeation. Albeit the most lipophilic prodrug exhibited a high stereoselectivity in permeation across full-thickness skin and underwent a higher percentage of concurrent stereoselective cutaneous hydrolysis, it did not show any stereoselectivity in permeation across skin treated with diisopropylfluorophosphate (DFP), an esterase inhibitor. Further, the concurrent cutaneous hydrolysis also ceased. In another but related study, mouse skin showed a very high stereoselectivity for propranolol isomers (R/S ratio: 6.7–18.4) as compared with liver (0.7–2.0) and plasma (1.7–4.7) [24]. The microsomal esterase activity was higher than cytosolic esterase in liver, while an opposite relation was found in the case of skin [24]. In the cases of skin and plasma, some esters exhibited smaller Michaelis–Menton constants (K_m) and larger rates of metabolism (V_{max}) values for the (R)-isomer than those of the (S)-isomer, while K_m was essentially the same between (R)- and (S)-isomers in liver. Interestingly, enzyme inhibition studies indicated that the carboxylesterases were primarily involved in prodrug hydrolysis in liver, whereas skin and plasma were found to be rich in both carboxylesterases and cholinesterases [24]. These results suggest that the hydrolyzing nature of skin esterases responsible for propranolol prodrug was sensitive to stereochemical configuration and more similar to those in plasma esterases than in liver esterases.

Another study, conducted by Udata et al. [25], focused on the stereoselective enzymatic hydrolysis and percutaneous permeation of four diasteromeric propranolol ester prodrug forms (1S,2S; 1R,2S; 1S,2R; 1R,2R) through full thickness shaved rat skin. The reported solubilities of 1S,2S and 1R,2R in acetate buffer pH 4 were 702 and 628 μg/mL, respectively. Accordingly, the transdermal flux of 1S,2S stereoisomer was approximately 3.5 times higher than that of 1R,2R. It was found that all four prodrugs were stereoselectively hydrolyzed faster by acetyl cholinesterase than butyryl cholinesterase, but 1S,2S diastereoisomer exhibited the highest resistance to enzymatic hydrolysis as reflected by its longer stability profile in liver and intestinal tissue preparations. This suggests that the design of prodrugs should involve not only the optimization of physicochemical characteristics but also biochemical stability by membranes containing degradating enzymes. The above reports clearly suggest that the chirality of a compound is critical to receptor-binding interactions. Thus consideration of stereoselectivity during prodrug design is of paramount importance.

Weston et al. studied the metabolism and activation of a polycyclic aromatic hydrocarbon, chrysene, in mouse, rat, and human skin, employing a short-term organ culture technique. Upon examination of the stereochemistry of the metabolic product, chrysene-1,2-diol was formed in each skin type; the (−)-enantiomer was found to be predominant [26]. Amin et al. have reported stereoselective metabolic activation of two carcinogens in mouse skin [27]. In their study, metabolic activation of a strong carcinogen (5-methylchrysene or 5-MeC) and a weak carcinogen (6-methylchrysene or 6-MeC) to dihydrodiols in vitro in rat and mouse liver and in vivo in mouse epidermis was investigated. For both carcinogens, only the R,R-enantiomers of each dihydrodiol predominated (greater than 90%). Further, when the dihydrodiol enantiomers in each case were tested for carcinogenic activity on mouse skin, the R,R-enantiomer was significantly more tumorigenic than the S,S-enantiomer [27].

Levin et al. have reported the differences in the carcinogenic activity between optical enantiomers. They investigated the differences in the tumor-initiating (carcinogenic) activity of optically pure (+)- and (−)-trans-7,8-dihydroxy-7,8-dihydrobenzo(a) pyrene on mouse skin using a two-stage tumorogenesis system [28]. They found that the (−)-enantiomer was 5- to 10-fold more potent than the (+)-enantiomer as a tumor initiator at the three dosage levels tested.

Fretland et al. reported the differential effects of the optical isomers of leukotriene B_4 antagonist on granulocyte diapedesis in the guinea pig dermis [29]. Diapedesis, the migration of cells through the walls of blood capillaries into the tissue spaces, is an important reaction of tissues to injury. When leukotriene B_4 (a proinflammatory product of arachidonic acid metabolism) was coadministered with racemate, (+)-enantiomer, and (−)-isomer of the leukotriene B_4 antagonist into the dermal site, the ED_{50} values for inhibition of granulocyte accumulation of 340 ± 30, 98 ± 5.6, and 1000 ± 142 ng, respectively, were reported [29].

4. SKIN TRANSPORT OF CHIRAL MOLECULES

4.1. Enantioselective Skin Permeation

Although there are many reports enumerating pharmacodynamic, pharmacokinetic, and toxicological differences between the stereoforms of chiral drugs, only a few studies have reported the permeability characteristics of enantiomers of a chiral molecule across the skin. Miyazaki et al. have investigated the difference on the permeation of R-(+)- and S-(−)-propranolol (PL) free base through excised rat skin in vitro [30]. Their data indicated that the flux of the S-(−)-enantiomer of PL across the normal skin was high compared with that of the R-(+)-isomer. The amount of S-(−)-enantiomer permeation was about 4 times higher at 12 h than that of the R-(+)-isomer (Fig. 4). However, when the stratum corneum was

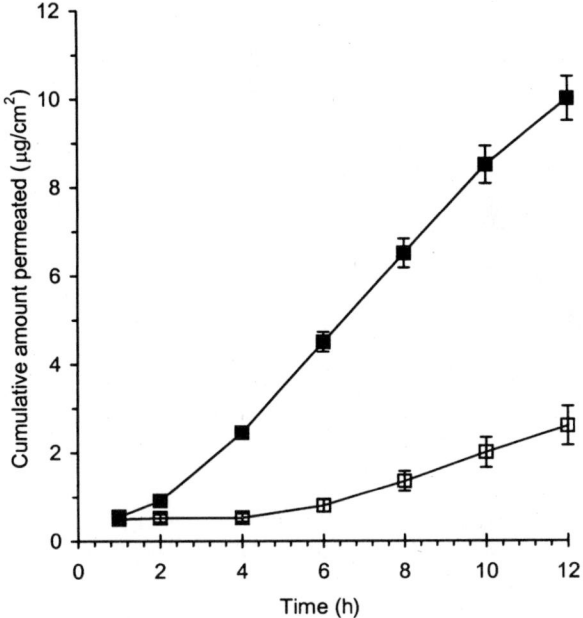

Figure 4 Permeation profiles of (□) R- and (■) S-propranolol free base across the intact rat skin. (Adapted from Ref. 30.)

removed by successive stripping using adhesive tape according to the method described by Washitake et al. [31], the flux values of both the enantiomers were found to be very close (Fig. 5) [30]. These results clearly suggest that there is a difference between the two enantiomers of propranolol in terms of their penetration through full thickness rat skin. Further, the factors responsible for such chiral discrimination appear to be residing in the stratum corneum of the skin.

Interestingly, similar transport experiments by Heard et al. [19] using human skin showed no differences in the permeation profiles of R- and S-propranolol from donor solutions of both the racemate and the pure enantiomers. Similar findings were further reported for the enantiomers of metoprolol [32] and ketoprofen [33] and for the diastereomers of ephedrine [18]; no significant differences were found in the permeation profiles of optical antipodes across skin from donor solutions containing the racemate. For example, when the permeation characteristics of individual enantiomers of metoprolol free base (MB) were investigated using hairless mouse skin, the permeation profiles of R- and S-enantiomers of MB from donor

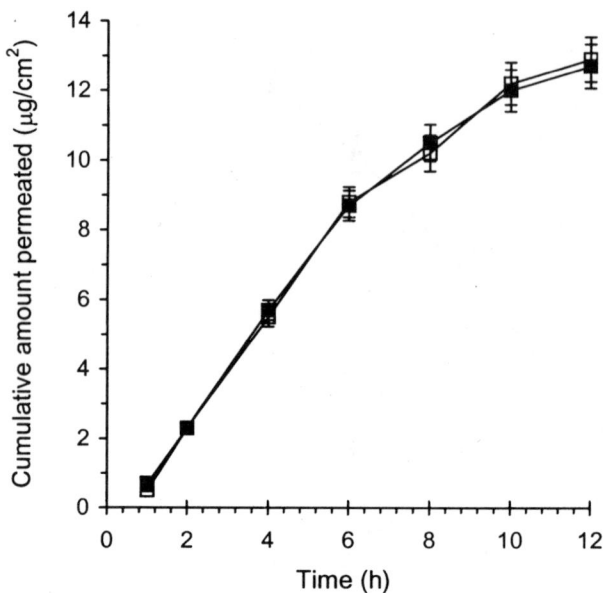

Figure 5 Permeation profiles of (□) R- and (■) S-propranolol free base across rat skin without stratum corneum. (Adapted from Ref. 30.)

solutions containing either racemate (RS-MB) or pure enantiomers were comparable (Fig. 6) [32].

Loschmann et al. proposed transdermal administration of the active enantiomer of a chiral dopamine D-2 receptor agonist N-0437 for sustained reversal of parkinsonian motor deficits in humans [34]. The selective dopamine D-2 receptor agonist produced a rapid and dose-dependent reversal of motor deficits lasting 90 to 120 min following intraperitonial or oral administration of the racemate to 1-methyl-4-phenyl-1,2,3,6-tetrahy-dropyridine (MPTP)-treated common marmosets. In contrast, application of racemate (±)-, (+)-, or (−)-enantiomers of N-0437 to the skin of MPTP-treated animals did not alter locomotor activity in the initial 4-hour period although other motor disabilities were reduced. However, 24 hours following application of the racemate or the (−)-enantiomer both locomotor activity and the other motor deficits induced by MPTP were improved. The increase in locomotor activity returned to basal values by 48–52 h following application of the racemate to the skin and by 72–76 h following administration of (−)-enantiomer of N-0437; the other motor deficits induced by MPTP were reduced for up to 72–76 h by both (±)- and

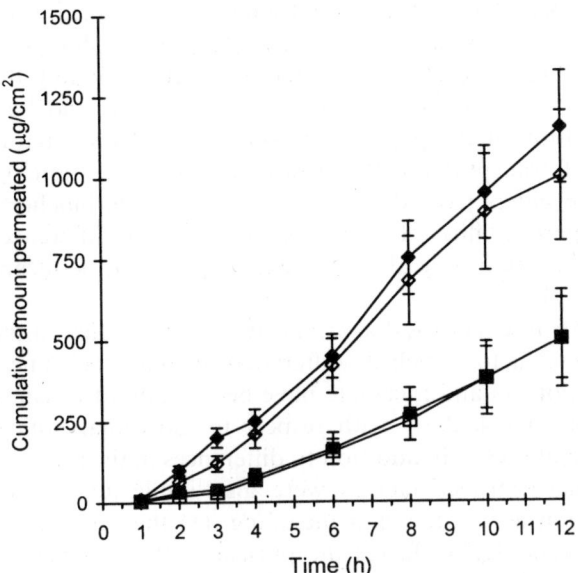

Figure 6 Permeation profiles of individual enantiomers of metoprolol (MB) free base across excised mouse skin from donor solutions containing either racemate or pure enantiomers. *Key*: (□) R-MB from RS-MB, (■) S-MB from RS-MB, (◇) R-MB from R-MB, and (◆) S-MB from S-MB enantiomer. (Adapted from Ref. 32.)

(−)-N-0437. Application to skin of the (+)-enantiomer produced no behavioral improvement or stimulation of locomotor activity. Based on these findings, transdermal administration of only the active enantiomer of N-0437 should be considered and further investigated for providing a prolonged reversal of parkinsonian motor deficits in man.

The lack of intrinsic enantioselectivity in the reported studies was attributed to the following factors [18,19,32–35]: (1) a high drug concentration which might overcome the enantioselective interaction, (2) the presence of organic solvent in the donor vehicle that causes conformational changes in keratin/ceramide molecules, and (3) a different pH of the donor solution from that of the skin. Further, it was observed that ethanol (a commonly used solvent in the donor solution) up to 60% in the donor vehicle increases the fluidity of the lipid domain and also forms new pores in the skin [36]. Under these conditions, therefore, enantiomers may overcome any possible enantioselectivity by diffusing through lipid and protein domains aided by the solvent flow.

Since percutaneous permeation studies are frequently conducted using laboratory animal models such as rats, mice, and guinea pigs, it should be understood that wide differences exist between these models, including the thickness of the stratum corneum, the number of sweat glands and hair follicles, and the distribution of the papillary blood supply. These factors affect both the routes of transport and the resistance to penetration. In addition, the human skin differs from different animal species in biochemical composition and permeability. The subtle biochemical differences between human and animal skin may alter the reaction between permeant molecules and the skin [37].

Apart from these intrinsic factors, due primarily to chiral nature of the skin components, extrinsic factors such as differences in physicochemical properties between enantiomers and racemate have been reported to cause stereoselective permeation, particularly with respect to individual optical isomers versus their racemates [38]. In addition to differences in the physicochemical properties, other extrinsic factors were also implicated in the enantioselective permeation across the skin including: (1) the presence of chiral permeation enhancers [32,33,39], (2) the presence of stereoselective retardants in the donor vehicle [40], (3) differences in the hydrolysis rates of the prodrugs of the enantiomers in epidermis/dermis [23–25], and (4) carrier mediated transport [41].

Stereoselective retardants have been used by some researchers in the donor vehicles as a means of blocking one enantiomer and thereby facilitating the permeation of the other enantiomer [19,40]. Stereoselective retardants such as cellulose tris (3,5-dimethyl phenyl carbamate), which are known for their ability to resolve enantiomers when used as chromatographic stationary phases, were examined. When pure propranolol enantiomers were used, the steady-state flux ratio of R/S was 1.70 across silastic membrane. No such differences in the permeation were observed in the absence of retardants.

Nakagawa et al. studied the chiral discrimination in the transport of ketoprofen and ibuprofen esters through an aqueous phase mediated by various serum albumins [41]. Serum albumins that act as carriers discriminated between enantiomers of alkyl esters of ketoprofen and ibuprofen in transport in the O/W/O (oil/water/oil) system using a U-shaped cell. The transport rate and the preferred enantiomer of the esters were substantially affected by pH, temperature, and species of albumin. Among five serum albumins studied, bovine serum albumin (BSA) showed the largest rate constant, and rat serum albumin (RSA) manifested the highest enantioselectivity. Regarding enantiomer selectivity in overall transport, it is anticipated that the ester uptake step plays an important role for BSA, whereas the ester release is the key step for RSA.

In the assessment of any enantioselective discrimination of chiral molecules with respect to skin permeation, the phenomena of chiral inversion should also be taken into account. While many investigations addressed the role of chiral inversion in the pharmacokinetic characterization of certain chiral agents including some recent reports on ibuprofen and fenoprofen [42,43], investigations on the chiral inversion within the skin are limited. Possible chiral inversion of ibuprofen following its topical administration has been investigated in vitro [44]. Incubation of ibuprofen with rat skin homogenates in the presence of coenzyme A, adenosine triphosphate (ATP), and magnesium provided no evidence for the formation of ibuprofenyl coenzyme A (the initial intermediate in the metabolic inversion of (R)- to (S)-ibuprofen). Moreover, similar incubation studies provided no indication of a change in the enantiomeric ratios of ibuprofen over the time course of the experiments. Percutaneous penetration studies of ibuprofen in a gel formulation through porcine skin indicated that the ibuprofen enantiomer levels in the reservoir solutions were consistent with racemic ibuprofen that permeated the skin without metabolic inversion, suggesting that, in the models studied, skin metabolism does not result in the chiral inversion of (R)- to (S)-ibuprofen [44].

In order to deliver drugs into and across the skin, several techniques are often utilized, including iontophoresis, phonophoresis, and electroporation. These delivery strategies are not expected to have any significant effect on the stereoselectivity in the skin permeation process. For example, Panus et al. examined the stereoselectivity of ketoprofen tissue permeation and drug delivery following cathodic iontophoresis (4 mA, 40 min) of pure enantiomers of ketoprofen onto the porcine medial thigh and found no transcutaneous stereoselective delivery [45].

4.2. The MTMT Concept

A simple relationship between the ratio of permeation fluxes and melting temperatures, referred to as melting temperature–membrane transport (MTMT), was proposed by Touitou et al. in the early 1990s [46]. In accordance with this mathematical model, the flux ratio of R or S enantiomer versus racemate (RS) through the skin (or other membrane) can be predicted, based on the thermodynamic characteristics of each stereoform.

Investigations focusing on thermal characteristics of chiral drugs have shown that melting temperature differences between racemate and enantiomers could be quite large owing to variations in their crystal structure [47–51]. Li et al. [50] performed thermal and thermodynamic analyses of a number of chiral molecules and concluded that thermo-dynamic prediction, thermal analysis, and structural data are in excellent agreement for identifying the nature of the racemic species. Besides, they found melting point differences (ΔT) between racemate and enantiomer of 50°, 38°, 25°, and 40°C for norephedrine, ephedrine, methylephedrine, and dexrazoxane, respectively.

An ideal solubility equation (Eq. 1) gives the relationship between solubility expressed as a mole fraction of dissolved drug in a particular vehicle (C_{max}), melting temperature (T_t), and the enthalpy of fusion (ΔH^f):

$$\ln C_{max} = \frac{-\Delta H^f (T_t - T)}{RT_t T} \qquad (1)$$

where R is a gas constant and T is the temperature of the solution.

Following Eq. (1) for stereoactive compounds, it is possible to choose the entity with the appropriate solubility by selecting the racemate or single enantiomer, when they differ in melting temperatures.

According to Fick's law, the flux of solute through a membrane (F or dQ/dt) is proportional to the concentration gradient that exists across the membrane (C) (Eq. 2).

$$F = \frac{dQ}{dt} = \frac{K_m C D}{h} \qquad (2)$$

where K_m is the partition coefficient, C is the concentration in the donor vehicle, D is the diffusion coefficient, and h is the thickness of the membrane. The donor vehicle should be saturated with diffusant (i.e., at maximum concentration in the donor compartment [C_{max}]) in order to obtain maximal flux (F_{max}).

The membrane fluxes of the enantiomers and their racemate having different melting temperatures will, therefore, not be equal, due to the solubility differences.

The dependence of the permeation fluxes ratio on thermodynamic parameters of pure enantiomer (S) and racemate (RS) of chiral compound can be expressed by Eq. (3).

$$\frac{\ln F_{max\,S}}{\ln F_{max\,RS}} = \ln \left(\frac{C_{max\,S}}{C_{max\,RS}} \right) = \frac{\Delta H_{RS}(T_{t\,RS} - T)}{RT_{t\,RS} T} - \frac{\Delta H_S (T_{t\,S} - T)}{RT_{t\,S} T} \qquad (3)$$

The basic assumption of the MTMT concept is that the diffusion coefficient (D) and the partition coefficient (K_m) are equal for each enantiomer and for different mixtures of enantiomers and there is no significant native enantioselectivity in transport across the membrane due to its chiral nature. The MTMT model was first used to predict the skin permeation behavior of propranolol enantiomers and racemate [46]. In this work, transport of propranolol freebase across human skin and tissue culture (Testskin[TM] LSE) has been reported. DSC measurements showed a difference of 20.8°C in the melting point between S-propranolol and racemate [46,50]. The S-enantiomer/racemate solubility ratios of 1.5, 2.5, 2.8, 2.6, and 4.0 in buffer pH 7.4, saline, mineral oil, 30% PEG aqueous solution, and propylene glycol, respectively, were detected in this study. By applying the MTMT concept, the calculated transdermal flux ratio of propranolol enantiomer versus racemate was 3.2. This predicted value corresponded with the experimentally obtained ratio of 3.3 and 3.1 for human skin and Testskin[TM], respectively. The melting point and solubility of R-propranolol in the listed vehicles were similar to values obtained with S-enantiomer [46]. Results of this study are summarized in Table 1. Since S-propranolol possesses a pharmacological effect two orders of magnitude higher than R-isomer [52] and a three-fold greater skin flux than racemate,

Table 1 Physicochemical Characteristics and Skin Permeation Fluxes of S-Enantiomer and Racemate (RS) of Propranolol Base

		S-Propranolol	RS-Propranolol	Ratio
				$C_{max\ S}/C_{max\ RS}$
Solubility,	Saline	0.55	0.22	2.5
C_{max} (mg/mL)	Buffer, pH7.4	0.70	0.47	1.5
	Propylene glycol (PG)	367.6	91.2	4.0 (experimental) 3.2 (theoretical)
	Mineral oil	6.22	2.19	2.8
	30% PEG400	5.99	2.35	2.5
Melting temperature, T_m (°C)		71.5	92.3	ΔT_m 20.8°C
Enthalpy, ΔH (cal/mol)		8665.3	10385.4	Not relevant
				$F_{max\ S}/F_{max\ RS}$
F_{max}, μg/cm^{2}h,	Testskin[TM]	335	108	3.1
(from PG)	Human cadaver skin	149	45	3.3
	Theoretical, predicted from MTMT model	—	—	3.2

Source: Adapted from Ref. 39.

the S-enantiomer may be proposed as the candidate of choice for transdermal administration [46]. Skin permeation results achieved in another related study showed that the transdermal flux was the same for both (R and S) enantiomers [19].

The MTMT concept is supported by the results obtained by Lawter and Pawelchak [53] and Wearley et al. [54] who investigated the transdermal permeation of nilvadipine (a calcium channel blocker) and genaconazole (an antifungal agent), respectively. Lawter and Pawelchak reported that enantiomers of nilvadipine have a melting point 34°C lower, and consequently a 10-fold higher saturation concentration, than the racemic compound [53]. The human skin permeation flux value of nilvadipine enantiomer was found to be approximately 7-fold higher than that of the racemic compound [53].

Genaconazole (SCH-39304), a potent antifungal agent that has topical as well as systemic efficacy, is available as a racemate of RR (SCH-42427) and SS (SCH-42426) enantiomers. The RR-isomer is the biologically active form that is responsible for most of the antifungal activity of genaconazole. Wearley et al. have investigated the relationship between the physicochemical properties, skin permeability, and topical antifungal activity of racemic genaconazole and its pure enantiomers [54]. The two enantiomers have melting points about 60°C lower than that of the racemate, therefore a higher solubility (~10-fold) as well as higher concentration in the donor vehicle can be achieved for enantiomers compared to that of racemic compound. Accordingly, the pure enantiomers (RR and SS) have exhibited approximately 9-fold higher flux values than that of the racemate across cadaver skin. The flux values measured from in vitro experimental studies correlated well with the solubility differences between the pure enantiomers and the racemic compound. However, when the lesions in the in vivo efficacy study using a guinea pig dermatophyte model were scored, only a minimal difference between racemate (SCH-39304) and the active isomer SCH-42427 was observed [54]. By fitting the data to the pharmacodynamic maximum observed effect parameter (E_{max}) model, the authors demonstrated that the maximum effect occurs at a concentration lower than the saturation concentration of the less soluble racemic compound. The authors concluded that the efficacy of certain topically active antifungal compounds such as genaconazole may not be linearly related to drug concentration in either the vehicle or the skin. It is possible that, under the experimental conditions, the in vivo efficacy of topically active antifungal formulation may not be a single-value function of the drug concentration.

The MTMT model was tested in additional works. Kommuru et al. [33] investigated the permeation behavior of the S-enantiomer vs. the racemate of ketoprofen, a nonsteroidal anti-inflammatory drug, from saturated solutions through nude mice skin. The melting temperature of racemic (RS)-ketoprofen is 94°C, and that of S-ketoprofen is 72°C, which results in ΔT of 22°C between the two entities. The authors reported an experimental flux ratio of 1.79 for ketoprofen enantiomer/racemate (Fig. 7). This value is very close to the theoretically calculated flux ratio of 1.97 using the MTMT relationship. These results show that the active S-enantiomer is more efficiently delivered through the skin. Moreover, since the therapeutic potency of ketoprofen resides mainly in the S-enantiomer [55], the development of a transdermal delivery system for this active S-enantiomer of ketoprofen is beneficial.

Interestingly, when the permeability of various chiral species of ketorolac (a nonnarcotic analgesic) through human cadaver skin was investigated, the observed skin flux of the racemic compound was approximately 1.5 times higher than those of both R- and S-enantiomers [56]. The reported melting point of racemic ketorolac mixture was 20°C lower than that of each pure enantiomer, resulting in about a twofold lower solubility of

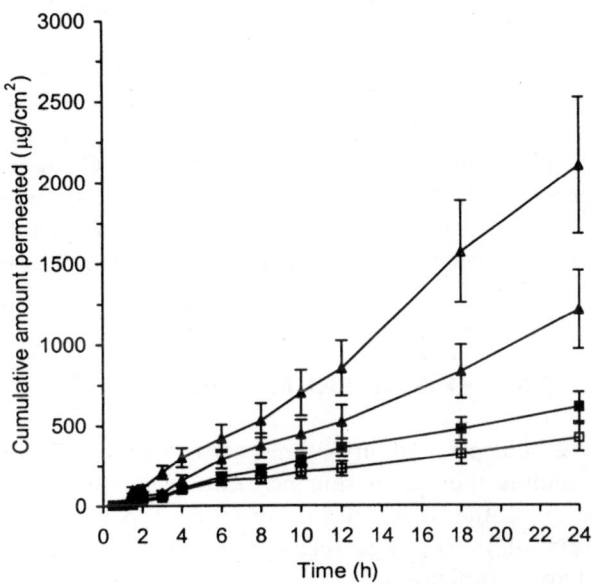

Figure 7 Permeation profiles of R- and S-enantiomers of ketoprofen (KP) across excised mouse skin from donor solution containing saturated ketoprofen solution. *Key*: (□)S-KP from RS-MP; (■)R-KP from RS-KP; (▲) S-KP from RS-KP, and (△) S-KP from S-KP enantiomer. (Adapted from Ref. 33).

enantiomers in water and isopropyl alcohol/water/isopropyl miristate mixture. The presence of two optical antipodes may enhance the skin permeation of each other, accounting for enhanced skin transport of the racemate as compared to that of single enantiomers.

Another recent study [57] focused on skin uptake of chiral terpenes. Stereoisomers of menthol have a 9°C lower melting point than racemic compound, while neomenthol enantiomers are present in the liquid state at room temperature, and its racemate melts at 51°C [50]. Changing the solvent composition had no significant effect on the racemate/enantiomers' solubility ratio of both terpenes. In further experiments no differences in skin uptake of (+)- or (−)-menthol were examined, but the difference in the melting temperature and solubility of enantiomers vs. racemic compound was reflected in a clear enhancement of racemate skin uptake when compared to pure stereoisomers. In the case of neomenthol, a more than 26°C difference in the melting point resulted in enantiomeric/racemic uptake ratios of between 2.5 and 8 (depending on the solvent composition).

As previously mentioned, the MTMT concept deals with a maximal flux of drug (F_{max}), and in all previously cited reports the donor compartment contained saturated drug solution exhibiting maximum thermodynamic activity of the drug in the vehicle. However, a number of studies have been performed where permeation of various stereoactive species of the drug was measured from nonsaturated drug solution.

For example, Kommuru et al. [32] reported that the melting temperature of the S-enantiomer of metoprolol free base was 6°C lower than that of the racemic compound. However, since the permeation studies were performed with donor solution not saturated with drug, no significant difference in the flux values between the S-form and the racemate was observed, as depicted in Fig. 6.

5. EUTECTICS AND CHIRAL TERPENE ENHANCERS

In the design of TDDS, the skin permeability often needs to be enhanced by various approaches including the use of skin permeation enhancers to reduce the barrier property of the skin. In the absence of permeation enhancers, systemic delivery of most drugs through the skin is limited, primarily because of the barrier function of the stratum corneum. Certain enhancers including some terpenes have a tendency to form eutectics when mixed with certain chemical entities. Terpenes are a group of chiral skin penetration enhancers that are derived from plant essential oils and are widely used as pharmaceutical excipients with various drugs. Terpenes are represented by members of various chemical classes including (1) alcohols such as carvacrol, eugenol, α-terpinol, linalool, nerolidol, menthol, thymol, eucalyptol and nerolidol, (2) ketones such as carvone, menthone and fenchone, (3) hydrocarbons such as limonene, α-pinene, dipentene (mixture of D-limonene and L-limonene), terpinolene, and cymene, and (4) oxides such as 1,8-cineole and ascaridole. The chemical structures of selected terpenes are depicted in Fig. 8. In recent years, many attempts have been made to investigate the use of terpenes as skin permeation enhancers, including menthol, linalool, limonene, and carvacrol to promote the transdermal transport of drugs, including chiral agents [57–78].

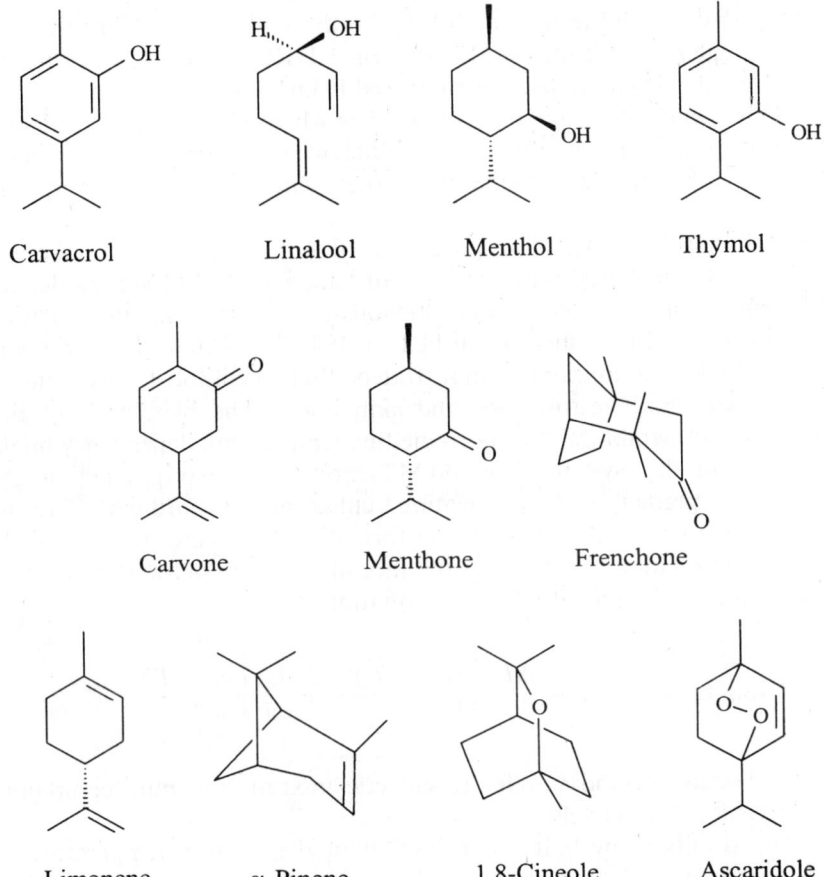

Figure 8   Chemical structures of selected terpenes.

The mechanism of action of terpenes involves increasing one or more of the following effects: (1) diffusion coefficient, (2) partition coefficient, (3) drug solubility (i.e., increasing the thermodynamic activity of the drug), (4) lipid extraction (i.e., disruption of lipid–protein domain), (5) macroscopic barrier perturbation, and (6) molecular orientation of terpene molecule within the lipid bilayer. However, the rate and extent of enhancement are dependent upon (1) type and physico-chemical characteristics (such as melting point, solubility) of terpene, (2) concentration of terpene used, (3) absence or presence of cosolvent or enhancer, and (4) concentration of cosolvent or enhancer, if present. Thus physicochemical parameters of terpenes including melting point depression must be carefully examined as they can influence the functional and therapeutic outcomes of drug products that contain terpene additives. To this end, many techniques such as wide-angle x-ray-diffraction, IR spectra, FT-IR, in vitro transepidermal water loss, and DSC are usually employed [63,64,76].

This section discusses only the studies where (1) the terpene affected the permeation of the enantiomer in a different way from that of racemate, and/or (2) an eutectic between the terpene and investigated drug was formed.

Differences in skin transport rates due to differences in thermal characteristics of stereoforms were reported, and the MTMT model for predicting such differences was previously discussed in this chapter [33,35,46,56,57]. In a study published in 1997 by Kaplun-Frischoff and Touitou [58], the authors demonstrated that menthol forms eutectic mixtures with both testosterone and skin lipids. The hypothesis of this work was that membrane transport–melting temperature dependency might also occur in this system. The MTMT model was extrapolated by the authors to predict skin permeation enhancement through lowering the melting point of the drug by the formation of a eutectic mixture. In this case the relationship between the flux of a compound and its melting temperature can be presented by the equation (Eq. 4)

$$\ln F_{\max E} - \ln F_{\max C} = \frac{\Delta H_C (T_{mC} - T)}{RT_{mC} T} - \frac{\Delta H_E (T_{mE} - T)}{RT_{mE} T} \tag{4}$$

where subscripts E and C refer to eutectic mixture and pure compound characteristics, respectively.

The results show that the reduction in the melting temperature of testosterone by 114°C, accompanied by an elevation in its solubility, brought about an increase in the transdermal flux of the drug through skin and silastic membrane. In permeation experiments of testosterone from saturated solutions through silastic membrane, a flux ratio of 2.9 was determined for eutectic mixture vs. testosterone alone, which is very close to a theoretically calculated ratio of 2.8 predicted by the MTMT model. An additional flux augmentation, to eight times the base line, in permeation of testosterone through mice skin was attributed to the influence of menthol on the stratum corneum lipid characteristics. This assumption was supported by DSC results demonstrating that menthol decreases the melting temperature of cholesteryl oleate and ceramides and affects the thermogram profile of isolated stratum corneum [58]

Subsequent studies have supported the above results. For instance, ibuprofen : terpene eutectic systems vs. a saturated aqueous solution of ibuprofen applied to untreated and to terpene pretreated skin were reported to cause a significant flux enhancement [59]. A system composed of ibuprofen : thymol 40 : 60 (% w/w) produced a flux 5.9 and 12.7 times higher than the flux values from a saturated aqueous solution with thymol pretreated skin and from a saturated aqueous solution across nonpretreated skin, respectively. Analogous data were achieved in permeation experiments through snake skin with lidocaine : menthol eutectic mixtures [60].

The same concept was extended to eutectic mixtures of propranolol with fatty acids [64]. Results of this recent study indicated that a good correlation was achieved between normalized predicted and experimental enhancement ratio (ER) values. ER values were normalized against the corresponding value of lauric acid, since the MTMT model does not include characteristics such as lipophilicity and molecular size. For example, predicted and experimental ER values for palmitic acid were 0.36 and 0.39, respectively. This study suggests that in some cases, in addition to disrupting stratum corneum lipids, fatty acids may promote permeation enhancement by the formation of eutectic mixture as well [64].

An interesting case in which both the drug permeant and the enhancers are chiral molecules was reported [32]. In this study the permeation characteristics of individual enantiomers and racemate of metoprolol free base (MB) as well as the influence of chiral permeation enhancers, l-menthol and (+/−)-linalool, on the permeation of MB were investigated using hairless mouse skin. In the absence of enhancers, the permeation profiles of R- and S-MB from donor solutions containing either racemate (RS-MB) or pure enantiomers (R- or S-) are comparable ($p < 0.05$). In the presence of enhancers, l-menthol and (+/−)-linalool, the flux values were increased 2.4- to 3.0-fold, respectively. The permeation profiles of R- and S-MB from donor solutions containing RS-MB were comparable. However, when the donor vehicle contained pure enantiomers, the permeation enhancing effect of l-menthol on S-MB was significantly

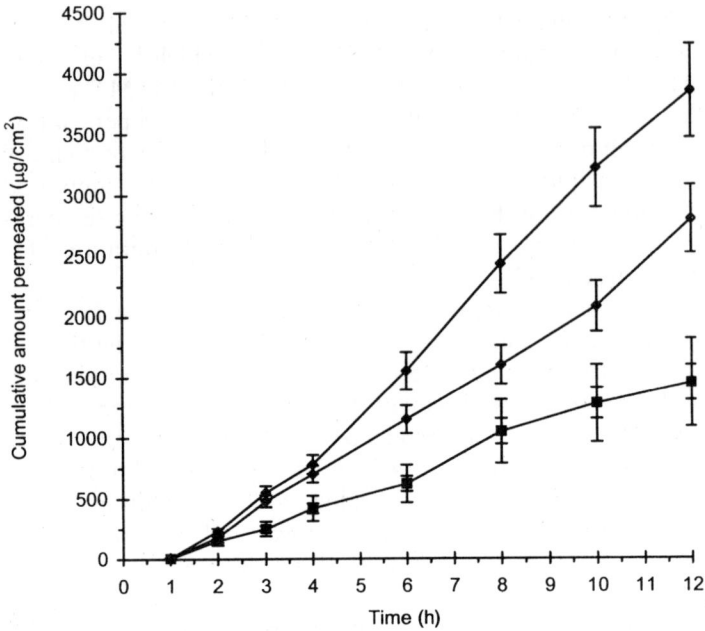

Figure 9 Permeation profiles of metoprolol (MB) enantiomers across mouse skin in the presence of 0.1% l-menthol. *Key*: (□) R-MB from RS-MB, (■) S-MB from RS-MB, (◇) R-MB from R-MB, and (◆) S-MB from S-MB enantiomer. *Note*: Both R-MB and S-MB from RS-MB permeation profiles are superimposed. (Adapted from Ref. 32.)

higher (by 25%) than on R-MB ($p < 0.05$) (Fig. 9). Further, in the presence of l-menthol, the flux of S-MB from donor solution containing pure S-MB was 35% higher than the flux of RS-MB from racemate. No such effect was seen with (+/−)-linalool. In all the investigations, no enantiomeric inversion was observed during the permeation process. The lag times were shorter in the case of l-menthol compared with (+/−)-linalool [32]. The effect of l-menthol on permeability of S-enantiomer (which is significantly more potent than the R-isomer) and racemate of propranolol across guinea pig skin was investigated [79]. In the presence of menthol, S-enantiomer of propranolol exhibited higher skin flux. The enhancement factors (EF, steady-state flux of S-propranolol with menthol/steady-state flux without menthol) for S-enantiomer and racemate of propranolol were 2.12 and 0.85, respectively. This suggests stereoselectivity in the permeation of propranolol

Table 2 Steady-State Flux, Lag Time, Permeability Coefficient, and Enhancement Factor Values for S-Enantiomer (S-PL) and Racemate (RS-PL) with and Without l-Menthol (M)

Drug	Steady-state flux $(\mu g\,cm^{-2}\,hr^{-1})$	Permeability coefficient $(cm^{-2}\,hr^{-1} \times 10^{-5})$	Enhancement factor
S-PL	6.90	69	—.
S-PL + M	14.60	146	2.12
RS-PL	14.44	144	—
RS-PL + M	12.22	122	0.85

Source: Adapted from Ref. 81.

enantiomer versus racemate across guinea pig skin in the presence of menthol (Table 2).

The effect of racemate and both enantiomers of 6-aminohexanoic acid 2-octylester as model enhancers with one chiral center on theophylline (achiral molecule) permeation through human skin in vitro was investigated by Vavrova et al. [80]. Their studies showed that there was no difference in the permeation enhancement ratios (ERs) between the (R)-(−) and (S)-(+) isomers of 6-aminohexanoic acid 2-octylester [the ERs were 2.72 ± 0.42 and 2.79 ± 0.60 for (R) and (S) enantiomers, respectively], suggesting that the enhancing properties of the compounds are not dependent on their spatial arrangement. Furthermore, a similar ER was found for the racemate [80].

The in vitro transport studies of enantiomers and racemate of timolol maleate (TM) across hairless mouse skin in the presence and absence of chiral terpene enhancers including menthol, limonene, and carvacrol have been recently investigated [81]. The steady-state flux values, apparent permeability coefficients, and enhancement factors for the two enantiomers of TM in the presence of selected terpenes are summarized in Table 3. Addition of l-menthol significantly increased the transport of individual enantiomers from pure isomers and racemate. The steady-state flux values and permeability coefficients increased significantly at all concentrations of l-menthol. While the flux values for the S- and R-enantiomers of TM were comparable in the presence of 0.5% l-menthol, the R-isomer had shown greater steady-state flux values than the S-TM at 0.2 and 1.0% concentrations of l-menthol. Further, the values of the enhancement factors for R-TM were significantly greater than those for S-isomer at all tested concentrations of l-menthol (Table 3). Similarly, l-limonene and carvacrol enhanced the transport of R-isomer to a significant extent over S-isomer, and the enhancement factors (EFs) were 4.6- and 13-fold, respectively, in the

Table 3 The Permeability Parameters of Pure S-Isomer and R-(from Racemic) Timolol Maleate (TM) Across Hairless Mouse Skin with and Without Chiral Enhancers

Formulation	Steady-state flux ($\mu g\,cm^{-2}\,h^{-1}$)	Permeability coefficient ($cm\,sec^{-1} \times 10^7$)	Enhancement factor
S-TM	5.6	4.3	1.0
S-TM + 0.2% w/v L-menthol	9.4	6.4	1.7
S-TM + 0.5% w/v L-menthol	13.2	8.7	2.4
S-TM + 1.0% w/v L-menthol	45.0	29.8	8.0
S-TM + 5.0% w/v L-menthol	68.2	45.1	12.2
R-TM	2.0	1.8	1.0
R-TM + 0.2% w/v L-menthol	8.1	7.3	4.1
R-TM + 1.0% w/v L-menthol	23.6	15.7	11.8
R-TM + 0.2% w/v D-limonene	12.5	8.3	4.6
S-TM + 0.2% w/v D-limonene	10.3	6.8	1.8
R-TM + 0.2% w/v carvacrol	35.1	23.2	13.0
S-TM + 0.2% w/v carvacrol	17.8	11.8	3.2

Source: Adapted from Ref. 81.

presence of 0.2% enhancer. Of the terpenes investigated, 0.2% carvocrol had the most significant enhancement effect (Fig. 10). The release profiles from transdermal patches containing pure enantiomers and racemates did exhibit stereoselectivity consistent with the results using the pure enantiomers [81]. The experimentally determined flux ratios in the presence of menthol were in agreement with theoretical values derived from the MTMT model.

6. CONCLUSIONS

This chapter provides a discussion on the skin's stereoselective permeation, metabolism, and binding of different stereoforms and racemates of chiral molecules. Not only is the published information in this field limited but also there are some controversial data that have been reported. Additional studies are required to answer the following questions: Is the skin an enantioselective membrane? Can this characteristic be confirmed by in vitro and in vivo experiments? Since wide differences exist between the laboratory models with regard to their biochemical composition and permeability, how can these issues be addressed in the preclinical investigations involving

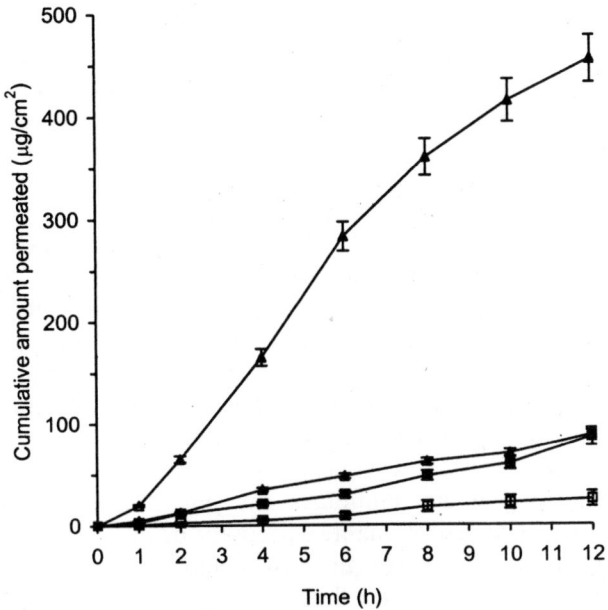

Figure 10 Permeation profiles of R-enantiomer of timolol maleate (R-TM) in the presence and the absence of selected chiral enhancers. (□) R-enantiomer, (■) R-enantiomer + 0.2% w/v D-limonene, (▲) R-enantiomer + 0.2% w/v L-menthol, and (◇) R-enantiomer + 0.2% w/v carvacrol. (Adapted from Refs. 39 and 81.)

chiral molecules? How can the preclinical data that is generated in laboratory animals be extrapolated to the clinical situation?

The existing knowledge in this rapidly evolving field clearly suggests that it is possible to modulate percutaneous fluxes of topically applied chiral drugs by a careful selection of the appropriate chemical entity, i.e., pure enantiomer, racemate, or a fixed combination of enantiomers for the intended therapeutic outcome. A variety of percutaneous absorption enhancers including some chiral molecules are available, which can be used in topical and transdermal formulations to achieve desirable flux values for optimal clinical outcomes. Since some chiral enhancers have the ability to affect the permeation of stereoforms of selected chiral agents in an enantioselective manner, enhancers of interest must be carefully screened for any possible stereoselectivity in their action. In the design of prodrugs and chemical derivatives of drugs for transdermal delivery that are expected to undergo metabolic activation, the stereoselectivity of the skin metabolizing enzymes must be carefully assessed.

Various physicochemical characteristics of several stereoisomers of the chiral molecule and the thermodynamic equations based on these properties provide an important tool in the evaluation of the stereoselective flux and skin permeation. A number of studies have shown that enantiomers with a lower melting point than their corresponding racemate exhibited higher solubility and consequently higher skin permeation profiles than the racemate. Using the MTMT model, the ratio of maximal fluxes of enantiomer/racemate can be predicted from the thermal characteristics of different stereoforms. This may be of significant importance, when one of the enantiomers has a more prominent pharmacological or toxic effect.

In the future, the choice between developing the racemate of a chiral drug versus single enantiomers will largely depend on a critical evaluation of their chiral characteristics with respect to their pharmacodynamic, pharmacokinetic, and toxicological effects. The potential for stereoselectivity in skin permeation due to stereochemical interactions between chiral drug, chiral excipients, and the biomembranes has generally been overlooked. Nevertheless, extrapolation of enantioselectivity in the permeation of enantiomers from in vitro data to in vivo behavior (permeation, disposition, and efficacy) is important and should be carried out carefully.

REFERENCES

1. Schaefer, H.; Redelmeier, T.E. *Skin Barrier – Principles of Percutaneous Absorption*; Karger AG: Basel, 1996.
2. Schaefer, H.; Zesch, A.; Stüttgen, G. *Skin Permeability*; Springer Verlag: Berlin, 1982.
3. Barry, W.B. *Dermatological Formulations: Percutaneous Absorption*; Marcel Dekker: New York, 1983.
4. Flynn, G.L. Cutaneous and transdermal delivery: processes and systems of delivery. In *Modern Pharmaceutics*; Banker, Rhodes, C.T., Eds.; Marcel Dekker: New York, 1995; 239.
5. Smith, W.P.; Christensen, M.S.; Nacht, S.; Gans, E.H. Effect of lipids on the aggregation and permeability of human stratum corneum. J. Invest. Dermatol. **1982**, *78*, 7–12.
6. Downing, D.T.; Stewart, M.E.; Wertz, P.W.; Colton, S.W.; Abraham, W.; Strauss, J.S. Skin lipids: an update. J Invest. Dermatol. **1987**, *Suppl. 88*, 2s–6s.
7. Yardley, H.J. Epidermal lipids. In *Biochemistry and Physiology of the Skin*; Goldsmith, L.A., Ed.; Oxford University Press: New York, 1983; 362–365.
8. Reiss, F. Therapeutics: percutaneous absorption, a critical and historical review. Am. J. Med. Sci. **1966**, *252* (5), 588–602.
9. Bell, G.H.; Davidson, J.N.; Scarbrough, H. *Textbook of Physiology and Biochemistry*, 5th Ed.; E. & S. Livingston: Edinburgh, 1963; Chap. 37.

10. Wilkes, G.L.; Brown, I.A.; Wildnauer, R.H.; The biomechanical properties of skin. CRC Crit. Rev. Bioeng. **1973**, *1* (4), 453.
11. Woodburne, R.T. Essentials of human anatomy. In: *Dermatological Formulations: Percutaneous Absorption*; Oxford University Press: New York, 1965; 1966.
12. Merk, H.F.; Jugert, F.K. Metabolic activation and detoxification of drugs and xenobiotica by the skin. In *Dermal and Transdermal Drug Delivery—New Insigts and Perspectives*; Gurny, R., Teubner, A., Eds.; Wiss.Verl. Ges.: Stuttgart, 1993; 91–100.
13. Kao, J.; Carver, M.P. Cutaneous metabolism of xenobiotics. Drug Metab. Rev. **1990**, *22*, 363–410.
14. Reiners, J.J.; Amador, R.C.; Pavone, A. Modulation of constitutive cytochrome P450 expression in vivo and in vitro in murine keratinocytes as a function of differentiation and extracellular Ca^{2+} concentration. Proc. Natl. Acad. Sci. USA **1990**, *87*, 1825–1829.
15. Zhu, Z.; Hotchkiss, S.; Boobis, A.; Edwards, R. Expression of P450 enzymes in rat whole skin and cultured epidermal keratinocytes. Biochem. Biophys. Res. Commun. **2002**, *297* (1), 65–68.
16. Gonzalez, M.C.; Marteau, C.; Franchi, J.; Migliore-Samour, D. Cytochrome P450 4A11 expression in human keratinocytes: effects of ultraviolet irradiation. Br. J. Dermatol. **2001**, *145* (5), 749–757.
17. Nasseri-Sina, P.; Hotchkiss, S.A.; Caldwell, J. Cutaneous xenobiotic metabolism: glycine conjugation in human and rat keratinocytes. Food Chem. Toxicol. **1997**, *35* (3–4), 409–416.
18. Heard, C.M.; Brain, K.R. Does solute stereochemistry influence percutaneous penetration? Chirality **1995**, *7*, 305–309.
19. Heard, C.M.; Watkinson, A.C.; Brain, K.R.; Hadgraft, J. In vitro skin penetration of propranolol enantiomers. Int. J. Pharm. **1993**, *90*, R5–R8.
20. Katz, E.P.; David, C.W. Unique side-chain conformations encoding for chirality and azimuthal orientation in the molecular packing of skin collagen. J. Mol. Biol. **1992**, *228* (3), 963–969.
21. Benezra, C.; Stampf, J.; Brbier, P.; Ducombs, G. Enantiospecificity in allergic contact dermatitis, Contact Dermatitis **1985**, *13*, 110–114.
22. Papageorgiou, C.; Stampf, J.L.; Benezra, C. Allergic contact dermatitis to tulips: an example of enantiospecificity. Arch. Dermatol. Res. **1988**, *280*, 5–7.
23. Ahmed, S.; Imai, T.; Otagiri, M. Evaluation of stereoselective transdermal transport and concurrent hydrolysis of several ester prodrugs of propranolol: mechanism of stereoselective permeation. Pharm. Res. **1996**, *13*, 1524–1535.
24. Ahmed, S.; Imai, T.; Yoshigae, Y.; Otagiri, M. Stereospecific activity and nature of metabolizing esterases for propranolol prodrug in hairless mouse skin, liver and plasma. Life Sci. **1997**, *61* (19), 1879–1887.
25. Udata, C.; Tirucherai, G.; Mitra, A.K. Synthesis, stereoselective enzymatic hydrolysis and skin permeation of diastereomeric propranolol ester prodrugs. J. Pharm. Sci. **1999**, *88*, 544–550.

26. Weston, A.; Hodgson, R.M.; Hewer, A.J.; Kuroda, R.; Grover, P.L. Comparative studies of the metabolic activation of chrysene in rodent and human skin. Chem. Biol. Interact. **1985**, *54* (2), 223–242.

27. Amin, S.; Huie, K.; Balanikas, G.; Hecht, S.S.; Pataki, J.; Harvey, R.G. High stereoselectivity in mouse skin metabolic activation of methylchrysenes to tumorigenic dihydrodiols, Cancer Res. **1987**, *47*, 3613–3617.

28. Levin, W.; Wood, A.W.; Chang, R.L.; Slaga, T.J.; Yagi, H.; Jerina, D.M.; Conney, A.H. Marked differences in the tumor-intiating activity of optically pure (+)- and (−)-trans-7,8-dihydroxy-7,8-dihydrobenzo(a)pyrene on mouse skin. Cancer Res. **1977**, *37*, 2721–2725.

29. Fretland, D.J.; Widomski, D.L.; Anglin, C.P.; Yu, S.; Djuric, S.W. Optical isomers of leukotriene B4 antagonist have differential effects on granulocyte diapedesis in the guinea pig dermis. Chirality. **1992**, *4* (6), 353–355.

30. Miyazaki, K.; Kaiho, F.; Inagaki, A.; Dohi, M.; Hazemoto, N.; Haga, M.; Hara, H.; Kato, Y. Enantiomeric difference in percutaneous penetration of propranolol through rat excised skin. Chem. Pharm. Bull. **1992**, *40*, 1075–1076.

31. Washitake, M.; Yajima, T.; Anmo, T.; Arita, T.; Hori, R. Studies on percutaneous absorption of drugs. 3. Percutaneous absorption of drugs through damaged skin. Chem. Pharm. Bull. **1973**, *21*, 2444–2451.

32. Kommuru, T.R.; Khan, M.A.; Reddy, I.K.; Effect of chiral enhancers on the permeability of optically active and racemic metoprolol across hairless mouse skin. Chirality **1999**, *11*, 536–542.

33. Kommuru, T.R.; Khan, M.A.; Reddy, I.K. Racemate and enantiomers of ketoprofen: phase diagram, thermodynamic studies, skin permeability and use of chiral permeation enhancers. J. Pharm. Sci. **1998**, *87*, 833–840.

34. Loschmann, P.A.; Chong, P.N.; Nomoto, M.; Tepper, P.G.; Horn, A.S.; Jenner, P.; Marsden, C.D. Stereoselective reversal of MPTP-induced parkinsonism in the marmoset after dermal application of N-0437. Eur. J. Pharmacol. **1989**, *166*, 373–380.

35. Reddy, I.K.; Kommuru, T.R.; Zaghloul, A.A.; Khan, M.A. Chirality and its implications in transdermal drug development. Crit. Rev. Ther. Drug Carrier Syst. **2000**, *17* (4), 285–325.

36. Manabe, E.; Sugibayashi, K.; Morimoto, Y. Analysis of skin penetration enhancing effect of drugs by ethanol-water mixed systems with hyrodynamic pore theory. Int. J. Pharm. **1996**, *129*, 211–217.

37. Marzulli, F.N.; Callahan, J.F.; Brown, D.W. Chemical structure and skin penetrating capacity of a short series of organic phosphates and phosphoric acid. J. Invest. Dermatol. **1965**, *44*, 339–344.

38. Li, Z.J.; Grant, D.J.W. Relationship between physical properties and crystal structures of chiral drugs. J. Pharm. Sci. **1997**, *86*, 1073–1078.

39. Afouna, M.I.; Fincher, T.K.; Khan, M.A.; Reddy, I.K. Percutaneous permeation of enantiomers and racemic mixture of chiral drugs and prediction of their flux ratios using thermal data: a pharmaceutical perspective. Chirality **2003**, *15*, 456–460.

40. Suedee, R.; Brain, K.R.; Heard, C.M. Differential permeation of propranolol enantiomers across human skin in vitro from formulations containing an enantioselective excipient. Chirality 1999, 11 (9), 680–683.

41. Nakagawa, H.; Shimizu, K.; Yamada, K. Chiral discrimination in the transport of ketoprofen and ibuprofen esters through an aqueous phase mediated by various serum albumins. Chirality 1999, 11 (5–6), 516–519.

42. Tan, S.C.; Patel, B.K.; Jackson, S.H.; Swift, C.G.; Hutt, A.J. Stereoselectivity of ibuprofen metabolism and pharmacokinetics following the administration of the racemate to healthy volunteers. Xenobiotica 2002, 32 (8), 683–697.

43. San Martin, M.F.; Soraci, A.; Fogel, F.; Tapia, O.; Islas, S. Chiral inversion of (R)-(−)-fenoprofen in guinea-pigs pretreated with clofibrate. Vet. Res. Commun. 2002, 26 (4), 323–332.

44. Millership, J.S.; Collier. P.S. Topical administration of racemic ibuprofen. Chirality 1997, 9 (3), 313–316.

45. Panus, P.C.; Ferslew, K.E.; Tober-Meyer, B.; Kao, R.L. Ketoprofen tissue permeation in swine following cathodic iontophoresis. Phys. Ther. 1999, 79 (1), 40–49.

46. Touitou, E.; Chow, D.D.; Lawter, J.R. Chiral β-blockers for transdermal delivery. Int. J. Pharm. 1994, 104, 19–28.

47. el-Arini, S.K.; Giron, D.; Leuenberger, H. Solubility properties of racemic praziquantel and its enantiomers. Pharm. Dev. Technol. 1998, 3, 557–564.

48. Yonemochi, E.; Yoshihashi, Y.; Terada, K. Quantitative relationship between solubility, initial dissolution rate and heat of solution of chiral drugs. Pharm. Res. 2000, 17, 90–93.

49. Leising, G.; Resel, R.; Stelzer, F.; Tasch, S.; Lanziner, A.; Hantich, G. Physical aspects of dexibuprofen and racemic ibuprofen. J. Clin. Pharmacol. 1996, 36, 3S–6S.

50. Li, Z.J.; Zell, M.T.; Munson, E.J.; Grant, D.J.W.; Characterization of racemic species of chiral drugs using thermal analysis, thermodynamic calculation and structural studies. J. Pharm. Sci. 1999, 88, 337–346.

51. Rustichelli, C.; Gamberini, M.C.; Feriolo, V.; Gamberini, G. Properties of the racemic species of verapamil hydrochloride and gallopamil hydrochloride. Int. J. Pharm. 1999, 178, 111–120.

52. Barett, A.M.; Cullum, V.A. The biological properties of the optical isomers of propranolol and their effects on cardiac arrhythmias. Br. J. Pharmacol. 1968, 34, 43–55.

53. Lawter, J.R.; Pawelchak, J. Transdermal permeation of chiral compounds. Proc. Int. Symp. Control. Rel. Bioact. Mater. 1989, 16, 1308.

54. Wearley, L.; Antonacci, B.; Cacciapuoti, A.; Assenza, S.; Chaudry, I.; Eckhart, C.; Levine, N.; Loebenberg, D.; Norris, C.; Parmegiani, R.; Sequeira, J.; Yarosh-Tomaine, T. Relationship among physicochemical properties, skin permeability, and topical activity of the racemic compound and pure enantiomers of a new antifungal. Pharm. Res. 1993, 10 (1), 136—140.

55. Jamali, F.; Brocks, D.R. Clinical pharmacokinetics of ketoprofen and its enantiomers. Clin. Pharmacokinet. 1990, 19, 197–217.

56. Roy, S.D.; Chatterjee, D.J.; Manoukian, E.; Divor, A. Permeability of pure enantiomers of ketorolac through human cadaver skin. J. Pharm. Sci. **1995**, *84*, 987–990.

57. Mackay, K.M.; Williams, A.C.; Barry, B.W. Effect of melting point of chiral terpenes on human stratum corneum uptake. Int. J. Pharm. **2001**, *228*, 89–97.

58. Kaplun-Frischoff, Y.; Touitou, E. Testosterone skin permeation enhancement by menthol through formation of eutectic with drug and interaction with skin lipids. J. Pharm. Sci. **1997**, *86*, 1394–1399.

59. Stott, P.W.; Williams, A.C.; Barry, B.W. Transdermal delivery from eutectic systems: enhanced permeation of a model drug, ibuprofen. J. Control. Release **1998**, *50*, 297–308.

60. Kang, L.; Jun, H.W.; McCall, J.W. Physicochemical studies of lidocaine-menthol binary systems for enhanced membrane transport. Int. J. Pharm. **2000**, *206*, 35–42.

61. Stott, P.W.; Williams, A.C.; Barry, B.W. Mechanistic study into the enhanced transdermal permeation of a model β-blocker propranolol, by fatty acids: a melting point depression effect. Int. J. Pharm. **2001**, *219*, 161–176.

62. Yamane, M.A.; Williams, A.C.; Barry, B.W. Terpene penetration enhancers in propylene glycol/water co-solvent systems: effectiveness and mechanism of action. J. Pharm. Pharmacol. **1995**, *47* (12A), 978–989.

63. Cornwell, P.A.; Barry, B.W.; Stoddart, C.P.; Bouwstra, J.A. Wide-angle x-ray diffraction of human stratum corneum: effects of hydration and terpene enhancer treatment. J. Pharm. Pharmacol. **1994**, *46* (12), 938–950.

64. Vaddi, H.K.; Ho, P.C.; Chan, Y.W.; Chan, S.Y. Terpenes in ethanol: haloperidol permeation and partition through human skin and stratum corneum changes. J. Control. Release **2002**, *81* (1–2), 121–133.

65. Cal, K.; Janicki, S.; Sznitowska, M. In vitro studies on penetration of terpenes from matrix-type transdermal systems through human skin. Int. J. Pharm. **2001**, *224* (1–2), 81–88.

66. Moghimi, H.R.; Williams, A.C.; Barry, B.W. Enhancement by terpenes of 5-fluorouracil permeation through the stratum corneum: model solvent approach. J. Pharm. Pharmacol. **1998**, *50* (9), 955–964.

67. Gao, S.; Singh, J. In vitro percutaneous absorption enhancement of a lipophilic drug tamoxifen by terpenes. J. Control. Release **1998**, *51* (2–3), 193–199.

68. Zhao, K.; Singh, J. Mechanism(s) of in vitro percutaneous absorption enhancement of tamoxifen by enhancers. J. Pharm. Sci. **2000**, *89* (6), 771–780.

69. Katayama, K.; Takahashi, O.; Matsui, R.; Morigaki, S.; Aiba, T.; Kakemi, M.; Koizumi, T. Effect of l-menthol on the permeation of indomethacin, mannitol and cortisone through excised hairless mouse skin. Chem. Pharm. Bull. (Tokyo). **1992**, *40* (11), 3097–3099.

70. El-Kattan, A.F.; Asbill, C.S.; Kim, N.; Michniak, B.B. The effects of terpene enhancers on the percutaneous permeation of drugs with different lipophilicities. Int. J. Pharm. **2001**, *215*, 229–240.

71. El-Kattan, A.F.; Asbill, C.S.; Michniak, B.B. The effect of terpene enhancer lipophilicity on the percutaneous permeation of hydrocortisone formulated in HPMC gel systems. Int. J. Pharm. **2000**, *198* (2), 179–189.
72. Jain, A.K.; Thomas, N.S.; Panchagnula, R. Transdermal drug delivery of imipramine hydrochloride. Effect of terpenes. J. Control. Release **2002**, *79* (1–3), 93–101.
73. Williams, A.C.; Barry, B.W. Terpenes and the lipid-protein-partitioning theory of skin penetration enhancement. Pharm. Res. **1991**, *8* (1), 17–24.
74. Cornwell, P.A.; Barry, B.W. Sesquiterpene components of volatile oils as skin penetration enhancers for the hydrophilic permeant 5-fluorouracil. J. Pharm. Pharmacol. **1994**, *46* (4), 261–269.
75. Magnusson, B.M.; Runn, P.; Koskinen, L.O. Terpene-enhanced transdermal permeation of water and ethanol in human epidermis. Acta Derm. Venereol. **1997**, *77* (4), 264–267.
76. Vaddi, H.K.; Ho, P.C.; Chan, S.Y. Terpenes in propylene glycol as skin-penetration enhancers: permeation and partition of haloperidol, Fourier transform infrared spectroscopy, and differential scanning calorimetry. J. Pharm. Sci. **2002**; *91* (7), 1639–1651.
77. Stott, P.W.; Williams, A.C.; Barry, B.W. Transdermal delivery from eutectic systems: enhanced permeation of a model drug, ibuprofen. J. Control. Release **1998**, *50* (1–3), 297–308.
78. Godwin, D.A.; Michniak, B.B. Influence of drug lipophilicity on terpenes as transdermal penetration enhancers. Drug Dev. Ind. Pharm. **1999**, *25* (8), 905–915.
79. Zahir, A.; Kunta, J.R.; Khan, M.A.; Reddy, I.K. Effect of menthol on permeability of an optically active and racemic propranolol across guinea pig skin. Drug Dev. Ind. Pharm. **1998**, *24* (9), 875–878.
80. Vavrova, K.; Hrabalek, A.; Dolezal, P. Enhancement effects of (R) and (S) enantiomers and the racemate of a model enhancer on permeation of theophylline through human skin. Arch. Dermatol. Res. **2002**, *294* (8), 383–385.
81. Madiraju, C. Master's thesis. Transport of timolol across hairless mouse skin: effect of chiral terpene enhancers. The University of Louisiana at Monroe, LA, 2001.

4

Stereoselective Drug Delivery Through Prodrug Approach

Teruko Imai and Masaki Otagiri
Kumamoto University, Kumamoto, Japan

1. INTRODUCTION

The goal in most drug development processes is the design of agents that are optimally effective in curing a defined illness with minimum or no side effects. For a traditional (nontargeted) delivery system, only a minor fraction of the total administered drug reaches the target site. Therefore a considerable portion of the administered therapeutic agent may interact with nontarget sites, resulting in inefficient drug delivery and/or undesirable side effects. A drug delivery system that has the ability to target a drug to a particular site (such as a specific tissue, organ, cell, enzyme, bacterium, or virus) would optimize therapy and minimize or eliminate possible drug toxicity. One approach to the site-specific delivery of a parent drug to a certain tissue involves the design of drug delivery systems targeted to specific drug carrier proteins and/or enzymes. Among molecular modification strategies, the prodrug approach would be useful for site-specific delivery of the drug.

The prodrug approach has been recognized as an integral part of the new drug design process. This approach matured as a branch of pharmaceutical research during the 1970s, and several prodrugs are now in clinical use. By definition, a prodrug is a pharmacologically inactive derivative of a parent drug molecule that requires an in vivo spontaneous biological process and/or enzymatic transformation in order to revert to the active parent drug molecule, and thus to improve, delay, prolong, control, and/or target the action of the parent drug molecule. The prodrug approach therefore involves chemical manipulation of parent drugs to help overcome problems associated with their general pharmacokinetics and to optimize their delivery. The use of prodrugs in this regard has undergone considerable expansion, largely as the result of an increased awareness and understanding of the physicochemical factors that affect drug delivery and of the efficacy of this approach. However, prodrugs must be stable in vitro during manufacturing processes and over their shelf life.

In the design of prodrugs based on site-specific transport or site-selective activation at a target site, it is important to ensure the affinity of the drug molecule to the corresponding protein transporter and/or enzyme so that site-specific delivery is achieved. The stereoconformation of a drug molecule is important for its recognition by a corresponding transporter and/or enzyme. Understanding the functional characteristics of the transporter proteins and/or enzymes would be helpful in utilizing these proteins and/or enzymes in the design of such site-specific prodrugs. A successful design of a site-specific prodrug through the consideration of enzyme–substrate specificity, therefore, requires an extensive knowledge of that particular enzyme or group of enzymes.

This chapter presents a discussion on prodrug design through targeting several transporters for oligopeptides, bile acids, and glucose, as well as enzymes such as γ-glutamyltransferase, aminopeptidase, and carboxylesterase. The recognition of substrate conformation by these proteins is described, and some successful examples of the synthesis of prodrugs which are recognized by targeted transporters and/or enzymes are also shown. In addition, antibody-directed enzyme prodrug therapy and gene-directed enzyme prodrug therapy will be introduced as new approaches to develop selectively targeted anticancer drugs that specifically target tumor cells. Finally, the stereoselective hydrolysis of a prodrug by carboxylesterases in the liver, small intestine, and skin as well as the stereoselective hydrolysis activity of skin esterases resulting in stereoselective transdermal penetration of drug are described.

2. DESIGN OF PRODRUGS

The prodrug approach using chemical modification of the parent molecular structure can be illustrated as shown in Fig. 1. After administration, the prodrug reverts to the original active compound (the parent drug) in vivo through an enzymatic or nonenzymatic biotransformation process. Albert [1] first introduced the term prodrug for compounds that have the above properties. Considerable attention has been focused on the prodrug approach since the early 1970s. It is now considered to be a new type of

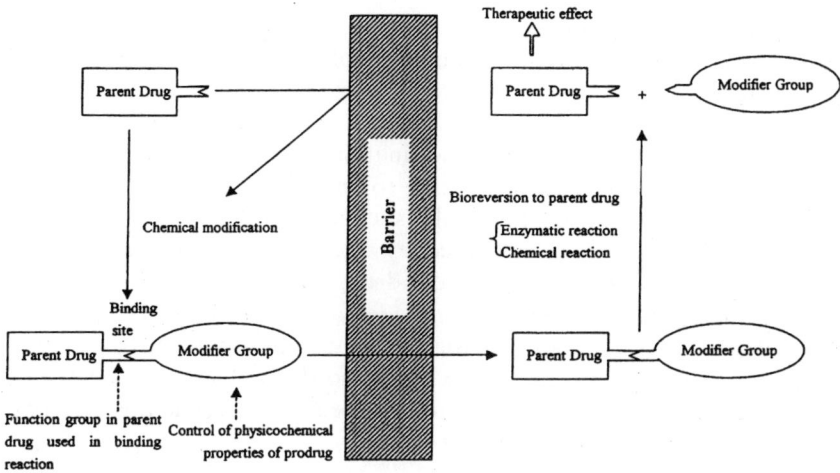

Figure 1 Schematic illustration of prodrug concept.

drug delivery system (DDS). Using some novel types of modifications, prodrugs can be designed with characteristics that confer site-specificity on the parent drug molecules, which would be important with respect to their therapeutic use.

Prodrugs should be designed depending on their respective uses. The rational goals for handling problems encountered in the absorption of a drug include one or more of the following approaches:

1. Solving difficulties in pharmaceutical formulation such as improving stability in the gastrointestinal tract and/or enhancing poor water solubility
2. Amelioration of in vivo behavior through improving membrane permeability or large intestine targeting
3. Modification of in vivo reactions to minimize side effects such as mucous membrane stimulation, unpleasant taste and/or smell, irritation, and/or avoiding first-pass metabolism

Based on differences in free radials, prodrugs can be divided into:

1. *Mutual prodrugs.* Prodrugs that contain free radials with pharmacological activity, which are effective in terms of decreasing side effects such as mucous membrane stimulation.
2. *Macromolecular prodrugs.* Prodrugs that are designed by combining low molecular weight drugs with macromolecular carriers, which have optimal physicochemical and biological properties.

Once past the barrier, owing to the properties of the macromolecular weight carrier, the low molecular drug (pharmacologically active moiety) is released from the macromolecular carrier to
exert its therapeutic action. If the parent drug molecule has poor
permeability in the gastrointestinal mucosal membrane, a macromolecular prodrug can be used to enhance the permeability at a
limited specific absorption site (such as the large intestine). There,
the prodrug can be activated by intestinal bacteria, resulting in
targeted efficacy for the parent drug.

3. *Prodrug with pharmacologically inactive promoiety.* This approach
   depends on the ability of the pharmacologically inactive promoiety to attach with the active parent compound so as to inactivate
   the drug (drug latentiation) and subsequently regenerate the drug
   effectively from the prodrug in vivo.

There are two primary questions emerging in the design of prodrugs:

1. What type of substituent structure is available for addressing a
   given problem through the prodrug approach?
2. What conditions in vivo are required for the parent drug to be
   regenerated?

The substrate–enzyme reaction is important in the conversion of enzymatically hydrolyzable prodrugs. Therefore knowledge of enzyme properties
such as substrate specificity, reactivity, organ distribution, and physiological
role is essential for the successful design of site-targeted prodrugs. In
addition to specifically designed prodrugs, a number of conventional drugs
have been found to serve as prodrugs, regenerating the active moiety after
metabolism. In these cases, active drugs are regenerated from their prodrugs
by reductases or oxidases but not esterases. Futraful and carmofur, for
example, are hydrated to the unstable 5-fluorouracil by the action of cytochrome P450. Loxoprofen is also converted to the cis or trans hydroxide
through the reduction of the keto group, but only the trans form is active.
The generation of the trans form is a major metabolic product in humans
and is also viable in monkeys and rats. However, most of these examples are
not predictable. In fact, the most common prodrugs are those that require
enzymatic hydrolysis in the presence of a hydrolase, which exists abundantly
in vivo. Free radicals limited to hydroxyl, carboxyl, and amino groups are
usually attached to the parent compounds, generating the ester and amide
bonds. The majority of clinically used prodrugs are ester derivatives
modified by hydroxyl and carboxyl groups of the parent drugs. This is
perhaps because in many cases an "ester bond" helps in enhancing membrane permeability and/or solubility of the prodrug. Different forms of

chemical modification, which are utilized in prodrug design, are shown in Table 1.

3. PROTEIN-RELATED STEREOSELECTIVE DISPOSITION OF PRODRUGS

Several attempts have been made to facilitate intestinal absorption and tissue distribution of less permeable compounds by utilizing several transport systems through the modification of parent compounds to give a specific conformational analog. In addition, many enzymes have been implicated in stereoselective bioconversion. Some of these include esterases, aminopeptidases, carboxypeptidases, γ-glutamyltransferases, β-glucronidases, and β-glucosidases. An enzymatically targeted prodrug approach might be achieved by chemical modification and subsequent confirmation of a specific moiety against a certain enzyme that is susceptible to a recognized catalytic reaction. In the case of such prodrugs, enzymes are regarded as the in vivo reconversion targets.

Some examples of site-specifically delivered prodrugs, based on site-specific transport and release of active drugs, are currently available. Transporters for glucose, oligopeptides, amino acids, and bile acids are related to specific membrane transporters of drug molecules into certain tissues, and to enzymes such as glucronidase and peptidase. Biotransformation processes of such prodrugs to their active parent compounds are related to specific tissues, where such enzyme systems are selectively available. In such cases, when the transporter or the enzyme recognizes the prodrug, it may promote its delivery to a specific site. Figure 2 shows several intestinal epithelial cells' transporters (carriers) that can recognize the stereospecific structure of a substrate.

4. UTILIZATION OF TRANSPORTERS FOR STEREOSELECTIVE DISPOSITION OF PRODRUGS

4.1. Utilization of Monosaccharide Transporters for Drug Absorption

Several attempts have been made to facilitate the intestinal absorption and tissue distribution of less permeable compounds by utilizing monosaccharide transport systems, through modification of the parent compounds to sugar analogs. The intestinal brush border membrane transport of hexose can be attributed to a Na^+/glucose cotransporter (SGLT1) for D-glucose,

Table 1 Prodrug Forms of Various Functional Groups in Parent Drug Substances

Functional group	Prodrug form
Hydroxyl -OH	$-O-\overset{\overset{\text{O}}{\|\|}}{C}-R$ — Carbonate esters
	$-O-PO_2-R$ — Phosphate esters
	$-O-CH_2-O-\overset{\overset{\text{O}}{\|\|}}{C}-R$ — Acyloxymethyl ethers
Carboxyl -COOH	$-\overset{\overset{\text{O}}{\|\|}}{C}-O-R$ — Carbonate esters
	$-\overset{\overset{\text{O}}{\|\|}}{C}-S-R$ — Thioesters
	$-\overset{\overset{\text{O}}{\|\|}}{C}-NH-R$ — Amides
	$-\overset{\overset{\text{O}}{\|\|}}{C}-O-CH_2-O-\overset{\overset{\text{O}}{\|\|}}{C}-R$ — Acyloxymethyl esters
Amino -NH$_2$	$-NH-\overset{\overset{\text{O}}{\|\|}}{C}-R$ — Amides
	$-NH-\overset{\overset{\text{O}}{\|\|}}{C}-O-R$ — Carbamates
	$-NH-CH_2-O-\overset{\overset{\text{O}}{\|\|}}{C}-R$ — Acyloxymethyl amines
	$-NH-CH_2-\overset{\overset{R_2}{\|}}{N}-R_1$ — N-Mannich bases
Thiol -SH	$-S-\overset{\overset{\text{O}}{\|\|}}{C}-R$ — Thioesters
	$-S-CH_2-O-\overset{\overset{\text{O}}{\|\|}}{C}-R$ — Acyloxymethyl thioethers

Intestinal Epithelial Cell

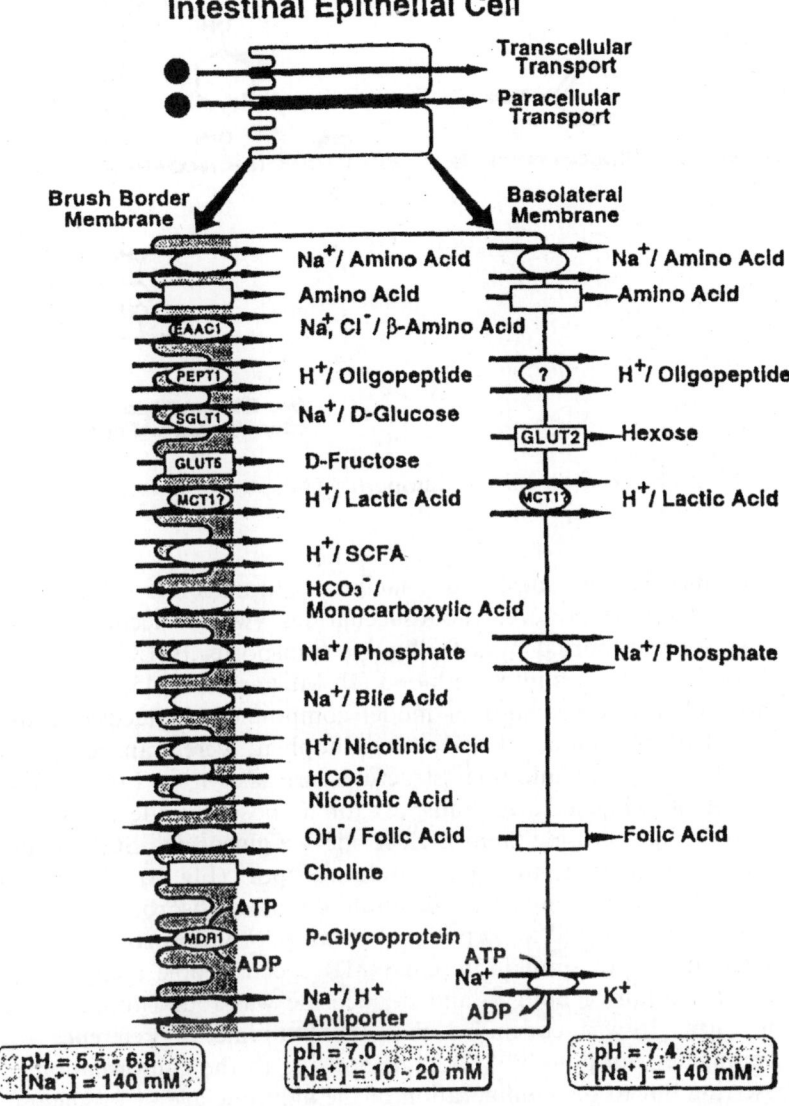

Figure 2 Summary of intestinal epithelial transporters. Transporters shown by square and oval shapes demonstrate active and facilitated transporters, respectively. The name of cloned transporters were shown within square or oval shapes. In the case of active transporters, the same direction of arrows represents *symport* of substrate and the driving force. Arrows going in the reverse direction mean the *antiport*. (Adapted from Ref. 1.)

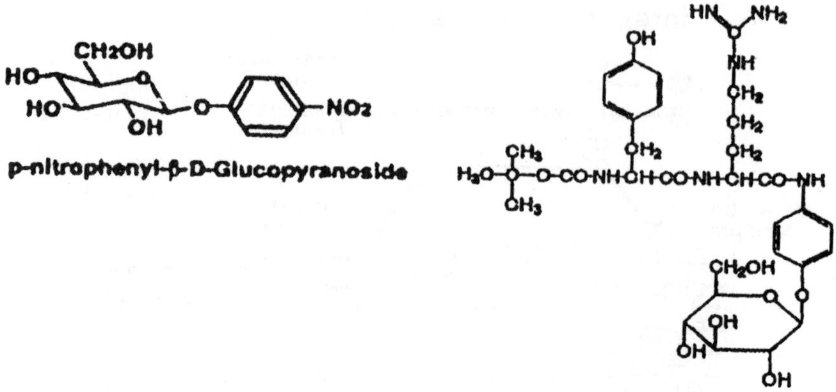

Figure 3   Drug absorbed via glucose transporter. (Adapted from Ref. 6.)

D-galactose, and their analogs, and a facilitative transporter GLUT5 for D-fructose [2,3]. Furthermore, monosaccharides (which accumulate in enterocytes) are transported by a facilitative transporter (GLUT2) across the basolateral membranes into the blood [4]. Mizuma et al. [5,6] reported that monosaccharide conjugates of model compounds, β-glucosides and β-galactosides of p-nitrophenol and acetaminophen, were transported by SGLT1 in the rat small intestine (structures are shown in Fig. 3). The preferences of SGLT1-mediated transport for monosaccharide conjugates were glucoside > galactoside and β-glucoside > α-glucoside; SGLT1 did not significantly contribute to α-galactoside transport (Fig. 4). Their data indicated that β-glucose conjugates constitute the most absorbed form and may improve the intestinal absorption of some poorly absorbable drugs. By molecular dynamics simulation and NMR spectroscopic analysis, the glucose conformation of both α- and β-glucoside was found to be in the $^4C_1$ chair form. It was concluded that the β-anomeric preference for glucose conjugate transport by SGLT1 is not due to the conformation of the glucose ring but to the configuration of the aglycone at C-1 of glucose. SGLT1 may have sufficient space to accommodate the β-position of glucose in the $^4C_1$ chair form, possibly larger than that in the α-position for glucoside transport.

A strategy for the enhancement of intestinal absorption by derivation into a monosaccharide analog has also been applied to peptides. The coupling of kyotorphin (an unstable peptide) with β-glucoside enhanced the absorption and clearance of kyotorphin in the rat-everted jejunum and decreased its metabolic clearance in rat intestinal homogenate [7]. A much larger peptide, insulin, was also modified with p-nitrophenyl-α-D-gluco-pyranoside and p-nitrophenyl-α-D-mannopyranoside, where the expected hypoglycemic effects were observed after intraintestinal administration in rats. The intestinal absorption of insulin apparently occurred following the modification of insulin with sugar and exhibited both improved hydrolytic stability and membrane permeability [8]. However, the p-nitrophenyl-α-L-arabino-pyranoside-insulin derivative was not effective.

4.2. Utilization of Oligopeptide Transporters

The apical membrane of epithelial cells of the small intestine express peptide transporter that mediates the electrogenic proton-coupled uptake of dietary di- and tripeptides from the intestinal lumen into the cell [9]. The peptide transporters are unique among solute transporters in that H^+ is the cotransported ion species, and these carriers are able to bind and translocate an enormous number of peptides. They also transport a variety of pharmacologically active peptidomimetic molecules, including β-lactam antibiotics [10], angiotensin converting enzyme inhibitors [11], and prodrugs, thus allowing their oral delivery with high bioavailability.

Dipeptides containing one or two L-isomers of amino acids possesse a high affinity for the transporter, compared to those carrying one or two D-isomers, as shown in Table 2. Based on a large series of dipeptides with different amino- and carboxy-terminal amino acid residues, as either L- and/or D-enantiomers, incorporation of D-amino acids into the carboxy-terminal position generally causes a more pronounced reduction in affinity to a peptide transporter than that in the N-terminal position; while dipeptides consisting solely of D-enantiomers show no detectable affinity [12,13]. However, the D-configuration in the N-terminus of a peptide substrate still allows for efficient transport.

All orally active β-lactams carry an unsubstituted or modified D-phenylglycine side group in the N-terminus position and therefore should have affinities similar to di- and tripeptides with an N-terminal D-amino acid [14]. The stereoselective transport of β-lactam antibiotics is of particular interest because the L-isomer of aminocephalosporin failed to be absorbed in vivo to a significant extent [15]. However, a study on D- and L-enantiomers of cephalexin and loracarbef demonstrated the higher affinity of the L-isomer to oligopeptide transporter than the D-isomer (Fig. 5) [16]. Consequently, the apparent low absorption rate of L-isomers of amino cephalosporines does not appear to be due to lack of transport by the peptide transporter, but, more likely, because of the rapid enzymatic

Table 2 Net Rates of Serosal Appearance of Phenylalanine and Dipeptide in Isolated Loops of Rat Jejunum Perfused by a Single Pass

Luminal Solute (50–120 min)	Conc. (mM)	Rate of appearance $(\text{nmol min}^{-1}(\text{g dry wt})^{-1})$	
		Phenylalanine	Dipeptide
L-P henylalanine	1.0	1136 ± 88	—
	0.5	582 ± 48	—
	0.1	111 ± 14	—
D-Phenylalanine	1.0	144 ± 20	—
	0.5	95 ± 14	—
L-Alanyl-L-phenylalanine	1.0	1136 ± 194	0
	0.5	355 ± 51	0
	0.1	121 ± 15	0
L-Alanyl-D-phenylalanine	1.0	1038 ± 118	0
	0.5	348 ± 31	0
	0.1	103 ± 30	0
D-Alanyl-L-phenylalanine	1.0	689 ± 10	124 ± 20
	0.5	228 ± 11	22 ± 5
	0.1	75 ± 15	2 ± 2
D-Alanyl-D-phenylalanine	1.0	16 ± 2	35 ± 3
	0.5	12 ± 2	0
	0.1	2 ± 0.1	0
L-Phenylalanyl-L-alanine	1.0	1230 ± 182	0
	0.5	521 ± 51	0
	0.1	111 ± 19	0
L-Phenylalanyl-D-alanine	1.0	247 ± 25	0
	0.5	126 ± 13	0
	0.1	21 ± 8	0
D-Phenylalanyl-L-alanine	1.0	17 ± 1.3	87 ± 3
	0.5	13 ± 2.7	46 ± 3
	0.1	6 ± 4	22 ± 2
D-Phenylalanyl-D-alanine	1.0	13 ± 7	9 ± 1
	0.5	0	4 ± 0.8

Mean rates (± S.E.M., $n = 4$) were calculated by covariance from all the time points between 80 and 120 min.

hydrolysis of the L-isomers (not the D-isomers) in the gastrointestinal tract. Furthermore, it has been reported that the minimal structure requirements for substrate binding and transport include two ionized head groups separated by at least four methylene groups based on a series of medium-chain fatty acids that contained an amino group as a head group (ω-amino

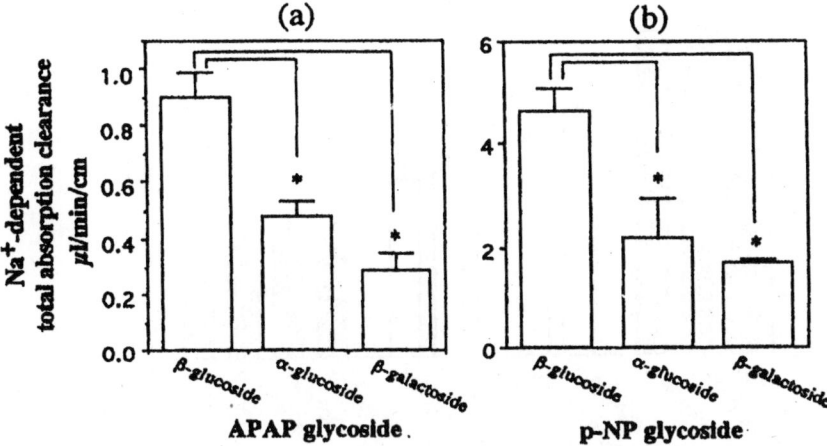

Figure 4 Comparison of Na$^+$-dependent total transport clearance of aceto-aminophen (APAP) glycosides (a) and p-nitrophenol (p-NP) glycosides (b). Data represent means ± S.E.M ($n = 3$–6). An asterisk represents a significant difference ($p < 0.05$) by Tukey–Kramer's test. (Adapted from Ref. 6.)

fatty acids) [17]. Consequently, a > 500 pm < 630 pm distance between the carboxyl carbon and the amino nitrogen will be sufficient for substrate recognition and transport.

Since peptide transporters have a broad spectrum of specificity, the intestinal peptide transport system can be employed to improve the intes-tinal absorption of certain drugs by converting them chemically to their corresponding di- or tripeptide prodrugs. The very low bioavailability of α-methyldopa (which is taken up by the Na$^+$-coupled neutral amino acid transporter) has been improved by the use of dipeptide prodrugs. Various dipeptide derivatives of α-methyldopa have been synthesized. These derivatives have shown significantly higher membrane permeability than the parent drug and are metabolized to their active moieties in mucosal tissue. As shown in Fig. 6, the use of α-methyldopa-phenylalanine increased the in vivo systemic availability of α-methyldopa to nearly 100% [11].

4.3. Utilization of Bile Acid Transporters

Bile acids are synthesized from cholesterol in the liver and then secreted into the bile and finally enter the lumen of the small intestine. Bile acids are

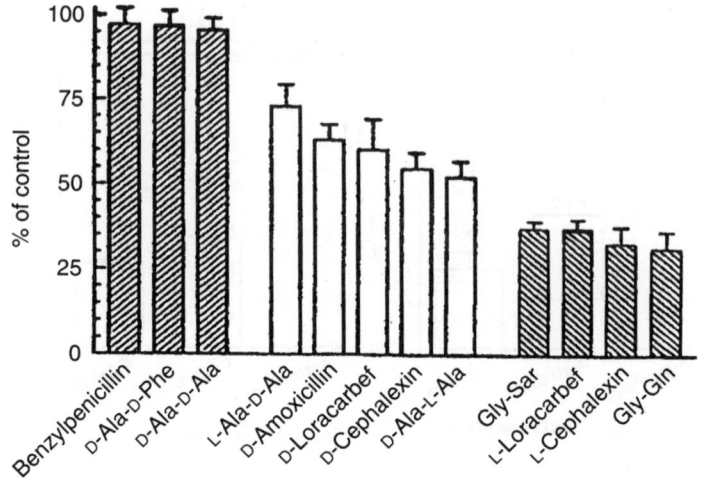

Figure 5 Inhibition of [³H]-cefadroxil uptake into Caco-2 cells by stereoisomers of alanyl peptides and selected β-lactam antibiotics. In addition, influx inhibition by the dipeptides glycyl-sarcosine (Gly-Sar) and glycyl-glutamine (Gly-Gln) and the β-lactams, amoxicillin and benzylpenicillin, is presented. Uptake of 1 mM cefadroxil into Caco-2 cells was measured at apical buffer pH 6.5 for 30 min at 37°C in the absence (control) or presence of 10 mM of the inhibitors. The aminopeptidase inhibitor, amasatin, was present at a concentration of 100 µM in each experiment. Values are expressed as the mean ± SD of 3–6 monolayers. Hatched columns (/): no significant inhibition; open columns: 50% inhibition; back hatched columns (\ ): 60% inhibition. (Adapted from Ref. 16.)

reabsorbed (up to 95%) in the small intestine by an Na⁺ gradient-driven transporter that is located in the brush border membrane of the ileum [18]. They are also taken up into the liver through an Na⁺-dependent manner [19]. The bile acid transporter is responsible for the uptake of drugs that are covalently bound to a bile acid [20]. Therefore bile acids are of particular interest in drug targeting for two main reasons:

1. Under normal physiological conditions, bile acids undergo an enterohepatic circulation that involves only the liver and small intestine.
2. During this biological recycling of bile acids in the enterohepatic circulation, the bile acid (1.5 to 4 g) is circulated 6 to 15 times per day, suggesting high overall transport capacity of the transport system for bile acids in the liver and the small intestine.

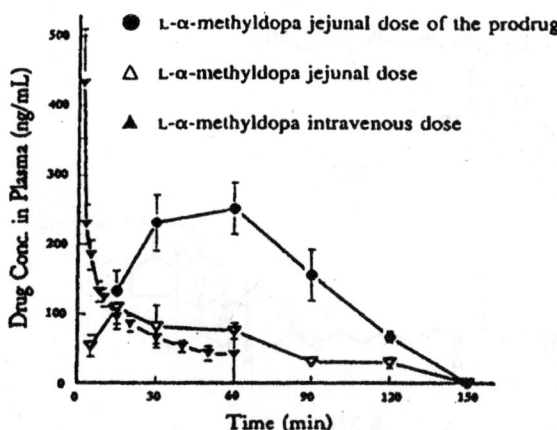

Figure 6 Plasma profile of L-α-methyldopa following intravenous dose of L-α-methyldopa and jejunal dose of L-α-methyldopa and L-α-methyldopa-phenylalanine ($n = 6$–7). (Adapted from Ref. 11.)

In a structure–activity investigation, the following structural elements necessary for the molecular recognition of a bile acid by the ileal Na^+/bile acid transport system were identified [21,22]:

1. A negatively charged side chain on the bile acid molecule
2. At least one α-oriented hydroxyl group in the steroid nucleus at positions 3, 7, or 12
3. A cis configuration of ring A and B of the steroid nucleus
4. A 12 α-hydroxy group at the steroid nucleus

Prodrugs, which are attached to bile acids as promoieties to improve the intestinal absorption of poorly absorbed parent drugs, including peptides, have been reported. The uptake of a peptide–bile acid conjugate via a bile acid transporter system requires the specific molecular interaction of a bile acid moiety by the ileal bile acid transporter system. The attachment of peptides to the bile acid side chain at position C-24 resulted in low ileal absorption of peptide. Furthermore, molecular modeling revealed that the attachment of a peptide to position 7 or 12 greatly changes the shape of the bile acid moiety [23]. Molecular modeling demonstrates the dramatic changes in a bile acid molecule after attaching the bulky peptide groups, even after attachment to the 3-position of the bile acid molecule, raising the question of whether such peptide–bile acid conjugates are preferably recognized by the bile acid or the oligopeptide transport system.

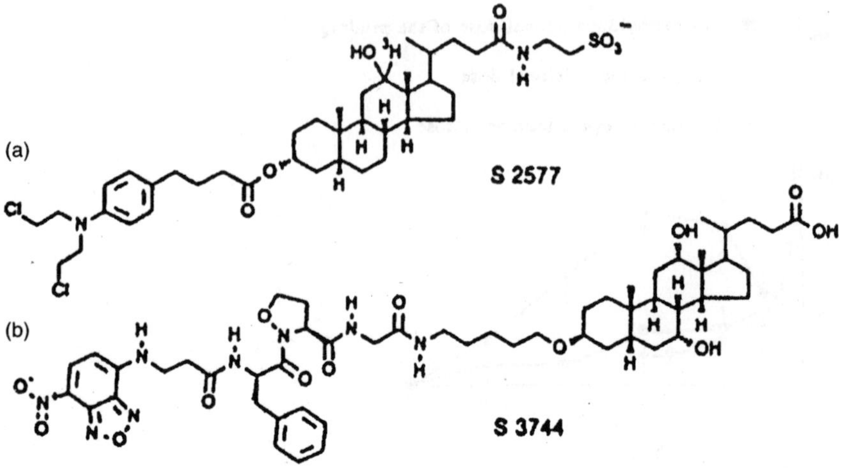

Figure 7 Structural formulas of bile acid conjugate with (a) chlorambcil and (b) NBD-oxalylpropylpeptide. (Adapted from Ref. 22.)

The peptide–bile acid conjugates significantly inhibited Na$^+$-dependent taurochorate uptake rather than H$^+$-dependent dipeptide uptake in rabbit ileal brush border membrane vesicles. This indicates the existence of a preferred molecular recognition of the peptide–bile acid conjugates by the intestinal Na$^+$/bile acid cotransporter. To measure the intestinal absorption of the peptide–bile acid conjugates, peptides or their conjugates were instilled into an ileal segment in an in vivo intestinal perfusion model after cannulation of the common bile duct. The peptide–bile acid conjugate, S3744 (Fig. 7), with a size of the peptide moiety corresponding to a hexa- or heptapeptide, showed an overall intestinal absorption of about 7%, compared with less than 0.5% for the parent peptide or its *t*-butylester [20].

The chlorambucil–bile acid conjugate inhibited the uptake of Na$^+$-dependent taurochorate into both the liver and the intestinal membranes in a concentration-dependent manner [24]. Kullak-Ublick et al. [25] demonstrated that a chlorambucil-taurocholic acid conjugate was transported by the bile acid transporter in cRNA-injected oocytes. Furthermore, the liver-selectivity of HMG-CoA reductase inhibitors conjugated to bile acids have been investigated [26,27]. The Na$^+$-dependent uptake of taurocholate into rat hepatocytes and into rat ileal brush border membrane vesicles was found to be inhibited by the bile acid prodrugs of HMG-CoA reductase inhibitors in a concentration-dependent manner. Bile acid–prodrugs, which

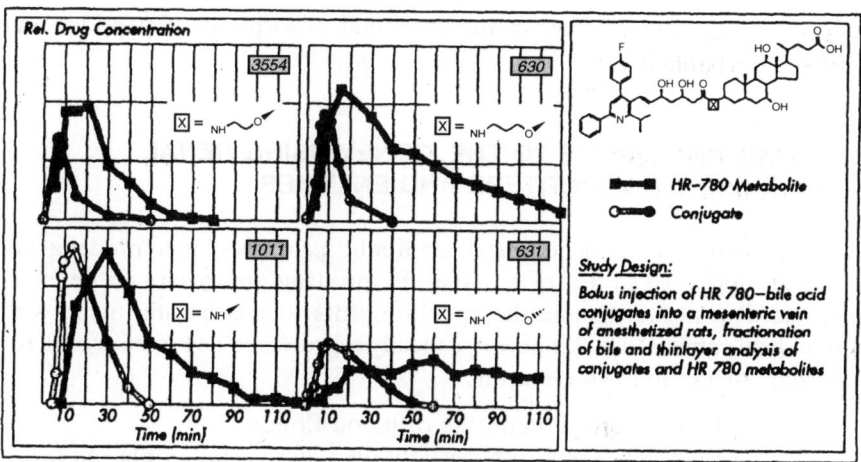

Figure 8 Influence of linker structure in HMG-CoA reductase inhibitor–bile acid prodrugs on the biliary secretion profile. The bile acid conjugates of the HMG-CoA reductase inhibitor HR 780 (1 mM) dissolved in 10 mM Tris/Hepes buffer (pH 7.4)/ 300 mM mannitol/5% ethanol were injected as bolus into a mesenteric vein of anesthetized rats. After cannulation of the common bile duct, bile was fractionated and analyzed for the prodrugs or metabolites by thin-layer chromatography. (Adapted from Ref. 27.)

contain a conserved natural bile acid side chain at position 17, showed a significantly higher affinity to the ileal bile acid transporter. In addition, cholesterol biosynthesis in HepG2 cells was inhibited by the prodrug S3554 at an IC_{50} of 68 nM, indicating that the active drug was released within the HepG2 cell from the inactive prodrug S3554. The influence of linker structure in HMG-CoA reductase inhibitor–bile acid prodrugs on the biliary secretion profile of the prodrug and regenerated drug is depicted in Fig. 8. The release profile of the drug within hepatocytes was greatly influenced by the structure of the linker between the drug and the bile acid moiety. A delayed and long lasting release of the HMG-CoA reductase inhibitor was achieved with a 3-carbon linker in the α-orientation at position 3 of the bile acid moiety. The tissue distribution, after its intravenous administration, demonstrated a clear preference of the drug–bile acid conjugate for the liver and the biliary system. Significantly lower concentrations of the drug–bile acid conjugate and/or the parent drug were found in extrahepatic organs such as the heart and the adrenal. In summary, modified bile acids can be useful for the synthesis of liver-specific prodrugs, for achieving liver-specific drug targeting, and improving the intestinal absorption of peptides and poorly absorbable drugs.

5. TARGETING DRUGS TO THE KIDNEY USING RENAL SPECIFIC TRANSPORTERS AND ENZYMES

In cases where the transport and metabolic processes of a prodrug are selectively localized in a target tissue, the prodrug approach can be effectively used in delivering the therapeutic agents to the specific tissue with minimal side effects. The kidneys are ideally represented as target organs for prodrugs for the following reasons:

1. Kidneys receive a high rate of blood flow.
2. The renal proximal tubules contain a large array of plasma membrane transport systems.
3. Renal cells contain unique drug metabolizing enzymes.

Glutathione, γ-glutamyl derivatives, and β-lyase substrates have been investigated as renal-specific prodrugs based on the presence of certain transporters such as plasma membrane transport systems for glutathione [28] and cysteine [29] conjugates and/or enzymes such as γ-glutamyltransferase (EC 2.3.2.2), dipeptidase, and β-lyase (EC4.4.1.13).

Lash et al. [30] have developed S-(6-purinyl)glutathione as a renal selective prodrug of 6-mercaptopurine. The delivery of S-(6-purinyl)glutathione to the kidney and the subsequent generation of 6-mercaptopurine is represented in Fig. 9. S-(6-purinyl)glutathione, which has a negative charge at physiological pH, can either undergo glomerular filtration or enter the tubular lumen. Luminal S-(6-purinyl)glutathione is rapidly hydrolyzed by γ-glutamyltransferase and dipeptidase on the proximal tubular brush border membrane to S-(6-purinyl)-L-cysteine, which is transported into the renal cell by a Na^+-dependent transporter. Intracellular S-(6-purinyl)-L-cysteine can be metabolized to 6-mercaptopurine by β-lyase enzyme. It has also been reported that S-(6-purinyl)glutathione can be transported across the basolateral plasma membrane into the renal cell by a Na^+- and membrane potential–dependent transporter. Intracellular S-(6-purinyl) glutathione is then efficiently transported out of the cell into the lumen by a membrane potential-dependent transporter on the brush border plasma membrane, where it is processed in the lumen as described above [30,31]. In addition, Drieman et al. [32] showed that several N-acetyl-γ-glutamyl derivatives are able to function as renal prodrugs. Andreadou et al. [33] synthesized several selenocysteine derivatives with aliphatic and benzylic selenium substituents that serve as excellent substrates for β-lyase enzyme.

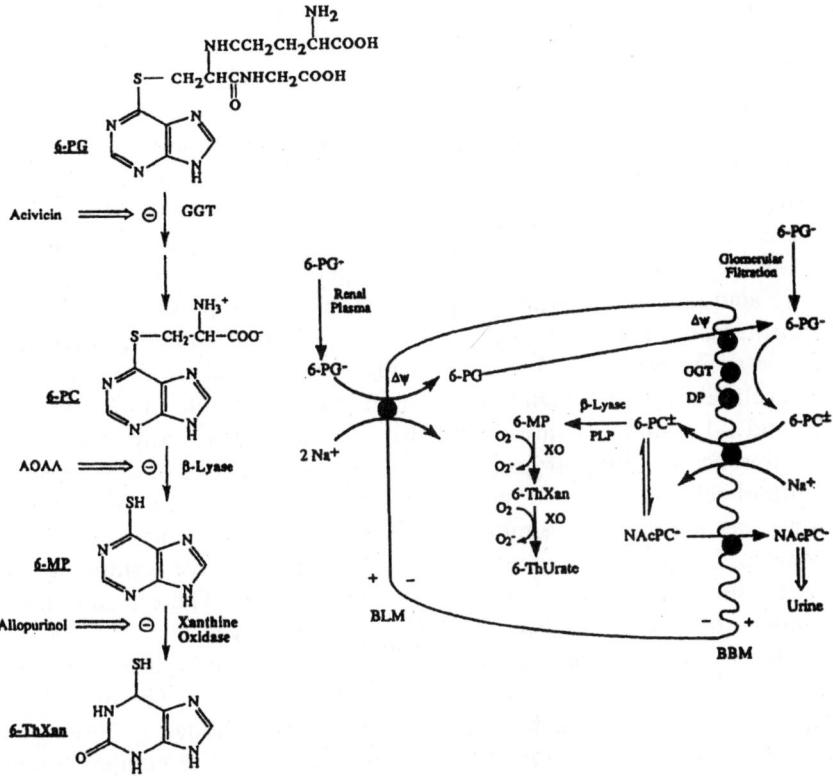

Figure 9 Handling of 6-PG in renal proximal tubular cells. Abbreviations: BBM, brush border plasma membrane; BLM, basolateral plasma membrane; 6-MP, 6-mercaptopurine; 6-PG, S-(6-purinyl) glutathione; 6-PC, S-(6-purinyl)-L-cysteine; NAcPC, *n*-acetyl-S-(6-purinyl)-L-cysteine; GGT, γ-glutamyltransferase; DP, dipeptidase; β-lyase, cysteine conjugate β-lyase; PLP, pyridoxal phosphate; XO, xanthine oxidase; 6-ThXan, 6-thioxanthine; 6-ThUrate, 6-thiourate; AOAA, aminooxyacetic acid; and Δψ, membrane potential. (Adapted from Ref. 30.)

6. UTILIZATION OF ENZYMES FOR THE STEREOSELECTIVE DISPOSITION OF PRODRUGS

6.1. Biodegradation of Amino Acid Prodrugs by Aminopeptidases

Amino acid prodrugs have been investigated in the past as promoieties to improve the aqueous solubility of poorly soluble drugs. Generally, the resulting amide bond is more chemically stable than an ester bond. Such stability of the amide bond often limits·its utility for hydrolytic degradation reactions as shown for the amino acid prodrug of prazosin. However, the amide bond of amino acid prodrugs could be enzymatically cleaved by an enzyme such as amino peptidase or γ-glutamyltranspeptidase.

Pochopin et al. [34] synthesized the alanine, glycine, leucine, lysine, and phenylalanine derivatives of dapsone (an antileprotic agent), which are water-soluble and chemically stable. Prodrugs that contain L-amino acids are all substrates for leucine aminopeptidase. The corresponding D-amino acid derivatives were completely stable in the presence of leucine aminopeptidase for up to 8 hours, which is consistent with the specificity of leucine aminopeptidase for amino acids in the L-configuration. This of course excludes glycine, which is not a chiral molecule. Table 3 indicates that L-Ala- and L-Leu-dapsone are good substrates for leucine aminopeptidase. In the presence of chymotrypsin, L-Ala-, L-Leu-, Gly, and L-Lys-dapsone were completely stable over a period of 4 h, and, as expected, only L-Phe-dapsone was a substrate. Similarly, only L-Lys-dapsone was a substrate for trypsin. Trypsin has a greater affinity for Lys and Arg residues while chymotrypsin preferentially hydrolyzes aromatic acids. These results have implications for the oral delivery of amino acid prodrugs.

The hydrolysis of L-amino acid prodrugs of dapsone occurrs in human plasma and blood, and their half-lives are listed in Table 3. The hydrolysis of dapsone derivatives, which contain D-amino acid residues; is limited by the extent of their solubility (less than 5%) in whole blood. A comparison of data on the plasma and whole blood in Table 3 indicates that whole blood

Table 3   Enzymatic Parameters for L-Aminoacyl Prodrug as Substrates of Leucine Aminopeptidase and Half-Lives of L-Aminoacyl Prodrugs in the Human Plasma and Blood

Compound	Degradation by leucine aminopeptidase		Half-life (in min; ± S.E., $n = 3$)	
	$K_m (\times 10^5)$ (M)	$V_{max} (\times 10^8)$ (Ms^{-1})	Human blood	Human plasma
L-Ala-dapsone	59.1	23.0	20.5 ± 0.7	39.6 ± 6.7
L-Leu-dapsone	8.96	3.83	1.7 ± 0.2	34.4 ± 9.6
L-Phe-dapsone	22.0	3.20	8.8 ± 0.7	190.0 ± 56.0
Gly-dapsone	37.6	2.58	14.6 ± 0.9	107.0 ± 5.6
L-Lys-dapsone	6.09	0.52	10.9 ± 0.7	29.7 ± 4.8

Source: Ref. 34.

has a greater catalytic activity on prodrug activation. Aminopeptidases are associated with erythrocytes and leukocytes of different species [35], and therefore the hydrolysis of amino acid amides would be expected to be greater in whole blood than in plasma. Similarly, the amino acid prodrug of amino-combretastatin, an antitumor compound, is also hydrolyzed by aminopeptidase in erythrocytes. L-amino acid derivatives were rapidly hydrolyzed and showed antitumor activity in vivo while the D-threonine derivative did not, owing to its limited hydrolysis in vivo [36].

6.2. Antibody- and Gene-Directed Enzyme Prodrug Therapies

A new approach for targeting anticancer drugs to tumor cells has been investigated. Antibody-directed enzyme prodrug therapy (ADEPT) and gene-directed enzyme prodrug therapy (GDEPT) systems can increase the delivery of drugs to cancer tissue selectively [37,38]. The ADEPT approach separates the targeting from the cytotoxic function in a two-step treatment. The selective component is an antibody within an enzyme conjugate. The antibody binds the antigen, which is preferentially expressed on the surface of tumor cells. In the first step, the antibody–enzyme conjugate is administered and time is allowed for it to accumulate at the tumor and to clear from blood and normal tissue. In the second step, a nontoxic prodrug is administered that is converted specifically by the enzyme at the tumor site. An antibody–enzyme conjugate molecule can catalyze the conversion of many molecules of the prodrug into a cytotoxic drug, thus enabling a higher concentration of drug at the tumor site. Another important factor is the bystander effect, which affects the killing of surrounding tumor cells even though they do not express tumor antigen and do not bind the antibody–enzyme conjugate.

In the GDEPT approach, foreign enzymes are delivered and expressed in target cells where they can activate subsequently administered nontoxic prodrugs to active drugs. Since the expression of foreign enzymes will not occur in all cells of a targeted tumor in vivo, a bystander cytotoxic effect is required whereby the prodrug is cleaved to generate an active drug.

Carboxypepetidase G2 (CPG2) and β-glactosidase are typical enzymes that recognize the steric conformation of a substrate among enzymes used in the ADEPT and GDEPT systems [38,39]. CPG2, a bacterial enzyme that is a zinc-dependent metalloproteinase, catalyzes the scission of amidic, urethanic, or ureidic bonds between a benzene nucleus and L-glutamic acid. Figure 10 depicts the cleavage of carboyl-L-glutamyl linkage by CGP2. A feasible ADEPT system has been developed using CPG2 linked to a monoclonal antibody for the targeting of tumor cells. In a clinical trial of ADEPT,

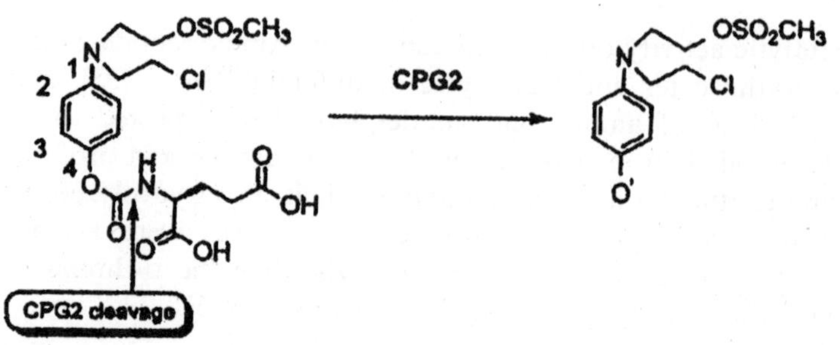

Figure 10 Cleavage of carbamoyl-L-glutamyl linkage by CPGs. (Adapted from Ref. 38.)

a galactosylated antibody-CPG2 conjugate was used in order to improve the clearance of the antibody–enzyme conjugate from the blood. At 24–48 hours after an administration of antibody-CPG2 conjugate to patients with a colorectal carcinoma of the lower intestinal tract, the prodrug (2-chloro-ethyl-(2-mesyloxyethyl)aminobenzoyl-L-glutamic acid (CMDA) was injected over 1–5 days [40]. Cyclosporin was orally coadministered. All patients produced IgG and IgM antibodies to mouse immnoglobulins and CPG2 in their body. From a total of eight patients evaluated, four partial responses and one mixed response were recorded.

A GDEPT strategy based on CPG2 has also been developed. This is versatile because CPG2 can be expressed either intracellularly [41] or extracellularly anchored to the outer membrane of tumor cells [42]. In order to test the efficacy of the externally expressed CPG2 systems, MDA MB361 xenograft models in which established solid tumors expressing either β-gal (control) or CPF2 [MDA MB 361-stCPG2(Q)3] were treated with the prodrug CMDA, and tumor growth was then measured. The results are shown in Fig. 11. CMDA treatment did not alter the rate of growth of the tumor expressing β-gal. In contrast, the regression of tumors was observed in all the mice bearing MDA MB 361-stCPG2(Q)3 xenografts following CMDA treatment, and 16 days later one prodrug-treated mouse was in remission. After 30 days, the tumors of the remainder of the treated mice began to grow and the animals were retreated with CMDA at day 43. Following the treatment the size of tumors approached a plateau. An approximate 80 day growth delay was calculated. The expression of enzyme in the prodrug treated animals was 0.56 ± 0.1 U/g tissue at day 105, in comparison with 0.60 ± 0.09 U/g tissue at day 39 in the case of controls, demonstrating that the expression of stCPG(Q)3 was maintained. These data demonstrate the feasibility of the GDEPT approach.

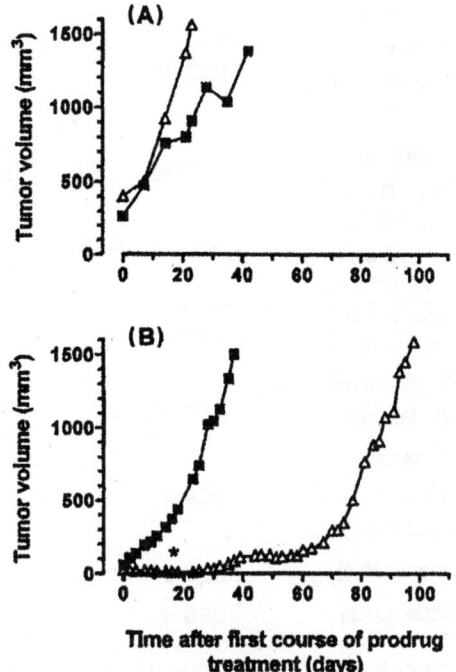

Figure 11   The antitumor activity of CMDA in xenografts. (A) MDA MB 361-β-gal. (B) MDA MB 361-stCPG2(Q)3. The control group (closed squares) received vehicle alone, and the treated group (open triangles) received CMDA. Symbols represent mean values. * indicates the occurrence of a cure. (Adapted from Ref. 38.)

6.3.   Bioconversion of Ester-Containing Prodrugs by Carboxylesterases

Prodrugs containing ester bonds are mainly hydrolyzed by various types of esterases. The enzymes for such hydrolysis include "A"-esterases which hydrolyze organophosphates and "B"-esterases which are inhibited by organophosphates and include butyrylcholinesterase (EC3.1.1.8), carboxylesterase (EC3.1.1.1), and cholesterol esterase (EC3.1.1.13). Among these esterases, carboxylesterase plays a major role in the metabolism of ester- and amide-containing compounds in the diet and drugs. The mammalian carboxylesterases represent a multigene family, the gene products of which are localized in the endoplasmic reticulum of several tissues, such as liver, plasma, muscle, kidney, lung, small intestine, brain and monocyte. Recently,

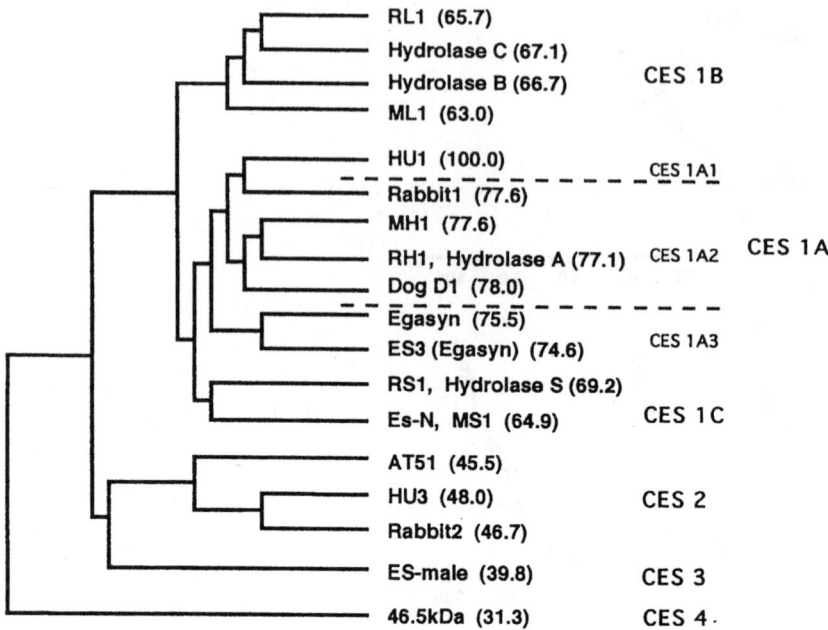

Figure 12 Phylogenetic tree of the carboxylesterase superfamily. (Adapted from Ref. 43.)

Satoh and Hosokawa [43], classified carboxylesterase isozymes into four families, referred to as CES1, CES2, CES3 and CES4 (Fig. 12) based on the homology and similarity of their characteristics. The CES1 family includes major forms of carboxylesterase isozymes (more than 60% homology of human liver carboxylesterase). The CES1 group is divided into three subfamilies designated CES1A, CES1B, and CES1C. The CES1A subfamily is further subdivided into three groups named CES1A1, CES1A2, and CES1A3. The CES 1A1 subfamily includes the major forms of human carboxylesterases, and the CES1A2 group includes the major isoforms of rat, dog, rabbit, and mouse carboxylesterase. The CES1B subfamily catalyzes long-chain acyl CoA hydrolysis. The CES1C family, the homology of which is similar to that of the CES1B family, is a secretory type of carboxylesterase that exists in plasma. The CES2 family mainly catalyzes N,O-acetyltransferase activity. The mammalian intestine includes only CES2 isozymes, although the liver includes both the CES1 and CES2 family. Therefore, the bioconversion of ester prodrugs is affected by the activity of the carboxylesterase in several tissues of each animal.

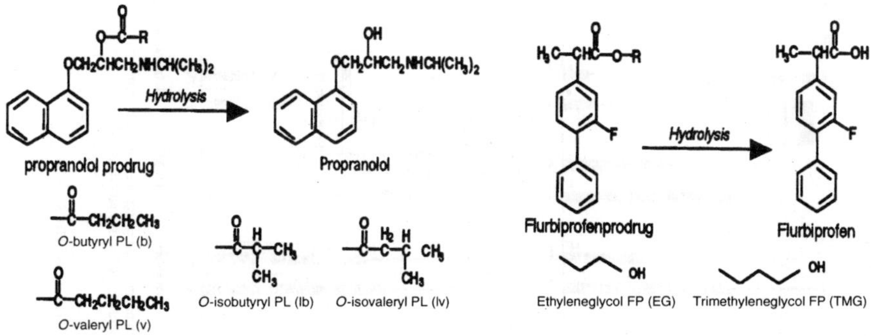

Figure 13 Structures of propranolol, flurbiprofen, and their ester prodrugs.

The hydrolysis of prodrugs of the chiral β-blocker propranolol and the nonsteroidal anti-inflammatory drug flurbiprofen (Fig. 13) by the hydrolases in the liver and small intestine microsomes is shown in Fig. 14 for different species [44,45]. Different propranolol prodrugs containing fatty acids of various chain lengths were hydrolyzed to different extents in both the liver and the small intestine of all the studied species except for the small intestine of the dog. Additionally, the hydrolysis was stereoselective in favor of the R-enantiomer for all the species and both tissues except for the liver microsomes of rats, which showed opposite stereoselectivity (Fig. 14). However, the differences in hydrolysis between the R- and S-isomers were larger in liver microsomes, compared with those in the microsomes of the small intestine, in all animal species examined (Fig. 14).

In order to clarify the contribution of hydrolytic activity in the small intestine to the systemic bioconversion of a prodrug, the plasma concentration of R- and S-propranolol and its prodrug in the mesenteric vein after administration of the prodrug in the rat jejunal loop was examined. When isovaleryl propranolol, which is slowly hydrolyzed in the small intestinal microsomes, was administered in the jejunal loop, only propranolol was detectable in the mesenteric vein, as shown in Fig. 15. In the rat, even the slowly hydrolyzed prodrug, isovaleryl-propranolol, was completely hydrolyzed to propranolol during the transport by the intestinal mucosa [46]. In addition, no stereoselective absorption profiles of propranolol prodrug and/or propranolol itself occured. Therefore the blood concentration of each isomer of propranolol after administration of the prodrug to the rat was the same as for the administration of propranolol. However, in contrast to a higher plasma concentration of (R)-propranolol after the administration of the parent drug propranolol to the dog, non-enantioselective plasma concentration of propranolol was observed after

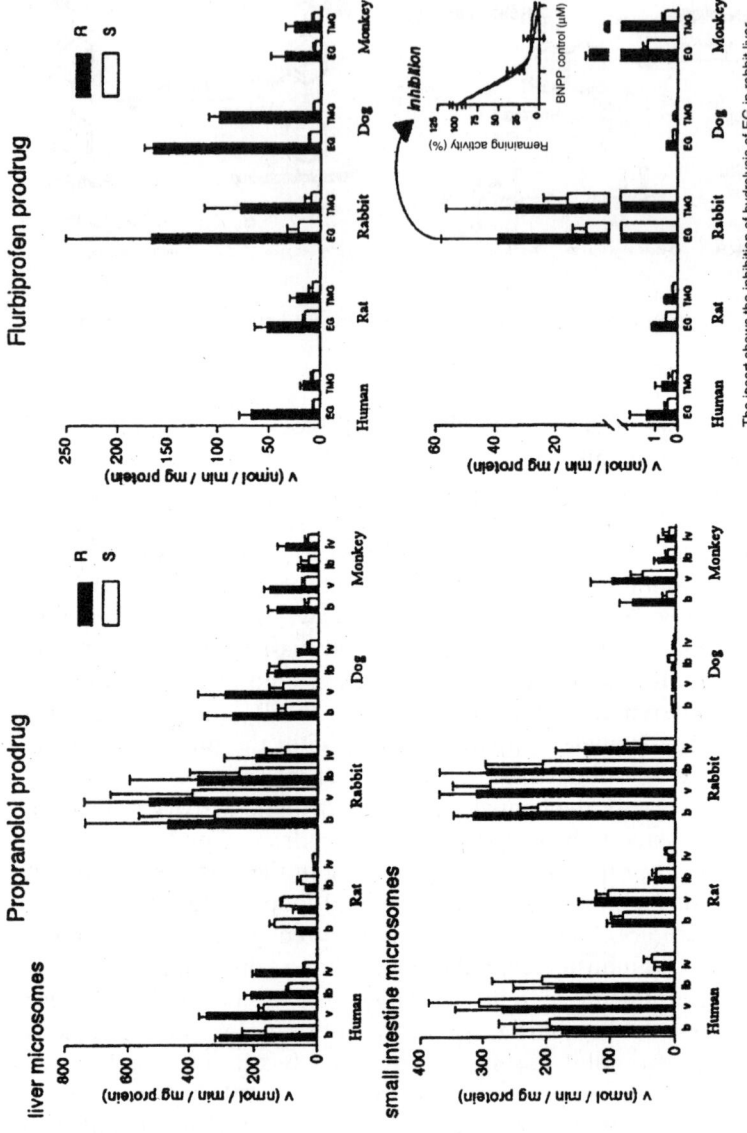

Figure 14 Species differences in hydrolase activity of liver and small intestine microsomes.

The insert shows the inhibition of hydrolysis of EG in rabbit liver microsomes by bis (p-nitro phenyl)-phosphate (BNPP).

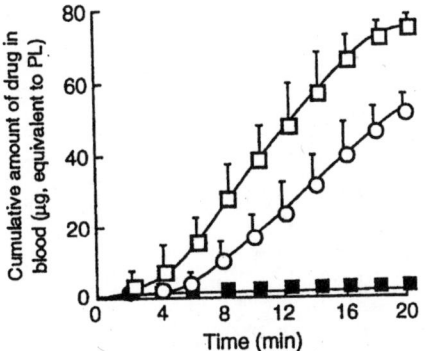

Figure 15 Cumulative amount of propranolol (PL: open symbols) and intact prodrug (closed symbol) in the mesenteric venous blood after administration of PL (open circle) and isovaleryl-PL (open square) in closed loop (8 cm) of rat jejunum (dose: 100 μg, equivalent to PL) under infusion of fresh blood (flow rate: 0.38 mL/ min). (Adapted from Ref. 46.)

an oral administration of isovaleryl-propranolol to the dog whose intestine was unable to hydrolyze the prodrug (Fig. 16) [47]. (S)-propranolol is preferentially eliminated by first-pass metabolism in the dog. An administration of propranolol prodrug to the dog can lead to a high concentration of the prodrug in the liver because of its relatively high lipophilicity, and the prodrug can then be rapidly hydrolyzed to propranolol in an (R)-preferential manner. (R)-propranolol might inhibit further metabolism of (S)-propranolol in the liver; consequently, the (S)-propranolol plasma concentration was also elevated.

In the case of the flurbiprofen prodrug (FP-PPA) that can be R-specifically hydrolyzed in the liver of all species (Fig. 14) to varied extents, preferential in vivo bioconversion for R-enantiomer after oral administration of the prodrug was predominant owing to the lack of hydrolytic activity of the small intestine [48]. As shown in Fig. 17, the R-flurbiprofen prodrug is completely converted to the corresponding parent drug, but only 75% of the S-isomer is converted to the S-flurbiprofen, which is the pharmacologically active form, after administration of the racemic prodrug to the rat. In contrast to rat liver, the (S)-flurbiprofen prodrug is rapidly hydrolyzed to the parent drug in the rat blood with a half-life of 20 seconds. However, the stereoselective bioconversion of the flurbiprofen prodrug in the rat is dependent on hepatic hydrolysis, which is R-preferential. Therefore hydrolase activity in the liver and small intestine is quite important in the design of prodrugs. Furthermore, the R-isomer of the flurbiprofen prodrug, which is substituted with an alcohol at the

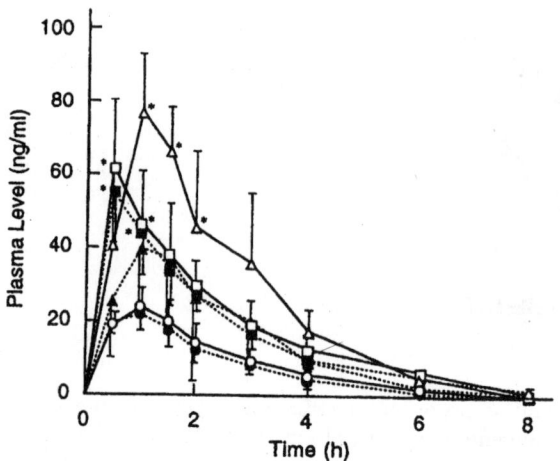

Figure 16 Mean plasma levels (±SD) of propranolol (PL) enantiomers after oral administration of PL (2 and 5 mg/kg) and isovaleryl-PL in equivalent molar doses of 2 mg/kg PL to four beagle dogs. Key: (○) (R)-PL from 2 mg/kg PL, (●) (S)-PL from 2 mg/kg PL, (△) (R)-PL from 5 mg/kg PL, (▲) (S)-PL from 5 mg/kg PL, (□) (R)-PL from isovaleryl-PL, (■) (S)-PL from 2.45 mg/kg isovaleryl-PL. *$p < 0.01$ vs. PL enantiomers after 2 mg/kg PL administration. (Adapted from Ref. 47.)

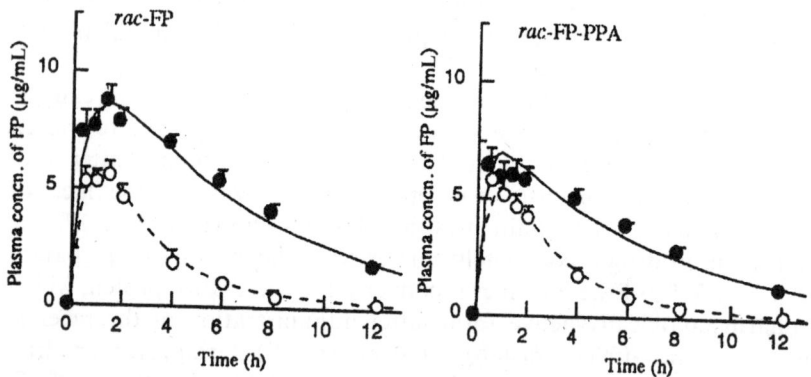

Figure 17 Plasma concentration of flurbiprofen (FP) enantiomers after oral administration of rac-FP and rac-FP-PPA (equivalent to 3 mg/kg of rac-FP) to rats. Data represent the mean ± ($n = 6$). (●) S(+)-FP. (○): R(−)-FP. Solid and broken lines are the simulation curves calculated for S(+)- and R(−)-FP, respectively. (Adapted from Ref. 48.)

carboxylic acid portion of flurbiprofen, is a much better substrate than the S-isomer in all animals species tested. Interestingly, the small intestinal esterases barely hydrolyze the flurbiprofen prodrug because of its bulky acyl group. In addition, the dog intestine shows an extremely low hydrolytic activity for both the propranolol and flurbiprofen prodrugs because of the absence of carboxylesterase isozymes in this species.

6.4. Stereoselective Transdermal Transport of Prodrugs

The advantages of transdermal drug delivery over other routes of administration are well documented, including the avoidance of variable absorption rate, the first-pass hepatic metabolism, and improved therapeutic activity. Skin penetration studies have usually focused only on the physicochemical factors affecting the transport of drugs across the skin. However, recently, various aspects of the metabolism of xenobiotics in the skin have been considered [49], and it has been demonstrated that skin contains various enzymes (such as esterases), which are capable of hydrolyzing a variety of esters [50,51]. Prodrugs containing an ester bond are widely used for enhancing percutaneous penetration. However, the stereoselective hydrolysis of prodrug in skin has been overlooked.

In Table 4, the stereoselective hydrolysis of a propranolol prodrug in skin is compared with that in the liver and plasma [52]. The greatest hydrolysis rate per mg of protein content was observed in the liver homogenate followed by plasma and the skin homogenate. The hydrolysis rate of the R-isomer was found to be faster than the S-isomer in all tissue preparations with all the prodrugs, except for acetyl propranolol in a liver homogenate (Table 4). Meanwhile, in the skin homogenate, very high stereoselective metabolism (R/S ratios fall between 6.7 and 18.4) was observed. Furthermore, K_m value of skin cytosol was close to the K_m value for plasma, indicating that the nature of skin esterases involved in the hydrolysis of propranolol prodrug are different from those of the liver but are more similar to plasma (Table 5). Based on inhibition studies, the hydrolysis of the propranolol prodrugs in hairless mouse skin might involve butyryl cholinesterase and carboxylesterase.

Propranolol did not show any stereoselectivity in penetration, but prodrugs with a higher degree of lipophilicity, e.g., valeryl-, caproyl-, isocaproyl-, and cyclohexanoyl-propranolol [53], showed stereoselectivity in terms of the total cumulative amount of propranolol and prodrug (Table 6) [53,54]. Interestingly, the R-isomer of this lipophilic prodrug was hydrolyzed to an extent greater than 90%. In order to investigate the possibility of stereoselective permeation of the prodrug, a permeation experiment was performed with caproyl-propranolol across full-thickness

Table 4 Stereoselective Hydrolysis Rate Constant (K) per mg of Protein Content, of *rac*-Propranolol Prodrugs in Different Tissue Preparations of Hairless Mouse and in Tris-HCl Buffer (pH 7.4) Containing 0.15 M KCl

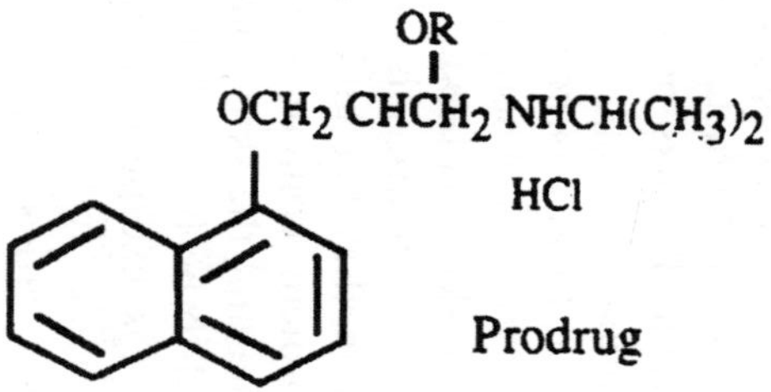

OR
|
$OCH_2 CHCH_2 NHCH(CH_3)_2$

HCl

Prodrug

R		K (× 10³ min⁻¹)				K (× 10³ min⁻¹ mg⁻¹)			
		Buffer (pH7.4)	R/S	Plasma (5%)	R/S	Liver homogenate 10,000 g	R/S	Skin homogenate 10,000 g	R/S
Acetyl	(R)	3.0±0.1	1.0	29.9±3.8*	2.9	73.5±17.8	0.7	3.3±0.1*	7.8
	(S)	3.0±0.1		10.3±3.5		103.2±19.7		0.4±0.1	
Propionyl	(R)	3.0±0.2	1.0	73.5±2.5*	4.7	418.2±11.2	1.3	7.5±0.8*	17.3
	(S)	3.0±0.5		15.8±5.1		316.0±16.7		0.4±0.2	

			Ratio		Ratio		Ratio		Ratio
Butyryl	(R)	2.1 ± 0.3	1.0	274.5 ± 28.9	2.0	1743.5 ± 211.9	1.6	8.5 ± 0.3*	6.7
	(S)	2.0 ± 0.3		134.6 ± 21.6		1125.6 ± 85.5		1.3 ± 0.1	
Valeryl	(R)	1.7 ± 0.1	1.0	350.8 ± 3.3*	1.8	4564.4 ± 141.6	1.3	10.8 ± 0.5*	7.7
	(S)	1.7 ± 0.1		196.7 ± 2.8		3492.4 ± 155.7		1.4 ± 0.5	
Caproyl	(R)	1.6 ± 0.3	1.0	295.3 ± 34.9	1.7	4829.8 ± 234.6	1.2	12.6 ± 0.8*	8.4
	(S)	1.5 ± 0.2		176.1 ± 26.7		3885.7 ± 658.7		1.5 ± 0.3	
Isobutyryl	(R)	3.4 ± 0.7	1.0	104.9 ± 4.0*	3.7	1218.8 ± 173.6	1.9	3.9 ± 0.3*	12.3
	(S)	3.4 ± 0.5		28.6 ± 1.3		641.4 ± 92.1		0.3 ± 0.0	
Isovaleryl	(R)	0.5 ± 0.0	1.0	47.9 ± 5.0	1.5	156.0 ± 19.1*	2.0	1.7 ± 0.3*	9.8
	(S)	0.5 ± 0.1		31.8 ± 5.8		78.5 ± 14.2		0.2 ± 0.1	
Isocaproyl	(R)	2.4 ± 0.1	1.0	221.4 ± 8.8	1.8	4883.7 ± 699.8	1.6	12.4 ± 0.2*	9.8
	(S)	2.3 ± 0.1		122.2 ± 4.0		3081.4 ± 294.0		1.3 ± 0.1	
Cyclopropanoyl	(R)	0.6 ± 0.0	1.0	45.6 ± 0.3*	3.5	42.5 ± 5.7	1.2	5.2 ± 0.7*	16.7
	(S)	0.6 ± 0.0		12.9 ± 0.1		35.0 ± 8.0		0.3 ± 0.1	
Cyclohexanoyl	(R)	0.9 ± 0.1	1.0	169.9 ± 8.9*	2.0	3588.1 ± 129.0	1.2	11.0 ± 0.3*	18.4
	(S)	1.0 ± 0.1		85.0 ± 2.0		3007.1 ± 188.2		0.6 ± 0.1	

Each value is the mean ± S.D. ($n = 4$).
Protein content: 5% plasma. 3.25 mg/mL; liver homogenate, 512 µg/mL; skin homogenate, 7.35 mg/mL.
*$p < 0.05$ versus (S) isomer.

Table 5 Enzyme Kinetics of Hydrolysis of the Enantiomers of Propranolol Prodrugs in Hairless Mouse Skin, Liver, and Plasma

Tissue preparations	Substrate	Enantiomer	K_m (µM)	V_{max} (nmol/min/mg)	V_{max}/K_m (µL/min/mg)
Skin	Butyryl-	(R)	183.4 ± 4.8	8.0 ± 0.4	43.8 ± 0.8*
cytosol	propranolol	(S)	N.D.	N.D.	
microsomes	Butyryl-	(R)	389.2 ± 17.3	3.0 ± 0.8	8.3 ± 0.9
	propranolol	(S)	N.D.	N.D.	
cytosol	Caproyl-	(R)	48.7 ± 4.8*	18.1 ± 3.5*	371.2 ± 7.8*
	propranolol	(S)	166.5 ± 18.3	10.3 ± 1.8	61.5 ± 3.6
Liver	Butyryl-	(R)	356.7 ± 17.2	32.3 ± 5.7*	90.8 ± 16.9*
cytosol	propranolol	(S)	357.2 ± 27.6	24.6 ± 1.2	69.1 ± 5.0
microsomes	Butyryl-	(R)	219.3 ± 45.4	470.6 ± 21.9*	2208.3 ± 459.7
	propranolol	(S)	197.0 ± 11.0	389.5 ± 36.7	1999.2 ± 297.9
Plasma	Butyryl-	(R)	45.1 ± 9.8*	15.5 ± 2.3*	351.4 ± 62.3*
	propranolol	(S)	106.6 ± 11.1	11.6 ± 1.4	110.3 ± 19.1

Data are expressed as means ± S.D. ($n = 3$). N.D.: Not detected. *$p < 0.05$ versus corresponding (S) enantiomer.
Source: Ref. 52.

skin and enzymatically inactive or less active full-thickness skin, prepared by treatment with, DFP; an inhibitor of esterases. Figure 18 shows nonstereoselective penetration of caproyl-propranolol in DFP-treated skin. Furthermore, isovaleryl-propranolol, a prodrug that is resistant to hydrolysis, did not show any stereoselectivity during permeation across enzymatically active full-thickness skin, as shown in Table 6. These results indicate that if the enzymatic activity of skin could be destroyed, hydrolysis-sensitive prodrug also failed to show enantiomeric differences in permeation.

The stereoselective penetration of a prodrug might be explained by different permeation rates between the derived propranolol in the skin and its prodrug. Although the permeation of propranolol and isovaleryl propranolol were the same in full-thickness skin, permeation of propranolol was markedly increased across stripped skin, while isovaleryl propranolol failed to show such enhancement (Fig. 19). Therefore it can be concluded that the stratum corneum is the main barrier to the permeation of propranolol, while the propranolol itself can pass very easily across the hydrophilic epidermis and dermis.

The enantioselective differences in skin-catalyzed hydrolysis may play an important role in the percutaneous fate of topically applied substances. On the basis of the above results, the mechanism of the stereoselective permeation of lipophilic prodrug is proposed and is shown schematically in

Table 6 Cumulative Penetrated Amount at 10h (Q_{10}), Steady State Flux (J_s), Permeability Coefficient (K_p), Lag Time (τ), and Hydrolyzed Percentage (after 10h) of the Isomers of Propranolol Prodrugs Through Full-Thickness Skin of Hairless Mouse

Compound		Q_{10} (nmol/cm^2)	R/S	J_s (nmol/cm^2·h)	R/S	$K_p \times 10^4$ (cm/h)	τ(h)	S/R	Hydrolyzed percentage	R/S
Propranolol	R	164.0 ± 34.9	1.00	30.57 ± 2.74	1.01	0.78 ± 0.07	4.85 ± 0.44	0.99	—	—
	S	164.0 ± 33.2		30.22 ± 2.74		0.77 ± 0.07	4.79 ± 0.40		—	
O-Acetyl PL	R	169.4 ± 24.5	0.99	24.22 ± 3.00	0.99	7.19 ± 0.89	3.13 ± 0.16	1.01	6.8 ± 0.9*	2.83
	S	170.6 ± 24.3		24.53 ± 3.10		7.28 ± 0.92	3.17 ± 0.16		2.4 ± 0.4	
O-Propionyl PL	R	85.6 ± 10.9	0.93	19.48 ± 2.31	0.87	6.24 ± 0.74	6.12 ± 0.22	0.91	37.1 ± 5.2*	15.46
	S	91.7 ± 10.4		22.44 ± 2.56		7.19 ± 0.82	5.63 ± 0.18		2.4 ± 0.4	
O-Butyryl PL	R	253.8 ± 41.3*	1.24	56.96 ± 5.37*	1.09	3.49 ± 0.33*	6.08 ± 0.38*	1.13	64.8 ± 4.2*	4.95
	S	205.5 ± 28.3		52.23 ± 5.37		3.21 ± 0.33	6.87 ± 0.20		13.1 ± 5.1*	
O-Valeryl PL	R	149.8 ± 1.1*	1.56	31.46 ± 4.26*	1.21	18.33 ± 2.48*	5.58 ± 0.70*	1.30	89.6 ± 1.6**	2.74
	S	95.9 ± 3.5		26.01 ± 2.45		15.14 ± 1.43	7.23 ± 0.89		32.7 ± 4.3	
O-Caproyl PL	R	181.0 ± 21.7*	1.74*	26.11 ± 1.98*	1.38	40.83 ± 3.10*	2.99 ± 0.31*	1.57	96.8 ± 3.0*	1.38
	S	104.5 ± 15.5		18.87 ± 2.18		29.49 ± 3.40	4.68 ± 0.89		69.9 ± 7.8*	
O-Isobutyryl PL	R	120.2 ± 41.2	1.00	21.85 ± 3.67	0.95	4.94 ± 0.83	4.71 ± 0.68	1.07	18.2 ± 7.8*	4.79
	S	120.0 ± 41.3		23.06 ± 3.45		5.21 ± 0.78	5.05 ± 0.82		3.8 ± 1.6	
O-Isovaleryl PL	R	158.7 ± 20.3	1.00	32.62 ± 5.84	1.01	24.07 ± 4.31	5.48 ± 0.25	0.99	10.7 ± 2.6*	9.73
	S	158.7 ± 12.7		32.38 ± 6.85		23.89 ± 5.05	5.45 ± 0.56		1.1 ± 0.7	
O-Isocaproyl PL	R	123.9 ± 13.8*	2.47	19.75 ± 1.35*	1.54	25.67 ± 1.76*	3.81 ± 0.28*	1.55	95.6 ± 0.4*	2.07
	S	50.2 ± 8.3		12.63 ± 0.62		16.42 ± 0.80	5.90 ± 0.42		46.2 ± 8.4*	
O-Cyclopropanoyl PL	R	153.3 ± 57.1	0.91	29.15 ± 4.22	0.95	12.62 ± 1.8	4.69 ± 1.29	0.98	8.2 ± 2.4*	13.67
	S	167.8 ± 52.1		30.60 ± 5.99		33.33 ± 2.22	4.78 ± 1.06		0.6 ± 0.2	
O-Cyclohexanoyl PL	R	49.3 ± 1.6*	4.81	7.77 ± 0.53*	2.14	21.59 ± 1.47*	3.75 ± 0.22*	2.25	96.9 ± 1.0	2.84
	S	10.3 ± 1.6		3.63 ± 0.66		10.09 ± 1.83	8.42 ± 1.09		34.1 ± 5.4	

Values are the mean ± S.D. ($n = 4$); *$p < 0.05$ vs. corresponding S-isomer; **$p < 0.01$ vs. corresponding S-isomer.
Source: Ref. 54.

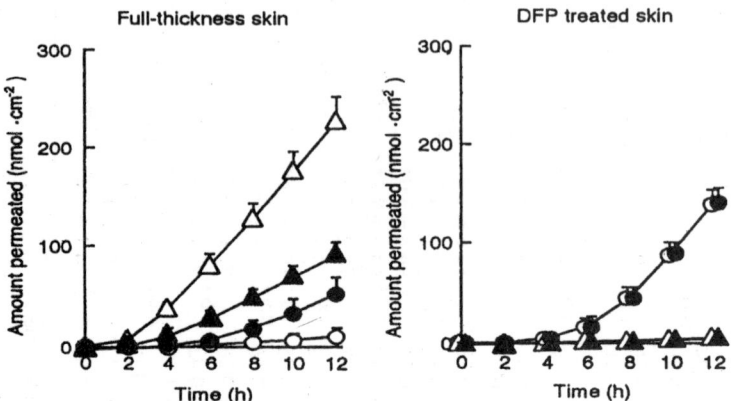

Figure 18 Permeation profiles of caproyl propranolol enantiomers across full-thickness skin and DFP-treated skin. Circles and triangles represent intact prodrug and converted propranolol, respectively. Open and closed symbols represent (R) and (S) isomers, respectively. (Adapted from Ref. 54.)

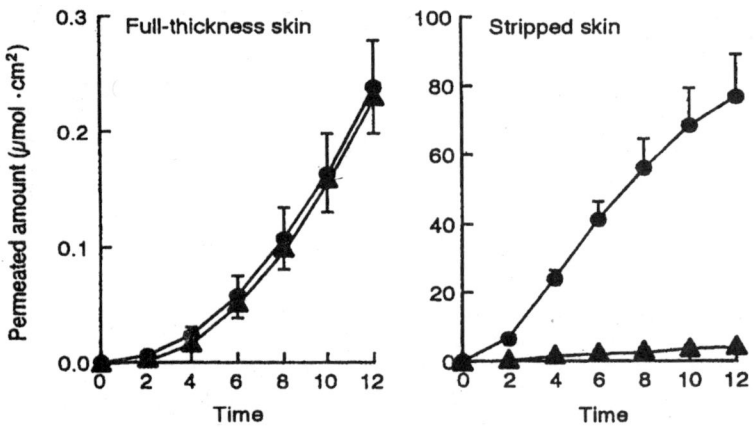

Figure 19 Cumulative penetrated amount of (S)-propranolol (circle) and (S)-isovaleryl propranolol (triangle) across hairless mouse skin. (Adapted from Ref. 54.)

Fig. 20 [54]. The stratum corneum is lipidlike in nature and favors the penetration of lipophilic drug; the viable epidermis and dermis, however, are hydrophilic in nature and present a barrier to the lipophilic drug. Therefore prodrugs of higher lipophilicities readily enter and diffuse within

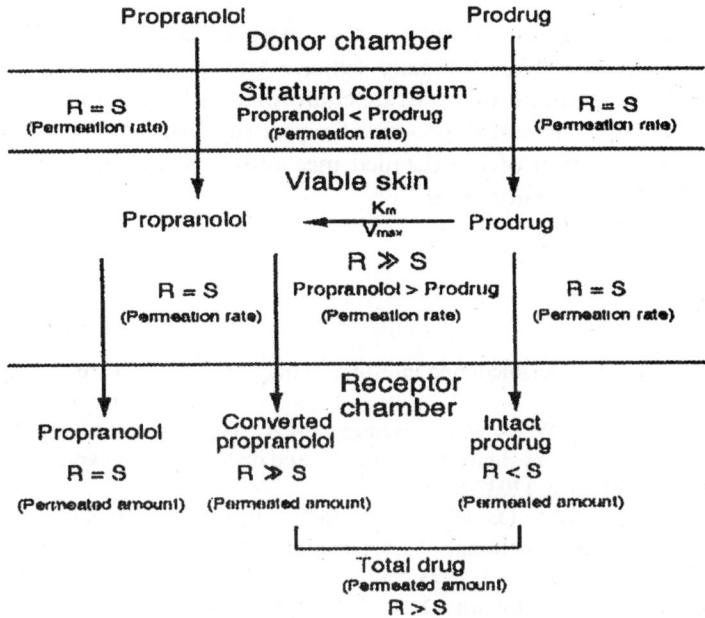

Figure 20 Possible mechanisms of stereoselective penetration of lipophilic propranolol prodrugs across hairless mouse skin. (Adapted from Ref. 54.)

the stratum corneum but are unable to penetrate easily through the viable epidermis and dermis. The skin esterases are mainly present in the viable epidermis where the prodrugs are more likely hydrolyzed. Since the R-isomer is rapidly hydrolyzed, the converted R-propranolol, which is hydrophilic, can easily pass the dermis layer. However, the S-isomer of the prodrug mainly remained as an intact form and faces difficulties in passing the dermal layer due to its lipophilicity. Thus stereoselective penetration was not due to the stereoselective transport of propranolol or prodrug but rather to substantial differences in hydrolytic properties $(R > S)$ (Fig. 20).

7. CONCLUSIONS

The body provides an intensely chiral environment in which the great majority of important processes exhibit stereospecificity. The disposition of xenobiotics is determined by their interaction with highly chiral endogenous molecules such as transporters and enzymes. Therefore the site-specific delivery of a drug can be achieved through the prodrug approach by targeting specific drug transporters and/or enzymes in certain tissues. However, only a few prodrugs actually achieve site-specific disposition in vivo because the recognition of substrate conformation by these proteins is quite complex. The successful site-specific prodrug approach can be realized by a demonstration of the detailed mechanism of the interaction of these proteins with their substrates.

REFERENCES

1. Tsuji, A.; Tamai, I. Carrier-mediated intestinal transport of drugs. Pharm. Res. **1996**, *13* (7), 963–977.
2. Lee, W.S.; Kanai, Y.; Wells, R.G.; Hediger, M.A. The high affinity Na^+/glucose cotransporter. Reevaluation of function and distribution of expression. J. Biol. Chem. **1994**, *269*, 12032–12039.
3. Rand, E.B.; Depaoli, A.M.; Davidson, N.O.; Bell, G.I.; Burant, C.F. Sequence, tissue distribution, and functional characterization of the rat fructose transporter GLUT5. Am. J. Physiol. **1993**, *264*, G1169–1176.
4. Thorens, B.; Cheng, Z.-Q.; Brown, D.; Lodish, H.F. Liver glucose transporter: a basolateral protein hepatocyte and intestine and kidney cells. Am. J. Physiol. **1990**, 259, C279–285.
5. Mizuma, T.; Ohta, S.; Awazu, S. The β-anomers and glucose preferences of glucose transport carrier for intestinal active absorption of monosaccharide conjugates. Biochim. Biophys. Acta **1994**, *1200*, 117–122.
6. Mizuma, T.; Nagamine, Y.; Dobashi, A.; Awazu, S. Factors that cause the β-anomeric preference of Na^+/glucose cotransporter for intestinal transport of monosaccharide conjugates. Biochim. Biophys. Acta **1998**, *1381*, 340–346.
7. Mizuma, T.; Sakai, N.; Awazu, S. Na^+-dependent transport of aminopeptidase-resistant sugar coupled tripeptides in rat intestine. Biochim. Biophys. Res. Commum. **1994**, *203*, 1412–1416.
8. Haga, M.; Saito, K.; Shimaya, T.; Maezawa, Y.; Kato, Y.; Kim, S.W. Hypoglycemic effect of insulin derivatives in rats. Chem. Pharm. Bull. **1990**, *38*, 1983–1986.
9. Liang, R.; Fei, Y.-J.; Prasad, P.D.; Ramamoorthy, S.; Han, H.; Yang-Feng, T.L.; Hediger, M.A.; Ganapathy, V.; Leibach, F.H. Human intestinal H^+/peptide cotransporter.: cloning functional expression and chromosomal localization. J. Biol. Chem. **1995**, *270*, 6456–6463.
10. Tamao, I.; Tomizawa, N.; Takeuchi, T.; Nakayama, K.; Higashida, H.; Tsuji, A. Functional expression of transport of intestinal dipeptide/β-lactam antibiotic transporter in *Xenopus laevis* oocytes. Biochem. Pharmacol. **1994**, *48*, 881–888.
11. Yee, S.; Amidon, G.L. Oral absorption of angiotensin converting enzyme inhibitors and peptide prodrugs. In *Peptide Based Drug Design*; Taylor, M.D., Amidon, G.L., Eds.; American Chemical Society: Washington, DC, 1995; 299–316.

12. Thwaites, D.T.; Hirst, B.H.; Simmons, N.L. Substrate specificity of the di/tripeptide transporter in human intestinal epithelia (Caco–2): identification of substrates that undergo H^+-coupled absorption Br. J. Pharmacol. **1994**, *113*, 1050–1056.

13. Lister, N.; Sykes, A.P.; Bailey, P.D.; Boyd, C.A.; Bronk, J.R. Dipeptide transport and hydrolysis in isolated loops of rat small intestine: effects of stereospecifity. J. Physiol. **1995**, *484*, 173–182.

14. Amidon, G.L.; Sinko, P.J. Characterization of the oral absorption of β-lactam antibiotics. II; Competitive absorption and peptide carrier specificity. J. Pharm. Sci. **1989**, *78*, 723–727.

15. Kramer, W.; Girbig, F.; Gutjaha, U.; Howalewski, S.; Adam, F.; Schiebler, W. Intestinal absorption ·of β-lactam antibiotics and oligopeptides—functional and stereoselective reconstitution of the oligopeptide transporter system from rabbit small intestine. Eur. J. Biochem. **1992**, *204*, 923–930.

16. Wenzel, U.; Thwaites, D.T.; Daniel, H. Stereoselective uptake ·of β-lactam antibiotics by the intestinal peptide transporter. Br. J. Pharmcol. **1995**, *116*, 3021–3027.

17. Döring, F.; Will, J.; Amasheh, S.; Clauss, W.; Ahlbrecht, H.; Daniel, H. Minimal molecular determinants of substrates for recognition by the intestinal peptide transporter. J. Biol. Chem. **1998**, *273*, 23211–23218.

18. Wilson, F.A. Intestinal transport of bile acids. Am. J. Physiol. **1981**, *241*, G83-G92.

19. Meier, P.J.; Eckhat, U.; Schroeder, A. Substrate specificity of sinusoidal bile acid and organic anion uptake systems in rat and human liver. Hepatol **1997**, *26*, 1667–1677.

20. Kramer, W.; Wess, G.; Enhsen, A. Falk, E.; Hoffmann, A.; Neckermann, G.; Schubert, G.; Urmann, M. Modified bile acids as carriers for peptides and drugs. J. Control. Rel. **1997**, *46*, 17–30.

21. Lack, L. Properties and biological significance of the ileal bile salt transport system. Environ. Health Perspect. **1979**, *33*, 79–90.

22. Kramer, W.; Gitbig, F.; Gutjahr, U.; Kowalewski, S.; Jouvenal, K.; Muller, G.; Tripier, D.; Wess, G. Intestinal bile acid absorption: Na^+-dependent bile acid transporter activity in rabbit small intestine correlates with the coexpression of an integral 93 kDa and a peripheral 14 kDa bile acid binding protein along the duodenum–ileum axis. J. Biol. Chem. **1993**, *268*, 18035–18046.

23. Wess, G.; Kramer, W.; Schubert, G.; Enhsen, A.; Barinhaus, K.-H., Glombik, H.; Müller, S.; Bock, K.; Kleine, H.; John, M.; Neckermann, G.; Hoffmann, A. Synthesis of bile acid drug conjugate: potential drug-shuttles for liver-specific targeting. Tetrahedron Lett. **1992**, *34*, 819–822.

24. Kramer, W.; Wess, G.; Schubert, G.; Bickel, M.; Girbig, F.; Gutjahr, U.; Kowalewski, S.; Baringhaus, K.-H.; Enhsen, A.; Glombik, H.; Müller, S.; Neckermann, G.; Schulz, S.; Petzinger, E.; Liver-specific drug targeting by coupling to bile acids. J. Biol. Chem. **1992**, *267*, 18598–18604.

25. Kullak-Ublick, G.A.; Glasa, J.; Boker, C.; Oswald, M.; Grutzner, B.; Hagenbuch, B.; Stieger, B.; Meier, P.J.; Beuers, U.; Kramer, W.; Wess, G.;

Paumgartner, G. Chlorambcil-taurocholate is transported by bile acid carriers expressed in human hepatocellular carcinomas. Gastroenterol. **1997**, *113*, 1295–1305.

26. Kramer, W.; Wess, G.; Enhsen, A.; Bock, K.; Falk, E.; Hoffmann, A.; Neckermann, G.; Gantz, D.; Schulz, L.; Nickau, L.; Petzinger, E.; Turley, S.; Dietschy, J.M. Bile acid derived HMG-CoA reductase inhibitors. Biochim. Biophys. Acta **1994**, *1227*, 137–154.

27. Kramer, W.; Wess, G.; Enhsen, A.; Falk, E.; Hoffmann, A.; Nekermann, G.; Schubert, G.; Urmann, M. Modified bile acids as carriers for peptides and drugs. J. Controlled Release **1997**, *46*, 17–30.

28. Lash, L.H.; Jones, D.P. Uptake of the glutathione conjugate S-(1,2-dichlorovinyl)glutathione by renal basal-lateral membrane vesicles and isolated kidney cells. Mol. Pharmacol. **1985**, *28*, 278–282.

29. Schaeffer, V.H.; Stevens, J.L. Mechanism of transport of toxic cysteine conjugates in rat kidney cortex membrane vesicles. Mol. Pharmacol. **1987**, *32*, 293–298.

30. Lash, L.H.; Shivnani, A.; Mai, J.; Chinnaiyan, P.; Kraouse, R.J.; Elfarra A. Renal cellular transport, metabolism, and cytotoxicity of S(-6-purinyl)glutathione, a prodrug of 6-mercaptopurine, and analogues. Biochem. Pharmacol. **1997**, *54*, 1341–1349.

31. Lash, L.H.; Jones, D.P. Renal glutathione transport: characteristics of the sodium-dependent system in the basal-lateral membrane. J. Biol. Chem. **1984**, *259*, 14508–14514.

32. Drieman, K.C.; Thijssen, H.H.W.; Struyker-Boudier, H.A.J. Renal selective N-acetyl-L-γ-glutamyl prodrugs: studies on the selectivity of some model prodrugs. Br. J. Pharmacol. **1993**, *108*, 204–208.

33. Andreadou, I.; Menge, W.M.P.B.; Commander, J.N.M.; Worthington, E.A.; Vermaeulen, N.P.E. Synthesis and novel Se-substituted seleocysteine derivatives as potential kidney selective prodrugs of biologically active selenol compounds: evaluation of kinetics of β-elimination reactions in rat renal cytosol. J. Med. Chem. **1996**, *39*, 2040–2046.

34. Pochopin, N.L.; Charman, W.N.; Stella, V.L. Amino acid derivatives of dapsone as water-soluble prodrugs. Int. J. Pharm. **1995**, *121*, 157–167.

35. Lochs, H.; Morse, E.L.; Adibi, S.A. Uptake and metabolism of dipeptides by human red blood cells. Biochem J. **1990**, *271*, 133–137.

36. Ohsumi, K.; Hatanaka, T.; Nakagawa, R.; Fukuda, Y.; Morinaga, Y.; Suga, Y.; Nihei, Y.; Ohishi, K.; Akiyma, Y.; Tsuji, T. Synthesis and antitumor activity of amino acid prodrugs of amino-combretastatins. Anti-Cancer Drug Design **1999**, *14*, 539–548.

37. Niculescu-Duvaz, I.; Spooner, R.; Marais, R.; Springer, C.J. Gene-directed enzyme prodrug therapy. Bioconjugate Chemistry **1998**, *9*, 4–22.

38. Niculescu-Duvaz, I.; Friedlos, F.; Niculescu-Duvaz, D.; Davies, L.; Springer, C.J. Prodrug for antibody- and gene-directed enzyme prodrug therapies (ADEPT and GDEPT). Anti-Cancer Drug Design **1999**, *14*, 517–538.

39. Bakina, E.; Farquhar, D. Intensely cytotoxic anthracycline prodrugs: galactosides. Anti-Cancer Drug Design **1999**, *14*, 507–515.

40. Sharma, S.K.; Bagshawe, K.D.; Burke, P.J.; Boden, J.A.; Rogers, G.T.; Antoniw, P.; Springer, C.J.; Melton, R.G.; Sherwood, R.F. Galactosylated antibodies and antibody-enzyme conjugates in ADEPT. Cancer **1994**, *73*, 1114–1120.

41. Marais, R.; Spooner, R.A.; Light, Y.; Martin, J.; Springer, C.J. Gene-directed enzyme prodrug therapy with a mustard prodrug/carboxypeptidase G2 combination. Cancer Res. **1996**, *56*, 4735–4742.

42. Marais, R.; Spooner, R.A.; Stribblinf, S.M.; Light, Y.; Martin, J.; Springer, C.J. A cell surface tethered enzyme improves efficiency in gene-directed enzyme prodrug therapy. Nature Biotechnology, **1997**, *15*, 1373–1377.

43. Satoh, T.; Hosokawa, M. The mammalian carboxylesterase: from molecules to functions. Annu. Rev. Pharmacol. Toxicol. **1998**, *38*, 257–288.

44. Yoshigae, Y.; Imai, T.; Horita, A.; Matsukane, H.; Otagiri, M. Species differences in stereoselective hydrolase activity in intestinal mucosa. Pharm. Res. **1998**, *15*, 626–631.

45. Yoshigae, Y.; Imai, T.; Horita, A.; Otagiri, M. Species differences for stereoselective hydrolysis of propranolol prodrugs in plasma and liver. Chirality **1997**, *9*, 661–666.

46. Yoshigae, Y.; Imai, T.; Aso, T.; Otagiri, M. Species differences in the disposition of propranolol prodrugs derived from hydrolase activity in intestinal mucosa. Life Sci. **1998**, *62*, 1231–1241.

47. Shameem, M.; Imai, T.; Yoshigae, Y.; Sparreboom, A.; Otagiri, M. Stereoselective hydrolysis of O-isovaleryl propranolol and its influence on the clearance of propranolol after oral administration. J. Pharm. Sci. **1994**, *83*, 1754–1757.

48. Fukuhara, A.; Imai, T.; Otagiri, M. Stereoselective disposition of flurbiprofen from a mutual prodrug with a histamine H2-antagonist to reduce gastrointestinal lesions in the rat. Chirality **1996**, *8*, 494–502.

49. Kappus, H. Drug metabolism in the skin. In *Pharmacology of the Skin II*; Greaves, M.W., Shuster, S., Eds.; Springer-Verlag: Berlin, 1989; 123–163.

50. McCracken, N.W.; Blain, P.G.; Williams, F.M. Nature and role of xenobiotic metabolizing esteras in rat liver, lung, skin, and blood. Biochem. Pharmacol. **1993**, *45*, 31–36.

51. Heymann, E.; Hoppe, W.; Krusselman, A.; Tshoetsel, C. Organophosphate sensitive and insensitive carboxylesterases in human skin. Chem-Biol. Interac **1993**, *87*, 217–226.

52. Ahmed, S.; Imai, T.; Yoshigae, Y.; Otagiri, M. Stereospecific activity and nature of metabolizing esterases for propranolol prodrug in hairless mouse skin, liver and plasma. Life Sci. **1997**, *61*, 1879–1887.

53. Ahmed, S.; Imai, T.; Otagiri, M. Influence of stereoselective hydrolysis on stereoselective penetration of O-acetyl- and O-cyclohexanoylpropranolol through hairless mouse skin. Pharm. Sci. **1995**, *1*, 429–432.

54. Ahmed, S.; Imai, T.; Otagiri, M. Evaluation of stereoselective transport and concurrent cutaneous hydrolysis of several ester prodrugs of propranolol: mechanism of stereoselective permeation. Pharm. Res. **1996**, *13*, 1524–1529.

5

Stereoselectivity in Drug Action and Disposition: An Overview

Bhavesh K. Patel and Andrew J. Hutt
King's College London, London, England

1. INTRODUCTION

In the foreword to the 1994 reprint of Edwin Abbott's classic eighteenth-century science-fiction novel concerning a two-dimensional world, "Flatland," Anjam Khursheed* comments, "Flatland will always be a profound inspiration to those who refuse to live in a world dominated by limited dimensionality." Until relatively recently, pharmacology has been very much a limited-dimensionality science, the three-dimensional nature of drug molecules being largely neglected. Stereoselectivity in drug action has been known since the early years of the last century [1], but apart from a relatively few instances, it was overlooked in what was almost a golden age of drug discovery and development between the 1950s and the early 1970s. As a result of this neglect, by the late 1980s racemates accounted for 25% of the products available in a survey of 1675 drugs [2]. However, over the last 15 to 20 years there has been a change in philosophy with respect to chiral pharmaceuticals.

This change has been brought about to a large extent by advances in methodology associated with the stereoselective synthesis and stereospecific analysis of chiral drug molecules, together with an increased appreciation of the potential significance of the differential biological properties of the

---

*Khursheed, A. (1994). Foreword. In: E.A. Abbott, *Flatland*. Oxford: Oneworld Publications, pp. 1–4.

enantiomers of chiral drugs administered as racemates. The use of racemates is regarded by some as polypharmacy with the proportions of the drug dose being determined by chemical rather than therapeutic or pharmacological criteria, and by others as drugs containing 50% impurity. As a consequence of the advances in technology and increasing safety concerns [3], drug chirality became an issue for both the regulatory agencies and the pharmaceutical industry [4–7], and in recent years there has been a move toward the development of single stereoisomer products [8–10]. In addition to the development of new drugs, the resurgence of interest in drug stereochemistry has resulted in a number of agents, previously marketed as racemates, being reevaluated as single stereoisomer products [11]. These so-called racemic or chiral switches have resulted in a number of agents being marketed as both single stereoisomer and racemic products at the same time.

In spite of the move towards single stereoisomer products, the number of agents currently available as racemates is considerable, and for the majority of these relatively little is known with respect to the pharmacodynamic, pharmacokinetic, or toxicological properties of the individual enantiomers present in such mixtures. This chapter is not intended to be an exhaustive compilation of stereoselectivity in drug action but aims to provide an overview, illustrate some of the complexities that may arise, and indicate the significance of stereochemical considerations in pharmacology. More detailed evaluation of individual classes of drugs are presented in Chaps. 6 and 7.

2. BIOLOGICAL DISCRIMINATION OF STEREOISOMERS

The differential biological activity of stereoisomers is not a new phenomenon in spite of the considerable interest over the last twenty years. Pasteur, in 1858, showed that the mold *Penicillium glaucum* metabolized (+)-tartrate more rapidly than the (−)-enantiomer and commented, "the idea of the influence of the molecular asymmetry of natural organic products is introduced into physiological studies, this important characteristic being perhaps the only distinct line of demarcation which we can draw today between dead and living matter" [12].

This was followed in 1886 by Piutti's observation that (+)-asparagine had a sweet taste whereas the (−)-enantiomer was insipid. Following Puitti's report, Pasteur made the remark that "the nervous tissue might itself be dissymmetric," an observation that Holmstedt regards as the first mention of stereoselectivity of a receptor [12]. These differences in taste have been established for other amino acids such that those of the D-configuration series are sweet whereas those of the L-series are either bitter or tasteless.

Claims that enantiomers may also differ in their odors were disputed for a number of years, and the observed differences were associated with contamination of samples. However, in the early 1970s, highly purified samples of the α,β-unsaturated terpenoid ketones (R)- and (S)-carvone were isolated and shown to have odors of spearmint and caraway, respectively [13–15]. (R)-Carvone was also shown to be a stronger odorant than the S-enantiomer on a threshold basis [14]. Similarly, the related terpine enantiomers (R)- and (S)-limonene have odors of orange and lemon respectively [13]. It is noteworthy that the odors of (R)- and (S)-amphetamine also differ, being described by individuals involved in their manufacture as musty and fecal respectively. This had been appreciated for a number of years before the 1970s but was not widely known, as it was not thought to be significant [13].

The differential pharmacological activity of a pair of drug enantiomers was reported in the first decade of the twentieth century when the British pharmacologist Cushny demonstrated differences in the activity of atropine [(±)-hyoscyamine] and (−)-hyoscyamine and (−)- and (+)-adrenaline. The potential significance of stereochemistry in pharmacology and medicinal chemistry was pointed out in a textbook on synthetic drugs published in 1918, which contains the statement: "In the case of stereo-isomerides, which are optically active, marked differences in the physiological action are very often encountered" [16]. Further, the author compares atropine with (−)-hyoscyamine, (+)- and (−)-nicotine, and the "striking example" of natural (−)-adrenaline with the (+)-enantiomer [16]. Cushny [1] went on to publish what is probably the first book devoted to the study of stereochemistry in pharmacology.

Discrimination between stereoisomers in biological environments should in fact not be unexpected, as at a molecular level such environments are composed of "handed" macromolecules, i.e., proteins, glycolipids and nucleic acids, from the chiral precursors of L-amino acids and D-carbohydrates. In addition, the macromolecular structures of these biopolymers also give rise to chirality as a result of helicity, e.g., the protein α-helix and DNA double helix. Such helical structures may have either a left or right handed turn in the same way that a spiral staircase may be left or right handed. In the case of the DNA double helix and the protein α-helix, the biopolymers have a right-handed turn.

As nature has made a preference in terms of its chirality it is not therefore surprising that enzymes and receptor systems exhibit stereochemical preferences, particularly as many of the natural substrates and ligands

for these systems are single stereoisomer chiral molecules, e.g., neurotrans-
mitters, endogenous opioids, and hormones.

   A significant conceptual advance toward the rationalization of the
biological discrimination of a pair of enantiomers in terms of drug action
was presented in the early 1930s by Easson and Stedman [17]. The interaction between a drug and a receptor is associated with bonding interactions
between the functionalities in the drug and complementary sites on the
receptor, the three-dimensional spatial arrangement of such functionalities
being of considerable significance. Easson and Stedman (17) postulated
that the differences in activity arose as a result of differential binding of the
pair of enantiomers to a common site. The more active enantiomer takes
part in a minimum of three intermolecular interactions with the receptor,
whereas the less active enantiomer could interact at two sites only (Fig. 1).
Thus the fit of the enantiomers to the receptor differs, as does the energy of
the interaction. An achiral analog of the drug could also interact at two sites
and may be expected to possess similar biological activity as the less active
stereoisomer (Fig. 2).

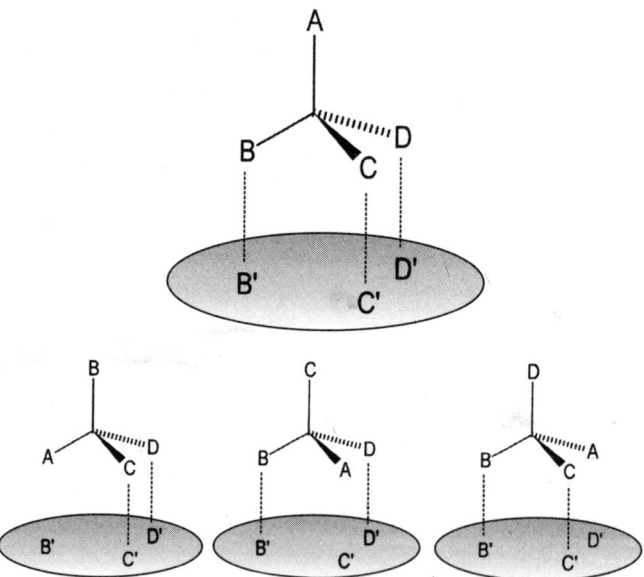

Figure 1   Easson–Stedman model of the drug–receptor interaction [17]. The more
active stereoisomer (top) is involved with three simultaneous complementary
bonding interactions with the receptor active site, B$\cdots$B′, C$\cdots$C′ and D$\cdots$D′;
its less active enantiomer (lower) may interact at two sites only irrespective of its
orientation to the active site.

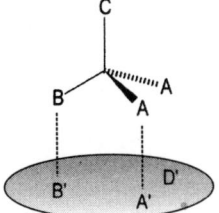

Figure 2 Achiral analog that may also take part in two complementary bonding interactions with the receptor similarly to the less active enantiomer (see Fig. 1) [17].

This hypothesis was supported by an examination of the activity of (*R*)-adrenaline, where the three interactive groups are the basic methylamino, the catechol ring, and the secondary alcohol. The corresponding deoxy compound, *N*-methyldopamine, where the secondary alcohol is removed, was found to have similar activity to (*S*)-adrenaline, where the hydroxyl group is oriented away from the complementary receptor binding site [18]. Subsequent comparisons of data between pairs of enantiomers and their corresponding deoxy analogs for a number of related adrenergic agents indicated that the Easson–Stedman hypothesis [17] did not always appear to hold; in some instances the achiral compounds were more potent than the "less" active enantiomers. However, additional investigations [18,19], following the development of theories regarding directly and indirectly acting amines and the use of catecholamine-depleted tissue preparations, indicated that the anomalies were due to variable indirect activities of the achiral and "less" active enantiomers. The (−)-enantiomers were found to be more potent than their (+)-enantiomers or deoxy analogs in both normal and reserpine-pretreated tissue preparations, whereas the (+)-enantiomers and achiral compounds were found to be equipotent in catecholamine-depleted preparations but of variable activity in normal tissue, resulting in the conclusion that the Easson–Stedman hypothesis only holds for sites of direct action [18,19].

Ogston [20,21], seemingly unaware of the Easson–Stedman model, proposed a similar three-point attachment model to rationalize the observed stereoselectivity in the enzymatic transformation of symmetrical prochiral substrates, e.g., citrate and aminomalonate (Fig. 3) [22]. Similarly, Dalgleish [23], also unaware of the Easson–Stedman model [17], rationalized his observations concerning the resolution of the enantiomers of a number of amino acids on paper chromatography by a "three-point attachment." In a subsequent telephone conversation with Bentley [24], Dalgleish stated that he was "terribly impressed by the Ogston hypothesis." It is therefore

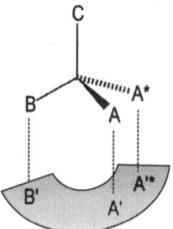

Figure 3 Three-point interaction between the symmetrical molecule C_{AA*BC} to an enzyme surface. Even though the substrate molecule is symmetrical, i.e., group A is identical to A*, these groups are not sterically equivalent and are said to be enantiotopic. The molecule is in fact prochiral. If site A'* is the catalytic site, then group A* in the substrate but not A will undergo transformation to the product [20].

unfortunate that he did not cite Ogston [20] in the manuscript. Bentley [24] also stated that the Easson–Stedman three-point model [17] was not generally known among biochemists at that time.

Sokolov and Zefirov [25] developed the Ogston enzyme model for prochiral substrates into what they described as a "rocking tetrahedron" model. In this model, enantioselectivity is examined in dynamic terms rather than the static terms of the Ogston proposal [20,21]. In the rocking tetrahedron model the substrate binds to the enzyme via two interactions and has conformational flexibility. The two enantiotopic groups (A and A*, Fig. 4) occupy overlapping but identical volumes, and the enantioselectivity of the enzymatic transformation is dependent on the orientation and interaction between the active site and the "volume." If the interaction is at right angles to the plane of the substrate (from point X, Fig. 4), then enantioselectivity is not observed, whereas if the interaction is in the plane (e.g., from point Y), the process is potentially highly enantioselective, or enantiospecific. Thus in compounds with "small" highly flexible enantiotopic functionalities, minimal stereoselectivity would be observed. In contrast, if the enantiotopic groups are large bulky substituents, which are conformationally less flexible, then greater selectivity, tending towards specificity, is expected [25].

Chiral recognition models continue to be a matter of considerable interest, not only with respect to biological activity but also with recognition phenomena in chemistry and particularly the separation sciences [26–29]. Booth et al. [30] extended the rocking tetrahedron concept to a conformationally driven chiral recognition mechanism, which they had previously discussed in terms of chromatographic recognition [31]. They proposed a four stage chiral recognition process. The initial step involved the formation

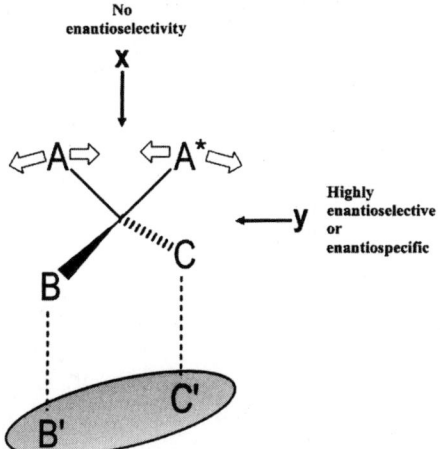

No enantioselectivity

Highly enantioselective or enantiospecific

Figure 4 The rocking tetrahedron model of Sokolov and Zefirov [25]. The substrate C_{AA^*BC} takes part in two bonding (B$\cdots$B′ and C$\cdots$C′) and has conformational flexibility in the site with movement of the enantiotopic groups A and A*. ($\Rightarrow$ indicates conformational flexibility of the bound substrate; $\rightarrow$ possible sites of catalytic activity; see text for discussion.)

of a selectand–selector complex (tethering), followed by conformational adjustment of the complex in order to optimize the molecular interactions and complex activation via additional interactions and finally an expression of molecular fit, i.e., chiral recognition [30]. They proposed that this general process describes enantiomeric discrimination for all cases of chiral selection.

More recently, Mesecar and Koshland [32], following an investigation of the enzyme isocitrate dehydrogenase, have found that the three-point model does not always hold. Examination of metal-free crystals of the enzyme structure reveals that only the 2S,3R-(L)-isocitrate binds, whereas in the presence of magnesium ions only the 2R,3S-(D)-enantiomer binds. Examination of x-ray structures of the two enzyme–substrate complexes reveals three common binding sites for both enantiomeric substrates that differ at a fourth site. Based on their observations, the authors proposed that the three-point model is only applicable if the assumption is made that the substrate can approach a planar surface from one direction. Thus a fourth location, either a direction requirement or an additional binding site, is essential to distinguish between a pair of enantiomers (Fig. 5).

In recent years there has been extensive discussion as to the minimal requirements for chiral recognition. Some authors say that only a diastereomeric intermediate is required, whereas others have stated that interactions

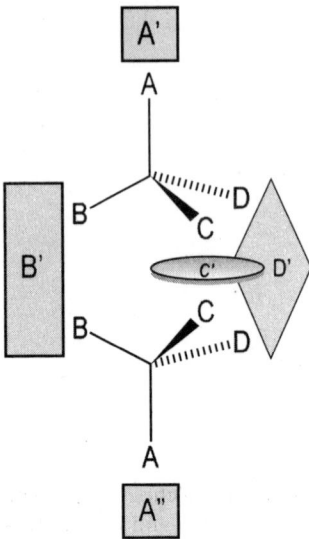

Figure 5   Four-point location model. If the target/binding site(s) protrude from a surface or are in a cleft in the macromolecular structure, then either enantiomer may bind at three sites (e.g., B····B′, C····C′ and D····D′), and the bonding interaction is determined by the approach direction. Alternatively a fourth interaction/location may be required (i.e., A····A′ or A····A″). The initial interaction may result in conformational changes in structure as proposed in the conformationally driven model.

must take place. The current consensus would appear to be that a diastereomeric intermediate must be involved, but that this alone may not be sufficient for recognition, and that interactions between eight centers [28], and hence a four-point attachment, are required. In such cases, where the drug/substrate may approach the target surface in one direction, three-point attachment involving six interactive centers may be sufficient, i.e., similarly to the Mesecar and Koshland [32] hypothesis; the direction provides the fourth location.

2.1. Stereochemical Terminology and Pharmacodynamic Activity

The differential pharmacodynamic activity of drug stereoisomers has resulted in additional terminology. The enantiomer with the greater affinity, or activity, is termed the eutomer, whereas that with the lower affinity or activity is the distomer. The ratio of affinities, or activities, eutomer to distomer, is known as the eudismic ratio and its logarithm as the eudismic index. The slope of a plot of eudismic index versus the logarithm of the affinity of the eutomer (ideally expressed as either pA_2 or pD_2 values in pharmacology or k_i and k_m values in enzymology) for a homologous or congenic series is known as the eudismic affinity quotient (EAQ). The EAQ is a quantitative measure of the stereoselectivity within a compound series for a particular biological effect [2,33,34]. A positive slope is indicative of Pfeiffer's rule, i.e., the greater the difference in pharmacological activity between a pair of enantiomers, the greater the specificity exhibited by the eutomer [35]. The original hypothesis is based on a plot of isomeric activity ratio versus average human dose for a diverse series of drugs. In hindsight, the linear relationship obtained is somewhat surprising, as no considerations were made for differences in enantiomeric disposition, receptor type, or drug action, i.e., agonist or antagonist. A number of subsequent investigations have provided supporting data for this hypothesis [33], though exceptions to the rule have also been observed [36].

It is important to appreciate that the above terminology and des-ignations refer to a single activity of a drug and that for a dual action drug the eutomer for one activity may be the distomer for the other. For example, amosulalol (Fig. 6) is an adrenoceptor antagonist at both α- and β-receptors. The $(-)$-enantiomer is a nonspecific β- and selective α_1-adrenoceptor antagonist, whereas $(+)$-amosulalol is a selective and more potent α_1-antagonist [37]. Thus, in the case of amosulalol, the $(-)$- and $(+)$-enantiomers are the eutomers for β- and α-adrenoceptor blockade, respectively.

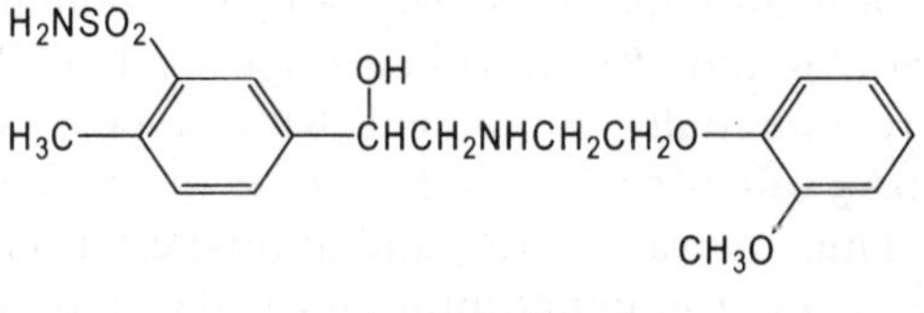

Amosulalol

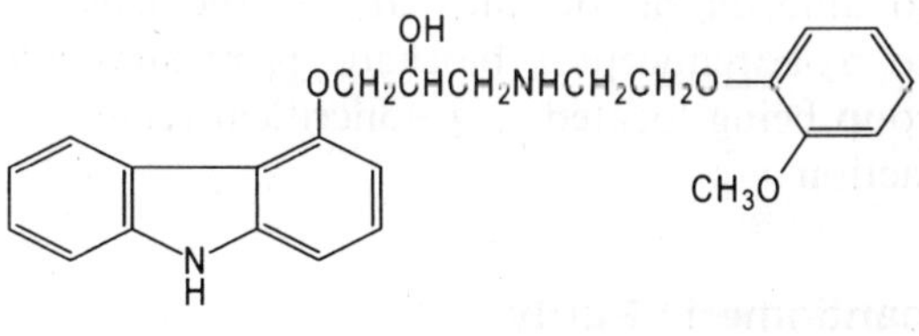

Carvedilol

Figure 6 Structures of the adrenoceptor antagonists, amosulalol and carvedilol.

Table 1   Pharmacodynamic Activity of the Enantiomers of Amosulalol

Receptor	Tissue	Eutomer pA$_2$ (enantiomer)	Eudismic ratio
β$_1$	Rat atrium	7.71 (−)	48
β$_2$	Guinea pig trachea	7.38 (−)	47
α$_1$	Rabbit aorta	8.31 (+)	14
α$_2$	Rat vas deferens	5.36 (+)	3

It is also of interest to note the difference in magnitude of the eudismic ratio for the receptor types (Table 1). Comparison of the affinity of the corresponding achiral desoxy derivative to that of the individual enantiomers results in a marked decrease with respect to β-blockade but relatively minor effects with respect to α-blockade; in the case of the α$_2$-adrenoceptor the affinity slightly increases (pA$_2$ 6.05) in comparison to the more potent (+)-amosulalol (pA$_2$ 5.36) [38]. These observations appear to contradict the Easson–Stedman hypothesis [17]; however, taken together with the decrease in eudismic ratio between the enantiomeric affinity for β- and α-adrenoceptors, they may well imply an alternative binding orientation of the drug enantiomers to the different receptors such that in the case of the α-adrenoceptor the hydroxy group is located in a less critical region for interaction [37].

In addition to a reversal in potency, a dual-action drug may show equipotency for one activity. Carvedilol (Fig. 6) is a nonselective β-adrenoceptor antagonist with vasodilator activity used in the treatment of hypertension and angina. (S)-Carvedilol is a potent competitive inhibitor at β$_1$-adrenoceptors, whereas the R-enantiomer is considerably less potent, with pA$_2$ values of 9.4 and 3.9 for (S)- and (R)-carvedilol, respectively [39]. In contrast, the enantiomers are essentially equipotent with respect to α$_1$-adrenoceptor blockade, with pA$_2$ values of 7.87 and 7.79 for the S- and R-enantiomers, respectively [39]. Thus the vasodilator and antihypertensive effects of the drug arise as a result of α$_1$-adrenoceptor blockade of both enantiomers and the β$_1$-blockade from the S-enantiomer which prevents reflex tachycardia. Similarly to amosulalol, outlined above, the lack of stereoselectivity with respect to α$_1$-adrenoceptor blockade is presumably associated with the hydroxy group being located in a noncritical region of the molecule for receptor interaction.

2.2.   Eudismic Ratio and Enantiomeric Purity

Determination of the eudismic ratio is obviously dependent on the stereochemical purity of the material under examination. This is particularly the case for the less active enantiomer, and as the eudismic ratio increases, the significance of trace quantities of the eutomer as an impurity of the distomer also increases. The degree of enantiomeric purity is frequently not reported in the pharmacological literature, or alternatively is presented in terms of optical rotation, particularly in the older literature, which is not a sensitive technique at levels of enantiomeric contamination of a few percent. Even if values of the specific rotation of a given analyte are known, then experimental factors such as temperature, solvent, and wavelength need to be reproduced with considerable care to ensure that accurate and comparable data can be obtained [40–42].

For example, initial investigations on the effect of the enantiomers of isoprenaline on blood pressure in cats and dogs yielded values for the eudismic ratio $(-)/(+)$ of 11.8 and 87.5, respectively, whereas later studies yielded ratios of 1000 and 450 [18;43]. The authors pointed out that purification to constant biological activity was a better criterion for enantiomeric purity than constant specific rotation or melting point [43], an argument confirmed by subsequent investigations by Barlow, who indicated the significance of trace quantities of enantiomeric impurities on the eudismic ratio. The interested reader should refer to the articles of Barlow [44] and Barlow et al. [45] for further details.

The significance of stereochemical purity may be illustrated by a consideration of the activity of the selective β_2-adrenoceptor agonist formoterol (Fig. 7) [46]. This compound contains two centers of chirality as part of its structure, and therefore four stereoisomers are possible. Initial investigations indicated that the β_2-agonist activity resided in the stereoisomer with the R,R absolute configuration and a rank order of potency $R,R > R,S > S,S \gg S,R$. Subsequent investigations reported much greater differences, the eudismic ratio $R,R/S,S$ increasing from 50 to 850 when the impurity of the eutomer in the distomer decreased from approximately 1.5% to < 0.1%. In addition, the increased stereoisomeric purity resulted in an altered potency ratio from that cited above to $R,R \ggg R,S > S,R \gg S,S$ [37].

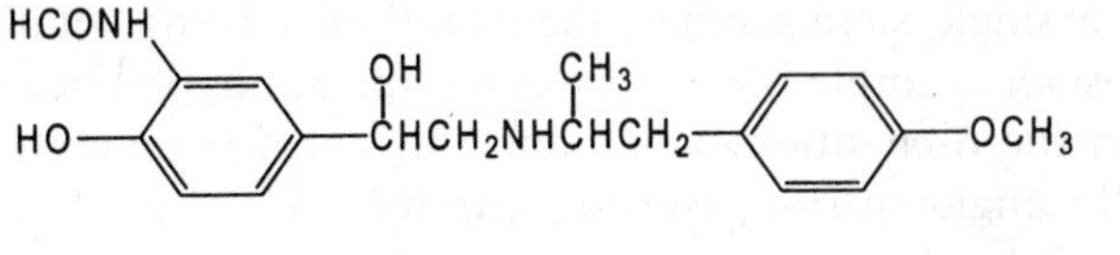

Formoterol

Figure 7 Structure of the selective β_2-adrenoceptor agonist formoterol.

With the current developments in stereospecific analytical method-ologies, particularly in the separation sciences, together with increased detector sensitivity, it should now be considered unacceptable to present pharmacological data on enantiomers without quoting their stereochemical purity.

3. PHARMACODYNAMIC COMPLEXITIES

The most important differences between enantiomers occur in drug–receptor interactions. Indeed, Lehmann [34] has stated, "the stereoselectivity displayed by pharmacological systems constitutes the best evidence that receptors exist and that they incorporate concrete molecular entities as integral components of their active-sites." In contrast to the pharmacoki-netic properties of a pair of enantiomers (Sec. 4), differences in pharma-codynamic activity tend to be more marked, and eudismic ratios of 100 to 1000 are not uncommon.

Examination of the pharmacodynamic properties of a pair of enan-tiomers may yield a number of possible scenarios:

> The required activity resides in a single stereoisomer, the enantiomer being biologically inert.
> Both enantiomers have similar pharmacodynamic profiles.
> The pharmacodynamic activity of a pair of enantiomers differs so that both may be marketed with different therapeutic indications.
> The enantiomers may have opposite effects at the same biological targets.
> One stereoisomer may antagonize the adverse effects of its enantiomer.
> The required activity resides in both stereoisomers, but the adverse effects are predominantly associated with one enantiomer.
> The adverse effects are associated with both enantiomers, but the required effect is predominantly associated with one enantiomer.

There are relatively few examples of drugs in which the pharmacodynamic activity is restricted to a single stereoisomer, the enantiomer being totally devoid of activity. One such example is α-methyldopa; the antihypertensive activity resides solely in the S-enantiomer [47] and this agent is marketed as a single isomer. The angiotensin-converting enzyme (ACE) inhibitor imidapril is another example: the inactive enantiomer is essentially devoid of activity, being greater than a millionfold less active than the eutomer [48]. Similarly, there are few examples where the required beneficial activity resides in a single stereoisomer and the adverse effects, or toxicity, reside in its enantiomer.

Instances are also known where the activity of a pair of enantiomers differs sufficiently that both are marketed with different therapeutic indications, e.g., dextromethorphan and levomethorphan, and levopropoxyphene and dextropropoxyphene. In the case of both these pairs of compounds, the first named is marketed as an antitussive whereas their enantiomers are analgesics.

That a pair of enantiomers can have opposite actions at the same receptor was at one time considered to be extremely rare. However, in recent years this phenomenon has become somewhat more common, which is presumably associated with the increased awareness of the potential significance of stereoselectivity in drug action. Whatever the reason, it is obvious that the evaluation of the activity of a racemate does not provide a clear indication of the properties of the drug.

The analgesic agent racemic picenadol (Fig. 8) is a partial μ-receptor agonist that arises as a result of the greater agonist potency of the (+)-3S,4R-stereoisomer than the weaker antagonist activity of (−)-picenadol at the same receptor [49]. Similarly, (R)-sopromidine (Fig. 8) is an H$_2$-agonist, whereas its S-enantiomer is an antagonist, the racemate being a partial agonist on guinea-pig atrium preparations [50]. A number of aporphine derivatives (Fig. 8) have also been shown to elicit opposite effects at the same receptor. For example, (R)-11-hydroxy-10-methylaporphine has been shown to be a selective 5-HT$_{1A}$ agonist [51], whereas its enantiomer is an antagonist [52]; (+)-(S)-apomorphine is an antagonist at D-1 and D-2 dopaminergic receptors, whereas the R-enantiomer is an agonist [53], and (R)-11-hydroxyaporphine activates dopamine receptors, whereas the S-enantiomer is an antagonist [54].

A more complex situation, illustrating that resolution and evaluation of the pharmacodynamic properties of the individual enantiomers present in a racemate may provide considerable insight into the observed activity of the racemate, is provided by an examination of 3-(3-hydroxyphenyl)-N-n-propylpiperidine (Fig. 8) and its analogs. The initial evaluation of the drug was carried out on the racemate, which was described as a highly selective agonist at dopaminergic autoreceptors. Following resolution, the R-enantiomer was found to selectively stimulate presynaptic dopaminergic receptors at high doses and postsynaptic receptors at lower doses [55]. In contrast, the S-enantiomer stimulated presynaptic dopamine receptors and blocked postsynaptic receptors. Thus following the racemate, the post-synaptic stimulation of the R-enantiomer is counteracted by the blockade due to the S-enantiomer resulting in a selective profile for the racemate. The S-enantiomer in this instance, together with the S-enantiomer of the

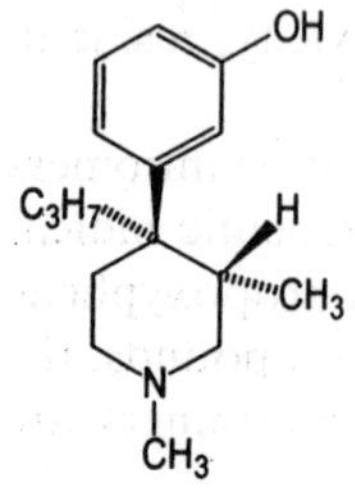

(3S,4R)-Picenadol

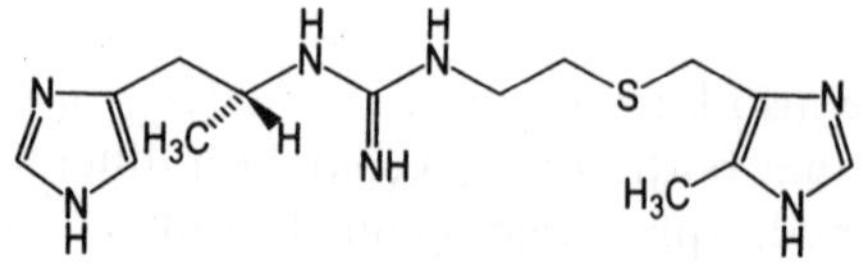

(R)-Sopromidine

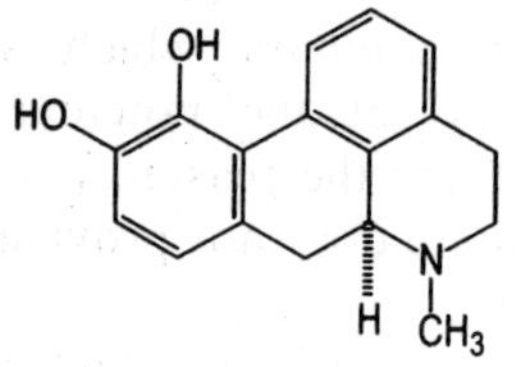

(S)-Apomorphine 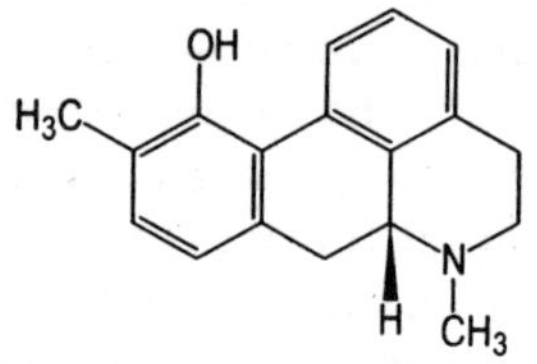   (R)-11-Hydroxy-10-methylaporphine

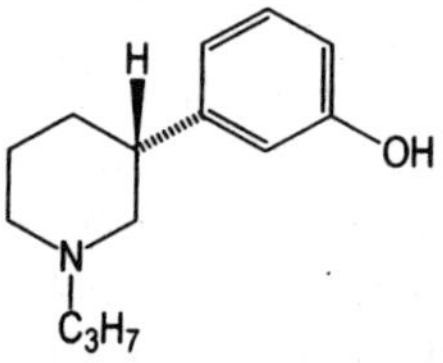

(R)-3-(3-Hydroxyphenyl)-N-n-propylpiperidine

Figure 8 Examples of agents the enantiomers of which have opposite effects at the same receptor site.

corresponding ethyl derivative, attenuates dopaminergic function by action at both presynaptic and postsynaptic receptors by agonism and antagonism, respectively [55].

Some of the more complex pharmacodynamic scenarios will be illustrated below with examples cited from selected pharmacological groups.

3.1. Dihydropyridine Calcium Channel Blocking Agents

Ion channels exist in a number of states, depending upon membrane or chemical potential, and these alternative states represent different conformations of channel proteins, which in turn may influence the binding site conformation or access pathway to the drug binding site [56]. In the case of the voltage-dependent calcium channel, the L-type is sensitive to

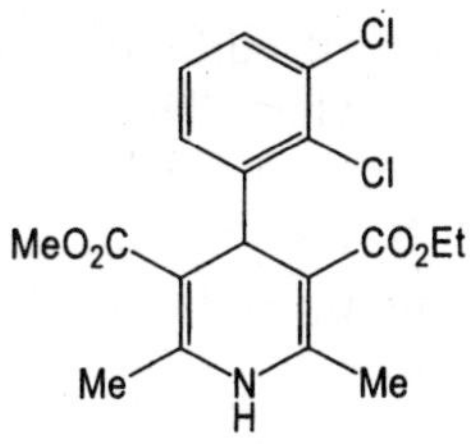

Felodipine

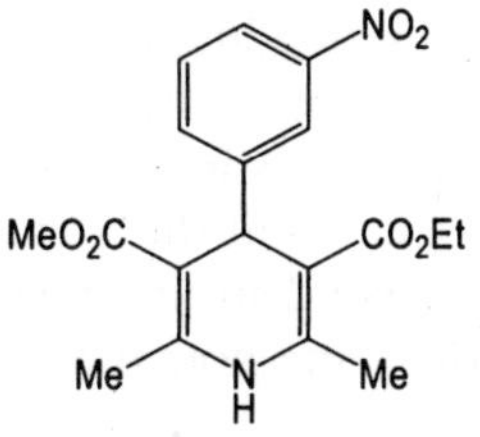

Nitrendipine

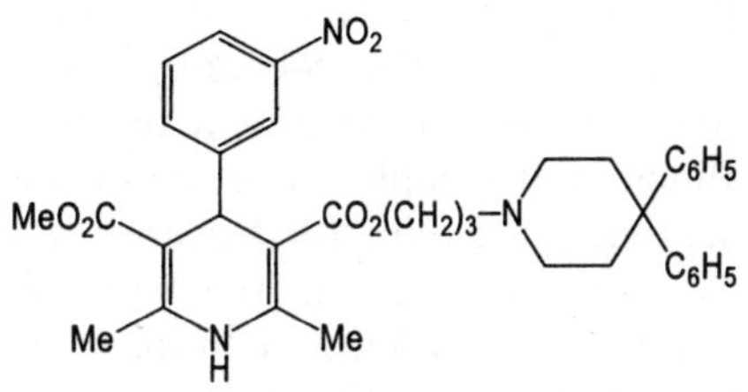

Niguldipine

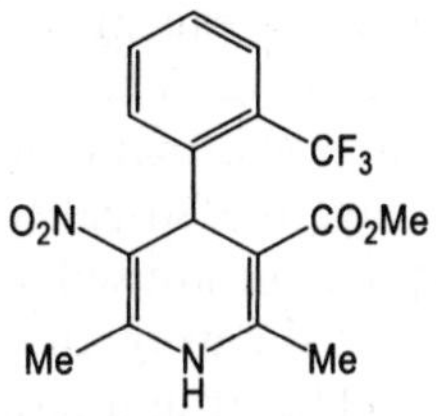

Bay K 8644

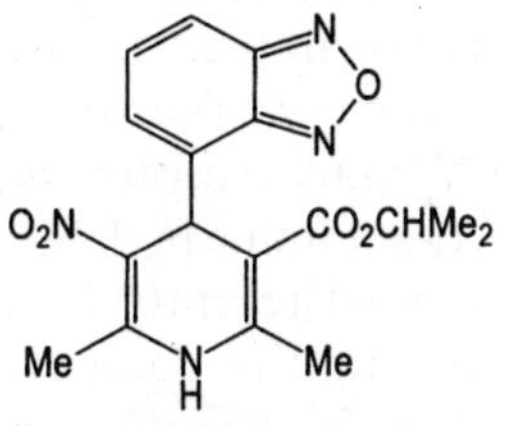

PN 202 791

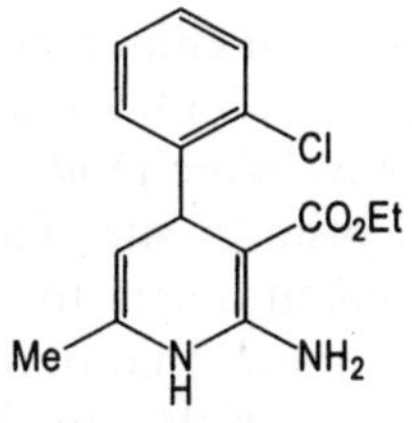

H 160/51

Figure 9 Structures of chiral 1,4-dihydropyridine calcium channel blocking agents.

dihydropyridine derivatives, and in those agents that exhibit chirality the resultant observed stereoselectivity may be complex. In some instances, one enantiomer of the dihydropyridines is more active than the other; in the case of felodipine, nitrendipine, and niguldipine (Fig. 9), the *S*-enantiomers are more potent with eudismic ratios (*S*/*R*) between 12 and 13, 7 and 8 and 28 and 45 respectively, depending on the test system used [57]. However, there are also well-known examples within this series of agents whose individual enantiomers have opposite effects on channel function. Thus the *S*-enantiomers of Bay K 8644, PN 202 791, and H 160/51 (Fig. 9) are calcium channel activators, whereas the *R*-enantiomers are antagonists [58–62].

It is thought that the individual enantiomers interact with different channel states, open and inactivated, the binding sites of which have opposite stereochemical requirements [63]. The situation becomes more complicated as the activating properties of the S-enantiomers of Bay K 8644 and PN 202 791 appear to be dependent on membrane potential, and they become antagonists at depolarizing potentials [64,65].

3.2. Ketamine

Ketamine is a general anaesthetic agent with analgesic activity and is commercially available, except in Germany where the S-enantiomer is available, as the racemate. The use of the drug is restricted as a result of postanaesthesia reactions, including hallucinations, nightmares, agitation, nausea and vomiting, the so-called emergence delirium; furthermore, the drug is also the subject of abuse [66]. In the case of ketamine, the beneficial and adverse effects of the drug are predominantly associated with the alternative enantiomers. The drug interacts with multiple binding sites including N-methyl-D-aspartate (NMDA) and non-NMDA glutamate receptors, nicotinic and muscarinic cholinergic and monoaminergic and opioid receptors [67]. The stereoselectivity of the individual enantiomers in terms of both their pharmacological and their clinical effects has been known since the late 1970s [68–70]. For example, the (+)-S-enantiomer has a three- to fourfold greater affinity for the phencyclidine binding site of the NMDA receptor, which is considered to be the primary site of action, than (R)-ketamine [67]. Similarly, the binding of the enantiomers to the μ- and κ-opioid receptors is stereoselective in favor of the S-enantiomer, two- to fourfold in comparison to (R)-ketamine, but with a 10- to 20-fold reduction in affinity compared to the NMDA receptor [71]. However, not all actions of the drug exhibit stereoselectivity, e.g., inhibition of noradrenaline release at therapeutically relevant concentrations is not stereoselective, and both enantiomers inhibit the neuronal uptake, but the S-enantiomer additionally inhibits the extraneuronal uptake of noradrenaline. Animal studies have indicated that (S)-ketamine exhibits a 1.5- to 3-fold greater hypnotic and threefold greater analgesic potency [68] and reduced CNS stimulation than the R-enantiomer, and investigations in man have revealed a 3.4-fold anaesthetic potency of (S)-ketamine [69]. Following comparative clinical investigations between (S)-ketamine and the racemate, the recovery phase was reduced, and the majority of patients felt more comfortable owing to a reduction in agitation, disorientation, and anxiety following administration of the single enantiomer [69,72]. Thus (S)-ketamine has been found to be a potent effective anaesthetic agent at reduced dosage, with a shorter recovery phase and fewer psychotomimetic emergence reactions than the racemate [67].

3.3. Dichlorophenoxyacetic Acid Diuretic Agents

Differences in pharmacodynamic activity between a pair of enantiomers may also be employed to yield an agent with an improved therapeutic profile compared to either individual stereoisomer. In the case of uricosuric-diuretic agent indacrinone (Fig. 10), the diuretic and natriuretic activity reside predominantly in the *R*-enantiomer, whereas the uricosuric effect is associated with (*S*)-indacrinone. However, following administration of the racemic drug the half-life of the *S*-enantiomer is too short at 2–5 hours compared to (*R*)-indacrinone (half-life 10–12 hours) to prevent the increase in serum uric acid [73]. Manipulation of the enantiomeric composition from the 1:1 ratio found in the racemate by increasing the content of the *S*-enantiomer to a ratio of 4:1 resulted in a mixture that was isouricemic, and an additional increase to 8:1 yielded a hypouricemic product [74]. Hence in the case of indacrinone the evaluation and exploitation of both the pharmacokinetic and the pharmacodynamic properties of the individual enantiomers resulted in an agent with potential for an improved therapeutic profile. The development of indacrinone was stopped in the mid-1980s [3], but from the investigations reported, the principle of the manipulation

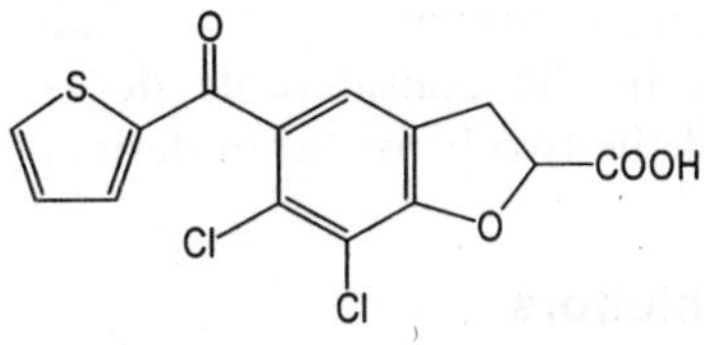

6,7-Dichloro-2,3-dihydro-5-(2-thienyl carbonyl)-2-benzofurancarboxylic acid

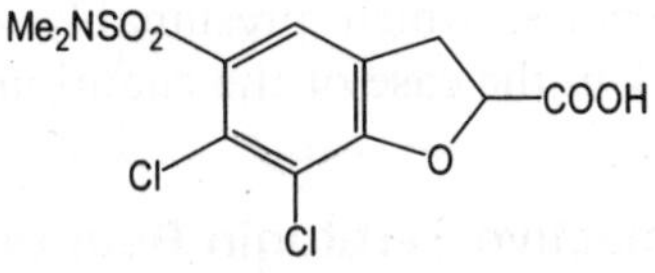

DBCA

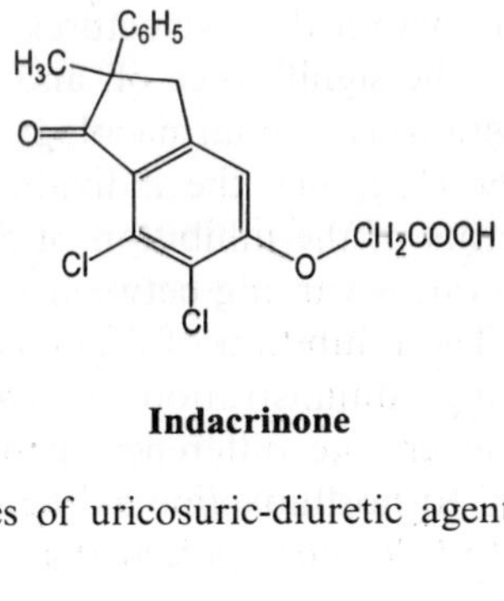

Indacrinone

Figure 10   Structures of uricosuric-diuretic agents that show stereoselectivity in action.

of stereoisomer composition for an improved therapeutic profile was established.

In the case of the related compound 6,7-dichloro-2,3-dihydro-5-(2-thienylcarbonyl)-2-benzofurancarboxylic acid (Fig. 10), the diuretic and uricosuric properties are essentially completely separated. The (+)-S-enantiomer, possessing the natriuretic activity, is twice as active as the racemate, with the uricosuric activity residing in the (−)-R-enantiomer [75,76]. It was postulated that a mixture of the two enantiomers could be prepared to yield an appropriate balance between the two actions [76]. Examination of nonracemic mixtures of the individual enantiomers in the chimpanzee resulted in the conclusion that the racemate possesses the optimal therapeutic ratio of the diuretic and uricosuric properties [75]. However, compared to humans the chimpanzee is a hyperresponder to uricosuric compounds and thus it may not be possible to extend this observation to man. A similar differentiation in enantiomeric activity has also been observed for the related compound 5-dimethylsulphamyl-6,7-dichloro-2,3-dihydrobenzofuran-2-carboxylic acid (DBCA) (Fig. 10), the uricosuric and diuretic activities being associated with the (+)-R- and (−)-S-enantiomers, respectively [77].

Both the above agents may be classified as dichlorophenoxyacetic acid derivatives, similarly to indacrinone. However, comparison of the structures (Fig. 10) indicates that the center of chirality occurs at "opposite ends" in the molecules, which presumably results in the considerable degree of selectivity in the case of the enantiomers of the two benzofuran derivatives.

3.4. Selective Serotonin Reuptake Inhibitors

The selective serotonin reuptake inhibitors (SSRIs) are an important class of antidepressant agents that appear to be better tolerated than the older tricyclic agents and have reduced drug/food interactions compared to the monoamine oxidase inhibitors. This group of agents includes compounds with one or two chiral centers within their structures, and several of these may be used to illustrate both the significance of, and the current thinking about, stereochemical considerations in pharmacology.

In the case of fluoxetine (Fig. 11), the individual enantiomers have similar potency in vitro in terms of the inhibition of 5-hydroxytryptamine (5-HT) uptake, with eudismic ratios varying between 1.1 and 1.9 depending on the test system used [78]. The inhibition of 5-HT reuptake lasted up to 24 hours and 8 hours following administration of (S)- and (R)-fluoxetine, respectively, to the rat. However, the difference in duration is probably associated with demethylation to norfluoxetine as the S-enantiomer of the metabolite is approximately 15-fold more potent than (R)-norfluoxetine as

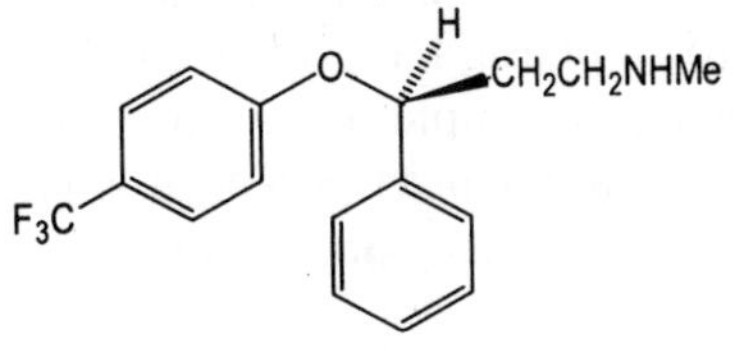

(*S*)-Fluoxetine

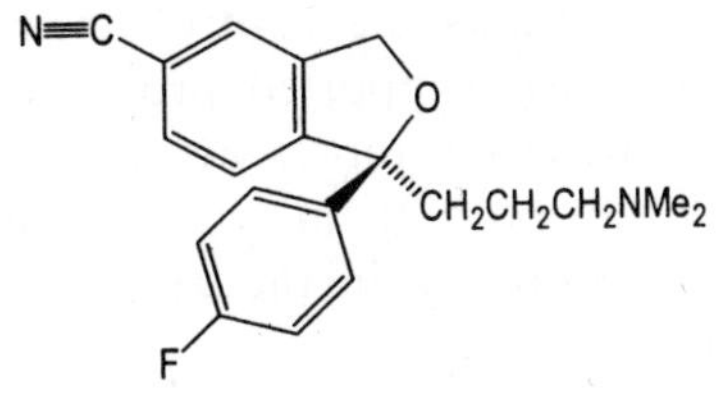

Escitalopram

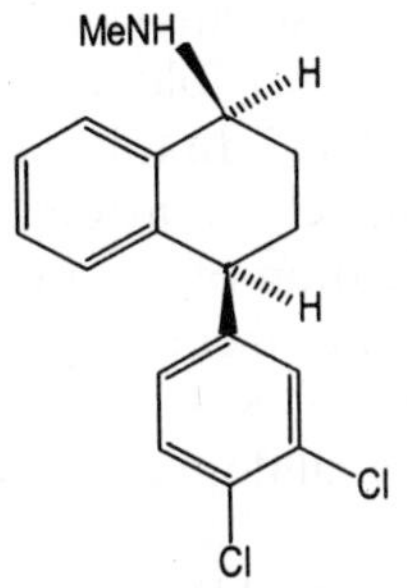

(1*S*,4*S*)-
Sertraline

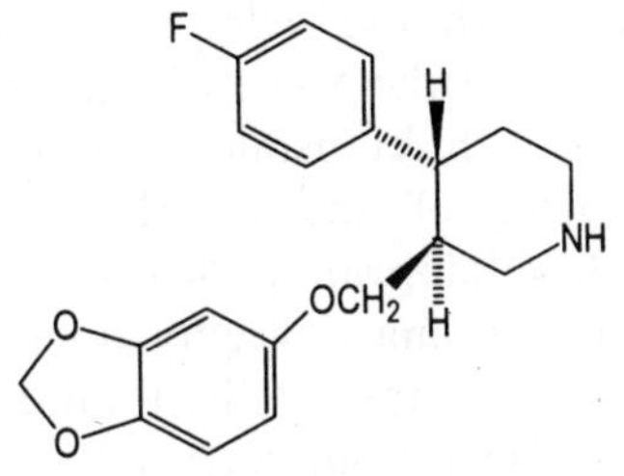

(3*S*,4*R*)-Paroxetine

Figure 11 Structures of the more active enantiomers of selective serotonin reuptake inhibitors (SSRI).

an SSRI and 1.5-fold more potent than (*S*)-fluoxetine [79]. As a result of the activity of the metabolite, it has been suggested that the concentrations of both enantiomers of the drug, together with those of (*S*)-norfluoxetine, should be taken into account in determining plasma concentration effect relationships [80]. The stereoselectivity of action of the drug and metabolite enantiomers also occurs in their inhibition of human liver microsomal cytochrome P450 (CYP) 2D6, the enantiomeric ratio (*S/R*) for the inhibition constant being 6.3 and 4.8 for the drug and metabolite, respectively [81].

As a result of the shorter washout period, the lack of a pharmacologically active metabolite, and the reduced accumulation and therefore increased flexibility for the treatment of depression, (*R*)-fluoxetine has been evaluated as a single enantiomer for the treatment of depression [11]. However, development of the drug was recently stopped owing to a small but significant increase in QT_C prolongation at the highest dose examined [82]. The single *S*-enantiomer has also been evaluated in a phase II placebo-controlled study for the prophylaxis of migraine. The results of the investigation indicated a decreased attack frequency earlier and greater in the treatment versus the control group, and the authors concluded that the data obtained supported progression to a phase III evaluation [83]. Fluoxetine therefore represents an interesting example of a marketed racemate where both enantiomers have been evaluated as potential single stereoisomer products, in the case of the S-enantiomer for an alternative indication.

A number of studies have indicated the greater potency of (S)-citalopram (Fig. 11) over that of the R-enantiomer in inhibition of 5-HT uptake, with a eudismic ratio between 130 and 160 [84,85]. As with fluoxetine, demethylation yields an active metabolite, desmethylcitalopram, which in the case of the S-enantiomer is approximately 6.7-fold less potent than the drug, but the eudismic ratio (S/R) decreases to 6.5. In this instance, the R-enantiomer of the metabolite is approximately fourfold more potent than (R)-citalopram [84]. Following administration of the racemate to patients, the plasma concentrations of the S-enantiomer are approximately one-third of those of the total drug, with a mean S/R ratio of 0.56 [86]. The single S-enantiomer, given the generic name escitalopram, has been marketed since 2002.

The related agents paroxetine and sertraline (Fig. 11) contain two centers of chirality in their structures, but both are marketed as single stereoisomers. The latter agent is interesting because the stereochemistry of the molecule has a marked influence on the selectivity of drug action (Table 2). In the case of the *trans* isomers, the (+)-enantiomer is a potent inhibitor of the uptake of serotonin, dopamine, and noradrenaline, the (−)-enantiomer being relatively selective for inhibition of noradrenaline uptake. In contrast, with the *cis* isomers, a separation of activity occurs with the (+)-1S,4S-stereoisomer, sertraline, retaining potent serotonin uptake inhibition activity [87,88]. The selectivity of action, expressed as a concentration ratio for the inhibition of dopamine and noradrenaline

Table 2 Relative Potency of the Stereoisomers of Sertraline for the Inhibition of Uptake of Biogenic Amines

	Inhibitory concentration (IC$_{50}$; µM)		
Stereoisomer	Serotonin	Dopamine	Noradrenaline
trans (+)-1R,4S	0.033	0.033	0.011
trans (−)-1S,4R	0.45	0.23	0.050
cis (+)-1S,4S*	0.06	1.1	1.2
cis (−)-1R,4R	0.46	0.29	0.38

*Sertraline is the *cis*-(+)-1S,4S-stereoisomer.

uptake compared to serotonin, being 18 and 20 for the cis-(+)-1S,4S-stereoisomer and 0.6 and 0.8 for the cis-(−)-enantiomer, whereas the corresponding values for the more potent trans-(+)-stereoisomer are 1 and 0.33 for dopamine and noradrenaline, respectively [87]. Thus in this instance stereochemical considerations resulted in an agent with approximately two-fold lower potency in terms of serotonin uptake inhibition but considerably greater selectivity of action.

4. STEREOSELECTIVITY IN PHARMACOKINETICS

Differentiation between stereoisomers also occurs in drug disposition and is of particular significance for those processes that depend upon a direct interaction between the drug and a chiral biological macromolecule, e.g., active transport processes, binding to plasma and tissue proteins, and drug metabolism [89–95].

4.1. Absorption

The absorption and transport of the majority of drugs across biological membranes occurs by passive diffusion, a process dependent upon physico-chemical properties, i.e., lipophilicity, ionization, and molecular size. Since enantiomers have identical physicochemical properties, stereoselectivity would not be expected; even though membrane phospholipids are chiral, the significance of lipophilicity appears to outweigh that of compound chirality. In contrast, differences between diastereoisomers may occur as a result of their differential solubility. However, in the case of compounds transported via carrier-mediated mechanisms, e.g., facilitated diffusion or active transport, processes involving a direct interaction between a substrate and a carrier macromolecule, stereoselectivity is expected. Preferential absorption of the L- compared to the D-enantiomers of dopa [96] and methotrexate [97,98] have been reported. In the case of the above examples, enantioselectivity in absorption is observed, whereas in the case of cephalexin, a cephalosporin antibiotic, diastereoselectivity for the L-epimer occurs. The L-epimer has shown a greater affinity than, and acted as a competitive inhibitor of D-cephalexin transport [99]. The L-epimer is also more susceptible to enzyme-mediated hydrolysis, with the result that it cannot be detected in plasma [99].

There is currently great interest in P-glycoprotein-mediated efflux mechanisms, and such transport systems are potentially stereoselective. However, there is limited data in the literature concerning this possibility, and some reports have been the subject of controversy. For example, in vitro data suggested that the *S*-enantiomer of the β-adrenoceptor antagonist, talinolol, is a slightly better substrate than (*R*)-talinolol for P-glycoprotein [100], which the authors suggested accounted for the lower plasma concentrations of the *S*- than the *R*-enantiomer in in vivo studies. However, a recent article suggests that modest presystemic metabolism via CYP 3A4, rather than P-glycoprotein efflux, is responsible for the minor but significant differences in (*R*)- and (*S*)-talinolol pharmacokinetics after oral administration [101].

4.2. Distribution

Stereoselectivity in drug distribution may occur as a result of binding to either plasma or tissue proteins and transport via specific tissue uptake and storage mechanisms. Differences between enantiomers in plasma protein binding have been reported for a number of drugs, examples of which are summarized in Table 3. Additional data for selected groups of drugs are presented in Table 3 of Chap. 6 and Table 2 of Chap. 7. Similarly to other pharmacokinetic parameters, the magnitude of the differences tends to be relatively modest, and may be less than 1%. The majority of drugs bind in a reversible manner to plasma proteins, notably to human serum albumin (HSA) and/or α_1-acid glycoprotein (AGP). Acidic drugs bind preferentially to HSA, with binding at site II (benzodiazepine site) on the protein generally displaying greater enantiomeric differences than at site I (warfarin site) [102], and basic drugs predominantely bind to AGP. It is noteworthy that stereoselectivity in binding may vary for different proteins, e.g., the protein binding of propranolol to AGP is stereoselective for the *S*-enantiomer, whereas binding to HSA favors (*R*)-propranolol [103]. In whole plasma the binding to AGP is dominant so that the free fraction of the *R*-enantiomer is greater than that of (*S*)-propranolol.

   Enantioselective tissue uptake, which is in part a consequence of enantioselective plasma protein binding, has been reported. For example, the transport of ibuprofen into both synovial and blister fluids is preferential for the *S*-enantiomer owing to the higher free fraction of this enantiomer in plasma [104]. In addition, the affinity of stereoisomers for binding sites in specific tissues may also differ and contribute to stereoselective tissue binding, e.g., (*S*)-leucovorin accumulates in tumor cells in vitro to a greater degree than the *R*-enantiomer [105]. The uptake of ibuprofen into lipids is stereoselective in favor of the *R*-enantiomer, but this is a result of stereospecific formation of the acyl-CoA thioester followed by incorporation as hybrid triglycerides (see Sec. 4.3 below and Chap. 8).

Table 3 Stereoselective Plasma Protein Binding

Drug	Free fraction (%)			Ratio
	S-enantiomer	R-enantiomer		
Acenocoumarol	2.0	1.8		1.11
Bupivacaine	4.5	6.6		1.47
Carvedilol	0.63	0.45		1.4
Chloroquine	33.3	51.3		1.7
Disopyramide	7.5	12.5	$\sim 1\,mg\,L^{-1}$	1.7
	20.7	33.8	$\sim 2.2\,mg\,L^{-1}$	1.6
Etodolac	0.85	0.47		1.8
Fenfluramine	2.8	2.9		1.03
Flurbiprofen	0.048	0.082		1.71
Gallopamil	5.7	4.0		1.4
Ibuprofen	0.64	0.42		1.5
Indacrinone	0.3	0.9		3.0
Mephobarbital	43.9	36.5		1.2
Methadone[a]	12.4 (−)	9.2 (+)		1.3
Mexiletine	28.3	19.8		1.4
Moxalactam	32	47		1.5
Pentobarbital	26	37		1.4
Propafenone	2.5	3.9		1.6
Propranolol	17.6	20.3		1.15
Sotalol	62	65		1.05
Tocainide[a]	86–91 (−)	83–89 (+)		1.0
Verapamil	11.5	6.3		1.8
Warfarin	2.2	3.6		1.6

[a]Parentheses indicate direction of optical rotation for the enantiomers. Data from Noctor [178] and Eichelbaum and Gross [179].

4.3. Metabolism

In drug metabolism, stereodifferentiation is the rule rather than the exception, and stereoselectivity in metabolism is probably responsible for the majority of the differences observed in enantioselective drug disposition. Stereoselectivity in metabolism may arise from differences in the binding of enantiomeric substrates to the enzyme active site and/or be associated with catalysis owing to differential reactivity and orientation of the target groups to the catalytic site [106]. As a result, a pair of enantiomers are frequently metabolized at different rates and/or via different routes to yield alternative products.

The stereoselectivity of the reactions of drug metabolism may be classified into three groups in terms of their selectivity with respect to the substrate, the product, or both [89]. Thus there may be substrate selectivity when one enantiomer is metabolized more rapidly than the other; product stereoselectivity, in which one particular stereoisomer of a metabolite is produced preferentially; or a combination of the above, i.e., substrate–product stereoselectivity, where one enantiomer is preferentially metabolized to yield a particular diastereoisomeric product [106,107]. An alternative classification involves the stereochemical consequences of the transformation reaction [92]. Using this approach, metabolic pathways may be divided into five groups.

1. *Prochiral to chiral transformations*: metabolism taking place either at a prochiral center or on an enantiotopic group within the molecule. For example, the prochiral sulphide cimetidine undergoes sulphoxidation to yield the corresponding sulphoxide (Fig. 12), the enantiomeric composition of which in human urine following oral drug administration is (+):(−): 71–75:29–25 [108,109]. Similarly, valproic acid preferentially yields (R)-2-n-propyl-4-pentenoic acid [110], and phenytoin undergoes stereoselective *para*-hydroxylation to yield (S)-4′-hydroxyphenytoin (Fig. 12) in greater than 90% enantiomeric excess following drug administration to man [111].

2. *Chiral to chiral transformations*: the individual enantiomers of a drug undergo metabolism at a site remote from the center of chirality with no configurational consequences. For example, (S)-warfarin undergoes

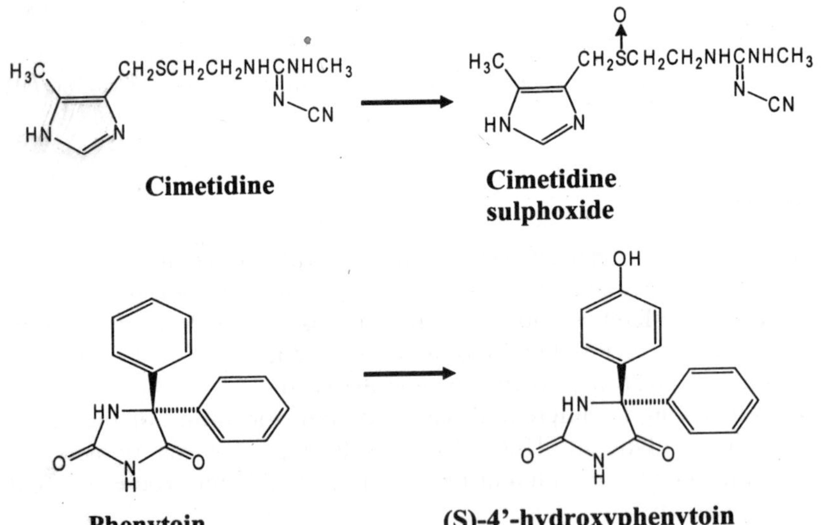

Figure 12   Examples of prochiral to chiral metabolic transformations.

aromatic oxidation mediated by CYP 2C9 in the 7- and 6-positions to yield (S)-7-hydroxy- and (S)-6-hydroxywarfarin in the ratio 3.5 : 1 [112].

3. *Chiral to diastereoisomer transformations*: a second chiral center is introduced into the drug either by reaction at a prochiral center or via conjugation with a chiral conjugating agent. Examples include the side-chain aliphatic oxidation of pentobarbitone and the keto-reduction of warfarin to yield the corresponding diastereoisomeric alcohol derivatives or the stereoselective glucuronidation of oxazepam [113].

4. *Chiral to achiral transformations*: the substrate undergoes metabolism at the center of chirality, resulting in a loss of asymmetry. Examples include the aromatization of the dihydropyridine calcium channel blocking agents, e.g., nilvadipine (Fig. 13), to yield the corresponding pyridine derivative [114,115] and the oxidation of the benzimidazole proton pump inhibitors, e.g., omperazole, which undergoes CYP 3A4–mediated oxidation at the chiral sulphoxide to yield the corresponding sulphone (Fig. 13). In the case of omeprazole, the reaction shows a tenfold selectivity for the S-enantiomer in terms of intrinsic clearance [116].

5. *Chiral inversion*: one stereoisomer is metabolically converted into its enantiomer with no other alteration in structure. The best known

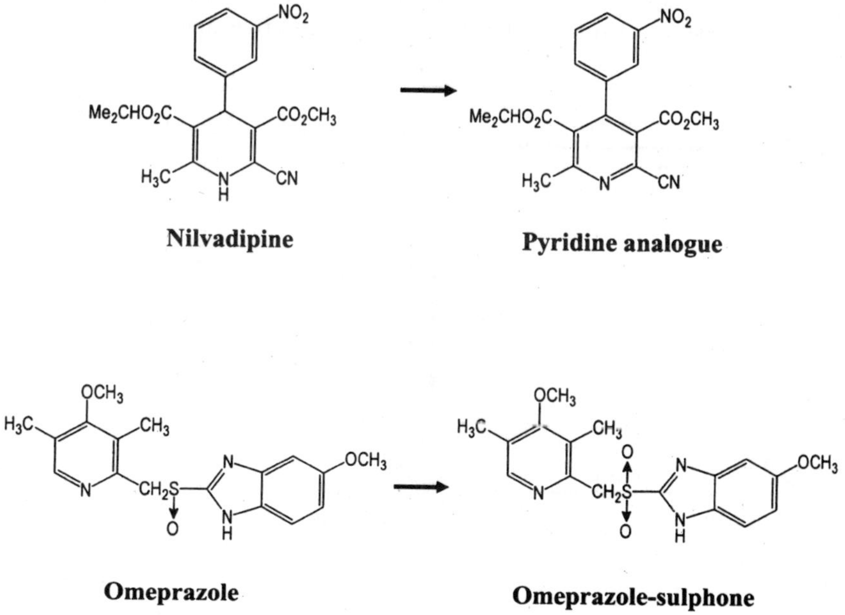

Nilvadipine   Pyridine analogue

Omeprazole   Omeprazole-sulphone

Figure 13 Examples of chiral to achiral metabolic transformations.

examples of agents undergoing this type of transformation are the 2-arylpropionic acid (2-APAs) nonsteroidal anti-inflammatory drugs (NSAIDs), e.g., ibuprofen, fenoprofen, flurbiprofen, ketoprofen [91,117], and the related 2-aryloxypropionic acid herbicides, e.g., haloxyfop [118]. In the case of the 2-APAs, the reaction is essentially stereospecific, the less active, or inactive, R-enantiomers undergoing inversion to the active S-enantiomers. With the 2-aryloxypropionic acid derivatives, the transformation is in the S- to R- direction as a result of the configurational designation according to the Cahn–Ingold–Prelog sequence rules, i.e., in terms of their three-dimensional spatial arrangement, the R-enantiomers of the 2-APAs correspond to the S-enantiomers of the 2-aryloxypropionic acids.

Chiral inversion of the 2-APAS is presented in detail as a whole chapter (Chap. 8) in this book. However, a brief overview of its mechanism is presented here. The mechanism of the inversion reaction (Fig. 14) has been extensively investigated both in vitro and in vivo and involves stereoselective formation of the acyl-coenzyme A (CoA) thioester of the (R)-2-APA, via an adenosine monophosphate intermediate. Once formed, the (R)-profenyl-CoA thioester undergoes epimerization at the center of chirality in the profen moiety, which is thought to proceed via an enolate intermediate, to yield the corresponding (S)-profenyl-CoA thioester. Hydrolysis of the acyl-CoA thioesters results in the liberation of the free acid. The enzyme

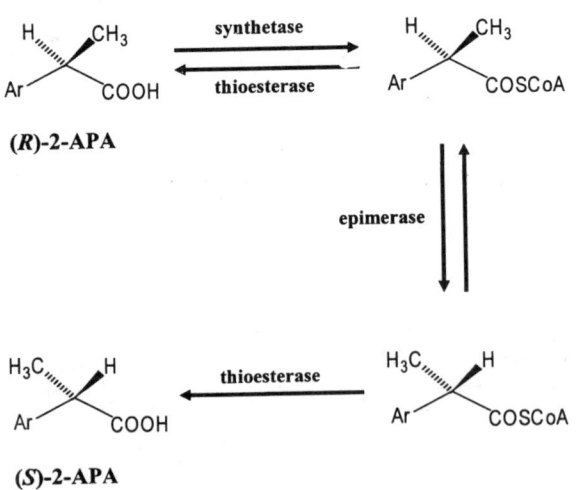

Figure 14 Mechanism of the metabolic chiral inversion of 2-arylpropionic acid NSAIDs.

mediating the initial step is thought to be the microsomal long-chain-acyl-CoA synthetase [119]. The epimerase enzyme catalyzes the interconversion of both the (R)- and (S)-profenyl-CoA derivatives. Acyl transfer of the profenyl moiety to glycerol results in the formation of hybrid triacylglycerols and the stereoselective tissue distribution of these agents referred to above [120,121].

Chiral allylic alcohols are known to undergo acid-catalyzed racemization and are therefore potentially stereochemically unstable in gastric acid following oral administration. The antiepileptic drug stiripentol undergoes acid-catalyzed racemization [122], and following oral administration of the individual enantiomers, both enantiomers are likely to be detected in plasma. Following administration of the R-enantiomer to rats, both enantiomers were readily detectable in blood [123]. However, following administration of (S)-stiripentol, only minor quantities of the R-enantiomer could be detected. This apparent unidirectional reaction was subsequently rationalized following an examination of the disposition of deuterium- or oxygen-18-labeled substrates. Following drug administration, both enantiomers undergo partial racemization in gastric acid, and both are absorbed, but the R-enantiomer undergoes glucuronidation in the liver that is both enantioselective and saturable. Following administration of (S)-stiripentol, the R-enantiomer produced by racemization undergoes conjugation with glucuronic acid and excretion in the bile, the S-enantiomer appearing in the systemic circulation. Whereas following administration of (R)-stiripentol, the glucuronidation pathway is saturated, and both enantiomers, (S)-stiripentol being formed in the gastric acid, are found in the systemic circulation [123,124].

Interconversion of other functionalities may also occur. For example, the reduction of chiral sulphoxides or N-oxides to the corresponding sulphides and tertiary amines and their subsequent reoxidation results in the loss and generation of a center of chirality and, depending on the stereoselectivity of the transformations involved, may result in the inversion of configuration. Following administration of either enantiomer of the peripheral vasodilator flosequinan (Fig. 15) to rats, the alternative enantiomer could be detected in plasma. The area under the plasma concentration versus time curve (AUC) accounted for approximately 8% for the S-enantiomer following administration of (R)-flosequinan, whereas that of the R-enantiomer following administration of (S)-flosequinan accounted for 25% [125]. Subsequent studies have indicated that the oxidation of the sulphide to (R)-flosequinan is mediated by both CYP and FMO, but that formation of the S-enantiomer is mediated by CYP [126], and it has been suggested that enzymes of the CYP 3A subfamily are involved [127].

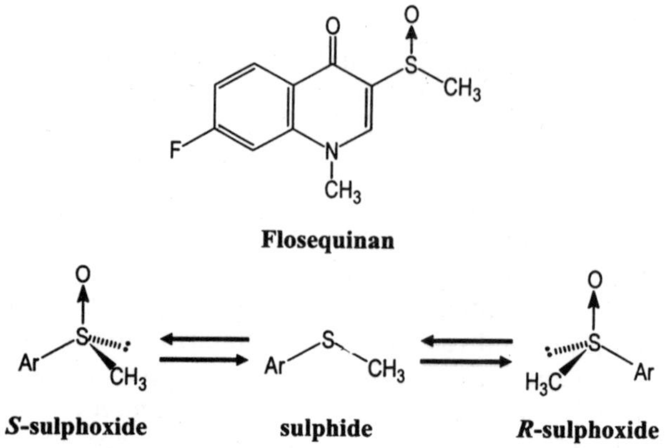

Flosequinan

S-sulphoxide   sulphide   *R*-sulphoxide

Figure 15   Flosequinan and the reduction–oxidation interconversion of sulphoxide enantiomers.

4.4. Renal Excretion

Stereoselectivity in renal clearance may arise as a result of either selectivity in protein binding, influencing glomerular filtration and passive reabsorption, or active secretion or reabsorption. Enantioselectivity in renal clearance has been reported for a number of drugs, and in the majority of instances the selectivity is relatively modest, with enantiomeric ratios between 1.0 and 3.0 (Table 4). In the case of the diastereoisomers quinine and quinidine, the difference is about fourfold, with values of 24.7 and 99 mL min^{-1} in man, respectively [128].

    However, for those agents that undergo active tubular secretion, interactions between enantiomers may occur such that their excretion differs following administration as single enantiomers versus the racemate. Administration of the quinolone antimicrobial agent (*S*)-ofloxacin with increasing amounts of the *R*-enantiomer to the cynomolgus monkey results in a reduction in both the total and the renal clearance of the *S*-enantiomer [129]. The renal excretion of ofloxacin involves both glomerular filteration and tubular secretion mediated by the organic cation transport system, and the enantiomeric interaction may be explained by competitive inhibition of the transport mechanism [129]. Similarly, enantiomeric interactions in renal tubular secretion have been suggested to account for differences in the total and the renal clearance of the enantiomers of the uricosuric diuretic 5-dimethylsulphamoyl-6,7-dichloro-2,3-dihydrobenzofuran-2-carboxylic acid (DBCA; Fig. 10) [130,131]. Administration of the racemic drug to the monkey results in a 25% reduction in the total and the renal clearance, and

Table 4 Renal Clearance of Drug Enantiomers in Man

Drug/metabolite	Enantiomer		Clearance	Units	Ratio	Reference
Acebutolol	R	CL_R	124	mL min^{-1}	1.03	180
	S		120			
Diacetolol	R	CL_R	70	mL min^{-1}	1.32	
(active metabolite	S		53			
of acebutolol)						
Atenolol	(+)	CL_R	109.7	mL min^{-1}	1.03	181
	(−)		112.5			
Chloroquine	(+)	CL_R	276	mL min^{-1}	1.03	182
	(−)		267			
	(+)	CL_{Ru}	824		1.59	
	(−)		519			
Disopyramide[a]	(−)-R	CL_R	0.75	mL min^{-1} kg^{-1}	1.17	135
	(+)-S		0.64			
	(−)-R	CL_{Ru}	6.26		1.40	
	(+)-S		8.75			
Mondesisopropyl-	(−)-R	CL_R	1.97	mL min^{-1} kg^{-1}	2.09	135
disopyramide	(+)-S		4.11			
(following	(−)-R	CL_{Ru}	3.21		2.19	
administration	(+)-S		7.02			
of the drug)						
Metoprolol	(−)	CL_R	69	mL min^{-1}	1.09	183
	(+)		75			
Mexiletine	(−)	CL_R	0.5	mL min^{-1} kg^{-1}	1.0	184
	(+)		0.5			
Ofloxacin	(+)-R	CL_R	7.53	L h^{-1}	1.05	185
	(−)-S		7.14			
Pindolol	(+)-R	CL_R	200	mL min^{-1}	1.20	133
	(−)-S		240			
	(+)-R	CL_{Ru}	453		1.18	
	(−)-S		534			
	(+)-R	CL_{TS}	157		1.25	
	(−)-S		196			
	(+)-R	CL_R	170	mL min^{-1}	1.31	134
	(−)-S		222			
	(+)-R	CL_{TS}	121		1.40	
	(−)-S		169			
Prenylamine	(−)-R	CL_R	1.3	mL min^{-1}	3.08	186
	(+)-S		4.0			
Terbutaline[a]	(−)	CL_R	1.5	mL min^{-1} kg^{-1}	1.80	187
	(+)		2.7			
Tocainide[a]	(−)	CL_R	55	mL min^{-1}	1.0	188
	(+)		55			

(continues)

Table 4   Continued

Drug/metabolite	Enantiomer		Clearance	Units	Ratio	Reference
Tranylcypromine	(−)	CL$_R$	15.3	mL min^{-1}	1.63	189
	(+)		24.9			
	(−)^a		8.1	mL min^{-1}	2.19	
	(+)		17.7			

CL$_R$, CL$_{Ru}$, and CL$_{TS}$ are the total, unbound, and tubular secretion clearances, respectively.
^{a}Indicates that the drug was administered as the individual enantiomers; in all other cases, the racemate was used.

a 30% reduction in the tubular secretion clearance, of the S-enantiomer in comparison to the values obtained following administration of the single enantiomer. In contrast, the corresponding reductions in the same parameters for (R)-DBCA did not achieve statistical significance [131]. Coadministration of the racemic drug with probenecid resulted in significant reductions in the tubular secretion of both enantiomers but was stereoselective for (S)-DBCA, the decrease in clearance being 53% and 14% for the S- and R-enantiomers, respectively [131].

Coadministration of probenecid has also been shown to reduce stereoselectively the renal clearance of (−)-sultopride, but not that of the (+)-enantiomer following administration of the racemic drug to rats [132]. In contrast, coadministration of the racemic drug with procainamide resulted in significant reductions in both total and renal clearance of both enantiomers [132].

Stereoselectivity in the renal clearance of the enantiomers of pindolol in man was initially reported by Hsyu and Giacomini [133], the tubular secretion of the (−)-S-enantiomer being 30% greater than that of (R)-pindolol. This observation has been confirmed by Somogyi et al. [134], who also showed that both the renal and the tubular secretion clearance of both enantiomers is inhibited by cimetidine, presumably by inhibition of the renal organic cation transport system. Interestingly, the renal clearance of (S)-pindolol, the enantiomer with the greater renal and the tubular secretion clearance, was reduced to a smaller extent (26%) than that of the R-enantiomer (34%), which may indicate that the secretion of the drug is mediated by more than one transporter, and that cimetidine has differential inhibitory properties [134].

Stereoselective renal clearance may also occur for metabolites. For example, following the repeated oral administration of the individual enantiomers of disopyramide, significant differences in both the total and the unbound renal clearances of the monodesisopropyl metabolite were observed, both processes being stereoselective for the (+)-*S*-enantiomer [135]. In contrast, the total renal clearance for the drug showed no stereoselectivity, whereas the unbound renal clearance of (*S*)-disopyramide was greater than that of the *R*-enantiomer [135]. The renal elimination of both enantiomers of both compounds was associated with tubular secretion, and the possibility exists that drug–metabolite–enantiomer interactions in renal tubular secretion may occur.

4.5. Pharmacokinetic Parameters

As a result of stereoselectivity in the processes of drug disposition, the pharmacokinetic profiles of the individual enantiomers of drugs administered as racemates frequently differ, and parameters derived from an analysis of "total drug" concentrations are of limited value and potentially misleading [136,137]. Similarly, pharmacokinetic–pharmacodynamic relationships, therapeutic drug monitoring, and bioequivalence data derived from investigations employing nonstereospecific analytical methodology provide information of limited value.

In comparison to the differences in pharmacodynamic parameters between enantiomers, the differences in pharmacokinetic parameters appear to be relatively modest, frequently one- to three-fold [93]. However, the magnitude of the difference may be appreciable, depending on the level of organization that the parameter represents. Pharmacokinetic parameters can be divided into three levels of organization:

1. Whole body, e.g., systemic clearance, volume of distribution, half-life
2. Organ, e.g., hepatic metabolic clearance, renal clearance
3. Macromolecular, e.g., fraction unbound in plasma, intrinsic metabolite formation clearance.

Whole body level parameters are determined by multiple organ parameters, which in turn are a reflection of multiple macromolecular parameters. Thus, in the case of a pair of enantiomers, the observed stereoselectivity for a particular parameter is dependent on the level of organization that the parameter represents and may be either amplified or attenuated [94]. For example, in the case of warfarin, the enantiomeric ratio (*R*/*S* 1.93) in half-life is determined primarily by differences in volume of distribution (*R*/*S* 1.83) rather than the modest difference in clearance (*R*/*S* 1.05). In contrast, examination of the alternative metabolic routes generates marked differences in stereoselectivity, e.g., aromatic oxidation at the 6-, 7-, and 8-positions of the coumarin ring yields enantiomeric ratios of *R*/*S* 1.19, *S*/*R* 6.33, and *R*/*S* 5.2, respectively. These, together with the corresponding values for the alcohol reduction products, are attenuated in summation. Pharmacokinetic parameters of selected chiral drugs are reviewed in Chaps. 6 and 7.

5. PRODRUGS

Certain stereochemical issues in drug delivery using a prodrug approach are discussed in Chap. 4. A brief discussion on prodrugs of some selected agents is presented below.

5.1. β-Lactam Antimicrobial Agents

Stereochemical considerations may also have significance in the design of prodrugs. In the case of the β-lactam antimicrobials, the esterification of the carboxyl group has been used extensively in order to improve their absorption following oral administration. In some instances, the addition of the promoiety results in the introduction of an additional center of chirality into the molecule, and, in the case of these single stereoisomer drugs, the formation of a pair of diastereoisomers. Cefuroxime axetil is the 1-acetoxyethyl ester prodrug of cefuroxime (Fig. 16), and differential chemical hydrolysis and photochemical stability of the two diastereoisomers has been reported [138]. The diastereomeric prodrug has also been reported to undergo differential hydrolysis using both serum and intestinal mucosal esterases isolated from both rat and dog tissue preparations [139]. The hydrolysis was stereoselective for the $1'S,6R,7R$-diastereoisomer, but the selectivity varied with enzyme source, the ratio $1'S/1'R$ being 13–14 and 2.5–3.4 for serum and intestinal esterases, respectively, with the rat tissue esterases being faster. Similarly, cefdaloxime (Fig. 16) has been esterified as the pivaloyloxyethyl derivative [140]. Pharmacokinetic studies in mice, rats, and dogs following administration of the individual stereoisomers and as a mixture indicated species variability in prodrug handling. Following administration to the dog, the $1'S$-epimer showed approximately three times the bioavailability of the $1'R$-epimer as measured by comparison of the areas under the serum cefdaloxime concentration-versus-time curves and urinary drug recovery [141]. In contrast, in the mouse all three forms of the prodrug, i.e., the individual epimers and the mixture, showed rapid and essentially complete absorption, whereas in the rat the bioavailability of the drug was reduced, but no differences between the prodrug epimers were observed [141]. Defossa et al. [140] have also stated that the absorption of the $1'S$-epimer was significantly higher in man; this stereoisomer was selected for further evaluation, and the pharmacokinetic properties

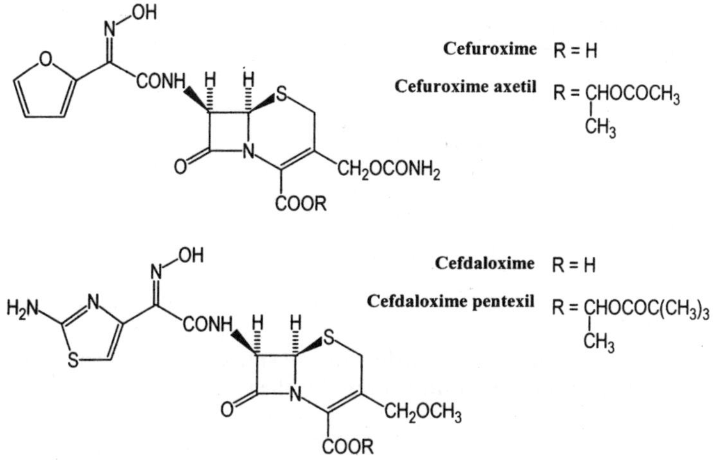

Figure 16 Structures of cefuroxime and cefdaloxime and their corresponding axetil and pentexil prodrug esters.

of cefdaloxime following prodrug administration to man have been reported [142].

5.2. Design of β-Adrenoceptor Antagonist Prodrugs

The stereoselectivity of enzymatic processes may also be utilized in the design of prodrugs. Following topical administration of ophthalmic drugs, the majority of the dose undergoes systemic absorption and may exert systemic effects. This is particularly the case with the β-adrenoceptor antagonists used, following local administration, for reduction of intra-ocular pressure in the treatment of glaucoma, where systemic cardiovascular effects have been reported. Following the observation that esters of adrenalone undergo hydrolysis and reduction to yield adrenaline only at the iris-ciliary body, it was hypothesized that ketone precursors of the aryloxypropanolamine derivative β-adrenoceptor antagonists would undergo a similar reduction at the site of action. Initial attempts to synthesize the required ketones were unsuccessful, presumably because of the stability of the amino ketone ethers, but the incorporation of a ketoxime function resulted in agents with the required activity and stability [143]. The lipophilic agents, e.g., the derivatives of propranolol and alprenolol, were found to be effective in reducing intraocular pressure and to be devoid of pharmacological activity when given orally [143,144]. Subsequent investigations have indicated that the ketoxime derivatives undergo

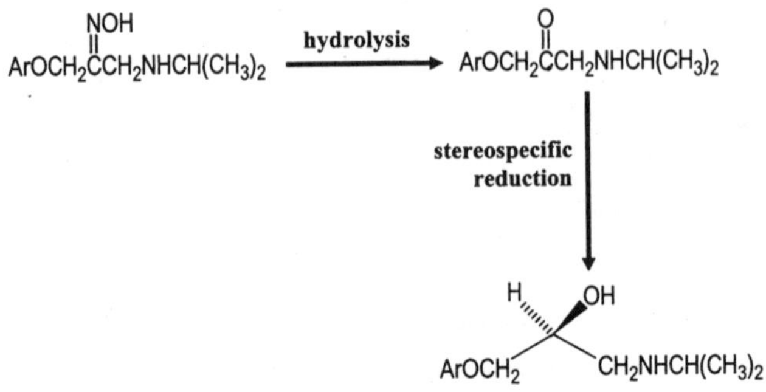

Figure 17 Hydrolysis and stereospecific reduction of ketoxime prodrugs of the aryloxypropanolamine β-adrenoreceptor antagonists in the eye.

hydrolysis to yield the corresponding inactive ketones followed by rapid stereospecific reduction of the prochiral ketone to the corresponding S-enantiomers of the active agents (Fig. 17) [145,146]. Thus the active agent is produced in situ, and the site of reduction is believed to be the iris-ciliary body, with the avoidance of the systemic effects of the drug.

6. TOXICOLOGY

There is relatively little information in the open literature on the comparative toxicity of individual enantiomers or between individual enantiomers and racemates. In addition, although a number of racemates have been withdrawn from the market as a result of adverse drug reactions, and arguments associated with the contribution of stereoselectivity in either disposition and/or action to the adverse events may be hypothesized, little data is available from studies in humans [3]. However, examples can be cited that are illustrative of aspects of the significance of stereochemical considerations in safety evaluation.

6.1. Thalidomide and Related Teratogenic Agents

A compound frequently cited, particularly in the popular lay and scientific press, as an example where the use of a single stereoisomer would have prevented a tragedy, is the teratogenic hypnotic-sedative agent, thalidomide (Fig. 18). An investigation in the late 1970s reported that following administration of the individual enantiomers to mice, both enantiomers possessed hypnotic activity, whereas the teratogenic activity resided solely in the S-enantiomer [147]. However, previous studies in a more sensitive

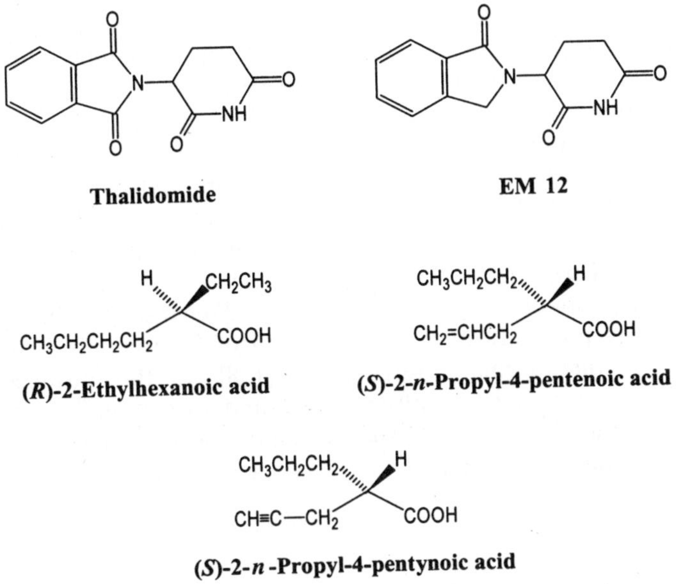

Thalidomide   **EM 12**

(**R**)-2-Ethylhexanoic acid   (**S**)-2-*n*-Propyl-4-pentenoic acid

(**S**)-2-*n*-Propyl-4-pentynoic acid

Figure 18   Teratogenic agents.

test species, New Zealand White rabbits, indicated that both enantiomers, and the racemate, are teratogenic [148], and additional investigations have indicated that the drug readily undergoes racemization in biological media [149–151]. Thus even if the single *R*-enantiomer had been commercially available in the early 1960s, patients would still have been exposed to both enantiomers of the drug as a result of its facile racemization. Thalidomide is therefore not a particularly useful example to cite in support of single stereoisomer products. Indeed, the senior author of the publication suggesting stereospecificity of the teratogenicity of the drug has dissociated himself from the biological studies presented [152].

A similar situation arises with the thalidomide analog EM 12 (2-(2,6-dioxopiperidine-3-yl)phthalimidine; Fig. 18). The enantiomers of EM 12 also undergo racemization both in vitro and in vivo following administration to female marmosets (*Callithrix jacchus*) [153,154]. In addition, following administration of the individual enantiomers to pregnant marmosets, the data obtained indicated modest stereoselectivity in the teratogenic potency for the *S*-enantiomer, but as a result of the extensive in vivo racemization of the drug, the significance of this observation is difficult to discern [155].

The facile racemization of both thalidomide and EM 12 obviously makes interpretation of the in vivo data difficult. However, that stereoselective teratogenicity may occur has been shown by an examination of the enantiomers of 2-ethylhexanoic acid (Fig. 18), a metabolite of the plasticizer di-(2-ethylhexyl)phthalate. (R)-2-Ethylhexanoic acid was teratogenic and embryotoxic following administration to mice, whereas the S-enantiomer was nontoxic, and the racemate produced an intermediate response [156]. Similarly, the S-enantiomer of 2-n-propyl-4-pentenoic acid (Fig. 18), a metabolite of valproic acid, has also been shown to be a more potent embryotoxin in mice [157]. The corresponding pentynoic acid derivative, 2-n-propyl-4-pentynoic acid (Fig. 18), shows similar enantioselectivity following administration to mice, the S-enantiomer being 7.5- and 1.9-fold more potent in terms of induction of exencephaly than the R-enantiomer or the racemate, respectively [158]. In this instance the pentynoic acid was found to be more potent and to exhibit greater enantioselectivity in the teratogenic response than the pentenoic acid derivative, thus providing an example of Pfeiffer's rule in teratogenicity.

6.2. Fenvalerate

The pyrethroid insecticide fenvalerate, (α-cyano-3-phenoxybenzyl-2-(4-chlorophenyl)isovalerate, contains two centers of chirality in its structure (designated as the 2 and α positions; Fig. 19) and therefore four stereoisomers, two pairs of enantiomers are possible. Initial evaluation of the mixture, by addition to the diet of a number of species, resulted in granulomatous changes in the liver, lymph nodes, and spleen. Separation and evaluation of the individual stereoisomers indicated that the toxicity was associated with one of the four, the 2R,αS-stereoisomer, and subsequent metabolic studies found the cause to be associated with the formation and disposition of a cholesterol ester of (R)-2-(4-chlorophenyl)isovalerate (Fig. 19). A metabolic transformation shown to be stereospecific in mice, only the 2R,αS-stereoisomer yielding the ester both in vitro and in vivo [159].

6.3. Male Antifertility Agents

Studies with the male antifertility agents 3-chloropropane-1,2-diol and 3-amino-1-chloropropane-2-ol have indicated that the antifertility activity in rat is due to the S-enantiomers, whereas the R-enantiomers are associated with nephrotoxicity [160,161]. Metabolic studies indicated that a common metabolite was the causative agent, which was postulated to be (R)-3-chlorolactate. Administration of the enantiomers of 3-chlorolactate to rats resulted in elevated urinary excretion of oxalate, presumably formed via oxidation of the intermediate 3-chloropyruvate, a known inhibitor of renal

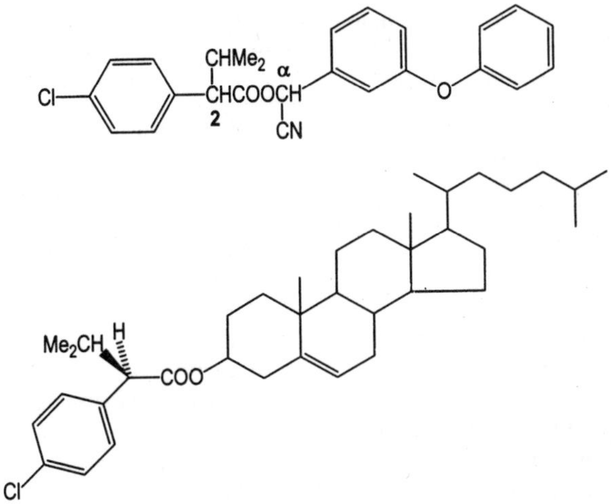

Figure 19   Fenvalerate and the structure of (R)-2-(4-chlorophenyl)isovaleric acid cholesterol ester.

mitochondrial metabolism and nephrotoxic agent [162] following the R- but not the S-enantiomer. Thus it would appear that the toxicity is associated with the metabolism of (R)-3-chlorolactate to 3-chloropyruvate.

7.   CHIRAL SWITCHES

In addition to new drug development, a number of established agents, previously marketed as racemates, have been reevaluated as single stereo-isomer products. This process, known as a chiral switch, has resulted in the remarketing of a number of compounds as single enantiomer products in a number of countries, examples of which are presented in Table 5 [11].

The concept of examining single enantiomers following an observation of unacceptable adverse effects of racemates in patients, or developments in technology that enable the production of a single enantiomer, is not new. For example, D-penicillamine, originally introduced into therapeutics for the treatment of Wilson's disease [163], has been used in rheumatology for decades. In the initial clinical studies for the treatment of Wilson's disease in the US, the use of the synthetic racemate resulted in optic neuritis [164]; but D-penicillamine obtained by the hydrolysis of penicillin was used in the UK, and the adverse effect was not observed at that time [165,166]. Similarly the initial use of racemic dopa for the treatment of Parkinson's disease resulted in a number of adverse effects, including

Table 5 Racemate to Single Enantiomer Switches

Drug	Action/indication	Country/status
Dexfenfluramine	Anoretic	Withdrawn
Levofloxacin	Antimicrobial	
Dilevalol	β-blocker	Withdrawn/development stopped
Dexibuprofen	NSAID	Availabile in Austria and Switzerland
Dexketoprofen	NSAID	UK, Spain
Levobupivacaine	Local anaesthetic	UK
(S)-Ketamine	Anaesthetic	Germany
(R)-Fluoxetine	Antidepressant	Development stopped
Esomeprazole	Proton-pump inhibitor	UK, USA
Cisatracurium	Neuromuscular blockade	UK, USA
(R)-Salbutamol	β₂-Agonist	USA
Levocertirizine	Allergy, H₁-antihistamine	UK
(R,R)-Methylphenidate	Attention-deficit hyperactivity disorder	USA
(S)-Citalopram (escitalopram)	SSRI	UK, USA

Additional agents currently in development include drug (indication): (S)-doxazosin (benign prostatic hyperplasia); (R,R)-formoterol (β₂-agonist); (S)-fluoxetine (migraine prophylaxis); (S)-oxybutinin (urinary incontinence); (+)-norcisapride (nocturnal heartburn); (−)-pantoprazole and (S)-lansoprazole (both are proton-pump inhibitors).

nausea, vomiting, anorexia, involuntary movements, and granulocytopenia [167]. In contrast, treatment with L-dopa resulted in halving the dose, and the adverse effects were either reduced or abolished; the granulocytopenia was not observed with the single enantiomer [168]. The oral contraceptive agent norgestrel, the activity of which resides in the levorotatory enantiomer, was initially marketed in the UK in 1974 as a racemate. However, following developments in the synthetic methodology, the single enantiomer levonorgestrel was marketed in 1979, and both single-enantiomer and racemic preparations of the drug are commercially available and the subject of individual monographs in the British Pharmacopoeia (2002).

However, the development and evaluation of single stereoisomer products from previously marketed racemic mixtures is not a trivial matter, and several instances may be cited in which drug development, or clinical studies, have been stopped owing to adverse effects [3,82,169].

The R,R-stereoisomer, dilevalol, the β-blocking stereoisomer of the combined α- and β-blocking agent labetalol, was initially marketed in Japan and Portugal and approved in a number of other countries; but it was rapidly withdrawn owing to hepatotoxicity in a small number of patients. The incidence of hepatotoxicity appeared to be greater than that observed with the mixture, and the magnitude of the adverse reaction could not have been predicted [3]. The development of (R)-fluoxetine has also been terminated recently as a result of adverse effects (see Sec. 3.4; Refs. 82, 169).

A nonselective β-blocking agent with class III antiarrhythmic activity, sotalol, is used as a racemate. The individual enantiomers are equipotent in terms of their antiarrhythmic activity, but the (−)-enantiomer is between 14 and 50-fold more potent, depending on the test system used, in terms of β-blockade [170,171]. (+)-Sotalol was evaluated as an antiarrhythmic agent in patients with depressed ventricular function following myocardial infarction in the SWORD trial (survival with oral d-sotalol). The investigation was terminated prematurely, following recruitment of less than 50% of the required number of patients, as a result of increased mortality (5%) in the drug treatment group compared to the placebo group (3.1%) [172]. It has been suggested that the results of the SWORD trial are associated with the therapeutic approach rather than the value of (+)-sotalol [173] and that the combined β-blocking and class III antiarrhythmic activity present in the racemate provides a more effective treatment. Shah et al. [3] have postulated that a nonracemic mixture of the enantiomers of sotalol could be developed in order to provide a balance between the two activities.

In the case of the anoretic agent fenfluramine, both the racemic drug and the single enantiomer dexfenfluramine were voluntarily withdrawn following an association with valvular heart disease [174,175]. However, from the available data the stereochemistry of the drug does not appear to be a contributory factor.

The examples cited above indicate that the removal of "isomeric ballast" from a mixture may have considerable financial consequences. For example, it has been estimated that the research and development costs of dilevalol represented an investment of $100 million [176]. In some instances, it could be argued that the development of single-enantiomer products from previously marketed racemates does not provide a genuine therapeutic benefit [177]. However, a number of commercially successful drugs are marketed as racemates, and the potential economic significance of single-enantiomer forms of these agents is obvious.

8. CONCLUDING COMMENTS

There can be little doubt that stereochemical considerations in pharmacology will continue, if only because of the scientific nonsense of evaluating the properties of two agents at the same time, and such investigations will continue to provide additional insights into drug action. For the large number of agents currently available as racemates, relatively little is known concerning the biological properties of the individual enantiomers. In terms of future drug development, single stereoisomers will be the norm, and racemates will require scientific justification. "Old" racemates will continue to be reevaluated and reintroduced as single-enantiomer products with cleaner pharmacological profiles and, in some instances, new indications resulting in therapeutic benefits. The move of pharmacology from being a science of "limited dimensionality" to one of spatial awareness has at times provoked considerable argument; indeed in some instances the shock to the-system appears to have been as traumatic as the revelation of a three-dimensional universe to the two-dimensional character in the novel mentioned in the introduction. However, if dimensional considerations result in improved drug safety and efficacy, then the double trouble involved will have been worthwhile.

REFERENCES

1. Cushny, A.R. *Biological Relations of Optically Isomeric Substances*; Bailliere, Tindall and Cox: London, 1926.
2. Ariëns, E.J.; Wuis, E.W.; Veringa, E.J. Stereoselectivity of bioactive xenobiotics. A pre-Pasteur attitude in medicinal chemistry, pharmacokinetics and clinical pharmacology. Biochem. Pharmacol. **1988**, *37*, 9–18.
3. Shah, R.R.; Midgley, J.M.; Branch, S.K. Stereochemical origin of some clinically significant drug safety concerns: lessons for future drug development. Adverse Drug React. Toxicol. Rev. **1998**, *17*, 145–190.
4. De Camp, W.H. The FDA perspective on the development of stereoisomers. Chirality **1989**, *1*, 2–6.
5. Cayen, M.N. Racemic mixtures and single stereoisomers: industrial concerns and issues in drug development. Chirality **1991**, *3*, 94–98.
6. Nation, R.L. Chirality in new drug development. Clinical pharmacokinetic considerations. Clin. Pharmacokin. **1994**, *27*, 249–255.
7. Rauws, A.G.; Groen, K. Current regulatory (draft) guidance on chiral medicinal products: Canada, EEC, Japan, United States. Chirality **1994**, *6*, 72–75.
8. Millership, J.S.; Fitzpatrick, A. Commonly used chiral drugs: a survey. Chirality **1993**, *5*, 573–576.
9. Shindo, H.; Caldwell, J. Development of chiral drugs in Japan: an update on regulatory and industrial opinion. Chirality **1995**, *7*, 349–352.
10. Branch, S. International regulation of chiral drugs. In *Chiral Separation Techniques, A Practical Approach*, 2nd Ed.; Subramanian, G., Ed.; Wiley-VCH: Weinheim, 2001; 319–342.

11. Tucker, G.T. Chiral switches. Lancet **2000**, *335*, 1085–1087.
12. Holmstedt, B. The use of enantiomers in biological studies: an historical review. In *Chirality and Biological Activity*; Holmstedt, B., Frank, H., Testa, B. Eds.; Alan R. Liss: New York, 1990; 1–14.
13. Friedman, L.; Miller J.G. Odour incongruity and chirality. Science **1971**, *172*, 1045–1046.
14. Leitereg, T.J.; Guadagni, D.G.; Harris, J.; Man, T.R.; Teranishi, R. Evidence for the difference between the odours of the optical isomers of (+)- and (−)-carvone. Nature **1971**, *230*, 455–456.
15. Russell, G.F.; Hills, J.I. Odour differences between enantiomeric isomers. Science **1971**, *172*, 1043–1044.
16. May, P. *The Chemistry of Synthetic Drugs*; Longman, Green & Co.: London, 1918; 34.
17. Easson, L.H.; Stedman, E. Studies on the relationship between chemical constitution and physiological action. V. Molecular dissymmetry and physiological activity. Biochem. J. **1933**, *27*, 1257–1266.
18. Patil, P.N.; LaPidus, J.B.; Tye, A. Steric aspects of adrenergic drugs. J. Pharm. Sci. **1970**, *59*, 1205–1234.
19. Patil, P.N.; Miller, D.D.; Trendelenburg, U. Molecular geometry and adrenergic drug activity. Pharmacol. Rev. **1975**, *26*, 323–392.
20. Ogston, A.G. Interpretation of experiments on metabolic processes using isotopic tracer elements. Nature **1948**, *162*, 963.
21. Ogston, A.G. Conditions for differential reaction of "identical" chemical groups. Nature **1958**, *181*, 1462.
22. Bentley, R. Ogston and the development of prochirality theory. Nature **1978**, *276*, 673–676.
23. Dalgliesh, C.E. The optical resolution of aromatic amino acids on paper chromatograms. J. Chem. Soc. **1952**, *137*, 3940–3942.
24. Bentley, R. Chiral recognition. Chem. Br. **1994**, *30*, 191–192.
25. Sokolov, V.I.; Zefirov, N.S. Enantioselectivity in two-point binding: rocking tetrahedron model. Dokl. Akad. Nank. **1991**, *319*, 1382–1383 (in Russian).
26. Pirkle, W.H.; Burke, J.A.; Wilson, S.R. X-Ray crystallographic support of a chiral recognition model. J. Am. Chem. Soc. **1989**, *111*, 9222–9223.
27. Topiol, S. A general criterion for molecular recognition: implications for chiral interactions. Chirality **1989**, *1*, 69–79.
28. Topiol, S.; Sabio, M. Interactions between eight centres are required for chiral recognition. J. Am. Chem. Soc. **1989**, *111*, 4109–4110.
29. Aires-de-Sousa, J.; Gasteiger, J. Prediction of enantiomeric selectivity in chromatography. Application of conformation-dependent and conformation-independent descriptors of molecular chirality. J. Mol. Graph. Model. **2001**, *53*, 1–16.
30. Booth, T.D.; Wahnon, D.; Wainer, I.W. Is chiral recognition a three-point process? Chirality **1997**, *9*, 96–98.
31. Booth, T.D.; Wainer, I.W. Investigation of the enantioselective separations of α-alkylarylcarboxylic acids on an amylose tris(3,5-dimethylphenylcarbamate)

chiral stationary phase using quantitative structure–enantioselective retention relationships (QSERR): identification of a conformationally driven chiral recognition mechanism. J. Chromatogr. A **1996**, *737*, 157–167.

32. Mesecar, A.D.; Koshland, D.E. A new model for protein stereospecificity, Nature **2000**, *403*, 614–615.

33. Lehmann, P.A.F.; Rodrigues de Miranda, J.F.; Ariëns, E.J. Stereoselectivity and affinity in molecular pharmacology. In *Progress in Drug Research*; Jucker, E., Ed.; Birkhauser Verlag: Basel, 1976; Vol. 20, 101–142.

34. Lehmann, P.A.F. Quantifying stereoselectivity or how to choose a pair of shoes when you have two left feet. Trend Pharmacol. Sci. **1982**, *3*, 103–106.

35. Pfeiffer, C.C. Optical isomerism and pharmacological action, a generalization. Science **1956**, *124*, 29–31.

36. Barlow, R.B. Enantiomers: how valid is Pfeiffer's rule? Trend Pharmacol. Sci. **1990**, *11*, 148–150.

37. Waldeck, B. Biological significance of the enantiomeric purity of drugs. Chirality **1993**, *5*, 350–355.

38. Honda, K.; Takenaka, T.; Miyata-Osawa, A.; Terai, M. Adrenoceptor blocking properties of the stereoisomers of amosulalol (YM-09538) and the corresponding desoxy derivative (YM-11133). J. Pharmacol. Exp. Ther. **1986**, *236*, 776–783.

39. Nichols, A.J.; Sulpizio, A.C.; Ashton, D.J.; Hieble, J.P.; Ruffolo, R.R. The interaction of the enantiomers of carvedilol with α_1- and β_1-adrenoceptors. Chirality **1989**, *1*, 265–270.

40. Rosin, J.; Williams, C.J. A note on specific rotation and temperature coefficients. J. Am. Pharm. Assoc. Sci. Ed. **1958**, *47*, 229.

41. Rosin, J.; Williams, C.J. Note on the effect of organic solvents upon specific rotation. J. Am. Pharm. Assoc. Sci. Ed. **1960**, *49*, 622–624.

42. Ceccarini, G.; Maione, A.M. Variations of optical rotation of naproxen: polarimetric determination in the presence of non chiral basic compounds. J. Pharm. Sci. **1989**, *78*, 1053–1054.

43. Lands, A.M.; Luduena, F.P.; Tullar, B.F. The pharmacologic activity of isopropylarterenol (Isuprel) compound with that of the optically inactive analog 1-(3,4-dihydroxyphenyl)-2-isopropylaminoethane HCl. J. Pharmacol. Exp. Ther. **1954**, *111*, 469–474.

44. Barlow, R.B. Differences in the stereospecificity of closely related compounds; a reinvestigation of the enantiomers of procyclidine, benzhexol and their metho- and etho-salts. J. Pharm. Pharmacol. **1971**, *23*, 90–97.

45. Barlow, R.B.; Franks, F.M.; Pearson, J.D.M. The relation between biological activity and degree of resolution of optical isomers. J. Pharm. Pharmacol. **1972**, *24*, 753–761.

46. Trofast, J.; Österberg, K.; Källström, B.-L.; Waldeck, B. Steric aspects of agonism and antagonism at β-adrenoceptors: synthesis of and pharmacological experiments with the enantiomers of formoterol and their diastereomers. Chirality **1991**, *3*, 443–450.

47. Gillespie, L.; Oates, J.A.; Crout, J.R.; Sjoerdsma, M. Clinical and chemical studies with α-methyldopa in patients with hypertension. Circulation **1962**, *25*, 281–291.

48. Kubota, H.; Nunami, K.; Hayashi, K.; Hashimoto, Y.; Ogiku, N.; Matsuoka, Y.; Ishida R. Studies on angiotensin converting enzyme inhibitors VI. Synthesis and angiotensin converting enzyme inhibitory activities of the dicarboxylic acid derivative of imidapril and its diastereoisomers. Chem. Pharm. Bull. **1992**, *40*, 1619–1622.

49. Carter, R.B.; Dykstra, L.A. Effects of picenadol (LY150720) and its stereo-isomers on electric shock titration in the squirrel monkey. J. Pharmacol. Exp. Ther. **1985**, *234*, 299–306.

50. Schunack, W.; Schwarz, S.; Gerhard, G.; Buyuktimkin, S.; Elz, S. Chiral agonists of histamine. In *Frontiers in Histamine Research*; Ganellin, C.R., Schwartz, J.C., Eds.; Pergamon: Oxford, 1985; 39–46.

51. Cannon, J.G.; Mohan, P.; Bojarski, J.; Long, J.P.; Bhatnagar, R.K.; Leoanrd, P.A.; Flynn, J.R.; Chatterjee, T.K. (*R*)-(−)-10-Methyl-11-hydroxyaporphine: a highly selective serotoneric agent. J. Med. Chem. **1988**, *31*, 313–318.

52. Cannon, J.G.; Moe, S.T.; Long J.P. Enantiomers of 11-hydroxy-10-methyl-aporphine having opposing pharmacological effects at 5-HT$_{1A}$ receptors. Chirality **1991**, *3*, 19–23.

53. Goldman, M.E.; Kebabian, J.W. Aporphine enantiomers. Interactions with D-1 and D-2 receptors. Mol. Pharmacol. **1984**, *25*, 18–23.

54. Gao, Y.; Zong, R.; Campbell, A.; Kula, N.S.; Baldessarini, R.J.; Neumeyer, J.L. Synthesis and dopamine agonist and antagonist effects of (*R*)-(−)- and (*S*)-(+)-11-hydroxy-N-n-propylaporphine. J. Med. Chem. **1988**, *31*, 1392–1396.

55. Wikström, H.; Sanchez, D.; Lindberg, P.; Hacksell, U.; Arvidsson, L.-E.; Johansson, A.M.; Thorberg, S.-O.; Nilsson, J.L.G.; Svensson, K.; Hjorth, S.; Clark, D.; Carlsson, A. Resolved 3-(3-hydroxylphenyl)-N-n-propylpiperidine and its analogues: central dopamine receptor activity. J. Med. Chem. **1984**, *27*, 1030–1036.

56. Triggle, D.J. On the other hand: the stereoselectivity of drug action at ion channels. Chirality **1994**, *6*, 58–62.

57. Eltze, M.; Boer, R.; Sanders, K.H.; Boss, H.; Ulrich, W.-R.; Flockerzi, D. Stereoselective inhibition of thromboxane-induced coronary vasoconstriction by 1,4-dihydropyridine calcium channel antagonists. Chirality **1990**, *2*, 233–240.

58. Franckowiak, G.; Bechem, M.; Schramm, M.; Thomas, G. The optical isomers of the 1,4-dihydropyridine Bay K 8644 show opposite effects on Ca channels. Eur. J. Pharmacol. **1985**, *114*, 223–226.

59. Hof, R.P.; Ruegg, U.T.; Hof, A.; Vogel, A. Stereoselectivity at the calcium channel: opposite action of the enantiomers of a 1,4-dihydropyridine. J. Cardiovascular. Pharmacol. **1985**, *7*, 689–693.

60. Williams, J.S.; Grupp, I.L.; Grupp, G.; Vaghy, P.L.; Dumont, L.; Schwartz, A. A profile of the oppositely acting enantiomers of the dihydropyridine 202–791 in cardiac preparations: receptor binding, electrophysiologic and pharmacolo-gical studies. Biochem. Biophys. Res. Commun. **1985**, *131*, 12–21.

61. Gjorstrup, P.; Harding, H.; Isaksson, R.; Westerlund, C. The enantiomers of the dihydropyridine derivative H 160/51 show opposite effects of stimulation and inhibition. Eur. J. Pharmacol. **1986**, *122*, 357–361.

62. Van Amsterdam, F.Th.M.; Punt, N.C.; Haas, M.; Van Amsterdam-Magnoni, M.S.; Zaagsma, J. Stereoisomers of Bay K 8644 show opposite activities in the normal and ischemic rat heart. A comparison with nifedipine. Naunyn-Schmied. Arch. Pharmacol. **1989**, *339*, 647–652.

63. Kwon, Y.-W.; Triggle, D.J. Chiral aspects of drug action at ion channels: a commentary on the stereoselectivity of drug actions at voltage-gated ion channels with particular reference to verapamil actions at the Ca^{2+} channel. Chirality **1991**, *3*, 393–404.

64. Kass, R.S. Voltage-dependent modulation of cardiac calcium channel current by optical isomers of Bay K 8644: implications for channel gating. Circulation Res. **1987**, *61* (suppl. 1), 1–5.

65. Kamp, T.J.; Sanguinetti. M.C.; Miller, R.J. Voltage and use dependent modulation of cardiac calcium channels by the 1,4-dihydropyridine (+)-202 791. Ciculation Res. **1989**, *64*, 338–351.

66. Eastman, D.; Hickey, M.; Hickey, F. Ketamine misuse identified. Pharm. J. **1992**, *248*, 444.

67. Kohrs, R.; Durieux, M.E. Ketamine: teaching an old drug new tricks. Anesth. Analg. **1998**, *87*, 1186–1193.

68. Ryder, S.; Way, W.L.; Trevor, A.J. Comparative pharmacology of the optical isomers of ketamine in mice. Eur. J. Pharmacol. **1978**, *49*, 15–23.

69. White, P.F.; Ham, J.; Way, W.L.; Trevor, A.J. Pharmacology of ketamine isomers in surgical patients. Anesthesiology **1980**, *52*, 231–239.

70. White, P.F.; Schuttler, J.; Stanski, D.R.; Horai, Y.; Trevor A.J. Comparative pharmacology of the ketamine isomers: studies in volunteers. Br. J. Anaesth. **1985**, *57*, 197–203.

71. Hustveit, O.; Maurset, A.; Oye, I. Interaction of the chiral forms of ketamine with opioid, phencyclidine, sigma and muscarinic receptors. Pharmacol. Toxicol. **1995**, *77*, 355–359.

72. Schuttler, J.; Stanski, D.R.; White, P.F.; Trevor, A.J.; Horai, Y.; Verotta, D.; Sheiner, L.B. Pharmacodynamic modelling of the EEG effects of ketamine and its enantiomers in man. J. Pharmacokin. Biopharm. **1987**, *15*, 241–253.

73. Vlasses, P.H.; Irvin, J.D.; Huker, P.B.; Lee R.B.; Ferguson, R.K.; Schrogie, J.J.; Zacchei, A.G.; Davies, R.O.; Abrams, W.B. Pharmacology of enantiomers and (−)-*p*-OH metabolite of indacrinone. Clin. Pharmacol. Ther. **1981**, *29*, 798–807.

74. Tobert, J.A.; Cirillo, V.J.; Hitzenberger, G. James, I.; Pryor, J.; Cook, T.; Buntinx, A.; Holmes, I.B.; Lutterbeck, P.M. Enhancement of uricosuric properties of indacrinone by manipulation of enantiomer ratio. Clin. Pharmacol. Ther. **1981**, *29*, 344–350.

75. Fanelli, G.M.; Watson, L.S.; Bohn, D.L.; Russo, H.F. Diuretic and uricosuric activity of 6,7-dichloro-2,3-dihydro-5-(2-thienylcarbonyl)-2-benzofuran-carboxylic acid and stereoisomers in chimpanzee, dog and rat. J. Pharmacol. Exp. Ther. **1980**, *212*, 190–197.

76. Hoffman, W.F.; Woltersdolf, O.W.; Novello, F.C.; Cragoe, E.J.; Springer, J.P.; Watson, L.S.; Fanelli, G.M. (Acylaryloxy)acetic acid derivatives. 3. 2,3-Dihydro-5-acyl-2-benzofurancarboxylic acids, a new class of uricosuric diuretics. J. Med. Chem. **1981**, *24*, 865–873.

77. Nakamura, M.; Kawabata, T.; Itoh, T.; Miyata, K.; Harada, H. Stereoselective saluretic effect and localization of renal tubular secretion of enantiomers of S-8666, a novel uricosuric antihypertensive diuretic. Drug Dev. Res. **1990**, *19*, 23–36.

78. Wong, D.T.; Fuller, R.W.; Robertson, D.W. Fluoxetine and its two enantiomers as selective serotonin uptake inhibitors. Acta. Pharm. Nord. **1990**, *2*, 171–180.

79. Wong, D.T.; Bymaster, F.P.; Reid, L.R.; Mayle, D.A.; Krushunski, J.H.; Robertson, D.W. Norfluoxetine enantiomers as inhibitors of serotonin uptake in rat brain. Neuropsychopharmacology, **1993**, *8*, 337–344.

80. Baumann, P.; Rochat, B. Comparative pharmacokinetics of selective serotonin reuptake inhibitors: a look behind the mirror. Int. Clin. Psychopharmacol. **1995**, *10* (suppl. 1), 15–21.

81. Stevens, J.C.; Wrighton, S.A. Interaction of the enantiomers of fluoxetine and norfluoxetine with human liver cytochrome P450. J. Pharmacol. Exp. Ther. **1993**, *266*, 964–971.

82. Thayer, A. Eli Lilly pulls the plug on prozac isomer drug. Chem. Eng. News **2000**, October 30, 8.

83. Steiner, T.J.; Ahmed, F. Findley, L.J.; MacGregor, E.A.; Wilkinson, M. (*S*)-Fluoxetine in the prophylaxis of migraine: a phase II double-blind randomised placebo-controlled study. Cephalagia **1998**, *18*, 283–286.

84. Hyttel, J.; Bøgesø, K.P.; Perregaard, J.; Sánchez, C. The pharmacological effect of citalopram resides in the (*S*)-(+)-enantiomer. J. Neural Trans. **1992**, *88*, 157–160.

85. Sánchez, C.; Hogg, S. The antidepressant effect of citalopram resides in the *S*-enantiomer (Lu 26-054). Poster presented at the Annual Meeting of the Society of Biological Psychiatry, May 11–18, 2000; Chicago, IL, USA, 2000.

86. Rochet, B.; Amery, M.; Baumann, P. Analysis of enantiomers of citalopram and its demethylated metabolites in plasma of depressive patients using chiral reverse-phase liquid chromatography. Ther. Drug Monit. **1995**, *17*, 273–279.

87. Koe, B.K.; Weissman, A.; Welch, W.M.; Browne, R.G. Sertraline, 1S,4S-N-methyl-4-(3,4-dichlorophenyl)-1,2,3,4-tetrahydro-1-naphthylamine, a new uptake inhibitor with selectivity for serotonin. J. Pharmacol. Exp. Ther. **1983**, *226*, 686–700.

88. Welch, W.M.; Kraska, A.R.; Sarges, R.; Koe, B.K. Nontricyclic antidepressant agents derived from *cis* and *trans*-1-amino-4-aryltetralins. J. Med. Chem. **1984**, *27*, 1508–1515.

89. Jenner, P.; Testa, B. The influence of stereochemical factors on drug disposition. Drug Metab. Rev. **1973**, *2*, 117–184.

90. Williams, K.; Lee, E. Importance of drug enantiomers in clinical pharmacology. Drugs **1985**, *30*, 333–354.

91. Caldwell, J.; Hutt, A.J.; Fournel-Gigleux, S. The metabolic chiral inversion and dispositional enantioselectivity of the 2-arylpropionic acids and their biological consequences. Biochem. Pharmacol. **1988**, *37*, 105–114.

92. Caldwell, J.; Winter, S.M.; Hutt, A.J. The pharmacological and toxicological significance of the stereochemistry of drug disposition. Xenobiotica **1988**, *18*, 59–70.

93. Tucker, G.T.; Lennard, M.S. Enantiomer specific pharmacokinetics. Pharmacol. Ther. **1990**, *45*, 309–329.

94. Levy, R.H; Boddy, A.V. Stereoselectivity in pharmacokinetics: a general theory. Pharm. Res. **1991**, *8*, 551–556.

95. Mason, J.P.; Hutt, A.J. Stereochemical aspects of drug metabolism. In *The Impact of Stereochemistry on Drug Development and Use*; Aboul-Enein, H.Y., Wainer, I.W.. Eds.; John Wiley: New York, 1997; 45–105.

96. Wade, D.N.; Mearrick, P.T.; Morris, J.L. Active transport of L-dopa in the intestine. Nature **1973**, *242*, 463–465.

97. Hendel, J.; Brodthagen, H. Entero-hepatic cycling of methotrexate estimated by use of the D-isomer as a reference marker. Eur. J. Clin. Pharmacol. **1984**, *26*, 103–107.

98. Itoh, T.; Ono, K.; Koido, K.-I.; Li, Y.-H.; Yamada, H. Stereoselectivity of the folate transporter in rabbit small intestine: studies with amethopterin enantiomers. Chirality **2001**, *13*, 164–169.

99. Tamai, I.; Ling, H.-Y.; Timbul, S.-M.; Nishikido, J.; Tsuji, A. Stereospecific absorption and degradation of cephalexin. J. Pharm. Pharmacol. **1988**, *40*, 320–324.

100. Wetterich, U.; Spahn-Langguth, H.; Mutschler, E.; Terhaag, B.; Rösch, W.; Langguth, P. Evidence for intestinal secretion as additional clearance pathway of talinolol enantiomers: concentration- and dose-dependent absorption *in vitro* and *in vivo*. Pharm. Res. **1996**, *13*, 514–522.

101. Zschiesche, M.; Lemma, G.L.; Klebingat, K.-J.; Franke, G.; Terhaag, B.; Hoffmann, A.; Gramatté, T.; Kroemer, H.K.; Siegmund, W. Stereoselective disposition of talinolol in man. J. Pharm. Sci. **2002**, *91*, 303–311.

102. Fehske, K.J.; Muller, W.E.; Wollert, U. The location of drug binding sites in human serum albumin. Biochem. Pharmacol. **1981**, *30*, 689–692.

103. Walle, U.K.; Walle, T.; Bai, S.A.; Olanoff, L.S. Stereoselective binding of propranolol to human plasma, α_1-acid glycoprotein and albumin. Clin. Pharmacol. Ther. **1983**, *34*, 718–723.

104. Seideman, P.; Lohrer, F.; Graham, G.G.; Duncan, M.W.; Williams, K.M.; Day, R.O. The stereoselective disposition of the enantiomers of ibuprofen in blood, blister and synovial fluid. Br. J. Clin. Pharmacol. **1994**, *38*, 221–227.

105. Mader, R.M.; Steger, G.G.; Risovski, B.; Sieder, A.E.; Locker, G.; Gnant, M.F.X.; Jakesz, R.; Rainer, H. Pharmacokinetics of rac-leucovorin versus (*S*)-leucovorin in patients with advanced gastrointestinal cancer. Br. J. Clin. Pharmacol. **1994**, *37*, 243–248.

106. Testa, B. Conceptual and mechanistic overview of stereoselective drug metabolism. In *Xenobiotic Metabolism and Disposition*; Kato, R., Estabrook, R.W., Cayen, M.N., Eds.; Taylor and Francis: London, 1989; 153–160.

107. Low, L.K.; Castagnoli, N. Enantioselectivity in drug metabolism. In *Annual Reports in Medicinal Chemistry*; Clarke, F.H., Ed.; Academic Press: New York, 1978; Vol. 13, 304–315.

108. Cashman, J.R.; Park, S.B.; Yang, Z.-C.; Washington, C.B.; Gomez, D.Y.; Giacomini, K.M.; Brett, C.M. Chemical, enzymatic and human enantio-selective S-oxygenation of cimetidine. Drug Metab. Dispos **1993**, *21*, 587–597.

109. Kuzel, R.A.; Bhasir, S.K.; Oldham, H.G.; Damani, L.A.; Murphy, J.; Camilleri, P.; Hutt, A.J. Investigations into the chirality of the metabolic sulfoxidation of cimetidine. Chirality **1994**, *6*, 607–614.

110. Porubek, D.J.; Barnes, H.; Theodore, L.J.; Baillie, T.A. Enantioselective synthesis and preliminary metabolic studies of the optical isomers of 2-n-propyl-4-pentenoic acid, a hepatotoxic metabolite of valproic acid. Chem. Res. Toxicol. **1988**, *1*, 343–348.

111. Poupaert, J.H.; Cavalier, R.; Claesen, M.H.; Dumont, P.A. Absolute con-figuration of the major metabolite of 5,5-diphenylhydantoin, 5-(4'-hydroxy-phenyl)-5-phenylhydantoin. J. Med. Chem. **1975**, *18*, 1268–1271.

112. Rettie, A.E.; Korzekwa, K.R.; Kunze, K.L.; Lawrence, R.F.; Eddy, A.C.; Aoyama, T.; Gelboin, H.V.; Gonzalez, F.J.; Trager, W.F. Hydroxylation of warfarin by human cDNA–expressed cytochrome P450: a role for P450 2C9 in the aetiology of (S)-warfarin-drug interactions. Chem. Res. Toxicol. **1992**, *5*, 54–59.

113. Sisenwine, S.F.; Tio, C.O.; Hadley, F.V.; Liu, A.-L.; Kimmel, H.B.; Ruelius, H.W. Species-related differences in the stereoselective glucuronidation of oxazepam. Drug Metab. Dispos. **1982**, *10*, 605–608.

114. Niwa, T.; Tokuma, Y.; Nakagawa, K.; Noguchi, H. Stereoselective oxidation of nilvadipine, a new dihydropyridine calcium antagonist, in rat and dog liver. Drug Metab. Dispos. **1989**, *17*, 64–68.

115. Tokuma, Y.; Fujiwara, T.; Niwa, T.; Hashimoto, T.; Naguchi, H. Stereo-selective disposition of nilvadipine, a new dihydropyridine calcium antagonist in the rat and dog. Res. Comm. Chem. Path. Pharmacol. **1989**, *63*, 249–262.

116. Äbelo, A.; Andersson, T.B.; Antonsson, M.; Naudot, A.K.; Skånberg, I.; Weidolf, L. Stereoselective metabolism of omeprazole by human cytochrome P450 enzymes. Drug Metab. Dispos. **2000**, *28*, 966–972.

117. Hutt, A.J.; Caldwell, J. The metabolic chiral inversion of 2-arylpropionic acids—a novel route with pharmacological consequences. J. Pharm. Pharmacol. **1983**, *35*, 693–704.

118. Bartels, M.J.; Smith, F.A. Stereochemical inversion of haloxyfop in the Fischer 344 rat. Drug Metab. Dispos. **1989**, *17*, 286–291.

119. Knights, K.M.; Talbot, U.M.; Baillie, T.A. Evidence of multiple forms of rat liver microsomal coenzyme A ligase catalysing the formation of 2-arylpropionyl-coenzyme A thioesters. Biochem. Pharmacol. **1992**, *44*, 2415–2417.

120. Williams, K.; Day, R.; Knihinicki, R.; Duffield, A. The stereoselective uptake of ibuprofen enantiomers into adipose tissue. Biochem. Pharmacol. **1986**, *35*, 3403–3405.

121. Sallustio, B.C.; Meffin, P.J.; Knights, K.M. The stereospecific incorporation of fenoprofen into rat hepatocyte and adipocyte triacylglycerols. Biochem. Pharmacol. **1988**, *37*, 1919–1923.

122. Tang, C.; Zhang, K.; Lepage, F.; Levy, R.H.; Baillie, T.A. Metabolic chiral inversion of stiripentol in the rat II. Influence of route of administration. Drug Metab. Dispos. **1994**, *22*, 554–560.

123. Zhang, K.; Tang, C.; Rashed, M.; Cui, D.; Tombret, F.; Botte, H.; Lepage, F.; Levy, R.H.; Baillie, T.A. Metabolic chiral inversion of stiripentol in the rat I. Mechanistic studies. Drug Metab. Dispos. **1994**, *22*, 544–553.

124. Baillie, T.A.; Schultz, K.M. Stereoselectivity in the drug development process: role of chirality as a determinant of drug action, metabolism and toxicity. In *The Impact of Stereochemistry on Drug Development and Use*; Aboul-Enein, H.Y., Wainer, I.W., Eds.; John Wiley: New York, 1997; 21–43.

125. Kashiyama, E.; Todaka, T.; Odomi, M.; Tanokura, Y.; Johnson, D.B.; Yokoi, T.; Kamataki, T.; Shimizu, T. Stereoselective pharmacokinetics and interconversions of flosequinan enantiomers containing chiral sulphoxide in rat. Xenobiotica **1994**; *24*, 369–377.

126. Kashiyama, E.; Yokoi, T.; Odomi, M.; Kamataki, T. Stereoselective S-oxidation and reduction of flosequinan in rat. Xenobiotica **1999**, *29*, 815–826.

127. Kashiyama, E.; Yokoi, T.; Odomi, M.; Funae, Y.; Inoue, K.; Kamataki, T. Cytochrome P450 responsible for the stereoselective S-oxidation of flosequinan in hepatic microsomes from rats and humans. Drug Metab. Dispos. **1997**, *25*, 716–724.

128. Notterman, D.A.; Drayer, D.E.; Metakis, L.; Reidenberg, M.M. Stereoselective renal tubular secretion of quinidine and quinine. Clin. Pharmacol. Ther. **1986**, *40*, 511–517.

129. Okazaki, O.; Kurata, T.; Hakusui, H.; Tachizawa, H. Species-related stereoselective disposition of ofloxacin in the rat, dog and monkey. Xenobiotica **1992**, *22*, 439–450.

130. Nakano, M.; Kawahara, S. Stereoselective renal tubular secretion of a new uricosuric diuretic, 6,7-dichloro-5-(N,N-dimethylsulfamoyl)-2,3-dihydro-2-benzofurancarboxylic acid (*S*-8666) in cynomolgus monkeys. Xenobiotica **1992**, *20*, 179–185.

131. Nakano, M.; Higaki, K.; Kawahara, S. Enantiomer-enantiomer interaction of a uriscosuric antihypertensive diuretic (DBCA) in renal tubular secretion and stereoselective inhibition by probenecid in the cynomolgus monkey. Xenobiotica **1993**, *23*, 525–536.

132. Kamizono, A.; Inotsume, N.; Fukushima, S.; Nakano, M.; Okamoto, Y. Inhibitory effects of procainamide and probenecid on renal excretion of sultopride enantiomers in rats. J. Pharm. Sci. **1993**, *82*, 1259–1261.

133. Hsyu, P.-H.; Giacomini, K.M. Stereoselective renal clearance of pindolol in humans. J. Clin. Invest., **1985**, *76*, 1720–1726.

134. Somogyi, A.A.; Bochner, F.; Sallustio, B.C. Stereoselective inhibiton of pindolol renal clearance by cimetidine in humans. Clin. Pharmacol. Ther. **1992**, *51*, 379–387.

135. Le Corre, P.; Gibassier, D.; Sado, P.; LeVerge, R. Stereoselective metabolism and pharmacokinetics of disopyramide enantiomers in humans. Drug Metab. Dispos. **1988**, *16*, 858–864.

136. Ariëns, E.J. Stereochemistry, a basis for sophisticated nonsense in pharmacokinetics and clinical pharmacology. Eur. J. Clin. Pharmacol. **1984**, *26*, 663–668.

137. Evans, A.M.; Nation, R.L.; Sansom, L.N.; Bochner, F.; Somogyi, A.A. Stereoselective drug disposition: potential for misinterpretation of drug disposition data. Br. J. Clin. Pharmacol. **1988**, *26*, 771–780.

138. Fabre, H.; Ibork, H.; Lerner, D.A. Photoisomerization kinetics of cefuroxime axetil and related compounds. J. Pharm. Sci. **1994**, *83*, 553–558.

139. Mosher, G.L.; McBee, J.; Shaw, D.B. Esterase activity toward the diastereomers of cefuroxime axetil in the rat and dog. Pharm. Res. **1992**, *9*, 687–689.

140. Defossa, E.; Durckheimer, W.; Fischer, G.; Jendralla, H.; Klesel, N.; Wollmann, T. Cefdaloxime pentexil tosilate (HR916K): a diastereomerically pure novel oral cephalosporinester: synthesis and antibacterial activity *in vivo*. In Program and Abstracts of the Twenty-Second Interscience Conference on Antimicrobial Agents and Chemotherapy, Anaheim 1992; American Society for Microbiology: Washington, 1992; 142.

141. Isert, D.; Fischer, G.; Klesel, N.; Limbert, M.; Markus, A.; Riess, G. Cefdaloxime pentexil tosilate (HR916K): a diastereomerically pure novel oral cephalosporinester with outstanding absorption characteristics. In Program and Abstracts of the Twenty-Second Interscience Conference on Antimicrobial Agents and Chemotherapy, Anaheim 1992; American Society for Microbiology: Washington, 1992; 142.

142. Mendes, P.; Meyer, B.H.; Müller, F.O.; Scholl, T.; Luus, H.; de la Ray, N. Pharmacokinetics of cefdaloxime pentexil tosilate (HR916K) after a single oral dose in healthy volunteers. In Program and Abstracts of the Twenty-Second Interscience Conference on Antimicrobial Agents and Chemotherapy, Anaheim 1992; American Society for Microbiology: Washington, 1992; 142.

143. Bodor, N.; El Koussi, A.; Kano, M.; Nakamura, T. Improved delivery through biological membranes. 26. Design, synthesis and pharmacological activity of a novel chemical delivery system for β-adrenergic blocking agents. J. Med. Chem. **1988**, *31*, 100–106.

144. Bodor, N.; El Koussi, A. Improved delivery through biological membranes. LVI. Pharmacological evaluation of alprenoxime—a new potential antiglaucoma agent. Pharm Res. **1991**, *8*, 1389–1395.

145. Bodor, N.; Prokai, L. Site- and stereospecific ocular drug delivery by sequential enzymatic bioactivation. Pharm. Res. **1990**, *7*, 723–725.

146. Prokai, L.; Wu, W.-M.; Somogyi, G.; Bodor, N. Ocular delivery of the β-adrenergic antagonist alprenolol by sequential bioactivation of its methoxime analogue. J. Med. Chem. **1995**, *38*, 2018–2020.

147. Blashke, G.; Kraft, H.P.; Fickentscher, K.; Köhler, F. Chromatographische racemattrennung von thalidomide und tertogene wirking der enantiomere. Arzneim. Forsch. **1979**, *29*, 1640–1642.

148. Fabro, S.; Smith R.L.; Williams, R.T. Toxicity and teratogenicity of optical isomers of thalidomide. Nature **1967**, *215*, 269.

149. Eriksson, T.; Björkman, S.; Roth, B.; Fyge, A.; Höglund, P. Stereospecific determination, chiral inversion in vitro and pharmacokinetics in humans of the enantiomers of thalidomide. Chirality **1995**, *7*, 44–52.

150. Eriksson, T.; Björkman, S.; Roth, B.; Fyge, A.; Höglund, P. Enantiomers of thalidomide: blood distribution and the influence of serum albumin on chiral inversion and hydrolysis. Chirality **1998**, *10*, 223–228.

151. Reist, M.; Carrupt, P.-A.; Francotte, E.; Testa, B. Chiral inversion and hydrolysis of thalidomide: mechanisms and catalysis by bases and serum albumin, and chiral stability of teratogenic metabolites. Chem. Res. Toxicol. **1998**, *11*, 1521–1528.

152. Meyring, M.; Chankvetadze, B.; Blaschke, G. Enantioseparation of thalidomide and its hydroxylated metabolites using capillary electrophoresis with various cyclodextrins and their combinations as chiral buffer additives. Electrophoresis **1999**, *20*, 2425–2431.

153. Schmahl, H.-J.; Nau, H.; Neubert, D. The enantiomers of the teratogenic thalidomide analogue EM12: 1. Chiral inversion and plasma pharmacokinetics in the marmoset monkey. Arch. Toxicol. **1988**, *62*, 200–204.

154. Schmahl, H.-J.; Heger, W.; Nau, H. The enantiomers of the teratogenic thalidomide analogue EM12: 2. Chemical stability, stereoselectivity of metabolism and renal excretion in the marmoset monkey. Toxicol. Lett. **1989**, *45*, 23–33.

155. Heger, W.; Klug, S.; Schmahl, H.-J.; Nau, H.; Merker, H.-J.; Neubert, D. Embryotoxic effects of thalidomide derivatives on the non-human primate Callithrix jacchus: 3. Teratogenic potency of the EM12 enantiomers. Arch. Toxicol. **1988**, *62*, 205–208.

156. Hauck, R.-S.; Wegner, C.; Blumtritt, P.; Fuhrhop, J.-H.; Nau, H. Asymmetric synthesis and teratogenic activity of (*R*)- and (*S*)-2-ethylhexanoic acid, a metabolite of the plasticizer di-(2-ethylhexyl)phthalate. Life Sci. **1990**, *46*, 513–518.

157. Hauck, R.-S.; Nau, H. Asymmetric synthesis and enantiomeric teratogenicity of 2-n-propyl-4-pentenoic acid (4-en-VPA), an active metabolite of the anticonvulsant drug, valproic acid. Toxicol. Lett. **1989**, *49*, 41–48.

158. Hauck, R.-S.; Nau, H. The enantiomers of the valproic acid analogue 2-n-propyl-4-pentynoic acid (4-yn-VPA): asymmetric synthesis and highly stereoselective teratogenicity in mice. Pharm. Res. **1992**, *9*, 850–855.

159. Takamatsu, Y.; Kaneko, H.; Abiko, J.; Yoshitake, A.; Miyamoto, J. *In vivo* and *in vitro* stereoselective hydrolysis of four chiral isomers of fenvalerate. J. Pesticide Sci. **1987**, *12*, 397–404.

160. Jones, A.R.; Mashford, P.M.; Murcott, C. The metabolism of 3-amino-1-chloropropan-2-ol in relation to its antifertility activity in male rats. Xenobiotica **1979**, *9*, 253–261.

161. Porter, K.E.; Jones, A.R. The effects of the isomers of α-chlorohydrin and racemic β-chlorolactate on the rat kidney. Chem.-Biol. Interact. **1982**, *41*, 95–104.

162. Dobbie, M.S.; Porter, K.E.; Jones, A.R. Is the nephrotoxicity of (*R*)-3-chlorolactate in the rat caused by 3-chloropyruvate? Xenobiotica **1988**, *18*, 1389–1399.

163. Walshe, J.M. Penicillamine, a new oral therapy for Wilson's disease. Am. J. Med. **1956**, *21*, 487–495.

164. Tu, J.-B.; Blackewell, R.Q.; Lee, P.F. D,L-Penicilliamine as a cause of optical axial neuritis. J. Am. Med. Ass. **1963**, *185*, 83–86.

165. Walshe, J.M. Chirality of penicillamine. Lancet **1992**, *339*, 254.

166. Lee, A.; Lawton, N.F. Penicillamine treatment of Wilson's disease and optical neuropathy. J. Neurol. Neurosurg. Psych. **1991**, *58*, 746.

167. Cotzias, G.C.; Van Woert, M.H.; Schiffer, L.M. Aromatic amino acids and modification of Parkinsonism. New Engl. J. Med. **1967**, *276*, 374–379.

168. Cotzias, G.C.; Papavasiliou, P.S.; Gellene, R. Modification of Parkinsonism-Chronic treatment with L-dopa. New Engl. J. Med. **1969**, *280*, 337–345.

169. Anon. Side effects kill "new Prozac." Chem. Br. **2000**, *36*, 11.

170. Kato, R.; Ikeda, N.; Yabek, S.; Kannen, R.; Singh, B.N. Electrophysiologic effects of the levo- and dextrorotatory isomers of sotalol in isolated cardiac muscle and their in vivo pharmacokinetics. J. Am. Coll. Cardiol. **1986**, *7*, 116–125.

171. Advani, S.V.; Singh, B.N. Pharmacodynamic, pharmacokinetic and antiarrhythmic properties of d-sotalol, the dextro-isomer of sotalol. Drugs **1995**, *49*, 664–679.

172. Waldo, A.L.; Camm, A.J.; de Ruyter, H.; Friedman, P.L.; MacNiell, D.J.; Pauls, J.F.; Pitt, B.; Pratt, C.M.; Schwartz, P.J.; Veltri, E.P. Effect of d-sotalol on mortality in patients with left ventricular dysfunction after recent and remote myocardial infarction. Lancet **1996**, *348*, 7–12.

173. Colatsky, T.J. Antiarrhythmic drugs: where are we going? Pharm. News **1995**, *2*, 17–23.

174. Connolly, M.H.; Cary, J.L.; McGoon, M.D. Valvular heart disease associated with fenfluramine-phentamine. New Engl. J. Med. **1997**, *337*, 581–588.

175. Anon. Fenfluramine and dexfenfluramine withdrawn. Curr. Problems in Pharmacovigilance **1997**, *23*, 13–14.

176. Anon. Dilevalol cost Schering $ 100 million. Scrip **1990**, *1543*, 19.

177. Maier, N.M.; Franco, P.; Linder, W. Separation of enantiomers: needs, challenges, perspectives. J. Chromatogr. A **2001**, *906*, 3–33.

178. Noctor, T.A.G. Enantioselective binding of drugs to plasma proteins. In *Drug Stereochemistry, Analytical Methods and Pharmacology*; 2nd Ed.; Wainer, I.W., Ed.; Marcel Dekker: New York, 1993; 337–364.

179. Eichelbaum, M.; Gross, A.S. Stereochemical aspects of drug action and disposition. In *Advances in Drug Research*; Testa, B., Meyer, U.A., Eds.; Academic Press: London, 1996; Vol. 28, 1–64.

180. Piquette-Miller, M.; Foster, R.T.; Kappagoda, C.T.; Jamali, F. Pharmacokinetics of acebutolol enantiomers in humans. J. Pharm. Sci. **1991**, *80*, 313–316.

181. Boyd, R.A.; Chin, S.K.; Don-Pedro, O.; Williams, R.L.; Giacomini, K.M. The pharmacokinetics of the enantiomers of atenolol. Clin. Pharmacol. Ther. **1989**, *45*, 403–410.

182. Ofori-Adjei, D.; Ericsson, O.; Lindstrom, B.; Hermansson, J.; Adjepon-Yamoah, K.; Sjoqvist, F. Enantioselective analysis of chloroquine and desethyl-chloroquine after oral administration of racemic chloroquine. Ther. Drug. Monit. **1986**, *8*, 457–461.

183. Lennard, M.S.; Tucker, G.T.; Silas, J.H.; Freestone, S.; Ramsay, L.E.; Woods, H.F. Differential stereoselective metabolism of metoprolol in extensive and poor debrisoquin metabolisers. Clin. Pharmacol. Ther. **1983**, *34*, 732–737.

184. Grech-Belanger, O.; Turgeon, J.; Gilbert, M. Stereoselective disposition of mexiletine in man. Br. J. Clin. Pharmacol. **1986**, *21*, 481–487.

185. Okazaki, O.; Kojima, C.; Hakusui, H.; Nakashima, M. Enantioselective disposition of ofloxacin in humans. Antimicrob. Agents Chemoth. **1991**, *35*, 2106–2109.

186. Gietl, Y.; Spahn, H.; Knauf, H.; Mutschler, E. Single and multiple-dose pharmacokinetics of (*R*)-(−)- and (*S*)-(+)-prenylamine in man. Eur. J. Clin. Pharmacol. **1990**, *38*, 587–593.

187. Borgstrom, L.; Nyberg, L.; Jonsson, S.; Lindberg, C.; Paulson, J. Pharmacokinetic evaluation in man of terbutaline given as separate enantiomers and as the racemate. Br. J. Clin. Pharmacol. **1989**, *27*, 49–56.

188. Edgar, B.; Heggelund, A.; Johansson, L.; Nyberg, G.; Regardh, C.G. The pharmacokinetics of (*R*)- and (*S*)-tocainide in healthy subjects. Br. J. Clin. Pharmacol. **1984**, *16*, 216P–217P.

189. Weber-Grandke, H.; Hahn, G.; Mutschler, E.; Möhrke, W.; Langguth, P.; Spahn-Langguth, H. The pharmacokinetics of tranylcypromine enantiomers in healthy subjects after oral administration of racemic drug and the single enantiomers. Br. J. Clin. Pharmacol. **1993**, *36*, 363–365.

6

Stereospecific Pharmacokinetics and Pharmacodynamics: Selected Classes of Drugs

Dion R. Brocks
University of Alberta, Edmonton, Alberta, Canada

Majid Vakily
TAP Pharmaceutical Products, Inc., Lake Forest, Illinois, U.S.A.

Reza Mehvar
Texas Tech University Health Sciences Center, Amarillo, Texas, U.S.A.

1. INTRODUCTION

The concept of chirality and its importance as a determinant of the pharmacological properties of drugs have been a part of scientific knowledge since the middle to late 1800s [1]. The property of enantiomerism is conferred on a molecule when it is present in two isomers that form nonsuperimposable mirror images of one another (Fig. 1). Most often these molecules possess a center of asymmetry focused on one specific atom in the molecule. With respect to these centers of asymmetry, chiral drugs may take several forms. In most cases, the chiral center is a tetrahedral carbon atom to which are attached four distinct substituents. Alternatively, chiral drugs may possess sulfur or phosphorous atoms at their centers of asymmetry (Fig. 2). Less commonly, chirality may be conferred on some drugs on the basis of the presence of a plane of asymmetry in the molecule (Fig. 2). Such compounds are known as geometric isomers.

In an achiral environment, enantiomers share the same physico-chemical properties, with the exception of the direction in which they rotate

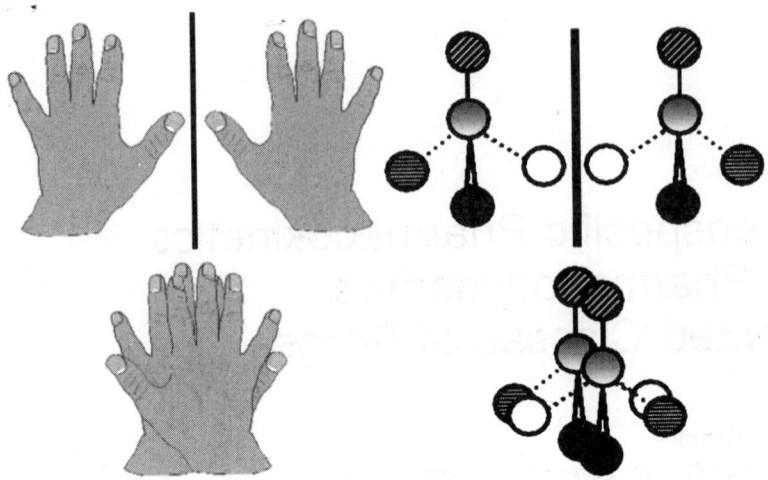

Figure 1 Enantiomers are two otherwise identical entities that form nonsuperimposable images of one another. Examples include a pair of hands (left) or a tetrahedral organic molecule to which are attached four different molecular substituents connected to a central atom (right). In both cases, although the respective images appear identical on both sides of the mirror, they cannot be superimposed upon one another. Enantiomers differ in the direction that they rotate plane polarized light.

plane polarized light when in solution form. However, it is well known that the enantiomers of chiral drugs may possess stereoselectivity in their pharmacodynamic properties. The basis for this difference between enantiomers lies in the three-dimensional nature of organic drug molecules and their target receptors, which are most often proteins. This provides a basis for spatial recognition of enantiomeric forms of a chiral molecule by a pharmacological receptor at its critical binding site for the drug–protein interaction. In a similar fashion, many drug disposition processes are facilitated or controlled by proteins for which enantioselectivity may also present itself. These processes may include absorption in the gastrointestinal tract, distribution, drug metabolism, and renal tubular secretion. As a consequence of this differential binding of enantiomers for proteins or other endogenous macromolecules, a wide range of differences may present themselves between enantiomers of a drug at the level of pharmacodynamics and/or pharmacokinetics.

Many of the drugs possessing chiral centers that were available in the earlier part of the 20th century were derived from natural product and were

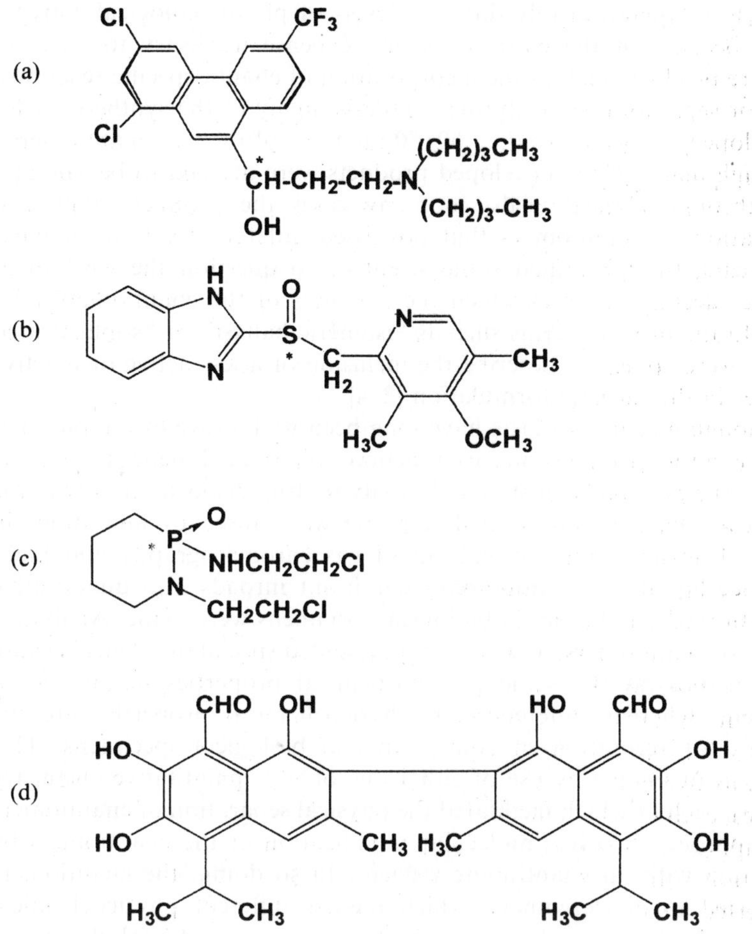

Figure 2 Centers of asymmetry may consist of one of several forms. (a) A carbon atom most commonly forms the chiral center, as is seen for halofantrine. (b) Sulfur atoms may form the center of asymmetry in some molecules, such as omeprazole. (c) In ifosfamide, a phosphorous atom is the center of asymmetry. (d) In some case, such as gossypol, a plane of asymmetry is formed not at a single atom, but rather between two atoms (the bond linking the two aromatic groups).

usually enantiopure as a result of the three-dimensional selectivity of enzymes for substrates. With the advancement of medicinal chemistry over the 20th century, the pharmacological treatment of disease saw a major shift in the available medicinal agents from natural to synthetic sources. Many of the enantiomers of these synthesized drugs were known to possess qualitatively or quantitatively different levels of pharmacological activity. However, because of the extra steps and expense necessary to yield an enantiopure product, such as the incorporation of enantiospecific reactions, reagents, or separation steps during synthesis, many of the synthetic drugs were developed as racemates (i.e., 50:50 mixtures of the two enantiomers). Even though many of the developed products were deemed to be safe and effective through clinical trials, in many cases the products contained a combination of enantiomers that possessed different levels of activity. In later years, this prompted some scientists to question the wisdom of developing racemic drugs in which the activities of the enantiomers were different. In the process, terms such as "isomeric ballast" or "sophisticated nonsense" were coined to describe the inclusion of a less active or inactive enantiomer in the racemic formulation [2–4].

Although pharmacologists have long been well aware of the pharmacological enantioselectivity in drug action, pharmacokineticists became focused on the possibility of stereoselectivity in drug disposition only in the past 25 years. Part of this belated response was caused by limitations in analytical chemistry. With the advent of gas chromatography and high-performance liquid chromatography, significant inroads into quantitative determination of xenobiotics in biological specimens were made. Analytical separation of enantiomers, however, still presented special problems because enantiomers possess the same physicochemical properties in an achiral environment, whereas differences in physicochemical properties are the usual basis for resolution of components in biological specimens. This problem was overcome by use of either one of two quantitative chemistry approaches, each of which facilitated the physical separation of enantiomers [5]. One approach involves molecular modification of the enantiomers by derivatization with an enantiopure reagent. In so doing, the enantiomers are converted to diastereomers, which possess different physicochemical properties and which can then be separated using conventional chromatography. The second method involves changing the environment in which the enantiomers are present during the separation procedure. This is facilitated by use of special chromatographic columns packed with materials that confer a three dimensional structure to the stationary phase. These packing materials may consist of protein molecules, or they may contain a special chemical in enantiopure form bonded to a particulate packing material. As the racemate passes through such a column, the enantiomers interact differently with the packing material, thus affording separation. These columns have become increasingly popular over the past decade as their reliability and durability have improved.

The purpose of this chapter is to provide an overview of the pharmacodynamic and pharmacokinetic properties of chiral drugs in humans used for disorders other than cardiovascular diseases, which are discussed in the next chapter. There have been literally hundreds of papers published just over the past 20 years dealing with stereoselectivity in pharmacokinetics and dynamics of chiral drugs. Due to the massive amount of information available, this review is not entirely comprehensive but is intended to provide the reader with an appreciation of the breadth of kinetic-dynamic differences between enantiomers of chiral drugs, and their possible therapeutic relevance. Because of their supportive nature toward the understanding of stereoselectivity, animal studies are cited where necessary in the discussion of aspects of pharmacodynamic and tissue distribution of enantiomers.

In order to provide a clinical focus, we have opted to write the chapter so that chiral drugs are grouped by therapeutic use. Although many of the chiral drug classes are discussed in this chapter, owing to the number of available studies into their stereoselective properties, the drugs used in the treatment of cardiovascular disease are presented separately in the next chapter. A more general overview of the principles of pharmacodynamic chiral recognition and stereoselectivity in the pharmacokinetics based on the individual processes (absorption, distribution, metabolism, and renal excretion) are presented in Chap. 5.

2. DRUGS USED IN THE TREATMENT OF INFECTIOUS DISEASE

2.1. Antibiotics

Several agents in the fluoroquinolone class of antibiotics are chiral and used clinically as the racemate. Varying degrees of stereoselectivity in antimicrobial activities have been reported. The enantiomers of ofloxacin possess a high degree of stereoselectivity in terms of antibacterial potency, with the S(-) enantiomer reportedly possessing a much greater antibacterial effect against gram-positive and gram-negative organisms than antipode [6]. The (+) enantiomer of tosufloxacin enantiomers shows a 10- to 100-fold higher in vitro level of antibacterial activity than the (−) enantiomer [7]. In contrast to these two agents, the R enantiomer of clinafloxacin displays only slightly more activity (smaller minimum inhibitory concentration) than antipode against gram-negative organisms and some gram-positive organisms. However, this difference in activity is probably clinically insignificant [8].

With respect to pharmacokinetic properties of fluoroquinolones, only a low to moderate degree of stereoselectivity has been reported for this class of drugs. The enantiomers of ofloxacin display a low but significant degree of stereoselectivity after single 200 mg oral doses of racemate to healthy subjects [6] (Table 1).The S:R ratios of $t_{1/2}$ and $AUC_{0-\infty}$ were both 1.1 (Table 1). Significant differences were also noted between enantiomers in renal and total CL, with values for the R-(+) enantiomer being 5–10% higher than the active S enantiomer. The renal CL was greater than that of renal filtration clearance for both enantiomers. Therefore, because there was no stereoselectivity in plasma protein binding, which ranged from 47 to 52%, the stereoselectivity in renal clearance was likely due to stereoselectivity at the level of active tubular secretion of the enantiomers. The oral Vd exceeded total body water but was not different between enantiomers [6]. These findings were largely repeated (S:R ratio = 1.09) in another study in preoperative patients undergoing cholecystectomy who were given 400 mg of the racemate as a 1 h intravenous infusion [9]. Gall bladder wall concentrations of the more active S-(−)-ofloxacin (levofloxacin) were well above the minimum inhibitory concentration against *E. coli* for over 36 hours after the infusion [9], which could confer on ofloxacin advantageous properties in the treatment of infections of the gall bladder.

The pharmacokinetics of clinafloxacin are nonstereoselective in human volunteers after both oral and intravenous doses (Table 2) [8]. Similarly, the oral pharmacokinetics of tosufloxacin are also nonstereoselective with respect to serum C_{max}, t_{max}, $t_{1/2}$, and AUC (Table 1). However, the renal clearance of the (−) enantiomer is significantly higher (12%) than that of the (+) enantiomer [7].

2.2. Antivirals

The pyrimidine antihuman immunodeficiency viral compound (±)-2′-deoxy-3′-oxa-4′-thiocytidine (dOTC) is administered as the racemate. In a stereoselective evaluation of the pharmacokinetics of dOTC in 12 healthy volunteers (Table 2) [10], it was found that the CL of the (−) enantiomer was 31% greater than antipode. In contrast, the Vss of the (−) enantiomer was lower than antipode [(+):(−) ratio of 1.4]. Plasma protein binding of dOTC enantiomers was not studied. The authors noted marked secondary peaks for both enantiomers after oral dosing, which was attributed to three absorption processes. The overall oral bioavailability was found to be over 75% for each enantiomer [10]. Nevertheless, because the enantiomers of dOTC have approximately the same level of antiviral activity [10], the stereoselectivity in pharmacokinetics may not have a clinical implication for this drug.

Table 1 Stereospecific Pharmacokinetic Data (Plasma or Serum Unless Stated Otherwise) of a Number of Chiral Drugs (Racemate Unless Specified) Given by Nonparenteral Routes (Oral Unless Specified)

Drug N (mean age/range)	Dose	Enantiomer	C_{max}, ng/mL	T_{max}, h	AUC, ng h/mL	$t_{1/2\lambda n}$, h	Ref.
Albuterol							
8 (21–40)	2 mg	R	0.46	2.0	2.1	—	114
		S	5.95	3.0	47.0	—	
	1.2 mg inhaled	R	1.6	0.12	3.0	—	
		S	3.1	1.0	25.0	—	
	1.2 mg + charcoal	R	1.5	0.08	2.5	—	
		S	2.5	0.17	8.0	—	
22 (22–64)	4 mg	R	1.0	2.5	3.2	—	291
		S	7.2	3.4	26.5	—	
	0.2 mg, endotracheal	R	0.4	0.9	0.7	—	
		S	0.7	1.1	1.6	—	
	0.8 mg, inhaled	R	1.2	0.35	1.9	—	
		S	2.0	0.63	7.0	—	
12 (20–29)	10 mg	R	3.5	1.8	13.5	—	
		S	23.2	1.8	87.2	—	
Aminoglutethimide							
6 (53) cancer patients	500 mg	(+)	3800	1.4	79900	16.8	242
		(−)	3300	1.3	47100	11.1	
Bicalutamide							
10 (57)	50 mg	R	848	30	225000	178	243
		S	84	5	1380	16	
10 (57 cirrhosis)	50 mg	R	750	24	182000	173	243
		S	81	2.5	996	23	
Chloroquine 6 (23–29)	150 mg enantiomer	R	155 (blood)	4.0	19300	294	23
		S	133	3.3	11300	237	

(continues)

Table 1 Continued

Drug N (mean age/range)	Dose	Enantiomer	C_{max}, ng/mL	T_{max}, h	AUC, ng h/mL	$t_{1/2\lambda n}$, h	Ref.
Chlorpheniramine							
6 (22–28) EM	8 mg maleate	R	5.4	3.5	124	20	219
		S	12.6	3.0	300	18	
24 (18–40)	4 mg	R	2.2	5.0	52.3	17.8	220
		S	5.2	3.0	119	18.3	
Etodolac							
6 (28)	200 mg	R	12000	1.8	71000	6.6	284
		S	2700	1.7	6200	4.8	
6 (73)	200 mg	R	15900	1.5	71000	7.7	284
		S	3800	1.3	10000	6.1	
Fluoxetine							
12 (21–50)	20 mg	R	2.7	9	126	29	135
		S	4.2	6	151	22	
6 (20–40, PM)	60 mg	R			1784	228	138
		S			10126	146	
6 (20–40, EM)	60 mg	R			749	62	138
		S			877	26	
Flurbiprofen							
6 (19–40)	100 mg	R	8100	1.7	40100	4.2	246
		S	8200	1.7	45400	4.2	
6 (26–68)	50 mg	R	4000	1.7	21400	5.1	292
		S	4200	1.7	27000	6.0	
8 uremic (26–68)	50 mg	R	2400	2.7	12200	5.3	292
		S	2700	2.4	15300	5.5	

Drug (ref)	Dose	Isomer					
Gossypol	20 mg (+)	(+)	909		15	133	230
11 (28–50)	20 mg (−)	(−)	414		3.2	4.6	
Halofantrine	500 mg HCl	(+)	345	18	9.92	—	31
6 (23–54)		(−)	278	14	6.13	—	
Hexobarbital							
6 (29)	500 mg	R	1440	1.8	3322	2.1	159
		S	3020	2.8	27300	6.0	
6 (71)	500 mg	R	1420	1.8	4840	2.8	159
		S	2510	2.3	27220	8.7	
Hydroxychloroquine							
72 (19–50)	155 mg	R	103 (blood)	4.1	10829	692	24
		S	70	3.2	5432	420	
24 (25)	155 mg	R	84 (blood)	3	3948	526	26
		S	52	2.5	2208	457	
E-10-hydroxynortriptyline10	75 mg hydrogen maleate	(+)			278	9.7	148
(23–43)		(−)			919	8.8	
Ibuprofen							
12 (24)	600 mg	R	14000	1.7	50500	3.2	293
		S	16000	2.1	78900	3.1	
12 (18–34)	400 mg	R	7700	4.4	42000	2.2	258
		S	8700	6.0	58000	2.8	
38 (6) cystic fibrosis	20 mg/kg	R	33900	~0.6	58900	1.4	288
		S	41100	~1	120000	1.7	
Ketoprofen							
5 (74)	50 mg	R	3730	1	6930	2.4	281
		S	3650	1	6340	2.0	
8 (24)	50 mg	R	3290	1.3	7000	3.8	280
		S	3140	1.3	5920	2.5	

(continues)

Table 1 Continued

Drug N (mean age/range)	Dose	Enantiomer	C_{max}, ng/mL	T_{max}, h	AUC, ng·h/mL	$t_{1/2\lambda n}$, h	Ref.
Ketorolac							
3 young adults	10 mg	R	464	1.03	2742	2.4	253
		S	210	1.08	708	1.2	
Lansoprazole							
6 (25)	30 mg	R	1040	2.9	7140	1.7	227
		S	360	1.9	1560	1.9	
Mefloquine							
8 (21–36)	250 mg HCl	(+)	120	18	19700	128	294
		(−)	360	30	189600	409	
9 (1.6)	25 mg/kg base	(+)	600		67200	62.4	39
		(−)	2420		528000	264	
9 (26–50)	750 mg HCl	(+)	460 (blood)	5.4	94000	206	37
		(−)	790 (blood)	4.8	402000	530	
Mephobarbital							
6 (34)	600 mg	R	1210	2.3	9740	7.5	157
		S	2710	3.5	228680	70	
MDMA ("Ecstasy")							
8 (22–32)	40 mg	R	33700	4		5.8	233
		S	21200	2		3.6	
Methylphenidate							
11 (18–30)	40 mg	d	18.1	2.36	120	5.7	151
		l	2.98	2.14	14.8	3.9	
9 (11)	10 mg	d	6.4	2.3	28	1.9	152
		l	1.3	2.4	4.6	1.4	

Drug (n)	Dose						Ref.
Ofloxacin 5 (33)	200 mg	R	1340	1.8	10620	6.3	6
		S	1350	1.8	11660	6.9	
Omeprazole 5 (27–65)	20 mg EM	(+)	107		244		226
		(−)	179		393		
	20 mg PM	(+)	511		1837		226
		(−)	388		1225		
Oxybutynin 18 (19–45)	5 mg	R	2.2	1.0 (median)	3.8	1.4	217
		S	4.1	0.8	5.5	0.9	
Pantoprazole 14 Japanese males	40 mg EM	(+)	1370	1.9	2550	0.94	225
		(−)	1450	2.0	3030	1.1	
	40 mg PM	(+)	2910	2.4	27300	7.8	225
		(−)	2240	2.4	7540	2.3	
Reboxetine 6 (24)	1 mg	(+)	7.3	2.2	113	11	145
		(−)	17	1.5	263	11	
9 (19–29)	4 mg	(+)	35	2.3	516	15	143
		(−)	77	1.8	1146	14	
6 (24)	3 mg	(+)	35	1.1	529	11	144
		(−)	77	0.94	1288	11	
Rolipram 6 (71)	1 mg (+)	(+)	15.7	0.4	29.2	4.6	146
	1 mg (−)	(−)	16.4	0.5	28.9	9.4	
Terbutaline 6 (22–23)	7.5 mg bid	R	17.8	4.0	135	—	120
		S	9.5	4.0	74.9	—	

(continues)

Table 1 Continued

Drug N (mean age/range)	Dose	Enantiomer	C_{max}, ng/mL	T_{max}, h	AUC, ng h/mL	$t_{1/2\lambda n}$, h	Ref.
Terodiline	177 mg	R	324	3.3		1.4	216
9 (26)		S	260	3.0		1.5	
Thioctic acid	600 mg, fasted	R	2654	1.0	2518	—	177
12 (26)		S	1415	1.0	1361	—	
	600 mg, fed	R	1735	2.5	1926	—	
		S	1003	2.5	1134	—	
Tiaprofenic acid							
4 (28)	300 mg	R	31200	1.5	81200	2.1	260
		S	29600	1.5	80800	2.2	
4 (28)	300 mg sustained-release	R	12800	4	72900		260
		S	13000	4	75500		
Tosufloxacin	204 mg	(+)	400	2.6	2780	3.6	7
6 (24–39)		(−)	440	2.4	2870	3.5	
Tranylcypromine	20 mg	(+)	19	0.8	28.6	0.75	147
6 (18–35)		(−)	49	1.0	130	1.6	
Vigabatrin	1500 mg	R	21762	0.8	6.5×10^5	8.1	169
6 (27)		S	11933	1.0	5.0×10^5	7.5	

6 (20 days)	40 mg/kg	R	34100	2.5	2.3×10^5	7.5	171
		S	14000	2.5	1.4×10^5	7.5	
6 (1)	50 mg/kg	R	20600	2.4	1.0×10^5	2.9	170
		S	13800	2.9	0.91×10^5	5.7	
6 (8.7)	50 mg/kg	R	34700	1.3	1.2×10^5	5.7	170
		S	19400	1.4	0.98×10^5	5.5	
Zileuton							
16 (33)	200 mg	R	0.83	2.30	3.49	1.88	130
		S	0.88	1.40	2.41	1.59	
	600 mg	R	1.80	2.30	10.8	2.49	130
		S	1.78	1.60	6.55	2.66	
Zopiclone							
12 (19–36)	15 mg	(+)	87	1.6	691	6.7	155
		(−)	44	1.5	210	3.8	
Warfarin							
8 (not stated)	15 mg R	R	—	—	0.96×10^5	48	47
	15 mg S	S	—	—	0.80×10^5	38	
	15 mg R + cimetidine	R	—	—	1.26×10^5	58	
	15 mg S + cimetidine	S	—	—	0.81×10^5	38	

EM, extensive metabolizer; PM, poor metabolizer.

Table 2 Stereospecific Pharmacokinetic Data of Chiral Drugs Given by Parenteral Routes (Intravenous Unless Otherwise Indicated)

Drug N (mean age/range)	Dose		$t_{1/2\lambda n}$, h	CL, L/h/kg	Vd, L/kg	Ref.
Albuterol						
7 (24)	1.6 mg	R	2.0	0.62	2.0	112
		S	2.9	0.39	1.8	
7 (23–40)	0.5	R	2.5	0.67	—	114
		S	4.7	0.21	—	
Bupivacaine	63–150 mg	R		0.55	3.6	295
12 surgical	intercostals	S		0.43	2.9	
Clinafloxacin	400 mg	R	5.6	0.25	1.74	8
13 (28)		S	5.7	0.24	1.69	
Cyclophosphamide	2.1 g/m^2	R	5.8	6.9 L/h	0.67	235
9 (13–19)		S	5.7	7.2	0.67	
dOTC	100 mg	(+)	19.2	0.18	1.2	10
12 (30)		(−)	8.9	0.24	0.84	
Ifosfamide						
14 (>18 y cancer)	3 g/m^2	R	7.6	2.5 L/h/m^2	26 L/m^2	238
		S	6.0	3.1 L/h/m^2	27 L/m^2	
12 (60)	3 g/m^2	R	7.1	0.060	0.61	240
		S	6.0	0.072	0.63	
5 (3–15)	2 g/m^2	R	4.5	3.8 L/h/m^2		241
		S	3.2	5.0 L/h/m^2		
Ketamine	2 mg/kg i.v.	R	2.6	0.99	3.0	203
25 (37) surgery patients		S	2.5	1.2	3.3	
Mepivacaine						
10 (20–35)	60 mg	R	1.9	~0.57	~1.5	211
		S	2.1	~0.25	~0.74	
10 (40)	731 mg nerve	R	2.8	0.73	3.1	212
	block	S	3.5	0.42	2.0	
Methadone	10–30 mg	R	38	0.13	6.7	179
8 (61, pain)		S	29	0.10	3.9	
Methylphenidate	10 mg	d	6.0	0.40	2.7	151
11 (18–30)		l	3.6	0.73	1.8	
Ofloxacin	400 mg	R	10	0.16	120	9
15 (30–80)		S	10	0.15	114	
Prilocaine	200 mg	R	1.5	2.0	3.6	207
10 (21–27)		S	2.1	1.5	3.8	

(continues)

Table 2 Continued

Reboxetine	0.3 mg	(+)	9.4	0.071	0.92	145
12 (24)		(−)	11	0.027	0.39	
Rolipram	0.1 mg	(+)	5.7	0.39	0.82	146
6 (71)		(−)	8.4	0.37	0.52	
Terbutaline	0.125	R	15.3	0.13	1.8	117
		S	12.7	0.19	1.9	
Thiopental	250–500 mg	R	9.6	0.25	1.9	198
7 (43–67) surgical		S	9.0	0.19	1.5	

2.3. Antimalarials

The chiral antimalarial drugs include primaquine, mefloquine, halofantrine (Fig. 2), quinacrine, lumefantrine, chloroquine, and hydroxychloroquine. Chloroquine and hydroxychloroquine are also used to treat patients with connective tissue disorders such as rheumatoid arthritis or systemic lupus erythematosis. Each drug in this class possesses an aromatic moiety and an aliphatic or alicyclic side chain containing a chiral carbon atom and an amine functional group. Most studies directed towards the elucidation of antimalarial activity of enantiomers have involved in vitro susceptibility tests of the enantiomers against strains of *P. falciparum*. In general, the antimalarial drugs display little to no stereoselectivity in their in vitro parasiticidal activities [11–14]. On the other hand, in vivo these drugs may exhibit stereoselectivity in their antimalarial effects, owing to stereoselective differences in pharmacokinetics [14]. Interspecies differences in susceptibility of parasites to enantiomers of antimalarial drug is demonstrated by mefloquine, which contains two asymmetric centers but is marketed as a racemate containing only (+)- and (−)-erythro stereoisomers. The individual enantiomers display no stereoselectivity in vitro against a chloroquine resistant strain of *P. falciparum* from Cameroon [12]. On the other hand, the enantiomers display in vitro stereoselectivity against strains of *P. falciparum* from Indochina and Sierra Leone (almost 2 : 1 in favor of the (+) enantiomer) [15]. By comparing stereoselectivity between several antimalarial drugs, Karle et al. have postulated that the more rigid the molecule, the greater the probability that stereoselectivity will be observed in antimalarial activity [15]. Although this observation may be valid, it does not explain the mechanism for in vitro stereoselectivity in antimalarial activity of mefloquine.

The adverse effects of some antimalarials can be stereoselective. This could be due to pharmacokinetic and/or pharmacodynamic differences between the enantiomers. For halofantrine, evidence has arisen that the (+)

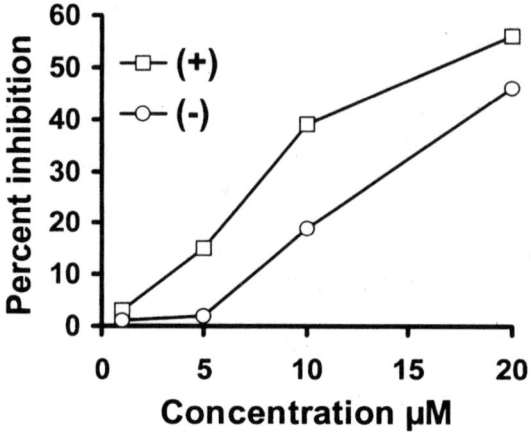

Figure 3 The enantiomers of halofantrine have similar antimalarial activities but may differ in their cardiotoxic effects. The figure illustrates the ability of varying concentrations of halofantrine enantiomers to inhibit the delayed rectifier potassium channel in feline ventricular myocytes. The (+) enantiomer appears to be more potent. (Figure plotted from data presented in Table II of Ref. 16).

enantiomer is more liable than antipode to cause inhibition of ventricular myocyte potassium channels in vitro, which may lead to the prolongation of the electrocardiographic Q-Tc interval and the development of serious cardiac arrhythmias (Fig. 3) [16]. In view of this finding, it may be of importance that after racemic intravenous halofantrine is administered to rats, the cardiac uptakes of halofantrine and its major chiral metabolite, desbutylhalofantrine, have been shown to be stereoselective, with concentrations of (+) enantiomers exceeding those of antipode [17]. On the other hand, in human liver and brain, the (+) enantiomer of chloroquine has 3- to 4-fold greater potency in its ability to inhibit the metabolism of histamine than does antipode [18], suggesting a pharmacodynamic difference between the enantiomers. An increase in tissue levels of histamine was postulated to be a risk factor for development of side effects associated with chloroquine therapy [18].

Stereoselectivity has been reported in the pharmacokinetic properties of some of the chiral antimalarial drugs (chloroquine, hydroxychloroquine, halofantrine, and mefloquine; Tables 1 and 3). Each of these drugs possesses low hepatic extraction ratios, large Vd and $t_{1/2}$ ranging from one day to months. Plasma protein binding has been reported to range from about 50% (chloroquine and hydroxychloroquine, Table 3) to over 99% (halofantrine [19]). Chloroquine and its hydroxylated congener,

Table 3 Unbound Fractions of Enantiomers of Chiral Drugs in Human Plasma for Drugs Discussed in this Chapter[a]

Drug	Unbound fraction (%)	Ratio	Ref.
Acenocoumarol	S, 2.1 R, 2.0	1.05 S:R	67
Aminoglutethimide	S, 26 R, 15	1.7 S:R	242
Bupivacaine	S, 4.5 R, 6.6	1.5 R:S	208
Chloroquine	S, 33.4 R, 57.3	1.7 R:S	23
	S, 33.4 R, 51.5	1.5 R:S	27
Etodolac	S, 0.85 R, 0.47	1.8 S:R	249
Flurbiprofen	S, 0.063 R, 0.050	1.3 S:R	184
Hydroxychloroquine	S, 47 R, 66	1.4 R:S	21
	S, 36 R, 63	1.8 R:S	28
E-10-hydroxynortriptyline	(+), 46 (−), 31	1.5 (+):(−)	148
Ketorolac	S, 0.61 R, 0.45	1.3 S:R	253
Lansoprazole	S, 5.8 R, 2.9	2.0 S:R	227
Mepivacaine	S, 25 R, 36	1.4 R:S	211
Phenprocoumon	S, 0.29 R, 0.56	1.9 R:S	69
Prilocaine	S, 73 R, 70	1.04 S:R	207
Thiopental	S, 15.6 R, 16.5	1.06 R:S	201
	S, 10.0 R, 12.4	1.2 R:S	198
Warfarin	S, 0.93 R, 1.2	1.3 R:S	54

[a]See Table 2 in Chap. 7 for data on cardiovascular drugs and Table 3 in Chap. 5 for some prototype chiral drugs.

hydroxychloroquine, both exhibit higher whole blood and serum concentrations of the R-(−) than of the S-(+) enantiomer [20–26].

Stereoselectivity has been reported in plasma protein binding of chloroquine and hydroxychloroquine (Table 3), with (+) enantiomers of both drugs being more highly bound to plasma proteins than the respective antipode [21,27,28]. A significant negative relationship is present between α_1 acid glycoprotein levels in plasma and unbound fraction of hydroxychloroquine enantiomers [21]. As a result, arthritic patients possess a higher degree of binding to plasma proteins than healthy subjects. This has not been assessed in patients with active malaria.

Whole blood and serum concentrations of chloroquine and hydroxychloroquine are much higher than those in plasma. Normally the binding of drug to blood cells is thought to be a reversible event. For these drugs, however, cellular uptake is thought to be due to sequestration of drug in blood cells by ion-trapping in acidic lysosomal organelles of lymphocytes, monocytes, and platelets, which is not a readily reversible event. Concentrations of these drugs in erythrocytes, which lack these organelles, are similar to those in plasma. Concentrations of enantiomers are higher in serum than plasma owing to platelets being activated during the clotting process, thus causing drug to be released from the cells. In terms of stereoselectivity, the R : S ratio is near unity in plasma and erythrocytes, in contrast to whole blood and serum in which the R : S ratio is close to 2 [21]. The higher R : S ratio in whole blood is likely due to the combination of a higher plasma unbound fraction of R-(−) hydroxychloroquine coupled with its lower plasma CL, which allows more R enantiomer to enter the white cells and platelets than antipode.

Over several months after the administration of a single dose, the S : R ratio of hydroxychloroquine in urine decreases, from an early post-dose value of >1 to much <1 over several months postdose [29,30]. The underlying cause of this reversal in enantiomer ratios was not addressed. Hydroxychloroquine follows a multicompartmental model. The reversal in urinary recovery enantiomer ratio may be due to stereoselectivity in extent of distribution and elimination, combined with redistribution, between the different compartments. What makes the observation unusual is the extended time period over which the change in ratios occurs.

(+)-Halofantrine attains higher plasma concentrations than (−)-halofantrine (Table 1) [31,32] and presumably it possesses a lower CL and Vd in humans, as is seen in rats [17,33,34]. Halofantrine exhibits stereo-selectivity in its binding to plasma lipoproteins, with the (−) enantiomer being present to a greater extent in lipoprotein-deficient plasma fractions than antipode [35]. In contrast, the (+) enantiomer has a greater affinity for lipoprotein-containing fractions than (−)-halofantrine.

Mefloquine is administered as a racemic mixture of the (−)-SR and (+)-RS diastereomers. Mefloquine exhibits much higher plasma and blood concentrations of the (−) enantiomer than the (+) enantiomer [36,37]. In plasma of subjects given 250 mg as single doses, the (−) : (+) ratios being 3 and almost 10 for C_{max} and AUC, respectively [36]. The level of stereo-selectivity was retained with repeated doses [36,38]. In Swedish Caucasian volunteers, poor and extensive metabolizers of debrisoquine apparently have similar pharmacokinetic properties of the enantiomers [38], suggesting that the metabolism of mefloquine is not governed by this pathway to a significant degree.

In healthy Caucasian volunteers given a single 250 mg oral dose of (±)-mefloquine [36], the AUC and C_{max} of mefloquine enantiomers in plasma appear to display a higher degree of stereoselectivity than in whole blood of adult Thai volunteers given single doses of 750 mg racemic meflo-quine (Table 1) [37]. However, based on plasma data, stereoselectivity in mefloquine pharmacokinetics in nine Thai children given 25 mg/kg of racemate were on the same level as those seen in adult Caucasians (Table 1) [39]. Therefore the observed differences in the degree of stereoselectivity between the adult Caucasian and Thai subjects is most likely due to the differences between whole blood and plasma levels of the enantiomers.

With respect to tissue uptake of mefloquine, in a postmortem analysis, concentrations of (+) mefloquine were higher than those of antipode in brain tissue [40]. This is consistent with the higher reported Vd/F of the (+) enantiomer of mefloquine relative to antipode. The uptake of mefloquine enantiomers into erythrocytes infected and uninfected with *P. falciparum* has been studied [41]. Over a wide range of concentrations, uptake of mefloquine enantiomers into uninfected red blood cells increased in a linear, nonstereoselective fashion. However, when incubated in the presence of erythrocytes infected with *P. falciparum*, stereoselectivity was noted in the uptake of the mefloquine enantiomers, with a (+):(−) ratio of 1.4 being observed after 3 h of incubation. It was not stated if the difference between enantiomers was significant in the infected vs. uninfected cells [41].

Some chiral metabolites of antimalarial drugs possess significant levels of antimalarial activity in comparison to parent drug. For example, the main circulating metabolite of halofantrine, (±)-desbutylhalofantrine, possesses an in vitro level of antimalarial activity similar to that of parent drug [42]. The enantiomers of this metabolite also share similar antimalarial activities in vitro. Stereoselectivity in pharmacokinetic properties of these chiral metabolites could influence an assessment of antimalarial activities. On the other hand, neither of the enantiomers of the major 4-carboxylic acid metabolite of mefloquine possesses antimalarial activity [43].

Stereoselective pharmacokinetic data on other antimalarial agents such as primaquine, quinacrine, and lumefantrine are not available for humans.

2.4. Antihelminthic Drugs

Albendazole is a broad-spectrum antihelminthic drug used to treat the tropical parasitic disease, human neurocysticercosis. Albendazole contains a sulfur group that is subject to oxidation to an active sulfoxide metabolite. The S-oxide metabolite forms a center of asymmetry at the sulfur atom, thus conferring chirality to the molecule. Albendazole sulfoxide steady-state pharmacokinetics have been studied in 18 patients with confirmed active neurocysticercosis after repeated dose administration of 5 mg/kg thrice daily for 8 days [44]. A high degree of stereoselectivity was observed in the plasma concentrations of the sulfoxide metabolite, with C_{max} and AUCss (+):(−) ratios of 5.5 and 9.2, respectively. It is not known if there is stereoselectivity in the antihelminthic activity of the (+)- and (−)-sulfoxide metabolites of albendazole.

2.5. Antileprosy Agents

The fetal malformations induced by thalidomide, and its historical significance in ensuring that the safety of drugs is well established during the drug development process, are well known. Originally developed for use as a sedative, thalidomide is a chiral drug that has made a resurrection in clinical use owing to its beneficial properties in leprosy, inflammatory conditions of the skin, and facilitation of immunosuppression. Thalidomide enantiomers undergo spontaneous bidirectional racemization in the presence of human serum albumin in plasma [45,46]. Compared to phosphate buffer, the presence of human serum albumin in the buffer (40 g/L) caused a 10-fold increase in the bidirectional rates of conversion of one enantiomer to the other [46]. In plasma, the rates of conversion were ~ 40% of that in solutions of human serum albumin, prompting the authors to hypothesize that endogenous components in plasma could competitively inhibit the albumin-induced conversion of enantiomers. Indeed, when capric acid, acetylsalicylic acid, or physostigmine were added to albumin solutions, the rates of interconversion of enantiomers declined substantially to levels equal to or lower than those observed in human plasma [46].

In humans given racemate or individual enantiomers of thalidomide by the oral route, stereoselective differences were noted in the plasma concentrations of the enantiomers [45]. In cases where the enantiomers were given alone, presence of antipode was detected in each group. After R(+) thalidomide or S(−) thalidomide alone, the R : S AUC ratios were 3.0 and 0.74, respectively. After racemate, the R : S ratio was 1.6. It was concluded that in addition to the interconversion of enantiomers, which was equal in both directions, the S enantiomer is preferentially metabolized in vivo [45].

3.   ORAL ANTICOAGULANTS

Several of the orally administered anticoagulants, including warfarin, phenprocoumon, and acenocoumarol are chiral and marketed as racemates. The coumarin derivatives inhibit the activity of vitamin K_1, an essential component in the carboxylation of glutamic acid residues in coagulation factors II, VII, X, anticoagulant protein C, and anticoagulant protein S. Specifically, the coumarin anticoagulants inhibit vitamin K_1 2,3-epoxide reductase, an enzyme necessary for the conversion of the inactive 2,3 epoxide species back to vitamin K_1 [47]. The pharmacological potencies of the enantiomers of oral anticoagulants may significantly differ (Fig. 4). A lower dose of S- than R-warfarin is needed to elicit an equivalent anticoagulant effect in Wistar rats [48]. A similar level of stereoselectivity

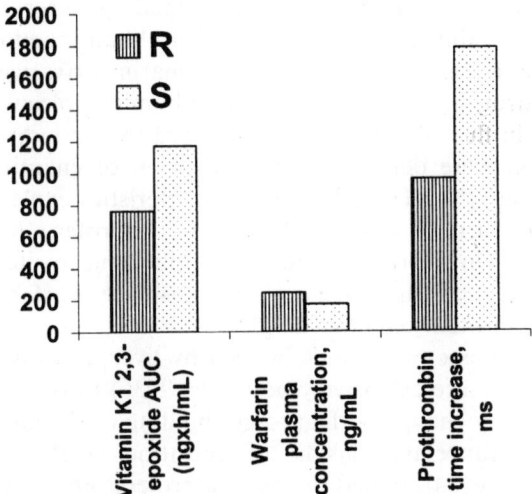

Figure 4 Pharmacokinetic and pharmacodynamic relationship of each warfarin enantiomer when 1 mg single warfarin enantiomer was administered to steady-state in five volunteers, followed by a short infusion of vitamin K1. It is apparent that the S enantiomer is more efficient in preventing conversion of the inactive vitamin K1 epoxide to active vitamin K1 and increasing prothrombin time. (Plot constructed from data presented in Ref. 290.)

in potency has been observed in humans [49]. Similarly, the anticoagulant potency of S-phenprocoumon is 4 to 5 times that for the antipode in rats [50].

Because each of the coumarin-type anticoagulants possesses a narrow therapeutic index, unexpected changes in unbound plasma concentrations of eutomer could result in a considerable risk of blood clotting or hemorrhage with potentially fatal consequences. Therefore, given stereoselectivity in pharmacodynamics, the importance of stereoselectivity in the disposition kinetics of the anticoagulants must be afforded special consideration.

3.1.   Stereoselective Pharmacokinetics/Dynamics of Oral Anticoagulants

Warfarin

Warfarin is specifically bound to drug binding site I (warfarin binding site) on human serum albumin [51]. Sudlow et al. [52] evaluated R,S-warfarin,

R- and S-warfarin, phenprocoumon, and acenocoumarin binding to serum albumin binding site I using 5-dimethylaminonaphthalene-1-sulfonamide (DNSA) as a site I specific fluorescent probe. All these coumarin derivatives displaced DNSA from the drug binding site I on the albumin molecule. Furthermore, no differences in fluorescence changes induced by the individual enantiomers of warfarin were detected, suggesting lack of enantioselectivity in binding to this site [52]. These binding characteristics explain the lack of displacement of warfarin by some NSAIDs (e.g., ibuprofen) that bind to site II on the albumin. On the other hand, drugs that bind to site I (phenylbutazone, tolbutamide, and sulfonamides) can displace warfarin enantiomers from the plasma protein binding site [52,53].

Warfarin enantiomers are extensively metabolized by liver, possess a low hepatic extraction ratio, and are extensively bound (> 99%) to plasma proteins (Table 3). Therefore any change in the protein binding of warfarin enantiomers may alter the clearance and plasma concentrations of R- and S-warfarin [54]. Yacobi and Levy [54] studied the plasma protein binding of racemic and individual enantiomers of warfarin in human blood. The free fraction of R-warfarin was significantly (32%) larger than that of S-warfarin (Table 3). The authors concluded that the difference in the potency of warfarin enantiomers could not be solely explained by the observed differences in the protein binding of the individual enantiomers but rather by the intrinsic ability of R- and S-warfarin for interactions with extravascular receptors.

The method by which the plasma protein binding of warfarin is determined is important in assessing stereoselectivity. For example, the free fractions of warfarin enantiomers are temperature dependent [55,56]. Using equilibrium dialysis, a similar degree of protein binding was determined for R- and S-warfarin at 27°C [57]. In contrast, protein binding of S-warfarin was significantly higher than that of antipode when the temperature was 37°C [58].

Tissue distribution studies in rats have indicated that the tissue : serum ratios of S-warfarin are concentration dependent. The liver : serum ratio was >10 at low concentrations and reached a limiting value of 0.25 at higher serum concentrations. The tissue : serum vs. concentration profile of S-warfarin indicated the presence of two tissue binding sites, one possessing high affinity and low capacity, the other low affinity and very high capacity [59]. The ramifications of nonlinear tissue binding may not be clinically relevant, as the higher dose (1 mg/kg) administered to the rats was much higher than the maximum dose typically used in patients (0.2 mg/kg). However, it is conceivable that in some patients some degree of nonlinear tissue binding of S-warfarin could confound the interpretation of the (±)-warfarin serum concentration vs. effect relationship.

In general, despite lower free fraction, S-warfarin is more rapidly eliminated than its antipode in humans, and hepatic metabolism accounts for almost all of the elimination of the drug [60]. Warfarin is metabolized to 6-hydroxywarfarin, 7-hydroxywarfarin, and two reduced derivatives, the diastereomers of warfarin alcohol [60]. The metabolism of R-warfarin involves oxidation to 6-hydroxywarfarin and subsequent reduction to (R,S)-warfarin alcohol. In contrast, S-warfarin is oxidized to 7-hydroxywarfarin and reduced to (S,S)-warfarin alcohol. The metabolic profiles of warfarin enantiomers in blood and urine were similar [53,60]. The ring-hydroxylated metabolites are inactive, and the activity of warfarin alcohols are substantially less than the parent molecule [53,60]. S-warfarin may also be metabolized to 6-hydroxywarfarin [60].

Human microsomes exhibit approximately 8 times greater 7-hydroxylation activity for the S enantiomer than that for antipode [61]. 7-Hydroxylation of (R,S)-warfarin by microsomes prepared from 35 human livers correlates more closely with that of S- ($r^2 = 0.95$) than with R-warfarin ($r^2 = 0.69$). A higher correlation ($r^2 = 0.73$) between 7-hydroxylation of R-warfarin and O-deethylation of 7-ethoxyresorufin indicated that R- and S-warfarin 7-hydroxylation reactions are catalyzed by separate CYP450 isoenzymes. Indeed, anti-CYP2C9 antibodies completely inhibited the 7-hydroxylation of S-warfarin. On the other hand, the 7-hydroxylation of R-warfarin was inhibited by about 70% in the presence of anti-CYP1A2 antibodies. Therefore it appears that the 7-hydroxylation of S-warfarin is mainly facilitated by CYP2C9, whereas R-warfarin is catalyzed by more than one CYP450 isoenzyme, including CYP1A2 [61].

Interestingly, microsomal 7-hydroxylation of racemic warfarin appears to be lower than that of S-warfarin, indicating the possibility of a metabolic interaction between warfarin enantiomers [61]. These in vitro studies showed that R-warfarin affected the catalytic activity of CYP2C9 by a noncompetitive mechanism [61]. However, in an earlier in vivo study a lack of enantiomeric interaction between R- and S-warfarin was suggested after single 1.5 mg/kg doses of the individual enantiomers and racemic warfarin [62]. The enantiospecific results obtained were successfully used to predict the pharmacokinetics and pharmacodynamics of racemic warfarin. The apparently contradictory findings between the two studies may be due to a number of experimental factors. One study used a supraclinical dose of warfarin in vivo, whereas the other involved human microsomes in vitro, the relevant hepatic concentrations of which are difficult to determine.

A number of drugs that act as substrates of CYP1A and CYP2C may interact with the metabolism of warfarin enantiomers. For example, sulphaphenazole and tolbutamide are competitive inhibitors of S-warfarin hydroxylation with Ki values of 100 and 0.5 μM, respectively [62].

Mutations in the CYP2C9 gene produces the expression of three allelic variants, CYP2C9*1, CYP2C9*2, and CYP2C9*3. The in vitro catalytic activities CYP2C9*2 and CYP2C9*3 are altered from that of the wild type enzyme. Steward et al. [63] genotyped a patient who had unusual sensitivity to warfarin therapy and could not tolerate more than 0.5 mg/day of the drug. Genotyping indicated that the patient was homozygous for CYP2C9*3. In addition, the plasma S:R ratio was about 0.5 in the control patient receiving 4–8 mg/day of warfarin, whereas the individual on very low warfarin dose had a plasma S:R ratio of 3.9. In contrast, the urinary 7-hydroxywarfrin S:R ratios were similar between the control and warfarin-sensitive patients. It was suggested that the expression of CYP2C9*3 was associated with diminished clearance of S-warfarin and a dangerously amplified therapeutic response to normal doses of the racemic warfarin [63].

The effect of genetic polymorphism of CYP2C9 or 2C19 on metabolism of warfarin enantiomers was examined in vivo in 86 Japanese patients with heart disease [64]. In patients with a mutation in the CYP2C9 gene, the metabolic clearance of S-warfarin via 7-hydroxylation was lower than with patients homozygous for wild type enzyme. In contrast, R-warfarin metabolic clearances did not differ between groups. Additionally, the metabolic clearances of both R- and S-warfarin were similar in poor and rapid metabolizers of CYP2C19, indicating that CYP2C19 does not play a significant role in the metabolism of warfarin enantiomers [64].

Stereoselective metabolism of warfarin is reflected in the plasma concentration vs. time profiles of warfarin enantiomers. Following administration of a single oral dose of 0.5 mg/kg of R- and S-warfarin on separate occasions to 7 normal healthy subjects, the $t_{1/2}$ of R-warfarin ranged from 19.9 to 69.8 hours [49], which was significantly longer than that of S-warfarin ($t_{1/2}$: 18 to 34.1 hours). After multiple oral doses, the $t_{1/2}$ of enantiomers were longer than those estimated after single doses. However, the volumes of distribution of enantiomers were similar after both single and multiple doses. The oral clearance of R-warfarin was significantly less than that of S-warfarin, resulting in higher plasma concentrations for R-warfarin. The ratio of R:S plasma concentration was on average 3.8 in four patients, due to stereoselective metabolism of warfarin enantiomers [49].

McAleer et al. [65] examined steady-state pharmacokinetics of racemic; R-, and S-warfarin in 10 anticoagulated patients receiving 2 to 8 mg/day doses of racemic warfarin. Mean steady-state oral clearances in these patients for racemic, R-, and S-warfarin were 0.040, 0.038, and 0.047 mL/min/kg, respectively. Chan et al. [66] investigated disposition of warfarin enantiomers and its metabolite in 36 patients receiving daily doses of racemic warfarin (on average 6.1 mg with the range of 2.5–12 mg) titrated to approximately the same anticoagulant response. The plasma concentrations of warfarin enantiomers and their metabolites exhibited considerable variability. Mean plasma concentrations of S-warfarin were lower than antipode. The overall oral clearance of S-warfarin was greater than that of R-warfarin with an approximately 27 times greater formation clearance for S-7-hydroxywarfarin (3.48 L/day) as compared to that of R-7-hydroxywarfarin (0.13 L/day) [66]. In this study, S : R enantiomeric ratios of (bound+unbound) warfarin plasma concentration ranged from 0.21 to 1.28. Although the unbound fractions of enantiomers in plasma were similar (0.5%), the clearance of free S-warfarin was approximately twice that of R-warfarin.

Acenocoumarol

After administration of 25 mg of each enantiomer and racemic acenocoumarol, S-acenocoumarol systemic clearance exceeds that of R-acenocoumarol by about tenfold [67]. In contrast, the steady-state volume of distribution of S enantiomer was only 1.5 to 2 times greater than that of R enantiomer. However, no difference is observed in the protein binding of acenocoumarol enantiomers, which was measured at concentrations higher than those observed in vivo (Table 3). It was concluded that the greater anticoagulant potency of R-acenocoumarol than that of its antipode may be attributed to stereoselectivity in metabolic clearance [67]. The observed greater clearance of S-acenocoumarol and hence lower plasma concentrations than antipode was consistent with the higher potency of the R enantiomer in vivo.

The 4-hydroxycoumarins undergo biotransformation similar to warfarin, forming hydroxylated metabolites at the 6 and 7 positions. These hydroxylated metabolites account for 63 to 99% of the metabolic clearance of 4-hydroxycoumarins, including acenocoumarol. The metabolic clearances of R- and S-acenocoumarol are 6 and 66 times higher than those of R- and S-warfarin, respectively [68]. The metabolism of acenocoumarol is also stereoselective in favor of the S enantiomer. Sulphaphenazole competitively inhibits the 7-hydroxylation of R- and S-acenocoumarol and the 6-hydroxylation of S-acenocoumarol. Omeprazole acts as a partial inhibitor of 6- and 7-hydroxylation of acenocoumarol enantiomers [68].

Phenprocoumon

Jahnchen et al. [69] investigated pharmacodynamics and pharmacokinetics of S- and R-phenprocoumon after a single oral dose of 0.6 mg/kg of racemic, S-, or R-phenprocoumon administered to 5 normal subjects. Based on the total area under the anticoagulant concentration vs. time curve, the S enantiomer was 1.6 to 2.6 times more potent than the

R enantiomer. The plasma protein binding of S-phenprocoumon is greater than that of antipode in human serum (Table 3) [69]. The protein binding of (R,S)-phenprocoumon is between that of the individual enantiomers. The difference in plasma protein binding of phenprocoumon enantiomers contributes to the lower apparent volume of distribution and clearance for S-phenprocoumon than those of its antipode [69].

3.2. Drug Interactions Affecting Pharmacokinetics/ Dynamics of Coumarin Anticoagulants

Interactions between the enantiomers of coumarin derivatives with commonly used drugs are presented in Table 4 [60,70–80]. A number of interactions have been reported at the level of displacement of enantiomers from plasma proteins and enzyme induction and inhibition. Although the addition of displacing drugs might increase the unbound fractions of the enantiomers of coumarins, only a transient increase in pharmacodynamic response is expected because of the restricted elimination of these drugs. Of a more serious nature, from the perspective of the clinician, are those interactions that influence the clearance of the unbound· anticoagulant enantiomer (Table 4).

Phenylbutazone (Table 4) provides an interesting example of a drug interaction that influences both unbound fraction and intrinsic clearance of unbound warfarin enantiomer in a stereoselective manner [60]. After multiple doses of phenylbutazone (100 mg tid) were administered to two subjects, the clearance of S-warfarin was reduced by half, whereas the clearance of R-warfarin was almost doubled. However, the plasma clearance of racemic warfarin was not significantly affected owing to the opposite direction of metabolic interaction for the individual enantiomers. Beyond this interaction, concomitant use of potent nonsteroidal anti-inflammatory agents such as phenylbutazone or piroxicam [79] with warfarin should be avoided owing to the ability of the combination to induce serious bleeding within the gastrointestinal tract.

Fluconazole provides an example of a drug that influences the intrinsic clearance of unbound enantiomer [70]. Following the administration of 400 mg/day of fluconazole to 6 healthy subjects, the CYP450-dependent metabolic clearances of both warfarin enantiomers were significantly reduced (Table 4) [70]. CYP2C9-catalyzed 6- and 7-hydroxylation of S-warfarin was inhibited by approximately 70%. This resulted in the prolongation of anticoagulant activity. In addition, CYP3A4-dependent 10-hydroxylation of R-warfarin was inhibited by 45%. Fluconazole also inhibited 6-, 7-, and 8-hydroxylation of R enantiomer by 61, 73, and 88%.

Table 4 In Vivo Drug Interactions with Enantiomers of Oral Anticoagulants

Coumarin analog	Interacting drug	Change in plasma concentration	Pharmacokinetic mechanism of interaction	Change in pharmacodynamics	Ref.
Warfarin	Cimetidine	R ↑ S ↔	Inhibition of hepatic CYP450 isoenzymes	Unclear	47
Warfarin	Phenylbutazone	R ↓ S ↑	↑ free fraction, ↑ clearance of R and ↓ clearance of S	↑	60
Warfarin	Fluconazole	R ↑ S ↑	Inhibition of metabolic clearances of R and S	↑	70
Warfarin	Benzbromarone	R ↔ S ↑	Inhibition of 7-hydroxylation of S	↑	71
Warfarin	Levofloxacin	R ↔ S ↔		↔	72
Warfarin	Ticlopidine	R ↑ S ↔	Possible inhibition of hepatic CYP450 isoenzymes	↔	73
Warfarin	Meloxicam	R ↔ S ↔	—	↔	74
Warfarin	Acetaminophen	R ↑ S ↑	Unknown	↔	75
Warfarin	Dicloxacillin	R ↓ S ↓	Unknown	→	76
Warfarin	Zileuton	R ↑ S ↔	Possibly altered metabolic clearance of R	↑	77
Warfarin	Terodiline	R ↔ S ↔	—	↔	78
Warfarin	Bucolome	R ↑ S ↔	Possible inhibition of CYP2C9	↑	80
Acenocoumarol	Piroxicam	R ↑ S ↔	Reduced metabolic clearance of R enantiomer	↑	79

Beyond those in vivo examples depicted in Table 4, there are some in vitro data suggesting stereoselective drug interactions with oral anticoagulants. For example, Hermans and Thijssen [68] investigated the potential of metabolic interactions in vitro between warfarin or acenocoumarol and cimetidine, propafenone, sulphaphenazole, or omeprazole using human liver microsomes. Sulphaphenazole competitively inhibited the 7- and in some experiments the 6-hydroxylation of S-warfarin and R- and S-acenocoumarol. Omeprazole partly inhibited the 6- and 7-hydroxylation of R-warfarin, R-acenocoumarol, and S-acenocoumarol [68].

4. DRUGS USED IN ASTHMA

The chiral β-agonists used in the treatment of asthma include albuterol (salbutamol), isoproterenol (isoprenaline), metaproterenol (orciprenaline), clenbuterol, formoterol, fenoterol, and terbutaline. Other chiral antiasthmatic drugs include the 5-lipooxygenase inhibitor zileuton, and the anticholinergic agent ipratropium bromide. Montelukast, a selective antagonist of cysteinyl leukotriene receptors, is marketed as a single R enantiomer [81]. For some of these drugs (e.g., albuterol and terbutaline), stereoselectivity in pharmacokinetic and pharmacodynamic properties has been extensively evaluated. For other drugs within and outside of the class of β-agonists, however, stereoselective information regarding pharmacodynamics and pharmacokinetics is scarce.

4.1. β2-Agonists

Pharmacodynamics

Stereoselectivity in the pharmacodynamics of β2-agonists has been extensively studied at both the receptor and the end-clinical response levels [82–95]. Except for trimethoquinol, the bronchodilator action of all β2-agonists is predominantly due to the R enantiomer. Using excised tissues from various animals and humans, it has been shown that β2-adrenoceptor agonist activity resides mainly with the R or (R,R) isomers of racemic albuterol, terbutaline, formoterol, and clenbuterol (Table 5) [82–95].

In addition to these in vitro studies, several clinical studies have clearly demonstrated that the bronchoprotective effect of albuterol is primarily due to the R enantiomer [96–98]. Ramsay et al. [97] investigated the protective effects of R- and S-albuterol and racemic albuterol in airway hyperresponsiveness induced by methacholine and adenosine monophosphate (AMP) in patients with bronchial asthma. R- and (R+S)-albuterol

Table 5 Stereoselective Pharmacodynamic Properties of Chiral Beta-Agonist Drugs

Drug	Experimental model	Pharmacodynamic response	Direction of stereoselectivity (eudismic ratio)	Ref.
Albuterol	Dog	Reduction of pulmonary resistance	R > S (50)	85
	Guinea pig	Inhibition of bronchospasm	R > S (40)	85
	Guinea pig trachea	Induced tone reduction	R > S (80)	86
	Guinea pig trachea	Trachea relaxation	R > R,S	87
Terbutaline	Guinea pig trachea	Trachea relaxation	R > S	88
Trimethoquinol	Guinea pig trachea	Bronchodilation	S > R,S (2)	89
	Guinea pig	Bronchodilation	S > R	90
	Human platelets	Antiaggregation	R > S	90
Clenbuterol	Mouse	Antidepressant activity	R > S	91
	Guinea pig trachea	Relaxation of trachea	R > S (1000)	92
	Guinea pig trachea	Blocking isoprenaline effect	R > S (100)	92
Formoterol	Guinea pig trachea and rabbit	Mucociliary activity	R > S	94

were equipotent in the bronchoprotective response 30 min after exposure to irritant [97,98]. After administration of 100 μg of R- or S-albuterol or 200 μg of the racemate using nebuliser and breath-activated dosimeter, the concentrations of AMP required to cause a 20% reduction in pulmonary function measured as forced expiratory volume in 1 second (FEV_1) increased by approximately 10.3, 1.1, and 9.3 times, respectively [97]. Similarly, the administration of R-albuterol, S-albuterol, and racemate increased the methacholine concentrations needed to produce a 20% reduction in pulmonary function by 6.7, 1.1, and 7.3 times, respectively. In addition, the R enantiomer and racemate increased FEV_1 on average by 12.4% and 12%, respectively. In contrast, S-albuterol did not significantly affect FEV_1 values [97]. Collectively, these studies indicate that the bronchodilator activity of racemic albuterol resides mainly in the R enantiomer.

After administration of R- and (R,S)-albuterol, linear relationships have been observed between plasma concentrations and pharmacodynamic responses for heart rate, QTc interval, T-wave amplitude, plasma glucose, and plasma potassium [96]. S-Albuterol resulted in a small but significant increase in the plasma potassium concentration. However, it did not change the other measured pharmacodynamic parameters. The small increase in plasma potassium after the administration of S-albuterol was attributed to the presence of R-albuterol formed by either chiral inversion in plasma, normal diurnal variation, or the possible direct effect of the distomer [96].

Tachyphylaxis is known to develop with chronic administration of sympathomimetic agents. For example, the regular use of racemic adrenaline was associated with unresponsiveness characterized by the lack of symptomatic relief from acute asthma. This unusual unresponsiveness to the bronchodilator effect of adrenaline has been referred to as "adrenaline fast" or as "locked lung syndrome" [99]. The bronchodilator potency of the next-generation sympathomimetic agent, racemic isoproterenol, also diminished upon regular use. In addition, it has been shown that with regular use of isoproterenol and other β2-agonists, a loss of protection from spasmogen exposure could occur. This paradoxical response to β2-agonists puzzled the medical community [99–104] and was largely blamed for the two asthma epidemics which occurred in the last 40 years [105].

It has been generally assumed that the S-enantiomers of sympathomimetics are practically inert and therefore lack desired or deleterious effects. However, the current data clearly indicate that the distomers do possess some activities, which may oppose those of the respective eutomers. For instance, S-isoprenaline (distomer) administration to asthmatic patients ($n = 10$) caused a substantial decrease in FEV_1 in 2 patients and increased reactivity to histamine 7h after inhalation [103]. Similarly, S-albuterol causes activation of human eosinophils, suggesting proinflammatory properties for this enantiomer. Both R-albuterol and racemic albuterol inhibited the IL-5 induced superoxide generation and eosinophil peroxidase release. In contrast, S-albuterol significantly enhanced the superoxide and peroxidase release by eosinophils [102,106].

S-Albuterol increases intracellular Ca^{2+} concentrations in tracheal smooth muscle cells, while its antipode causes a decrease. This effect of S-albuterol may cause worsening of the spasmogen-induced bronchoconstriction, or mask the desired effects of the active R enantiomer [102,106]. The effect of S-albuterol on intracellular calcium is not mediated by β2 receptor activation because this effect was abolished by atropine or 4-diphenylacetoxy-N-methylpiperidine, suggesting the possible interaction of S-albuterol with muscarinic receptors [102,107].

Following levalbuterol (R-albuterol) or (R,S)-albuterol as nebulized doses 3 times daily for 4 weeks to 328 asthmatic patients [108], FEV_1 was determined after the first dose, and at week 5 (1 week after the cessation of the double blinded part of the study). After the first dose, the improvements in FEV_1 for the combined levalbuterol groups (0.625 mg and 1.25 mg dose levels) were statistically greater than that observed for the combined racemic albuterol group (1.25 mg and 2.5 mg dose levels). However, the improvement in FEV_1 following multiple doses of levalbuterol and racemic albuterol did not reach statistical significance after 4 weeks of treatment. The differences in the overall improvement in pulmonary function were also not statistically significant between the individual doses of levalbuterol and racemic albuterol [108].

In 1999, the R-enantiomer of albuterol, levalbuterol, received approval from the US Food and Drug Administration for the treatment or prevention of bronchospasm. A number of recently published clinical studies have demonstrated the effectiveness of inhaled levalbuterol solution, delivered using a nebulizer, in the treatment of patients with mild to moderate asthma [109–111]. The dose of levalbuterol required to produce a significant bronchodilatory response is equivalent to 50% of the dose of the racemate. Consequently, the effect of levalbuterol on heart rate and potassium levels is substantially lower than that observed after the administration of therapeutic doses of the racemate, presumably owing to the absence of the S enantiomer [109].

Long acting β2-agonists are commonly recommended for patients who continue to experience bronchospasm despite corticosteroid therapy. Advantages include a rapid onset of action, sustained control of exercise-induced asthma, and nighttime control of asthma symptoms. (R,R)-Formoterol, the active isomer of racemic formoterol, is currently in Phase IIB clinical development in the United State as a long-acting anti-asthmatic agent. Results obtained in the early Phase IIA studies suggest that (R,R)-formoterol produces an immediate response that lasts for 24 hours. The clinical use of the active isomer of formoterol may provide a rapid relief of bronchospasm with the advantage of once daily dosing without tachyphylaxis [111].

Pharmacokinetics

The major pharmacokinetic parameters of albuterol and terbutaline are listed in Tables 1 and 2. Irrespective of the route of administration, the pharmacokinetics of racemic albuterol are stereoselective with a faster disappearance of the active R enantiomer from plasma (Tables 1 and 2). This is due to stereoselective metabolism and renal clearance of albuterol in favor of the R enantiomer. Both the total body and renal clearances of R-albuterol are approximately 2 to 3 times higher than those of the distomer, resulting in a $\geq$3-fold higher maximum plasma concentrations of

S enantiomer (Table 1) [112–114]. Furthermore, the degree of stereo-selectivity was confirmed to be highly dependent on the route of administration with the lowest concentrations of (R)-albuterol after oral administration (Table 1) [112,114].

For the majority of drugs, absorption from the gastrointestinal tract occurs via passive processes that cannot distinguish between the enantiomers of a racemate. Additionally, there is no evidence that any of the antiasthma drugs are absorbed by a carrier mediated process; thus stereoselectivity in the absorption of antiasthma agents currently used in clinical practice is not expected [115].

The β2-agonists may be administered by the oral, inhalation, subcutaneous, intramuscular, or intravenous routes, although inhalation is by far the most common. Although absorption of β2-agonists from biological membranes appears to be nonstereoselective, these drugs may be subject to extensive stereoselective first-pass metabolism in the gastrointestinal tract and in the liver [116]. This can result in a difference in the proportion of the active enantiomer entering the systemic circulation after oral compared to after parenteral routes of administration. For example, after albuterol orally (Table 1), stereoselective first-pass metabolism in favor of the active R enantiomer results in lower R : S AUC ratios than ratios observed after the IV injection (Table 2) [112]. Hence a given total drug concentration after oral administration would be expected to yield a less marked pharmacodynamic response than that of an equivalent total concentration after IV administration.

Ward et al. [114] investigated enantiomeric kinetics of albuterol following inhaled, IV, and oral doses of racemic albuterol. The inhaled dose was also studied after the coadministration of an oral dose of activated charcoal. Systemic exposure of S-albuterol was consistently higher than that of antipode after all three routes of administration. Furthermore, the R : S AUC ratio was significantly higher after IV dose (0.32) than that after oral administration (0.05). The R : S AUC ratio obtained after the inhaled dose of albuterol (0.13) was in between the values obtained for the oral and IV administrations. This is due to the low lung bioavailability of albuterol enantiomers coupled with a stereoselective metabolism of the partially ingested actuated dose in the gastrointestinal tract. Indeed, concomitant administration of charcoal, which impedes oral absorption of the ingested dose, significantly increased the R : S AUC ratios (0.29 vs. 0.13) after inhalation [114]. Therefore caution must be exercised in interpreting the bioavailability and/or bioequivalence data obtained for the β2-agonists based on the concentrations of total (R+S)-drug. It cannot be assumed that the direction and degree of stereoselectivity is the same for all structurally related compounds. For example, in contrast to albuterol, the oral bioavailability of the R enantiomer of terbutaline (14.8%) is higher than that of its antipode (7.5%) [117,118]. This is due to a more extensive sulfation of S-terbutaline presystemically as compared to antipode, resulting in a higher plasma concentration of eutomer (Tables 1 and 2) [117–120].

In general, β2-agonists are hydrophilic compounds with a low degree of binding to plasma proteins (< 5 to 55%). One exception is provided by the lipid-soluble β2-agonist clenbuterol, which is 97% bound to plasma proteins [121]. For albuterol, plasma protein binding is nonstereoselective and low, ranging from 8% to 55%, depending on the method used. After intravenous administration to humans, no statistical differences were apparent in the steady-state volume of distribution of albuterol enantiomers [113]. This suggests that perhaps neither plasma nor tissue protein binding of albuterol is stereoselective [112,114,122]. Similarly, the volume of distribution of terbutaline is nonstereoselective, being 1.90 and 1.76 L/kg for the S and R enantiomers, respectively [117].

The β2-agonists are eliminated by metabolism and renal excretion. However, the relative importance of these pathways is highly dependent on the route of administration. Following parenteral administration, renal clearance is the predominant pathway of elimination. In contrast, metabolism becomes more important after their oral administration. Drastically lower intact drug concentrations in plasma and urine after the oral administration of these antiasthma agents reflect an extensive first-pass metabolism in the gastrointestinal tract [121].

Owing to their hydrophilicity and presence of functional groups amenable to conjugation, β2-agonists are mainly subject to phase II metabolism. Some β2-agonists (e.g., albuterol) are exclusively biotransformed by sulfation, whereas others such as fenoterol form both sulfate and glucuronide conjugates. Conversely, β2-agonists such as formoterol are entirely glucuronidated. In contrast to these hydrophilic β2-agonists, more lipophilic long-acting β2-agonists (e.g., salmeterol) are generally metabolized by phase I oxidation [121,123,124].

Sulfation of enantiomers of albuterol and other β2-agonists is catalyzed by a monoamine form of phenol sulphotransferase. In general, the inactive S enantiomers of β2-agonists exhibit significantly greater affinity to phenol sulphotransferase than the eutomer. However, two exceptions are albuterol and salmeterol, which have greater affinity of their active R enantiomers towards this enzyme [113]. Formoterol, a chiral β2-agonist with two asymmetric carbon atoms, is eliminated by glucuronidation, which is stereoselective and highly variable in favor of the (S,S)-isomer [125,126]. Glucuronidation of formoterol enantiomers exhibits a large degree of variability among microsomes prepared from liver tissues obtained from different individuals. For example, the (S,S)/(R,R) ratio for Vm/km (intrinsic clearance) varied more than tenfold [127]. Therefore the authors suggested that the variable stereoselectivity in the glucuronidation of formoterol may lead to variability in the in vivo ratio of active:inactive plasma concentrations [127]. This may partially explain the interindividual variability in the duration of bronchodilation produced after the inhaled doses of this long-acting β_2-agonists.

Enantioselectivity in the renal clearance of chiral β2-agonists has been reported for albuterol and terbutaline [112,114,117]. After the IV administration of albuterol enantiomers and racemic albuterol, the renal clearance of the individual enantiomers and racemic drug exceeded creatinine clearance, thus implicating active tubular secretion in the urinary elimination of the drug. Renal clearance of R-albuterol was reported to be two- or threefold higher than that of S-albuterol [114]. Since plasma protein binding of albuterol is low [112], the differences in the renal clearance of albuterol enantiomers may be attributed to stereoselectivity in active tubular secretion.

As with albuterol, terbutaline renal clearance also exceeds creatinine clearance, indicating the involvement of active tubular secretion in the renal excretion of the drug. However, the direction of stereoselectivity in the renal clearance of terbutaline is opposite to that of albuterol. After IV doses of racemic terbutaline or the individual enantiomers, renal clearance of the S-enantiomer is approximately twice that of its antipode [117].

Assuming complete elimination through hepatic metabolism and renal pathways, the estimated hepatic clearances of R- and S-albuterol after IV injections are 0.40 and 0.036 L/min, respectively [114]. Assuming hepatic plasma flow of 0.8 L/min, these values correspond to hepatic extraction ratios of 0.50 and 0.045 for R- and S-albuterol, respectively. Therefore oral bioavailabilty based on hepatic first-pass metabolism should be 50% and 96% for the active and inactive enantiomers of albuterol, respectively. However, the observed oral bioavailabilities of the R and S enantiomers of albuterol are only 9.4% and 68.7%, indicating a substantial first-pass metabolism in the gastrointestinal tract. This can be attributed to approximately 80% and 30% sulfation of R- and S-albuterol, respectively, in the intestinal wall [112,114].

In contrast to albuterol, plasma concentrations of R-terbutaline are significantly higher than those of its inactive S enantiomer (Table 1). Borgstrom et al. [117] investigated the plasma pharmacokinetics of terbutaline enantiomers after single IV and oral doses of single enantiomers or the racemate. The IV doses were 0.125 mg of S or R enantiomer or 0.5 mg of the racemate, and oral doses were 5 mg of R- or S-terbutaline or the racemate. Mean steady-state volumes of distribution of enantiomers were similar (1.9 L/kg), although mean clearance was 0.19 and 0.13 L/h/kg for

S-terbutaline and R-terbutaline, respectively. The differences in clearance of enantiomers were primarily due to stereoselectivity in renal clearance. However, the nonrenal clearances of enantiomers were similar, suggesting a lack of stereoselectivity in the formation of sulfate conjugates in the liver. Indeed, urinary recoveries of conjugates after IV administration were similar for both enantiomers. After oral doses, the plasma concentrations of the R-enantiomer were approximately twice those of S-terbutaline. Intestinal metabolism in favor of the S enantiomer was considered to be the primary cause of stereoselectivity in terbutaline pharmacokinetics and bioavailability after its oral administration [117].

Special Use of Stereoselective Pharmacokinetic Data of β-Agonists

According to the International Olympic Committee's guidelines, only therapeutic doses of inhaled β2-agonists can be used for the treatment of exercise-induced asthma. The use of oral, parenteral, or very large inhaled doses of β2-agonists is considered to be illegal doping owing to the significant adrenergic and anabolic effects of the drugs. Stereoselectivity in albuterol pharmacokinetics has been explored as a means of discriminating between the authorized inhalational and the prohibited oral use of albuterol [128]. A criterion was established based on ratios of concentration of unconjugated albuterol enantiomers in a urine sample taken after inhaled or oral routes of administration. Due to the much larger (ten- to100-fold) oral doses, the urinary concentrations of unconjugated (R+S)-albuterol after oral administration were generally higher than those observed after inhaled albuterol. However, because of the variability in urine volume, the use of urinary concentrations alone was not sufficiently specific. Another discriminating parameter was the S:R ratio of the unconjugated albuterol excreted in urine. It was shown that the S:R ratios were >2.5 or <0.25 for oral or inhaled administration, respectively. This was attributed to significant presystemic stereoselective metabolism (sulfation) of the oral dose in favor of the R enantiomer, resulting in higher urinary concentrations of the unconjugated S-albuterol [128].

A combination of concentrations of unconjugated (R+S)-albuterol, and the corresponding S:R ratios in urine, were also used to distinguish between the· allowed (low inhaled doses) and prohibited (oral or high inhaled doses) use of the drug [128]. The accuracy of the method was demonstrated in situations mimicking the permitted use or abuse of the drug. All data points for the oral dosing were clearly separated from those after the inhalation using a plot of total (R+S) intact albuterol vs. their respective S:R ratios in urine.

4.2. Zileuton

It has been suggested that zileuton enantiomers are equipotent in inhibiting 5-lipooxygenase [129]. However, no study is available evaluating the pharmacodynamics of the individual enantiomers of zileuton. Stereoselective pharmacokinetics of zileuton have been observed [130], with concentrations of the R enantiomer exceeding those of antipode (Table 1).

Zileuton reportedly undergoes stereoselective glucuronidation at a N-hydroxy group. Sweeny et al. [131] showed that the human microsomal glucuronidation rates for the S enantiomer were 3.6 to 4.3 times greater than for antipode. The consequence of stereoselective metabolism is apparent in vivo with 49 to 76% faster clearance of S-zileuton after single oral doses of 200 to 800 mg in normal healthy subjects [130]. Furthermore, it has been demonstrated that each enantiomer of zileuton competitively inhibits the phase II metabolism of respective antipode [131].

In contrast to the β2-agonists, the plasma protein binding of zileuton is reportedly stereoselective, with an average binding of 96.3% and 87.8% for the S and R enantiomers, respectively (a threefold greater free fraction for R-zileuton) [132].

5. DRUGS USED TO TREAT PSYCHIATRIC DISEASE

5.1. Antipsychotics

The chiral antipsychotic drug thioridazine, which possesses a chiral carbon atom at position 2 of its piperidyl ring, is administered as the racemate. Thioridazine is metabolized to two sulfoxidated metabolites, 2-sulfoxide thioridazine (mesoridazine) and 5-sulfoxide thioridazine. These metabolites contain a second chiral center at their respective mono-oxidized sulfur atoms. The sulfoxide group of mesoridazine can be further metabolized to the sulfone metabolite, sulforidazine. In this step the chiral sulfoxide moiety is rendered achiral, thus leaving the R and S enantiomers with the single chiral carbon atom at position 2 of its piperidyl ring. The stereoselective steady-state plasma concentrations of thioridazine have been studied in 32 severely depressed patients that were coadministered the antidepressant agent, moclobemide [133]. Stereoselectivity was seen in the plasma thioridazine concentrations, with an R:S ratio (mean±SD) of 3.48 ± 0.93 being observed.

Reports of stereoselective tissue concentrations of chiral antipsychotic drugs in humans are scarce. Tissue concentrations of thioridazine have been reported to be stereoselective in a deceased patient who had received the drug chronically [134]. The (−):(+) concentration ratios in blood, bile, brain, and liver were 2.9, 2.2, 2.0, and 1.9, respectively. The enantiomer ratio in blood of the deceased patient paralleled that observed in the plasma of depressed patients [133].

5.2. Antidepressants

Fluoxetine

Fluoxetine is a widely used chiral antidepressant drug that acts as an inhibitor of serotonin reuptake. The pharmacologic activities of fluoxetine are discussed in detail in Chap. 5 (Sec. 3.4). Basically, the enantiomers of fluoxetine are equipotent with respect to therapeutic activity. In humans, the S-enantiomer has slower total body clearance than antipode [135]. The major products of fluoxetine biotransformation is the N-demethylated metabolites, R- and S-norfluoxetine, mediated by CYP2D6 [135]. It is known that while the S-metabolite enantiomer retains the same level of potency as R- and S-fluoxetine, R-norfluoxetine is much less potent [136,137]. In 12 healthy volunteers given single 20 mg doses of racemic fluoxetine (Table 1) [135], the ratio of S:R AUC and C_{max} were 1.2 and 1.5, respectively. Norfluoxetine kinetics after administration of parent drug were more stereoselective. The plasma AUC and C_{max} S:R ratios of norfluoxetine were 2.4 and 2.5, respectively. The ratios of norfluoxetine enantiomers to fluoxetine enantiomers were 9.8 and 5.5 for the S and R enantiomers, respectively. It would appear that the contribution of the S-norfluoxetine metabolite to pharmacological activity would be significant. In this study, alosetron, a drug indicated for treatment of irritable bowel syndrome, was not found to affect the pharmacokinetics of fluoxetine enantiomers in these subjects [135].

   In order to determine if the enantiomers of fluoxetine and its main metabolite norfluoxetine are substrates for CYP2D6, the drug (60 mg) was administered to six extensive (EM) and six poor metabolizers (PM) of sparteine, a marker for CYP2D6 activity [138]. In the PM and EM volunteers, the S:R ratios of AUC were 5.7 and 1.2, respectively (Fig. 5). The AUC of R and S enantiomers were 2.4-fold and 11.5-fold higher, respectively, in PM than in EM subjects. The authors concluded [138] that CYP2D6 is an important enzyme in the biotransformation of fluoxetine in humans, and that PM and EM of sparteine display different degrees of plasma concentrations and stereoselectivity in serum. The difference in the S:R ratios in poor and extensive metabolizers by itself is not likely to be clinically significant, because the enantiomers possessed the same level of pharmacological potency, and the plasma concentrations and enantiomer ratios of norfluoxetine were similar in poor and extensive metabolizers.

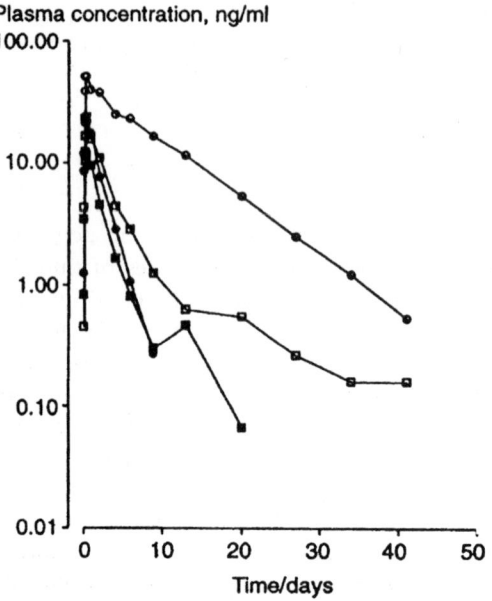

Plasma concentration, ng/ml

Time/days

Figure 5 The degree of stereoselectivity in plasma concentrations of individuals may be greatly impacted by polymorphic drug metabolism. For example, the stereoselectivity exhibited by fluoxetine in healthy volunteers after oral dosing of 60 mg racemate is much lower in extensive (closed symbols, $n = 6$) than in poor metabolizers (open symbols, $n = 6$) of the drug. R and S enantiomers denoted by square and round symbols, respectively. (From Ref. 138, with permission).

Citalopram

The stereoselective pharmacokinetics of the selective serotonin reuptake inhibitor, citalopram, have been studied in 10 healthy young subjects [139]. After administration of 40 mg racemate daily for 21 days by the oral route, stereoselectivity was found in citalopram plasma concentrations. The $R(-):S(+)$ ratios of C_{max} and AUCss were 1.5 and 1.6, respectively. No difference was noted in t_{max}, suggesting similar absorption rates of the enantiomers. The terminal phase $t_{1/2}$ values were 47 and 35 h for the R and S enantiomers, respectively. Renal clearance comprised 20% of each enantiomer's total clearance and was nonstereoselective. Although both enantiomers are extensively metabolized, both possess low hepatic extraction ratios [139]. Serotonin reuptake inhibition of citalopram primarily resides with the S enantiomer (see Chap. 5, Sec. 3.4 for more details), which attains lower concentrations in plasma. In eight geriatric patients (mean 77 y) given repeated doses [140], the mean R : S steady-state plasma concentrations of citalopram and desmethylcitalopram were 1.5 and 0.93, respectively. These values were similar to those present in younger patients [139].

Mianserin

Mianserin is a chiral tetracyclic antidepressant for which the S enantiomer is the eutomer. The influence of carbamazepine, an enzyme inducer of CYP3A4, on mianserin enantiomer concentrations was examined in 12 inpatients treated for depression [141]. The patients received at least 7 days of therapy with mianserin (60 mg per day) prior to study. In the control situation (before carbamazepine), the S : R ratios of 12 h postdose mianserin plasma concentrations were approximately twofold. At that point, the patients were coadministered carbamazepine 400 mg/day for 4 weeks along with their daily dose of mianserin. By week 4 of carbamazepine treatment, concentrations of the S and R enantiomers had declined by 49% and 38%, respectively. This corresponded with 2.0- and 1.6-fold increases, respectively, in the S- and R- desmethylmianserin : mianserin ratios. The findings confirmed the involvement of CYP3A4 in the metabolism of mianserin enantiomers to desmethylmianserine. Because of the possibility of a therapeutic window for S-mianserin, it was suggested that the dose of mianserin would need to be increased by a factor of two in the presence of carbamazepine [141].

Mianserin enantiomer concentrations are also influenced, in a stereoselective fashion, by coadministration of the chiral antipsychotic drug, thioridazine, an inhibitor of CYP2D6 [142]. Steady-state plasma concentrations of the S(+)-, but not the R(−)-enantiomer, were significantly increased in depressed Japanese patients (about twofold) by the coadministration of thioridazine. At the same time, there were significant increases in plasma concentrations of both R and S enantiomers of desmethylmianserin, which also possess pharmacological activity, after addition of thioridazine. The authors [142] concluded that the S-enantiomer of the parent drug and both enantiomers of desmethylmianserin are likely substrates for CYP2D6.

Reboxetine

Reboxetine is a new antidepressant that works at least in part by inhibiting norepinephrine reuptake. The drug possesses two chiral centers, but is marketed as the (−)-RR and (+)-SS enantiomers. The potency of the SS enantiomer is greater than the RR enantiomer. In healthy subjects given oral doses, the C_{max} and AUC of the RR enantiomer is approximately twofold higher than antipode. In examining effects on heart rate, it was found that the RR enantiomer causes fewer hemodynamic effects than racemate, suggesting that it is less potent in that respect than the SS enantiomer [143].

In a dose–rising study involving reboxetine in six healthy male volunteers given the racemate, stereoselectivity was observed in plasma concentrations (Table 1) [144]. With a single oral solution dose of 4.5 mg, the $(-):(+)$ ratios of plasma C_{max}, AUC, and $t_{1/2}$ were 2.3, 2.3, and 1.0, respectively. These findings were replicated after single doses of 1.5 and 3 mg. Given the high extent of plasma protein binding of reboxetine, the proportionately higher oral CL and oral Vd of the $(+)$-SS enantiomers relative to antipode appeared to suggest a stereoselective difference at the level of plasma protein binding. In a later study, the pharmacokinetics of reboxetine were established after intravenous dosing in male and female subjects [145]. The $(+):(-)$ ratios of Cl and Vd were approximately 2.7 and 2.4, respectively, in men and women given intravenous doses of 0.3 mg (Table 2). There were no apparent gender differences in the pattern of stereoselectivity in Cl or Vd. The bioavailability of both enantiomers was calculated to be near 1.

Rolipram

Rolipram is a racemic selective inhibitor of cAMP phosphodiesterase that has been used in the treatment of endogenous depression. The R(−)-enantiomer possesses the majority of the therapeutic effect. After separate administration of the enantiomers by the oral and intravenous routes (Tables 1 and 2), no stereoselectivity appeared in plasma concentrations or in calculated pharmacokinetic [146]. However, it is not known whether this is also true after the administration of the racemate, because enantiomer–enantiomer interactions may affect the enantiomeric plasma concentration ratios of a number of chiral drugs.

Tranylcypromine

Tranylcypromine is a chiral monoamine oxidase inhibitor used in the treatment of depression. The drug is similar to mefloquine in that it is contains a diastereomeric structure but is only administered as the 50:50 combination of the $(+)$-1S,2R and $(-)$-1R,2S species. The enantiomers possess differences in their pharmacological properties in that $(+)$ tranylcypromine is much more effective than its antipode in MAO inhibition, but the $(-)$ enantiomer causes greater diminution of catecholamine reuptake and release than $(+)$ enantiomer [147]. With respect to its pharmacokinetics (Table 1), the $(+)$ enantiomer seemed to be cleared via the oral route 4 to 8 times more rapidly than antipode based on significantly lower AUC values when administered in both the presence and the absence of antipode. A significantly higher renal CL (two-fold) was observed for (+) tranylcypromine than antipode, and an enantiomer–enantiomer interaction observed at this level in that renal CL of the (+) enantiomer was significantly increased in the presence of antipode [147]. However, renal CL appeared to play a minor role in the overall CL, so the clinical relevance of the finding is insignificant.

E-10-Hydroxynortriptyline

E-10-hydroxynortriptyline is the main chiral metabolite of the tricyclic antidepressant nortriptyline. Although the compound is less (50%) effective in inhibiting noradrenaline reuptake, it possesses a reduced capacity to elicit anticholinergic effects than nortriptyline, thus allowing it a possible therapeutic advantage [148]. After oral administration of single 75 mg capsules to 10 healthy volunteers, plasma concentrations of the (−) enantiomer were much greater than antipode, with (−):(+) AUC ratios of 2.6 being observed (Table 1). The metabolic clearance by conjugation was significantly higher for the (+) enantiomer than for the antipode. This was consistent with a higher free fraction of the (+) isomer, compared with (−) isomer (Table 3). Between 71 and 79% of the dose of each enantiomer of E-10-hydroxynortriptyline was recovered in the urine as conjugated plus intact enantiomers. Whereas the (+) enantiomer was dominant between the two conjugated enantiomers, the urinary excretion of (−) enantiomer exceeded that of the (+) enantiomer in terms of intact drug [148].

Trimipramine

The plasma concentrations of the chiral antidepressant trimipramine have been studied in a group of 27 depressed patients that had been phenotyped for CYP2D6 and CYP2C19 [149]. The patients were given 300–400 mg of the racemate daily for 5 weeks prior to measurement of plasma trimipramine enantiomer concentrations. Of these patients, 25 were extensive metabolizers for both isoenzymes, one was a poor metabolizer for CYP2D6, and one was a poor metabolizer for CYP2C19. Only a modest degree of stereoselectivity (mean L:D ratio of 1.15) was found in the mean plasma concentrations of trimipramine enantiomers in extensive metabolizers. In the patients phenotyped as being poor metabolizers of CYP2D6 or CYP2C19, the corresponding ratios were 1.66 and 0.75, respectively. L-trimipramine exhibits a higher degree of binding to dopamine and 5-hydroxytriptamine receptors of rat brain than does the D enantiomer

[150]. Therefore, it is possible that the effects of trimipramine can be altered in the poor metabolizers of CYP2D6 or CYP2C19.

5.3. Attention Deficit Disorder: Methylphenidate

Methylphenidate is used in the treatment of attention deficit hyperactivity disorder, mostly in pediatric and adolescent patients. The drug is chiral and is thought to act by inhibiting the reuptake of dopamine into neurons. The d enantiomer of methylphenidate is more potent than the l enantiomer. The CL ($d:l$ ratio 0.55) and Vd ($d:l$ ratio 1.5) of the enantiomers are stereoselective when 10 mg of the racemate is given intravenously to healthy volunteers (Table 2) [151]. After oral administration, the degree of stereoselectivity increases markedly, with $d:l$ ratios of AUC being between 8 and 10. The C_{max} $d:l$ ratio (between 6 to 10) was likewise higher than expected based on the intravenous CL values. The oral absolute bioavailabilities of the d and l enantiomers were approximately 23% and 5%, respectively, after administration of immediate or sustained-release dosage forms [151]. Both these values were much lower than expected based on the CL of the drug, which would predict the d and l enantiomers to be extracted by liver to a low and moderate extent, respectively. Presystemic metabolism by gastrointestinal tract might be an explanation for the less than expected oral enantiomer bioavailabilities.

In children given oral doses of the drug, the relative AUC and C_{max} of d and l enantiomers appears to be similar to that seen in adults given the drug (Table 1) [152,153]. Using the scanning reaction time of 7 boys, it was suggested that clockwise hysteresis was present in the plasma concentration vs. effect relationship of d-methylphenidate, although the loop was small [152]. No explanation, such as the formation of an inhibitory metabolite or the depletion of a substrate necessary for effect, was offered to explain the apparent clockwise hysteresis in those patients given the drug.

5.4. Sedative/Hypnotics

Zopiclone is a chiral hypnotic agent that possesses stereoselectivity in its pharmacodynamics and pharmacokinetics. The in vitro affinity of the S(+) enantiomer for binding to the benzodiazepine receptor is much higher (50-fold) than that of R(−)-zopiclone [154]. After oral doses of 15 mg of racemate to 12 healthy subjects, plasma AUC and C_{max} values are significantly higher for the S(+) enantiomer compared to antipode (Table 1) [155]. The (−):(+) ratio of oral clearance and Vd was 3.4. In human plasma, both enantiomers are highly bound to plasma. Although the unbound fraction of the S(+) enantiomer is 1.7-fold higher than the R(−) enantiomer [156], the oral volume of distribution of the R enantiomer has been reported to be 1.9-fold higher than that of antipode [155].

The stereoselective pharmacokinetics of hexobarbital and mephobarbital have been reported (Table 1). Both drugs display a very large degree of stereoselectivity in their plasma concentrations after racemic doses. In six healthy male volunteers, mephobarbital R : S ratios of C_{max} and oral clearance were reported to be 0.45 and 27.6, respectively, after oral 200 or 300 mg doses [157]. A significant amount of the dose of S-mephobarbital was thought to have been metabolized to phenobarbital. Presystemic metabolism of the S enantiomer was thought to play a role in the large difference in the AUC of the enantiomers, although the difference in C_{max} was not as great. In young healthy subjects receiving 500 mg of hexobarbital, $l:d$ ratios of C_{max} and oral clearance were reported to be 0.47 and 8.9, respectively [158]. Concentrations of the more active d enantiomer were almost 10-fold higher as assessed using AUC, which might confer its greater level of effect than antipode. An age-related stereoselective effect was found for hexobarbital (Fig. 6). In an older group of subjects (mean 68 y) given the drug, it was found that the oral clearance of the l enantiomer was significantly (49%) lower than in the younger (mean 23 y) subjects [158].

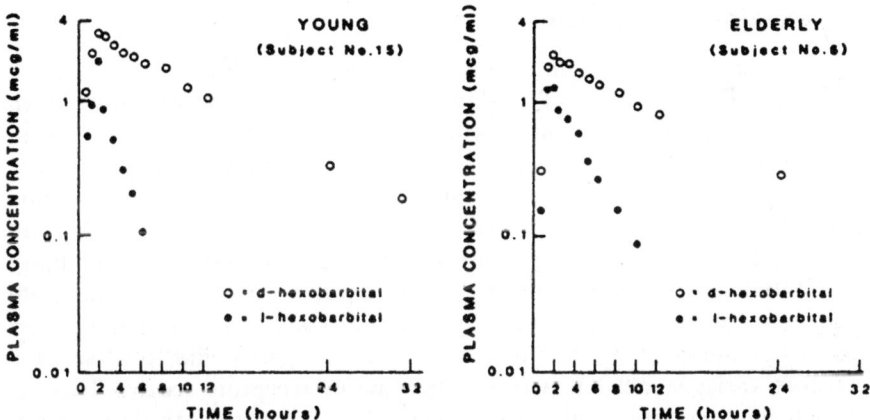

Figure 6   The degree of stereoselectivity in hexobarbital plasma concentrations is significantly lower in elderly subjects than in younger subjects. Concentrations of d-hexobarbital are higher in the elderly, although concentrations of the l enantiomer are the same in both groups of subjects. (From Ref. 158, with permission).

In contrast, there was no significant age-related impact on the oral clearance of the active *d* enantiomer (Fig. 6). The preferential age-related decline in oral clearance of the *l* enantiomer was thought to be due to a specific change in one or more of the pathways preferentially responsible for elimination of the *l*, but not the *d*, enantiomer.

Rifampin has been shown to increase stereoselectively the oral clearances of both hexobarbital enantiomers after administration of the racemate, although the degree of stereoselectivity in the interaction was age dependent [159]. In both young and elderly volunteers, the oral CL of the S(+) enantiomer increased with rifampicin pretreatment, by about tenfold in each group. The oral CL of the R(−) enantiomer, however, increased by 74-fold and 19-fold in young and elderly subjects, respectively. Therefore, the inductability of metabolism of R-hexobarbital is diminished with aging [159].

Some of the benzodiazepines are chiral. Oxazepam has been shown to undergo polymorphic stereoselective glucuronidation in Caucasian subjects, in favor of the more active S enantiomer [160]. In most subjects, the S:R ratio of the glucuronide conjugate was approximately 3.9. In about 10% of a sample size of 66 subjects, a lower ratio was found as a result of increased km for the S enantiomer.

6. DRUGS USED IN NEUROLOGICAL DISORDERS

6.1. Parkinsons Disease

There are a number of chiral anticholinergic drugs that are used in the treatment of mild forms of Parkinson's disease as monotherapy, or in combination with other antiparkinsons drugs for use in more severe presentations of the disease. These drugs are especially useful in treating tremor, and in alleviating extrapyramidal effects caused by antipsychotic drugs. For procyclidine and trihexyphenidyl, the R enantiomers are significantly and substantially more potent than antipode in their ability to bind to muscarinic receptors [161]. (+)-Biperiden was found to bind more strongly to muscarinic receptors than the corresponding (−) enantiomer [162]. The affinity of the enantiomers of these drugs to bind to muscarinic receptors varies somewhat with the subtype of receptor studied. There is no such binding data available for stereospecific antimuscarinic properties of some other chiral anticholinergic agents, such as orphenadrine and ethopropazine. With the exception of ethopropazine in the rat, in which stereoselectivity is absent, there is no pharmacokinetic data for the enantiomers of this class of drugs in humans [161].

6.2. Epilepsy

Hydantoin Derivatives

Mephenytoin is a chiral compound that is metabolized to its active chiral metabolite, phenylethylhydantoin. After administration of 300 mg orally to extensive metabolizers mephenytoin displays a massive degree of stereoselectivity in the plasma concentrations of the enantiomers of both parent drug and chiral metabolite [163]. In these patients the concentrations of R-mephenytoin and R-phenylethylhydantoin greatly exceeded those of their respective antipodes in plasma. In poor metabolizers, the direction of stereoselectivity in plasma concentrations of both parent drug and metabolite is similar to that in extensive metabolizers. However, the degree of stereoselectivity is significantly reduced in poor metabolizers, with the S:R oral CL ratios in extensive and poor metabolizers being 174 and 1.48, respectively. In poor metabolizers, urinary recovery of phenylethylhydantoin was 3.4-fold greater than in extensive metabolizers, suggesting that sequential metabolism of the metabolite was diminished in poor metabolizers.

Although phenytoin is not chiral itself, its major metabolite undergoes product stereoselectivity in the generation of a chiral compound, 5-(4-hydroxyphenyl-5-phenylhydantoin (p-HPPH), the R enantiomer of which has been suggested to be more strongly associated with gingivial hyperplasia than antipode [164]. The metabolism of phenytoin to p-HPPH is catalyzed by CYP2C9/19, for which four major genotypes have been identified. In one study involving 134 epileptic Japanese patients receiving phenytoin [165], the serum concentration ratios of R-pHPPH:phenytoin and S-pHPPH:phenytoin were examined in each of the different four genotypes. The ratios were much higher for S-pHPPH than for R-pHPPH in all four genotypes. Significant differences in the ratio for R-pHPPH were observed between two of the genotypes.

Ethotoin (3-ethyl-5-phenylhydantoin) is a chiral anticonvulsant drug with utility in generalized tonic-clonic and complex partial seizures. In six healthy volunteers given single 1 g oral doses, the only significant difference between enantiomers was in the AUC, which was 580 and 510 µg × h/mL for the R(−) and S(+) enantiomers, respectively [166].

Ethosuximide

Ethosuximide is a racemic drug used in the treatment of absence type seizure disorders. In an examination of plasma samples from 33 patients over a wide age range (2–67 y), there was no stereoselectivity in the plasma concentrations of the drug [167]. Racemization during the assay procedure could not be assessed, as separated quantities of individual enantiomers were not available to the authors. Nevertheless, the results suggest that there is no difference in pharmacokinetics of the enantiomers. In a subsequent study the authors found that the enantiomeric ratio of ethosuximide in umbilical cord plasma (obtained immediately postpartum) and in breast milk of three mothers was approximately 1 [168].

Vigabatrin

Vigabatrin is an irreversible inhibitor of the transaminase enzyme responsible for metabolism of the inhibitory neurotransmitter, GABA. The drug is used in the treatment of epilepsy, and therapeutic activity resides with the S-(+) enantiomer. After oral doses of 1500 mg of racemic vigabatrin, the plasma AUC is approximately 30% higher for the inactive R enantiomer relative to the S enantiomer in healthy men [169]. The authors stated that the pharmacokinetics of the drug are of minor importance because the drug irreversibly binds to the transaminase enzyme, and consequently enzyme regeneration rather than drug concentrations are the rate-limiting step in cessation of drug action [169].

As well as in adults, vigabatrin also displays stereoselectivity in its pharmacokinetics in pediatric patients (Table 1) [170], although there were some differences between infants and older children. In six infants (mean 1 y), the R : S ratios of C_{max}, AUC, and elimination rate constant were 1.5, 1.2, and 2.0, respectively. In older children (mean 8.7 y) the corresponding ratios were 1.8, 1.3, and 1.0, respectively. The AUCs of both enantiomers were higher in the older children than in the infants, suggesting an age-related decrease in clearance or increase in oral bioavailability. In five of the studied patients, the calculated renal CL was higher for the R than for the S enantiomer [170]. Pharmacokinetic data from neonatal patients (mean 21.3 days) [171] were comparable to those of adults, children, and infants. In neonates, the R : S C_{max} and AUC ratios were 2.4 and 1.6, respectively [171]. It was suggested that since the pharmacokinetics of vigabatrin enantiomers was similar across the age groups from neonates to adults, age-specific dosing recommendations may not be required for this drug.

In two mothers, the placental transfer at time of birth, and breast milk concentrations after birth, of vigabatrin enantiomers were examined [172]. Both women had measurable concentrations of vigabatrin in the cord and maternal venous blood. The umbilical vein : maternal vein concentrations ranged widely from 7% to 139% in the two women, although the ratios were greater for the R enantiomer than for the S enantiomer in both women. In the breast milk, the level of stereoselectivity was similar to that found in cord blood. The milk : plasma concentration ratios of the enantiomers were consistently higher for the R than for the S enantiomer. Nevertheless, the results suggested that the exposure of the infant to both vigabatrin enantiomers by breast milk was negligible [172].

In the presence of end-stage renal failure, the pharmacokinetics of vigabatrin enantiomers can become profoundly stereoselective in comparison to patients with normal renal function [173]. Under steady-state conditions, the R : S C_{max} and AUC ratios of vigabatrin were 7.9 and 6.2 in a 15-year-old female patient with end-stage renal disease.

Oxcarbazepine

Oxcarbazepine is an achiral antiepileptic drug structurally similar to carbamazepine. It is used for partial onset and generalized tonic-clonic seizures. The parent compound has little antiepileptic activity, but in vivo the drug is metabolized by 10-keto-reduction to yield a chiral metabolite, 10-hydroxycarbazepine, that possesses the majority of the antiepileptic activity of the administered oxcarbazepine [174]. In 12 healthy Chinese volunteers given 600 mg of the drug, there was a significant degree of product stereoselectivity, in that plasma concentrations of S-10-hydroxy-carbazepine were 5- or 6-fold greater than antipode. Because the enantiomers possess similar antiepileptic activities [174], this stereoselective metabolism may not be of any clinical significance.

Analogs of Valproic Acid

Valproic acid (VPA) itself is achiral and possesses a moderate propensity to cause teratogenicity in mice. The S(−) and R(+) enantiomers of the structural analogs, 4-yn-VPA and 4-en-VPA, have been synthesized. It has been shown that there was a progression from lowest to highest degree of teratogenic potential from the R enantiomers of these compounds to the S enantiomers [175]. The order of teratogenicity was R(+)-4-yn-VPA < R(+)-4-en-VPA < VPA < S(−)-4-en-VPA < S(−)-4-yn-VPA. In terms of sedation, valproic acid was most sedative, whereas the enantiomers of the chiral analogs were less sedative. The differential teratogenic properties of the enantiomers of the studied analogs were thought to be due to either an intrinsic ability of the enantiomers to cause teratogenicity, or pharmacokinetic properties allowing more of the S enantiomers to access the fetal circulation [175].

Valnoctamide is a chiral drug that has been examined for its utility as an antiepileptic drug. Valnoctamide possesses two chiral centers, and thus four stereoisomers. In seven healthy male volunteers given single oral 400 mg doses of the drug, the four isomers of valnoctamide had varying levels of CL/F, ranging from 4.2 to 8.7 L/h [176]. The Vd/F ranged from 59 to 74 L, and $t_{1/2}$ ranged from 5.8 to 10.3 h. Three of the isomers possessed similarities in CL and Vd, although one isomer had a significantly higher CL and shorter $t_{1/2}$. In epileptic patients given the same dose level, much higher CL/F and Vd/F values were reported than in healthy subjects. With repeat dose administration to the epileptic patients (200 mg thrice daily for 7 days), the CL/F was apparently lower for each of the four stereoisomers The authors believed that the higher CL/F of the stereoisomers in epileptic patients compared to healthy subjects was due to the inducing effects of carbamazepine on P-450 mediated metabolism of the drug [176]. Nonlinear elimination of the drug due to saturation of metabolism was proposed as an explanation for the decreased CL with repeated dosing to the epileptic patients.

6.3. Drugs Used to Treat Neuropathy

The effect of a high-fat meal on the stereoselective pharmacokinetics of thioctic acid, a drug used to treat peripheral and autonomic neuropathy in diabetic patients, has been studied [177]. Significant stereoselectivity was noted in the fasted condition (Table 1), with R : S ratios of ~ 1.86 being observed for C_{max} and AUC. After a high-fat meal, C_{max} and AUC of both enantiomers decreased, and t_{max} increased (Table 1). The R : S ratios of C_{max} and AUC after a high fat meal were slightly lower (~ 1.7) than those observed in the fasted state, suggesting that the food effect was basically nonstereoselective [177].

7. ANALGESICS

Most of the analgesic effect of methadone has been ascribed to the R enantiomer [178]. In ten patients suffering from chronic pain, methadone displays enantioselective pharmacokinetics after oral and intravenous doses [179]. R-Methadone was shown to have a significantly higher CL (1.2-fold), Vdss (1.7-fold) and $t_{1/2}$ (1.3-fold) than antipode. The higher CL and Vd were consistent with the higher reported unbound fraction of the R enantiomer in plasma [178]. After oral dosing, C_{max} and AUC of the R enantiomer were lower than those of the S enantiomers. Oral bioavailability was high for both enantiomers and was not significantly different. The R enantiomer of methadone has a plasma unbound fraction and renal clearance almost twice that of antipode [180]. The R : S plasma concentration ratios in another group of patients receiving methadone for treatment of opioid addiction were highly variable, ranging from 0.6 to 2.4. It was suggested that the presence of the S enantiomer in the racemic compound might autoinduce the metabolism of the drug, thereby lowering the plasma concentrations of the more active R enantiomer [180].

The use of methadone by addicts may have important consequences when the patient is breastfeeding her infant. In this case, if the infant is suffering from neonatal abstinence syndrome, exposure to methadone through breast milk might be advantageous. In a recent report [181], the concentrations of methadone enantiomers were determined in eight lactating mothers administered moderate to high doses of racemate. By using a paired Student t test on the data for immature milk, the milk : plasma ratio and estimated relative infant exposure (corrected for body weight) of R-methadone were significantly higher than those of antipode [181]. The authors believed that the infant exposure to the drug would be insufficient to prevent the neonatal abstinence syndrome.

The relative activities of the enantiomers of the chiral opioid analgesic drug tramadol have been studied in postsurgical gynecological patients. The authors monitored and quantified the use of the enantiomers or racemate, delivered by patient controlled analgesia, to determine the relative differences in enantiomer potency [182]. Patients needed to use significantly more (−) enantiomer than (+) enantiomer in order to alleviate pain, indicating that the (+) enantiomer possessed greater potency. The (+) enantiomer, however, appeared to cause side effects more frequently than antipode and racemate. It was suggested that the racemate was superior to the (+) enantiomer alone owing to its combination of effectiveness and low incidence of side effects. Plasma concentration monitoring was not addressed, so the possible impact of stereoselectivity in pharmacokinetics on the findings could not be determined. Recently, the same investigators reported [183] that in this same group of patients the mean effective serum concentrations for the (+) and (−) enantiomers after their sole administrations were 470 ± 323 and 771 ± 451 ng/mL, respectively. After racemate, the therapeutic concentrations of (±)-tramadol were 590 ± 410 ng/mL. This corresponded to (+) and (−) enantiomers serum concentrations of 307 ± 211 and 281 ± 198 ng/mL after racemate [183]. These in vivo data are consistent with in vitro studies demonstrating that the (+) enantiomer of tramadol has markedly higher binding affinity for the μ opioid receptor than antipode [185].

A recent study has found that the (+) enantiomer of the O-demethylated metabolite of tramadol, which possesses a μ opioid effect, is also a contributor to pain relief afforded by tramadol [186]. The formation of this metabolite is facilitated by CYP2D6, which could be of importance in considering the analgesic effect of the drug in poor metabolizers of sparteine. This is in agreement with the finding that the analgesic effects of tramadol are less intense in poor metabolizers of debrisoquine [187].

8. DRUGS USED IN ANESTHESIA

8.1. General Anesthesia

Several of the agents used as general anesthetics are chiral and used as the racemate. This group includes parentally administered agents such as thiopental and ketamine, and a number of fluorinated agents administered by inhalation, including halothane, enflurane, isoflurane, and desflurane.

A number of physical theories have been developed to help explain the mechanism(s) behind general anesthesia. For many years it was believed that general anesthesia was achieved as a result of perturbation of lipids within cell membranes. The potency of the anesthetics was related to their oil:water partition coefficients [188] and possibly the resultant ability of the general anesthetics to penetrate and disrupt electrophysiological activity in cell membranes [188,189]. Others have postulated that the anesthetic agent could form hydrated microcrystals in the central nervous system and thus interfere with water–lipid interactions, resulting in anesthesia [190]. Because physicochemical properties of enantiomers are the same, these physical-based mechanisms of general anesthesia imply that enantiomers should be equipotent as anesthetic agents. Recent evidence has arisen that has forced a reexamination of the issue [189,191]. It has become evident that the enantiomers of the chiral inhalational volatile anesthetics possess different abilities to bind to receptor proteins. Thus the study of stereo-specific drug action has allowed for the determination of the molecular mechanisms of actions of this class of drugs.

Volatile Anesthetics

The (+) enantiomer of isoflurane appears to be more potent as an anesthetic agent than is its antipode [192,193]. It appears that the (+) enantiomer causes a greater inhibition of neuronal activity by stimulation of $GABA_A$ receptors in the central nervous system [194]. Compared to isoflurane, halothane enantiomers do not appear to differ as much in their effects on $GABA_A$ responses [195].

With respect to toxicities, the volatile anesthetics do not appear to affect stereoselectively the electrophysiological function of the heart. Because of their narrow therapeutic window, it has been proposed that the use of a single enantiomer of these agents, the one possessing greater anesthetic effect but similar toxicity to antipode, could have significant therapeutic advantages over the racemate. As an example, isoflurane enantiomers share a similar ability to bind to L-type calcium channels in murine heart, although the (+) enantiomer possesses a greater anesthetic potency [196].

Pharmacokinetic data related to the volatile anesthetics are sparse, perhaps because of onerous analytical challenges. Recently, a number of sensitive gas chromatographic assays have been published that may enhance our knowledge of the pharmacokinetics of these chiral drugs.

Injectable Anesthetics

Thiopental. A number of studies have been published involving the injectable anesthetic thiopental. Thiopental is a very fast acting barbiturate and is one of the most widely used of the anesthetic agents. Preclinical studies involving the enantiomers have indicated that the S enantiomer is more potent and has a lower safety threshold than antipode. S-thiopental causes a greater potentiation of $GABA_A$ receptors of human $\alpha_1\beta_1\gamma_2$ subtype expressed in Xenopus oocytes injected with the human mRNA (S:R EC50 ratio = 1.8) [197]. Pentobarbital, which is also a chiral drug, was found to be much less potent in racemic form than racemic thiopental in accentuating $GABA_A$ activities. Although pentobarbital is one of the major metabolic by-products of thiopental, the enantiomers of pentobarbital were not studied individually in this paper [197].

Nguyen et al. [198] have studied the pharmacokinetics of thiopental enantiomers in presurgical subjects with normal hepatic and renal function (Table 2). The subjects were given thiopental as intravenous bolus doses or constant infusions. Plasma concentrations of the enantiomers of the metabolite pentobarbital were also monitored by stereospecific HPLC assay. The plasma CL (R:S = 1.28) and Vdss (R:S = 1.22) of R-thiopental were significantly higher than those of antipode. This level of stereoselectivity is much less than that seen for some other chiral barbiturate derivatives [157–159]. The extent of plasma protein binding of this acidic drug was high for both enantiomers (87–92%), and there was a significantly higher unbound fraction in plasma of the R enantiomer (R:S = 1.24) (Table 3). Hence it appeared that the pharmacokinetic differences observed in total (bound + unbound) plasma concentrations of the enantiomers was a result of stereoselectivity in plasma protein binding. Consequently, it appeared that the intrinsic CL of unbound thiopental enantiomers was not stereoselective. The plasma concentrations of pentobarbital measured in these patients were approximately 10% those of thiopental and were nonstereoselective [198].

Thiopental exhibits enantioselectivity in its EEG effects [199,200]. After the continuous infusion of thiopental enantiomers administered separately to rats [199], the lethality : anesthetic potency ratio of doses was significantly lower for S-thiopental (2.1) than for antipode (3.2). The authors also found that initial electroencephalographic changes were greater for

S- than for R-thiopental. After the administration of racemate, tissue concentrations at time of death were very similar in segments of the central nervous system and in peripheral tissues, with few significant differences. It was of note that after administration of the separate enantiomers to time of death, concentrations of R-thiopental were substantially and significantly higher in all brain tissues and most peripheral tissues [199]. This was as expected, since the kinetics of the enantiomers were the same after racemate, and the greater lethality of the S enantiomer when given alone abbreviated its length of infusion compared to antipode. It was of special interest that the heart concentrations of S enantiomer were not significantly different from those of R-thiopental, this yielding a lower heart:brain concentration ratio for the R enantiomer. These findings prompted the authors to suggest that the R enantiomer might be safer for clinical use than the racemate [199].

In 12 back surgery patients, thiopental exhibited no significant stereoselectivity in bound or unbound CL and Vd of the enantiomers when given as racemate [201]. In six of these patients, plasma protein binding measurements were high (> 80%) and nonstereoselective (Table 3). The only significant difference between enantiomers was in the percentage of drug recovered in urine. However, since the amount recovered was less than 0.7% of racemate, the finding was clinically not relevant.

Ketamine. A discussion of the pharmacodynamic complexities of ketamine is provided in Chap. 5 (Sec. 3.2). Ketamine is another injectable chiral general anesthetic, and the S enantiomer possesses approximately three-fold greater anesthetic potency than antipode, possibly by binding to N-methyl-D-aspartate receptors. Ketamine can cause adverse psychological effects, and these have been thought to be caused in part by inhibition of neuronal nicotinic acetylcholine receptors. Binding of ketamine enantiomers to these receptors has been shown to be nonstereospecific [202]. In a study involving surgical patients, 50 subjects were randomly allocated to receive either S-ketamine or racemic ketamine intravenously [203]. Using a stereospecific assay, it was found that S-ketamine had a significantly higher CL (16%) and Vd (9%) than antipode. The authors found that 1 mg/kg of S-ketamine had the same magnitude of increase on blood pressure and heart rate as did 2 mg/kg of racemate. The authors suggested that this might indicate that most of the hemodynamic changes are due to the S enantiomer. This suggestion infers linear relationships between dose or plasma concentrations of S- and/or racemic ketamine and increases in blood pressure and heart rate, which was not studied. There were no differences detected between plasma adrenaline and noradrenaline levels after administration of the S enantiomer alone or after racemate [203].

Due to its greater anesthetic potency and more favorable side effects profile, S-ketamine has supplanted racemate in several European countries [204]. In a study examining the utility of the S enantiomer preoperative medication in pediatric patients [204], the enantiomer was given by the rectal route to 40 children with or without midazolam. Another 22 children were premedicated only with midazolam. The authors found that S-ketamine had no advantage over midazolam alone as a preoperative sedative, and that S-ketamine was much less effective than had been observed previously in a similar cohort of subjects given racemic ketamine. The authors speculated [204] that perhaps the enantiomers are absorbed differently by the rectal route of administration.

8.2. Local Anesthetics

In recent years there has appeared some longer acting amide-type local anesthetics. These drugs include bupivacaine, mepivacaine, and prilocaine, each of which possesses a chiral center. For at least one of these drugs, bupivacaine, the S enantiomer has been found to cause less toxicity than antipode [205]. One of the potentially serious stereoselective toxicities associated preferentially with R-bupivacaine is cardiac arrhythmias. The S enantiomer of bupivacaine has been shown to possess a similar level of efficacy to that of racemate after epidural administration [206], which has led to the suggestion that the S enantiomer could make for a safer anesthetic agent.

The stereoselective pharmacokinetics of prilocaine (Table 2) and bupivacaine enantiomers have been studied after intravenous doses [207, 208]. For both drugs, the R enantiomers are cleared more rapidly than the respective S enantiomers. The R:S clearance ratios of both drugs lie between 1.25 and 1.34 [207,208].

Significant differences also are present between the Vdss of bupivacaine enantiomers (R:S = 1.56), but not between prilocaine enantiomers. The difference in unbound fraction in plasma between prilocaine enantiomers was statistically significant but negligible (S:R = 1.04) (Table 3) [207], which explains the lack of difference in Vd of the enantiomers. On the other hand, the unbound fractions of bupivacaine enantiomers differed significantly both statistically and numerically (R:S ratio of 1.5). When the unbound CL of bupivacaine enantiomers was determined, the value for the R enantiomer was actually less than that of antipode, indicating the influence of plasma protein binding in determining total (bound + unbound) plasma concentrations of the enantiomers [208].

When administered by the epidural route to elderly [209] and young [210] surgical patients, bupivacaine shows a stereoselectivity pattern similar to that after intravenous administration [208], with S enantiomer concentrations being significantly greater than those of R enantiomer in both age groups [209,210]. The S : R AUC and C_{max} ratios were 1.3 and 1.15, respectively in young patients [209]. For both enantiomers, mean residence time was determined to be about threefold higher after epidural as compared to intravenous administration. The mean absorption time was roughly the same for both enantiomers. In the elderly subjects, the AUC was significantly higher for S-bupivacaine than the R enantiomer in both arterial and pulmonary arterial plasma [209]. The unbound fraction in plasma in young subjects [210] was higher for R- than for S-bupivacaine (R : S ratio = 1.5) (Table 3) and was similar to that observed after intravenous administration [208]. Interestingly, within 5 min after epidural dose administration in elderly patients, plasma concentrations of both bupivacaine enantiomers were higher in pulmonary arterial plasma than in radial arterial plasma, indicating pulmonary uptake of both enantiomers of about 15% [209].

Mepivacaine has close structural similarity to bupivacaine, and its stereoselective pharmacokinetics have also been studied after parenteral administration (Table 2). In healthy volunteers given short intravenous infusions of mupivacaine, the R : S ratio of CL is 2.25 for total (bound + unbound) drug [211]. The CL of both mepivacaine enantiomers appears to be greater than its structural analog, bupivacaine, and the Vd was also stereoselective, in favor of the R enantiomer (R : S ratio = 1.79). The difference in Vd was attributed to a 30% greater unbound fraction in plasma of the R enantiomer (Table 3). After correcting the CL for unbound concentrations, the CL of unbound enantiomer was found to have an R : S ratio of 1.58. Thus the higher total (bound + unbound) CL of the R enantiomer is attributable to stereoselectivity primarily at the level of intrinsic CL, with some contribution from enantioselective differences in unbound fraction in plasma [211].

In addition to bupivacaine, Groen et al. [210] also studied the pharmacokinetics of mepivacaine after epidural administration in young surgical patients. The S : R ratios of AUC (1.89) and C_{max} (1.29) were similar to that observed after intravenous dosing by Burm et al. [211] in healthy volunteers. In a combined psoas compartment/sciatic nerve block technique involving two sequential injections of mepivacaine [212], similar stereoselectivity in pharmacokinetics was observed as for the epidural procedure [210]. With this combined technique [212], S : R ratios of AUC and C_{max} were 1.93 and 1.52, respectively. In five patients given varying total doses of bupivacaine by epidural infusion, the urinary excretion of bupivacaine was variably stereoselective, with R : S ratios ranging from 1 to over 4 [213]. The percentage of each enantiomer excreted into urine unchanged ranged from 8 to 43%.

A drug interaction has been identified between the antimycotic agent itraconazole and bupivacaine enantiomers. Coadministration of itraconazole orally with intravenous bupivacaine caused the clearance of both bupivacaine enantiomers to decrease by 25% [214]. This interaction is of concern for the drug, since the R enantiomer has been linked with the development of cardiac arrhythmias.

9. DRUGS AFFECTING MUSCLE FUNCTION

In a single healthy volunteer, the chiral skeletal muscle relaxant methocarbamol was found to possess stereoselectivity in plasma concentrations after intravenous administration. Based on the plasma concentration vs. time curve, the S enantiomer appeared to be cleared more rapidly than antipode, although pharmacokinetic data were not reported [215].

Terodiline is an antimuscarinic drug with calcium antagonist properties that was used to treat urinary incontinence in Europe until an unacceptable risk of Torsades de pointes, secondary to QTc prolongation, prompted its withdrawal from the market. It is known that the R enantiomer possesses the antimuscarinic properties, whereas the S enantiomer possesses most of the calcium antagonist properties [216]. It was also known that the R-terodiline possessed the majority of the beneficial effect on the detrusor muscle of the bladder. In healthy male and female volunteers given racemic, R-, or S-terodiline, stereoselectivity has been reported in the pharmacokinetics of terodiline enantiomers, with plasma concentration of the R enantiomer exceeding those of antipode (Table 1) [216]. It was hoped that the R enantiomer would possess less QTc prolongation than the S enantiomer, which would provide a rationale for the development of the pure R enantiomer for treatment of urinary incontinence. Unfortunately, it was the R and not the S enantiomer that caused more prolongation of the QTc interval, effectively precluding the drug as a safe detrusor relaxing agent.

Most of the articles reviewed for this chapter have examined the pharmacokinetics of drugs after their parenteral or oral administration. Reports of stereoselectivity in plasma concentrations of chiral drugs after their topical administration are scarce. Oxybutynin is an anticholinergic drug that is used to treat overactive bladder. Anticholinergic activity of the R enantiomer is approximately 100-fold greater than that of the S enantiomer [217]. After oral administration, it undergoes extensive first-pass metabolism to N-desethyloxybutynin, a compound that may contribute significantly to the anticholinergic side effect profile of the drug. Recently, the plasma concentrations of oxybutynin enantiomers after oral (Table 1) and topical administration (as a transdermal patch) of the racemate were reported [217]. Stereoselectivity was present after both oral and topical administration, with concentrations of S enantiomer exceeding those of antipode (AUC S:R ratios of ~1.4). Plasma concentrations of metabolite enantiomers were also stereoselective but formulation specific, with concentrations of R metabolite being greater than antipode after oral administration, but in the opposite direction after topical administration. Oral administration resulted in much higher AUC ratios of N-desethyloxybutynin enantiomers to oxybutynin enantiomers (S, 6.7 and R, 15) than did topical administration (S, 0.79 and R, 0.95). This difference was attributed to the lack of a significant first pass metabolism of oxybutynin by the transdermal route [217].

An important drug in the symptomatic treatment of spasticity associated with multiple sclerosis, baclofen, is chiral and used as the racemate. Although there is no stereospecific pharmacokinetic data available, it is known that the (−) enantiomer is more potent than antipode in alleviating spasticity and in eliciting antinociceptive reaction [218].

10. DRUGS USED TO TREAT ALLERGIES

The S enantiomer of the highly used H_1 antagonist chlorpheniramine is 100-fold more potent in vivo and in vitro than the R enantiomer in its antihistaminic effects [219]. In healthy volunteers, plasma concentrations of the more active S enantiomer are about two times higher than those of antipode [219,220]. For chlorpheniramine, stereoselectivity in pharmacokinetics and dynamics has been proposed to be a factor in the time discordance observed between effect and plasma concentrations attained using achiral analytical methods [219]. In extensive metabolizers for CYP2D6 given 8 mg of (±)-chlorpheniramine maleate, oral CL of the R enantiomer was over twofold higher than S-chlorpheniramine (Table 1). When the same dose of chlorpheniramine was administered concurrently with quinidine, a known inhibitor of CYP2D6, the plasma concentrations of both chlorpheniramine enantiomers increased, by an approximately equivalent magnitude. This may indicate that the fraction of dose of each enantiomer metabolized by CYP2D6 isoenzymes is similar. The $t_{1/2}$ for the enantiomers did not increase as much as the decrease in oral CL might have predicted. This might suggest a change in Vd of chlorpheniramine enantiomers by quinidine. In two poor metabolizers for CYP2D6, plasma concentrations were higher than those in extensive metabolizers, but the degree of stereoselectivity, indicated by the S:R ratio of AUC, was similar (2.7) to that observed in extensive metabolizers (2.4)

[219]. In a pharmacokinetic assessment, the H_2 antagonist, ranitidine, was shown not to interact with either enantiomer after administration of $(\pm)$-chlorpheniramine [220].

Terfenadine is a chiral nonsedating antihistamine that has been withdrawn from the market owing to its ability to cause increased QTc interval and ventricular arrhythmias, particularly when drugs that inhibit its metabolism are coadministered. The drug undergoes a high hepatic first pass extraction $(E = 0.99)$ after oral administration and is largely converted upon first pass to a carboxylic acid metabolite, terfenadine acid (fexofenadine). In five healthy male volunteers given 60 mg terfenadine, stereoselectivity was observed in the plasma concentrations of terfenadine acid [221]. The $(+):(-)$ ratios of C_{max} and AUC were 2.3 and 2.2, respectively, with no differences in t_{max} (~ 2.5 h) or $t_{1/2}$ (4 h). When fexofenadine was given by itself by the oral route, it was found to have a $(+):(-)$ concentration ratio of 1.7 [222]. The difference in plasma concentrations could not be explained by differences in renal excretion rate, as urine concentrations were the same in all urine fractions collected after administration of the oral dose. Fexofenadine has now replaced terfenadine in clinical use.

In a subset of samples obtained from a pharmacokinetic study involving venlafaxine, enantiomer concentrations were determined in pooled plasma samples from 9 subjects given single 50 mg doses of the HCl salt [223]. Four time points were examined after dosing, from 2 to 4 hours. In each of the pooled plasma specimens, the R:S ratio of venlafaxine enantiomers was approximately 1.35.

11. DRUGS USED IN THE TREATMENT OF GASTROINTESTINAL DISORDERS

Omeprazole (Fig. 2), pantoprazole, and lansoprazole are proton pump inhibitors used for hydrogen atom hypersecretory disorders of the stomach. Most chiral drugs used clinically possess a chiral carbon atom. Rather than a chiral carbon atom, each of these drugs possesses a tricoordinated mono-oxidized sulfur atom that forms a pyramidal optically active center of asymmetry. Each is administered as the racemate.

As part of a nonstereoselective dose ranging study involving 40 subjects, two subjects, a poor and an extensive metabolizer of pantoprazole, were identified, and their plasma samples were reassayed by stereospecific HPLC [224]. In the extensive metabolizer, the $(+):(-)$ AUC ratios of pantoprazole in each of the three doses administered ranged from 0.58 to 0.89. In the poor metabolizer subject, the $(+):(-)$ ratios were much higher, ranging from 2.7 to 3.5. The $t_{1/2}$ and C_{max} values followed the same trend, with much higher levels of stereoselectivity noted in the poor metabolizer subject [224]. Recently, it has been demonstrated that the metabolism of pantoprazole cosegregated with S-mephenytoin, a substrate for CYP2C19 [225]. As previously observed, poor metabolizers displayed much greater stereoselectivity in plasma concentrations than did extensive metabolizers [225]. In seven Japanese poor metabolizer subjects given single 40 mg doses of the racemate (Table 1), the $(+):(-)$ mean C_{max} and AUC ratios were 1.3 and 3.6, respectively. In comparison, in seven extensive metabolizers, the $(+):(-)$ mean C_{max} and AUC ratios were 0.94 and 0.84, respectively. Although plasma concentrations of both enantiomers were lower in extensive metabolizers, the metabolism of the $(+)$ enantiomer was apparently more affected than antipode by the difference in phenotype status [225].

The stereoselective pharmacokinetics of racemic omeprazole have been studied in healthy Swedish volunteers who had been phenotyped to be poor ($n = 5$) or extensive ($n = 5$) metabolizers of S-mephenytoin (Table 1) [226]. Similar to pantoprazole, phenotype status for CYP2C19 had a significant impact on the pharmacokinetics of the enantiomers. Concentrations of both omeprazole enantiomers were decreased in extensive metabolizers. In extensive metabolizers, the $(+):(-)$ ratios for C_{max} and AUC were 0.60 and 0.62, respectively. In poor metabolizers, however, the $(+):(-)$ ratio for C_{max} and AUC were 1.3 and 1.5, respectively. As observed for pantoprazole [226], the AUC of the $(+)$ enantiomer was increased to a greater extent in poor metabolizers than was the $(-)$ enantiomer AUC. The ratio of $(+)$-hydroxyomeprazole to $(+)$-omeprazole was greater than that of antipode in extensive metabolizers, and the ratio for $(+)$ enantiomer was reduced by 30-fold in poor metabolizers compared to only 3.4-fold for $(-)$ enantiomer, both of which suggested that the $(+)$ enantiomer is preferentially metabolized to $(+)$-hydroxyomeprazole by CYP2C19 in extensive metabolizers [226].

In six healthy Japanese subjects given 30 mg of racemate, marked stereoselectivity was found in the pharmacokinetics of lansoprazole (Table 1) [227]. The mean C_{max} and AUC ratios of R$(+)$:S$(-)$ lansoprazole were 2.9 and 4.5, respectively. The elimination $t_{1/2}$ of the enantiomers were similar, but a 4.8-fold larger Vd/F was determined for the S$(-)$ enantiomer. The twofold higher unbound fraction of the S$(-)$ enantiomer in plasma (Table 3) explains, in part, its higher Vd/F.

Ondansetron is a chiral antiemetic drug used to prevent nausea and vomiting associated with use of antineoplastics. Although chiral assays have been developed for the drug [228,229], in our search we could not find any literature that described the pharmacokinetic properties of the enantiomers of the drug in humans.

12. ANTIFERTILITY DRUGS

Gossypol (Fig. 2) is a naturally occurring compound that is administered as the racemate consisting of two geometric stereoisomers, and which has been examined for its utility as a male antifertility drug. The pharmacokinetics of gossypol enantiomers have been examined in healthy men after administration of either 20 mg racemate or 20 mg of (+) or (−) enantiomers separately (Table 1) [230]. A nonstereospecific assay was used to assay the samples. The mean (+) : (−) AUC and terminal phase half-life ratios were 4.8 and 29, respectively. Because the (−) enantiomer is pharmacologically active, the authors expressed concern that minimal accumulation would occur for the (−) enantiomer with repeated dosing. It would also follow that with repeated dosing, significant accumulation of the inactive (+) enantiomer would occur in plasma [230]. Gossypol has more recently been proposed for use as an anticancer agent, with the (−) enantiomer displaying a greater cytotoxic potential to tumor cell cultures than (+)-gossypol [231].

Interestingly, the (+) enantiomer of propranolol, which is largely devoid of β-blocking activity, has been found to show promise as an antifertility agent [232]. When used topically in combination with nonoxynol-9, sperm motility was significantly reduced compared to use of nonoxynol-9 alone. The basis of the effect was thought to be due to the membrane stabilizing effect of propranolol, which is shared between the two enantiomers.

13. DRUGS OF ABUSE

In recent years, there has been increased misuse of 3,4-methylenedioxymethamphetamine (MDMA; "Ecstasy") among young people. The drug causes euphoria and gives the user a sense of empathy. MDMA is a chiral compound used as the racemate, and the S-(+) enantiomer is reportedly more potent at causing the intended feelings of euphoria and empathy than antipode [233]. The plasma and urinary pharmacokinetic properties of MDMA enantiomers have recently been described in eight healthy male volunteers given oral doses (Table 1) [233]. Stereoselectivity was found in plasma concentrations of the enantiomers after its oral administration, with plasma concentrations of the less active R enantiomer exceeding those of S-MDMA. The oral CL values of the S enantiomer exceeded those of antipode by a factor of 2.4. Stereoselectivity was also noted in the Vd/F, with the S enantiomer possessing larger values than R-MDMA. Renal CL/F of the enantiomers did not appear to be stereoselective and accounted for a small proportion of overall CL (10–25% of R and <10% of S enantiomer CL_R, respectively) [233].

14. DRUGS USED TO TREAT CANCER

14.1. Cyclophosphamide and Ifosfamide

Two chiral drugs commonly used in the treatment of cancer are the alkylating agents cyclophosphamide and ifosfamide (Fig. 2). From a chemical perspective, both drugs are distinctive in that their chiral centers are phosphorous rather than carbon atoms. Another feature common to both drugs is that they are really prodrugs, as pharmacological action is attributable to metabolites formed by a specific pathway.

Cyclophosphamide is converted by cytochrome P-450 isoenzymes into a hydroxylated metabolite that can be further hydrolyzed into the active product phosphoramide mustard. This pathway also generates the urotoxic metabolite acrolein, although coadministration of a uroprotective agent can minimize this problem. In one study, concentrations obtained from nine patients given IV (±)-cyclophosphamide indicated that there was little to no stereoselectivity present in human plasma [234]. Further, in young, teenaged patients with various malignancies (Table 2), the pharmacokinetics of cyclophosphamide enantiomers were nonstereoselective (Table 2) [235]. The total clearance, volume of distribution, and half-life values were all virtually the same for both enantiomers. The formation clearance of the R enantiomer of a noncytotoxic metabolite (N-dechloroethyl-cyclophosphamide) was found to be twice that of the S metabolite, but there was no enantioselectivity in the formation of active 4-hydroxycyclophosphamide [235]. The latter metabolite is an intermediate in the formation of therapeutically beneficial phosphoramide mustard.

Busulfan is an antineoplastic agent that is sometimes administered with cyclophosphamide. Because of the convulsant potential of busulfan, it may in some cases be administered with an antiepileptic drug such as phenytoin or diazepam. Because phenytoin can induce the metabolism of some drugs, a study was conducted to examine the effects of phenytoin on cylophosphamide enantiomer pharmacokinetics in cancer patients [236]. The total body clearances of both cyclophosphamide enantiomers were higher in three patients given phenytoin than in three patients given diazepam. The induction of cyclophosphamide metabolism by phenytoin appeared to have a greater effect on the clearance of the S enantiomer. The pathway converting cyclophosphamide enantiomers to the active 4-hydroxylated metabolite seemed to be more affected by phenytoin than the pathway to the noncytotoxic metabolite.

The stereoselectivity in metabolism of ifosfamide enantiomers may have toxicological significance. Similar to cyclophosphamide, ifosfamide is biotransformed by two pathways mediated by P-450 isoenzymes.

The 4-hydroxylation pathway leads to the formation of phosphoramide mustard, which is the active byproduct, and acrolein. The alternate route is dechloroethylation, which is accompanied by formation of a neurotoxic and urotoxic metabolite, chloroacetaldehyde. Although qualitatively similar, the metabolism of ifosfamide differs from cyclophosphamide quantitatively. Whereas N-dechloroethylation is a minor pathway for cyclo-phosphamide, it accounts for about 50% of the elimination of ifosfamide [237]. The metabolic products of R-ifosfamide are R-2-dechloroethylifosfa-mide and S-3-dechloroethylifosfamide, whereas S-2-dechloroethylifosfamide and R-3-dechloroethylifosfamide are the products of S-ifosfamide metabolism [238].

With respect to the activation of ifosfamide by 4-hydroxylation, human liver microsomes indicated that the rate of conversion of the R enantiomer to the activated metabolite occurred at the same or a higher rate than did the S enantiomer [237]. S-Ifosfamide also displays higher in vitro maximal rates of formation (V_{max}) and lower affinity (higher km) of dechloroethylated metabolites than does the R enantiomer [239]. Formation of the R dechloroethylated metabolites is apparently catalyzed by CYP3A4, whereas formation of the corresponding S metabolites is due to CYP2B6 activity [230,231]. It has been suggested that the neurotoxic metabolite of ifosfamide is R(−)-3-dechloroethylifosfamide rather than chloroacetalde-hyde, based on higher urinary excretion of this metabolite in a patient exhibiting signs of ifosfamide-related neurotoxicity [239]. Given all of these findings, it has been proposed that the R enantiomer would be preferential to racemate owing to its slower rate of conversion to potentially toxic N-dechloroethylated metabolites [237,239].

Ifosfamide appears to possess a modest degree of stereoselectivity in pharmacokinetics of the parent drug. In studies involving cancer patients, including adult men and women, and in children, the CL of the S enantiomer is consistently higher than antipode, and the terminal elimina-tion phase $t_{1/2}$ is longer for the R enantiomer (Table 2) [238,240,241]. In one study [238], Vdss of the R enantiomer was slightly, but significantly, larger than that of the S enantiomer in female patients with squamous pelvic carcinoma. In another study involving male and female patients with small lung cell carcinoma [240], there were no enantiomeric differences in Vdss. In children receiving the drug for 72 h as three daily short infusions, or as constant IV infusion for 72 h [241], it was suggested that autoinduction of both enantiomers of ifosfamide occurred. This was based on the observation that in all five patients the CL values of S enantiomer were higher on day 3 than on day 1, the same being true for the R enantiomer in all but one of the patients.

Granvil et al. [238] found that in their female patients, the AUC ratio of dechloroethyl metabolites of ifosfamide to AUC of parent drug was higher for the S enantiomer than R-ifosfamide. Consistent with the in vitro results outlined above, this may suggest a higher extent of S enantiomer converted to these metabolites and/or a lower CL of these dechloroethylated metabolites formed from S-ifosfamide, compared with that of metabolites formed from the R enantiomer.

14.2. Other Chiral Antineoplastic Agents

The chiral anticancer drug aminoglutethimide acts as an inhibitor· of nonsteroidal aromatase. It is used as the racemate in the treatment of advanced breast cancer in postmenopausal women who have estrogen positive receptor tumors. The (+) enantiomer possesses greater antitumor activity than the antipode [242]. After oral dosing, the drug follows enantioselective pharmacokinetics, with greater C_{max} and AUC being attained for the (+) enantiomer than the antipode, reflecting higher CL of (−)-aminoglutethimide (Table 1) [242]. Stereoselectivity in renal CL was a reason for the enantiospecific differences in total CL. Specifically, the renal CL of (−)-aminoglutethimide was about half that of its oral total CL, whereas the renal CL of the (+) enantiomer was less than 25% of its oral total CL. The oral Vd of the enantiomers are not significantly different [(−):(+) ratio = 1.16], however, even though the (+) has a lower unbound fraction than antipode [(−):(+) ratio = 1.7] (Table 3). The reason for this apparent anomaly was not addressed [242].

Bicalutamide is a chiral nonsteroidal antiandrogen used in the treatment of prostate cancer. Pharmacological activity resides mostly with the R enantiomer. The drug displays marked stereoselectivity in plasma concentrations, with much faster clearance of the S form in plasma (Table 1) [243]. The R:S ratios of AUC after oral doses of 50 mg ranged from 74 to 290 in healthy subjects, and 97 to 290 in patients with hepatic impairment. The terminal $t_{1/2}$ is several days for the R enantiomer versus 6 to 37 hours for the S enantiomer [243].

15. DRUGS USED TO TREAT CONNECTIVE TISSUE DISORDERS: THE NONSTEROIDAL ANTI-INFLAMMATORY DRUGS

The nonsteroidal anti-inflammatory drugs (NSAIDs) are effective anti-inflammatory, antipyretic, and analgesic agents. Although not all of the NSAIDs are chiral, all of the drugs in its major chemical classes, the

2-arylpropionic acids (2-APA), possess a chiral center. As a collective group, the chiral NSAIDs are perhaps one of the most studied classes for enantioselectivity in pharmacokinetics. This class includes the prototype compound ibuprofen, as well as ketoprofen, fenoprofen, flurbiprofen, tiaprofenic acid, carprofen, pirprofen, benoxaprofen, and naproxen. Other non-APA chiral NSAIDs include etodolac and ketorolac.

The pharmacological action of the NSAIDs is primarily mediated by inhibition of prostaglandin synthetase [244]. For NSAIDs, it is the enantiomer possessing the S configuration that almost exclusively possesses the ability to inhibit prostaglandin activity. Of the chiral NSAIDs, all are administered in racemic form except naproxen, which is given as the single S enantiomer. Studies have been performed to examine the effectiveness of other single enantiomers of NSAIDs, such as S-ketoprofen, although no clear-cut advantage of the single enantiomer over racemate is apparent [245]. By far, most of the NSAIDs remain clinically used as the racemate.

The NSAIDs have been intensively studied from the perspective of stereoselectivity in pharmacokinetics. One of the compelling features of some of the 2-APA derivatives is that they can undergo a unidirectional chiral inversion from the inactive R to the active S enantiomer. This unique metabolic pathway is examined in detail in another chapter in this book.

15.1. Stereoselectivity in Plasma Concentrations of NSAIDs

As a class, the chiral NSAIDs display a wide range of stereoselectivity in plasma concentrations in humans (Table 1, Fig. 7). This is mostly due to variations in metabolic clearances of the enantiomers. Flurbiprofen and ketoprofen possess S : R AUC ratios of close to one (Table 1) [246,247]. In contrast, drugs such as fenoprofen, etodolac, and ketorolac possess S : R ratios that differ substantially from unity (Table 1) [248–250]. In all cases, concentrations of NSAID eutomer exceed those of distomer. Etodolac and ketorolac provide notable exceptions to this commonality, in that concentrations of distomer are much higher than those of the eutomer.

S-Fenoprofen concentrations greatly exceed those of the R enantiomer as a result of extensive chiral inversion [248]. An earlier report in a single subject indicated that the plasma concentrations of ketorolac enantiomers were nonstereoselective [251], but it was recognized later that a derivatization step in the assay [252] caused racemization of the enantiomers [253]. Assays using chiral columns have shown that plasma concentrations of inactive R-ketorolac are significantly higher than antipode [250,253]. This is consistent with a higher free fraction of S-ketorolac in plasma than that of the antipode (Table 3).

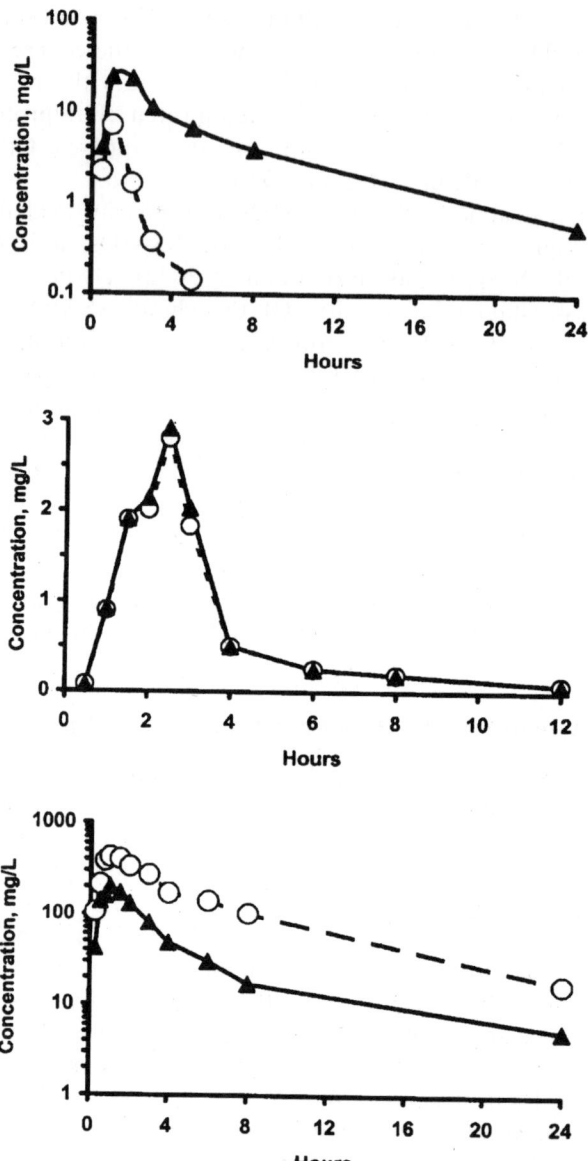

Figure 7 Plasma concentration vs. time profiles may significantly differ between drugs in the same therapeutic and chemical class. Upper panel, fenoprofen; middle panel, ketoprofen; lower panel, ketorolac. Circles, R enantiometer; triangles, S enantiometer (From Refs. 248, 280, and 253, respectively, with permission).

15.2. Pharmacokinetic Properties of NSAIDs

Absorption

Some 2-arylpropionates (e.g., ibuprofen and fenoprofen) undergo a unique unidirectional biotransformation from R to S enantiomer in the gastrointestinal tract and/or liver [254,255]. This topic, which is discussed in detail in another chapter, can affect the absorption parameters of the enantiomers. The absorption of NSAIDs across biological membranes occurs via passive diffusion, a process that cannot differentiate between enantiomers. On the other hand, presystemic unidirectional metabolism of R to S enantiomer may lead to an apparent stereoselectivity in the oral absorption of 2-arylpropionates. It has been suggested that the absorption rate may influence the degree of stereoselectivity in the plasma concentration of enantiomers [254–256]. Jamali et al. [257] suggested that an increase in the gastrointestinal residence time of ibuprofen could enhance the extent of inversion. Indeed, a significant positive correlation between t_{max} and S : R concentration ratios was observed after the oral administration of ibuprofen tablets with varying rate of release [257]. Since the pharmacological activity of 2-arylpropionates mainly resides with their S enantiomers, the effect of formulation on the enantiomeric inversion could influence the therapeutic effect. Interpretation of data by Cox et al. [258] suggested that presystemic inversion of ibuprofen can affect not only the S : R ratio of plasma concentrations but also their t_{max} [259].

Because many of the pharmaceutical excipients used in the formulation of racemic drugs are chiral and optically pure, there is a potential for the stereoselective interaction of the enantiomers' with the chiral matrix included in the formulation. For example, enantioselective pH-dependent release of tiaprofenic acid enantiomers from a sustained release formulation containing microcrystalline cellulose has been reported [260]. The differential release of tiaprofenic acid enantiomers, however, did not alter the pharmacokinetics of the individual enantiomers in rats. The possible effects of chiral excipients on the stereoselective release of racemates are discussed in a separate chapter in this book.

Distribution

The in vitro plasma protein binding has been reported to be stereoselective for many NSAIDs such as ibuprofen, flurbiprofen, ketoprofen, and etodolac (Table 3) [261–265]. Ibuprofen protein binding is stereoselective and nonlinear within the therapeutic concentration range (1 to 50 mg/L) [259]. The S enantiomer reportedly has a higher unbound fraction than antipode in human plasma [261,266]. An apparent decrease in the dose-normalized AUC values for ibuprofen is mainly due to nonlinear plasma protein binding [266]. The possibility of competitive plasma protein binding between enantiomers of ibuprofen has also been suggested [267]. However, use of specific fluorescent probes for binding sites I and II on human serum albumin did not demonstrate significant differences in the binding of individual enantiomers [267].

Enantiomeric competition for protein binding has also been reported for flurbiprofen enantiomers in rats, with displacement of R-flurbiprofen by its antipode from plasma proteins [268]. Jamali et al. [246] postulated that the interaction between enantiomers of flurbiprofen at protein binding level may explain the increase in the clearance of R-flurbiprofen. However, a more recently published study indicates that the extent of pharmacokinetic interaction between flurbiprofen enantiomers is not significant in humans [269].

Similar S:R concentration ratios in synovial fluid and plasma have been reported for ibuprofen [270,271], ketoprofen [272], flurbiprofen [273], and tiaprofenic acid [274]. On the other hand, Brocks et al. [264] reported that the stereoselectivity in the pharmacokinetics of etodolac enantiomers differs in synovial fluid from those observed in plasma. Following the administration of a 200 mg single oral dose of etodolac to 6 arthritic patients, the S to R concentration ratio was 0.074 in plasma and 0.17 in synovial fluid. The concentrations of (R)-etodolac (distomer) were similar in plasma and synovial fluid. In contrast, the synovial concentration of (S) enantiomer (eutomer) was 1.7-fold higher than that of plasma [264]. The same researchers also reported considerable levels of acyl-glucuronides of etodolac enantiomers in the synovial fluid. The presence of conjugated etodolac in the synovial fluid was attributed to the altered biological membrane properties and high circulating levels of acyl-glucuronidated etodolac conjugates [264].

Studies investigating stereoselective binding and distribution of NSAIDs to extrasynovial tissues are scarce. In rats, etodolac displays less stereoselectivity in tissue concentrations than in plasma and was thought to be due to a greater unbound fraction of S enantiomer in plasma relative to antipode [275]. For 2-aryl propionates, which undergo in vivo unidirectional inversion from R to S enantiomer (e.g., ibuprofen), the presence of these NSAIDs in the adipose tissue as hybrid triglycerides has been reported [276].

Metabolism

Most of the available chiral NSAIDs appear to be extensively metabolized in the body, and all possess a low hepatic extraction ratio. Because of the low hepatic extraction, stereoselectivity in metabolism of NSAIDs plays an important role in determination of the enantiomer plasma concentrations. The contribution of high plasma protein binding to the metabolism of these low-clearance drugs is also a critical determinant in their hepatic and total body clearance.

Chiral inversion is a unique metabolic pathway that is known to influence the plasma concentrations of some of the 2-APAs, such as ibuprofen and fenoprofen, but less so for other drugs such as ketoprofen and flurbiprofen (see Chap. 8). In addition to chiral inversion, these drugs are subject to drug-dependent stereoselectivity in microsomal oxidation and/or glucuronidation.

Ibuprofen enantiomers are mostly metabolized by P-450 mediated oxidation [259,277]. The formation clearances of the major metabolites of ibuprofen are reportedly higher for S- than for R-ibuprofen [278]. Ketoprofen enantiomers, which show little stereoselectivity in plasma concentrations, are mostly eliminated by acyl-glucuronidation. Renal clearance of the S enantiomer glucuronide is reported to be higher than that of the R enantiomer [279–281], which may be due to chiral inversion [282]. For tiaprofenic acid and flurbiprofen, little difference has been reported between enantiomers in amount of conjugated species recovered in urine, with S : R ratios near one [246,260]. For both drugs, some degree of stereoselectivity is apparent in plasma with S concentrations exceeding those of distomers. For these drugs, it is likely that there exists chiral inversion or stereoselectivity in oxidative or glucuronide metabolism.

Etodolac displays stereoselectivity in clearance of the enantiomers through glucuronidation, with S-etodolac having a 13-fold higher value than antipode [283,284]. Although not specifically studied, stereoselective clearance through other metabolic pathways, such as oxidation, account for most of the total body CL of the enantiomers. These pathways must also be highly stereoselective in order to account for the large magnitude of the differences between the enantiomers in plasma AUC, where inactive R enantiomer is about 10-fold higher than active S enantiomer [284]. Recently, it was shown by Boni et al. [285] that bioequivalence assessments of etodolac formulation must consider the concentrations of active S enantiomer. In the absence of stereospecific data, there was no detectable difference in C_{max} among three regimens differing in input rate of drug, and hence they were deemed to be bioequivalent. When a stereospecific assessment was followed, however, it was possible to identify bioinequivalence between the regimens based on the peak concentrations of the pharmacologically relevant S enantiomer.

Ketorolac apparently does not undergo chiral inversion. The CL/F and Vd/F of the enantiomers seemed to be similar whether given by im or oral doses [250,253], suggesting that bioavailability of each enantiomer is similar between the two routes of elimination.

Excretion

None of the NSAIDs appear to be significantly eliminated into urine directly in the unchanged form. Usually when significant amounts of unchanged drug are reported in urine, they are attributable to spontaneous hydrolysis of glucuronide conjugates. For ketoprofen, renal clearance of conjugated enantiomers was also nonstereoselective [281]. The clearance of glucuronidated etodolac enantiomers is nonstereoselective and appears to be facilitated by renal excretion [284]. In humans, bile is a minor pathway of excretion of glucuronidated etodolac and ketoprofen enantiomers [284,286].

15.3. Clinical Factors Affecting Plasma Concentrations of NSAIDs

Consideration of age-related changes in pharmacokinetics is especially relevant for the NSAIDs owing to the higher incidence of arthritic disease in the elderly population. Such physiological changes include decreases in albumin concentration, and in metabolic and renal efficiency, each of which could affect plasma concentrations of NSAIDs. A poignant example of the need to consider age-related changes in NSAIDs was provided by the chiral NSAID, benoxaprofen, which proved to be exceptionally toxic in elderly patients due to elevated and sustained plasma levels of the drug [287].

With respect to stereoselectivity, healthy elderly patients share a similar profile to healthy young subjects given single oral doses of etodolac (Table 1) [284]. For another NSAID, ketoprofen, young healthy subjects show circulating glucuronide conjugates near or below the level of detection. However, in elderly subjects the conjugate concentrations are readily measurable and stereoselective, with AUC of the S conjugates 2- or 3-fold higher than antipode [280]. In another study, Skeith et al. [281] have shown that elderly arthritic patients have same ketoprofen enantiomer AUCs but higher levels of conjugated ketoprofen enantiomers than do nonarthritic elderly volunteers. The renal clearance of the S-ketoprofen conjugate was significantly higher in the nonarthritic than the arthritic group. These arthritic patients had a significantly lower creatinine clearance than the nonarthritic volunteers, which may have contributed to the altered renal clearance of conjugates.

Cystic fibrosis patients may suffer from pulmonary inflammation that is treatable with NSAIDs. In 38 pediatric patients suffering from cystic fibrosis, the enantioselective pharmacokinetics of ibuprofen have been studied (Table 1) [288]. The S:R mean ratio of concentrations was 2.0 in this patient population, which is higher than that reported in most studies involving adults. Infants receiving ibuprofen have been reported to have a "flipped" ratio of ibuprofen enantiomer AUCs, as the R concentrations are higher than the S concentrations [289].

Some of the factors known to affect the chiral inversion of NSAIDs are discussed in the chapter by Davies.

16. CONCLUSION

As can be appreciated by this review of the literature, which is by no means comprehensive, there is a massive quantity of information related to stereoselectivity in pharmacodynamics and pharmacokinetics of drugs. As long as small molecules continue to be synthesized by medicinal chemists and used in patients, chirality will continue to be an important consideration in the pharmacological management of disease. Although the practice of using racemates has been questioned, to date there are relatively few examples where the use of a single enantiomer has documented clinical advantage.

Regulatory agencies and the pharmaceutical industry are well aware of the advances made in understanding the impact of stereochemistry on drug action and disposition. Because of possible issues related to development of the racemate, such as the need to understand the pharmacology and pharmacokinetics of two components in a dosage form rather than one, development of the racemate may be more complex than that of a single enantiomer or achiral drug. This will inevitably prompt the development of the eutomers of chiral drugs, or direct development efforts will be made preferentially toward achiral compounds. Hopefully this will not detract from the knowledge that many of the racemic drugs used currently have enjoyed therapeutic success in millions of patients receiving them. Avoidance of developing chiral drugs simply to circumvent issues related to the development of racemates per se may be disadvantageous, as chiral drugs may be the most pharmacologically active species in a series of structurally similar analogs. Development of the racemate may still have merit, in that production costs are less.

Many of the chiral drugs described in this and other chapters are well studied. Notable examples include β-blockers and the nonsteroidal anti-inflammatory drugs. Advances in the chiral aspects of some drugs, such as the general anesthetics, are providing exciting new insights into their mechanisms of action. In other cases, such as the β-adrenergic agonists, an appreciation of chirality may assist us in better understanding the phenomenon of tachyphylaxis with repeated use. On the other hand, it is apparent that for some drugs a paucity of published information persists in the literature. Examples of poorly understood chiral drugs include, at least from a pharmacokinetic aspect, the antiparkinson's anticholinergic drugs, and many of the diuretic agents (discussed in the following chapter). An explanation for the lack of scientific interest in some of these older drugs may be attributed to their loss of patent protection, which has diminished the influx of research secondary to a lack of financial interest. Nevertheless, since these drugs continue to be used for treatment of various diseases, an understanding of their stereoselective properties may increase their safety and clinical effectiveness. Undoubtedly, the scientific and medical community, and ultimately patients themselves, will continue to benefit from works, such as those reviewed here, that concentrate on the stereoselective aspects of chiral drugs.

REFERENCES

1. Riddell, F.G.; Robinson, M.J.T. J. H. Van't Hoff and J. A. Le Bel—their historical context. Tetrahedron **1974**, *30*, 2001–2007.
2. Ariens, E.J. Stereochemistry, a basis for sophisticated nonsense in pharmacokinetics and clinical pharmacology. Eur. J. Clin. Pharmacol. **1984**, *26*, 663–668.
3. Ariens, E.J.; Wuis, E.W.; Veringa, E.J. Stereoselectivity of bioactive xenobiotics. A pre-Pasteur attitude in medicinal chemistry, pharmacokinetics and clinical pharmacology. Biochem. Pharmacol. **1988**, *37*, 9–18.
4. Ariens, E.J. Racemic therapeutics—ethical and regulatory aspects. Eur. J. Clin. Pharmacol. **1991**, *41*, 89–93.
5. Pasutto, F.M. Mirror images: the analysis of pharmaceutical enantiomers. J. Clin. Pharmacol. **1992**, *32*, 917–924.
6. Okazaki, O.; Kojima, C.; Hakusui, H.; Nakashima, M. Enantioselective disposition of ofloxacin in humans. Antimicrob. Agents Chemother. **1991**, *35*, 2106–2109.
7. Minami, R.; Inotsume, N.; Nakamura, C.; Nakano, M. Stereoselective analysis of the disposition of tosufloxacin enantiomers in man. Eur. J. Clin. Pharmacol. **1993**, *45*, 489–491.
8. Humphrey, G.H.; Shapiro, M.A.; Randinitis, E.J.; Guttendorf, R.J.; Brodfuehrer, J.I. Pharmacokinetics of clinafloxacin enantiomers in humans. J. Clin. Pharmacol. **1999**, *39*, 1143–1150.
9. Gascon, A.R.; Campo, E.; Hernandez, R.M.; Calvo, B.; Errasti, J.; Pedraz Munoz, J.L. Pharmacokinetics of ofloxacin enantiomers after intravenous administration for antibiotic prophylaxis in biliary surgery. J. Clin. Pharmacol. **2000**, *40*, 869–874.
10. Smith, P.F.; Forrest, A.; Ballow, C.H.; Martin, D.E.; Proulx, L. Absolute bioavailability and disposition of (−) and (+) 2'-deoxy-3'-oxa-4'-thiocytidine (dOTC) following single intravenous and oral doses of racemic dOTC in humans. Antimicrob. Agents Chemother. **2000**, *44*, 1609–1615.

11. Baker, J.K.; McChesney, J.D. Differential metabolism of the enantiomers of primaquine. J. Pharm. Sci. **1988**, *77*, 380–382.

12. Basco, L.K.; Gillotin, C.; Gimenez, F.; Farinotti, R.; Le Bras, J. In vitro activity of the enantiomers of mefloquine, halofantrine and enpiroline against Plasmodium falciparum. Br. J. Clin. Pharmacol. **1992**, *33*, 517–520.

13. Fu, S.; Bjorkman, A.; Wahlin, B.; Ofori-Adjei, D.; Ericsson, O.; Sjoqvist, F. In vitro activity of chloroquine, the two enantiomers of chloroquine, desethylchloroquine and pyronaridine against Plasmodium falciparum. Br. J. Clin. Pharmacol. **1986**, *22*, 93–96.

14. Webster, R.V.; Craig, J.C.; Shyamala, V.; Kirby, G.C.; Warhu st, D.C. Antimalarial activity of optical isomers of quinacrine dihydrochlori e a inst chloroquine-sensitive and -resistant Plasmodium falciparum in vitro. Bio m. Pharmacol. **1991**, *42* (suppl), S225–S227.

15. Karle, J.M.; Olmeda, R.; Gerena, L.; Milhous, W.K. Plasmodium falciparum: role of absolute stereochemistry in the antimalarial activity of synthetic amino alcohol antimalarial agents. Exp. Parasitol. **1993**, *76*, 345–351.

16. Wesche, D.L.; Schuster, B.G.; Wang, W.X.; Woosley, R.L. Mechanism of cardiotoxicity of halofantrine. Clin. Pharmacol. Ther. **2000**, *67*, 521–529.

17. Brocks, D.R. Stereoselective halofantrine and desbutylhalofantrine disposition in the rat: cardiac and plasma concentrations, and plasma protein binding. Biopharm. Drug Dispos. **2002**, *23*, 9–15.

18. Donatelli, P.; Marchi, G.; Giuliani, L.; Gustafsson, L.L.; Pacifici, G.M. Stereoselective inhibition by chloroquine of histamine N-methyltrans-ferase in the human liver and brain. Eur. J. Clin. Pharmacol. **1994**, *47*, 345–349.

19. Cenni, B.; Meyer, J.; Brandt, R.; Betschart, B. The antimalarial drug halofantrine is bound mainly to low and high density lipoproteins in human serum. Br. J. Clin. Pharmacol. **1995**, *39*, 519–526.

20. Ofori-Adjei, D.; Ericsson, O.; Lindstrom, B.; Hermansson, J.; Adjepon-Yamoah, K.; Sjoqvist, F. Enantioselective analysis of chloroquine and desethylchloroquine after oral administration of racemic chloroquine. Ther. Drug Monit. **1986**, *8*, 457–461.

21. Brocks, D.R.; Skeith, K.J.; Johnston, C.; Emamibafrani, J.; Davis, P.; Russell, A.S.; Jamali, F. Hematologic disposition of hydroxychloroquine enantiomers. J. Clin. Pharmacol. **1994**, *34*, 1088 1097.

22. McLachlan, A.J.; Tett, S.E.; Cutler, D.J.; Day, R.O. Disposition and absorption of hydroxychloroquine enantiomers following a single dose of the racemate. Chirality **1994**, *6*, 360–364.

23. Augustijns, P.; Verbeke, N. Stereoselective pharmacokinetic properties of chloroquine and de-ethyl-chloroquine in humans. Clin. Pharmacokinet. **1993**, *24*, 259–269.

24. Midha, K.K.; Hubbard, J.W.; Rawson, M.J.; McKay, G.; Schwede, R. The roles of stereochemistry and partial areas in a parallel design study to assess the bioequivalence of two formulations of hydroxychloroquine: a drug with a very long half life. Eur. J. Pharm. Sci. **1996**, *28*, 283–292.

25. Tett, S.E.; McLachlan, A.J.; Cutler, D.J.; Day, R.O. Pharmacokinetics and pharmacodynamics of hydroxychloroquine enantiomers in patients with rheumatoid arthritis receiving multiple doses of racemate. Chirality **1994**, *6*, 355–359.
26. Ducharme, J.; Fieger, H.; Ducharme, M.P.; Khalil, S.K.; Wainer, I.W. Enantioselective disposition of hydroxychloroquine after a single oral dose of the racemate to healthy subjects. Br. J. Clin. Pharmacol. **1995**, *40*, 127–133.
27. Ofori-Adjei, D.; Ericsson, O.; Lindstrom, B.; Sjoqvist, F. Protein binding of chloroquine enantiomers and desethylchloroquine. Br. J. Clin. Pharmacol. **1986**, *22*, 356–358.
28. McLachlan, A.J.; Cutler, D.J.; Tett, S.E. Plasma protein binding of the enantiomers of hydroxychloroquine and metabolites. Eur. J. Clin. Pharmacol. **1993**, *44*, 481–484.
29. Brocks, D.R.; Pasutto, F.M.; Jamali, F. Analytical and semi-preparative high-performance liquid chromatographic separation and assay of hydroxychloroquine enantiomers. J. Chromatogr. **1992**, *581*, 83–92.
30. Fieger, H.; Iredale, J.; Wainer, I.W. Enantioselective determination of hydroxychloroquine and its major metabolites in urine and the observation of a reversal in the (+)/(−)-hydroxychloroquine ratio. Chirality **1993**, *5*, 65–70.
31. Gimenez, F.; Gillotin, C.; Basco, L.K.; Bouchaud, O.; Aubry, A.F.; Wainer, I.W.; Le Bras, J.; Farinotti, R. Plasma concentrations of the enantiomers of halofantrine and its main metabolite in malaria patients. Eur. J. Clin. Pharmacol. **1994**, *46*, 561–562.
32. Abernethy, D.R.; Wesche, D.L.; Barbey, J.T.; Ohrt, C.; Mohanty, S.; Pezzullo, J.C.; Schuster, B.G. Stereoselective halofantrine disposition and effect: concentration-related QTc prolongation. Br. J. Clin. Pharmacol. **2001**, *51*, 231–237.
33. Terefe, H.; Blaschke, G. Direct determination of the enantiomers of halofantrine and its pharmacologically active metabolite N-desbutylhalofantrine by high-performance liquid chromatography. J. Chromatogr. **1993**, *615*, 347–351.
34. Brocks, D.R.; Toni, J.W. Pharmacokinetics of halofantrine in the rat: stereoselectivity and interspecies comparisons. Biopharm. Drug Dispos. **1999**, *20*, 165–169.
35. Brocks, D.R.; Ramaswamy, M.; MacInnes, A.I.; Wasan, K.M. The stereoselective distribution of halofantrine enantiomers within human, dog, and rat plasma lipoproteins. Pharm. Res. **2000**, *17*, 427–431.
36. Gimenez, F.; Pennie, R.A.; Koren, G.; Crevoisier, C.; Wainer, I.W.; Farinotti, R. Stereoselective pharmacokinetics of mefloquine in healthy Caucasians after multiple doses. J. Pharm. Sci. **1994**, *83*, 824–827.
37. Martin, C.; Gimenez, F.; Bangchang, K.N.; Karbwang, J.; Wainer, I.W.; Farinotti, R. Whole blood concentrations of mefloquine enantiomers in healthy Thai volunteers. Eur. J. Clin. Pharmacol. **1994**, *47*, 85–87.
38. Hellgren, U.; Berggren-Palme, I.; Bergqvist, Y.; Jerling, M. Enantioselective pharmacokinetics of mefloquine during long-term intake of the prophylactic dose. Br. J. Clin. Pharmacol. **1997**, *44*, 119–124.

39. Bourahla, A.; Martin, C.; Gimenez, F.; Singhasivanon, V.; Attanath, P.; Sabchearon, A.; Chongsuphajaisiddhi, T.; Farinotti, R. Stereoselective pharmacokinetics of mefloquine in young children. Eur. J. Clin. Pharmacol. **1996**, *50*, 241–244.

40. Pham, Y.T.; Nosten, F.; Farinotti, R.; White, N.J.; Gimenez, F. Cerebral uptake of mefloquine enantiomers in fatal cerebral malaria. Int. J. Clin. Pharmacol. Ther. **1999**, *37*, 58–61.

41. Vidrequin, S.; Gimenez, F.; Basco, L.K.; Martin, C.; LeBras, J.; Farinotti, R. Uptake of mefloquine enantiomers into uninfected and malaria-infected erythrocytes. Drug Metab. Dispos. **1996**, *24*, 689–691.

42. Basco, L.K.; Peytavin, G.; Gimenez, F.; Genissel, B.; Farinotti, R.; Le Bras, J. In vitro activity of the enantiomers of N-desbutyl derivative of halofantrine. Trop. Med. Parasitol. **1994**, *45*, 45–46.

43. Basco, L.K.; Gillotin, C.; Gimenez, F.; Farinotti, R.; Le Bras, J. Absence of antimalarial activity or interaction with mefloquine enantiomers in vitro of the main human metabolite of mefloquine. Trans. R. Soc. Trop. Med. Hyg. **1991**, *85*, 208–209.

44. Marques, M.P.; Takayanagui, O.M.; Bonato, P.S.; Santos, S.R.; Lanchote, V.L. Enantioselective kinetic disposition of albendazole sulfoxide in patients with neurocysticercosis. Chirality **1999**, *11*, 218–223.

45. Eriksson, T.; Bjorkman, S.; Roth, B.; Fyge, A.; Hoglund, P. Stereospecific determination, chiral inversion in vitro and pharmacokinetics in humans of the enantiomers of thalidomide. Chirality **1995**, *7*, 44–52.

46. Eriksson, T.; Bjorkman, S.; Roth, B.; Fyge, A.; Hoglund, P. Enantiomers of thalidomide: blood distribution and the influence of serum albumin on chiral inversion and hydrolysis. Chirality **1998**, *10*, 223–228.

47. Choonara, I.A.; Cholerton, S.; Haynes, B.P.; Breckenridge, A.M.; Park, B.K. Stereoselective interaction between the R enantiomer of warfarin and cimetidine. Br. J. Clin. Pharmacol. **1986**, *21*, 271–277.

48. Pratt, S.K.; Winn, M.J.; Park, B.K. The disposition of the enantiomers of warfarin following chronic administration to rats: relationship to anticoagulant response. J. Pharm. Pharmacol. **1989**, *41*, 743–746.

49. Breckenridge, A.; Orme, M.; Wesseling, H.; Lewis, R.J.; Gibbons, R. Pharmacokinetics and pharmacodynamics of the enantiomers of warfarin in man. Clin. Pharmacol. Ther. **1974**, *15*, 424–430.

50. Schmidt, W.; Jahnchen, E. Stereoselective drug distribution and anticoagulant potency of the enantiomers of phenprocoumon in rats. J. Pharm. Pharmacol. **1977**, *29*, 266–271.

51. Yamasaki, K.; Maruyama, T.; Kragh-Hansen, U.; Otagiri, M. Characterization of site I on human serum albumin: concept about the structure of a drug binding site. Biochim. Biophys. Acta **1996**, *1295*, 147–157.

52. Sudlow, G.; Birkett, D.J.; Wade, D.N. Further characterization of specific drug binding sites on human serum albumin. Mol. Pharmacol. **1976**, *12*, 1052–1061.

53. Kelly, J.G.; O'Malley, K. Clinical pharmacokinetics of oral anticoagulants. Clin. Pharmacokinet. **1979**, *4*, 1–15.

54. Yacobi, A.; Levy, G. Protein binding of warfarin enantiomers in serum of humans and rats. J. Pharmacokinet. Biopharm. **1977**, *5*, 123–131.

55. Oester, Y.T.; Keresztes-Nagy, S.; Mais, R.F.; Becktel, J.; Zaroslinski, J.F. Effect of temperature on binding of warfarin by human serum albumin. J. Pharm. Sci. **1976**, *65*, 1673–1677.

56. O'Reilly, R.A. The binding of sodium warfarin to plasma albumin and its displacement by phenylbutazone. Ann. N Y Acad. Sci. **1973**, *226*, 293–308.

57. O'Reilly, R.A. Interaction of several coumarin compounds with human and canine plasma albumin. Mol. Pharmacol. **1971**, *7*, 209–218.

58. Koch-Weser, J.; Sellers, E.M. Drug interactions with coumarin anticoagulants. New Eng. J. Med. **1971**, *285*, 547–558.

59. Cheung, W.K.; Levy, G. Comparative pharmacokinetics of coumarin anticoagulants. XLIX: Nonlinear tissue distribution of S-warfarin in rats. J. Pharm. Sci. **1989**, *78*, 541–546.

60. Lewis, R.J.; Trager, W.F.; Chan, K.K.; Breckenridge, A.; Orme, M.; Roland, M.; Schary, W. Warfarin. Stereochemical aspects of its metabolism and the interaction with phenylbutazone. J. Clin. Invest. **1974**, *53*, 1607–1617.

61. Yamazaki, H.; Shimada, T. Human liver cytochrome P450 enzymes involved in the 7-hydroxylation of R- and S-warfarin enantiomers. Biochem. Pharmacol. **1997**, *54*, 1195–1203.

62. Levy, G.; O'Reilly, R.A.; Wingard, L.B., Jr. Comparative pharmacokinetics of coumarin anticoagulants XXXV: Examination of possible pharmacokinetic interaction between (R)-(+)- and (S)-(−)-warfarin in humans. J. Pharm. Sci. **1978**, *67*, 867–868.

63. Steward, D.J.; Haining, R.L.; Henne, K.R.; Davis, G.; Rushmore, T.H.; Trager, W.F.; Rettie, A.E. Genetic association between sensitivity to warfarin and expression of CYP2C9*3. Pharmacogenetics **1997**, *7*, 361–367.

64. Takahashi, H.; Kashima, T.; Nomizo, Y.; Muramoto, N.; Shimizu, T.; Nasu, K.; Kubota, T.; Kimura, S.; Echizen, H. Metabolism of warfarin enantiomers in Japanese patients with heart disease having different CYP2C9 and CYP2C19 genotypes. Clin. Pharmacol. Ther. **1998**, *63*, 519–528.

65. McAleer, S.D.; Foondun, A.S.; Feely, M.; Chrystyn, H. Steady-state clearance rates of warfarin and its enantiomers in therapeutically dosed patients. Chirality **1997**, *9*, 13–16.

66. Chan, E.; McLachlan, A.J.; Pegg, M.; MacKay, A.D.; Cole, R.B.; Rowland, M. Disposition of warfarin enantiomers and metabolites in patients during multiple dosing with rac-warfarin. Br. J. Clin. Pharmacol. **1994**, *37*, 563–569.

67. Gòdbillon, J.; Richard, J.; Gerardin, A.; Meinertz, T.; Kasper, W.; Jahnchen, E. Pharmacokinetics of the enantiomers of acenocoumarol in man. Br. J. Clin. Pharmacol. **1981**, *12*, 621–629.

68. Hermans, J.J.; Thijssen, H.H. Human liver microsomal metabolism of the enantiomers of warfarin and acenocoumarol: P450 isozyme diversity determines the differences in their pharmacokinetics. Br. J. Pharmacol. **1993**, *110*, 482–490.

69. Jahnchen, E.; Meinertz, T.; Gilfrich, H.J.; Groth, U.; Martini, A. The enantiomers of phenprocoumon: pharmacodynamic and pharmacokinetic studies. Clin. Pharmacol. Ther. **1976**, *20*, 342–349.
70. Black, D.J.; Kunze, K.L.; Wienkers, L.C.; Gidal, B.E.; Seaton, T.L.; McDonnell, N.D.; Evans, J.S.; Bauwens, J.E.; Trager, W.F. Warfarin-fluconazole. II. A metabolically based drug interaction: in vivo studies. Drug Metab. Dispos. **1996**, *24*, 422–428.
71. Takahashi, H.; Sato, T.; Shimoyama, Y.; Shioda, N.; Shimizu, T.; Kubo, S.; Tamura, N.; Tainaka, H.; Yasumori, T.; Echizen, H. Potentiation of anticoagulant effect of warfarin caused by enantioselective metabolic inhibition by the uricosuric agent benzbromarone. Clin. Pharmacol. Ther. **1999**, *66*, 569–581.
72. Liao, S.; Palmer, M.; Fowler, C.; Nayak, R.K. Absence of an effect of levofloxacin on warfarin pharmacokinetics and anticoagulation in male volunteers. J. Clin. Pharmacol. **1996**, *36*, 1072–1077.
73. Gidal, B.E.; Sorkness, C.A.; McGill, K.A.; Larson, R.; Levine, R.R. Evaluation of a potential enantioselective interaction between ticlopidine and warfarin in chronically anticoagulated patients. Ther. Drug Monit. **1995**, *17*, 33–38.
74. Turck, D.; Su, C.A.; Heinzel, G.; Busch, U.; Bluhmki, E.; Hoffmann, J. Lack of interaction between meloxicam and warfarin in healthy volunteers. Eur. J. Clin. Pharmacol. **1997**, *51*, 421–425.
75. Kwan, D.; Bartle, W.R.; Walker, S.E. The effects of acetaminophen on pharmacokinetics and pharmacodynamics of warfarin. J. Clin. Pharmacol. **1999**, *39*, 68–75.
76. Mailloux, A.T.; Gidal, B.E.; Sorkness, C.A. Potential interaction between warfarin and dicloxacillin. Ann. Pharmacother. **1996**, *30*, 1402–1407.
77. Awni, W.M.; Hussein, Z.; Granneman, G.R.; Patterson, K.J.; Dube, L.M.; Cavanaugh, J.H. Pharmacodynamic and stereoselective pharmacokinetic interactions between zileuton and warfarin in humans. Clin. Pharmacokinet. **1995**, *29* (Suppl 2), 67–76.
78. Hoglund, P.; Paulsen, O.; Bogentoft, S. No effect of terodiline on anticoagulation effect of warfarin and steady-state plasma levels of warfarin enantiomers in healthy volunteers. Ther. Drug Monit. **1989**, *11*, 667–673.
79. Bonnabry, P.; Desmeules, J.; Rudaz, S.; Leemann, T.; Veuthey, J.L.; Dayer, P. Stereoselective interaction between piroxicam and acenocoumarol. Br. J. Clin. Pharmacol. **1996**, *41*, 525–530.
80. Matsumoto, K.; Ishida, S.; Ueno, K.; Hashimoto, H.; Takada, M.; Tanaka, K.; Kamakura, S.; Miyatake, K.; Shibakawa, M. The stereoselective effects of bucolome on the pharmacokinetics and pharmacodynamics of racemic warfarin. J. Clin. Pharmacol. **2001**, *41*, 459–464.
81. Balani, S.K.; Xu, X.; Pratha, V.; Koss, M.A.; Amin, R.D.; Dufresne, C.; Miller, R.R.; Arison, B.H.; Doss, G.A.; Chiba, M.; Freeman, A.; Holland, S.D.; Schwartz, J.I.; Lasseter, K.C.; Gertz, B.J.; Isenberg, J.I.; Rogers, J.D.; Lin, J.H.; Baillie, T.A. Metabolic profiles of montelukast sodium (Singulair),

a potent cysteinyl leukotriene1 receptor antagonist, in human plasma and bile. Drug Metab. Dispos. **1997**, *25*, 1282–1287.

82. Waldeck, B. Biological significance of the enantiomeric purity of drugs. Chirality **1993**, *5*, 350–355.

83. Dobbins, D.E.; Buehn, M.J.; Dabney, J.M. Stereospecificity of the anti-inflammatory actions of terbutaline. Microcirc. Endothelium Lymphatics **1990**, *6*, 3–20.

84. Dobbins, D.E.; Buehn, M.J.; Dabney, J.M. Bradykinin-mediated edema formation is blocked by levorotatory but not dextrorotatory terbutaline. Microcirc. Endothelium Lymphatics **1988**, *4*, 377–397.

85. Brittain, R.T.; Farmer, J.B.; Marshall, R.J. Some observations on the -adrenoceptor agonist properties of the isomers of salbutamol. Br. J. Pharmacol. **1973**, *48*, 144–147.

86. Hartley, D.; Middlemiss, D. Absolute configuration of the optical isomers of salbutamol. J. Med. Chem. **1971**, *14*, 995–996.

87. Hawkins, C.J.; Klease, G.T. Relative potency of (−)- and (plus-)-salbutamol on guinea pig tracheal tissue. J. Med. Chem. **1973**, *16*, 856–857.

88. Jeppsson, A.B.; Johansson, U.; Waldeck, B. Steric aspects of agonism and antagonism at beta-adrenoceptors: experiments with the enantiomers of terbutaline and pindolol. Acta Pharmacol. Toxicol. (Copenh) **1984**, *54*, 285–291.

89. Kiyomoto, A.; Iwasawa, Y.; Harigaya, S. Studies on tetrahydroisoquinolines (THI). VI. Effects of trimetoquinol on tracheal and some other smooth muscles. Arzneimittelforschung **1970**, *20*, 46–52.

90. Dalton, C.; Crowley, H.J.; Czyzewski, L.B. Trimethoquinol-different pharmacological properties of optical isomers. Biochem. Pharmacol. **1976**, *25*, 2209–2210.

91. Martin, P.; Puech, A.J.; Brochet, D.; Soubrie, P.; Simon, P. Comparison of clenbuterol enantiomers using four psychopharmacological tests sensitive to beta-agonists. Eur. J. Pharmacol. **1985**, *117*, 127–129.

92. Waldeck, B.; Widmark, E. Steric aspects of agonism and antagonism at beta-adrenoceptors: experiments with the enantiomers of clenbuterol. Acta Pharmacol. Toxicol. (Copenh) **1985**, *56*, 221–227.

93. Lindberg, S.; Khan, R.; Runer, T. The effects of formoterol, a long-acting beta 2-adrenoceptor agonist, on mucociliary activity. Eur. J. Pharmacol. **1995**, *285*, 275–280.

94. Penn, R.B.; Frielle, T.; McCullough, J.R.; Aberg, G.; Benovic, J.L. Comparison of R-, S-, and RS-albuterol interaction with human beta 1- and beta 2-adrenergic receptors. Clin. Rev. Allergy Immunol. **1996**, *14*, 37–45.

95. Handley, D.A.; Walle, T.; Fang, K.O.; Hett, R.; Gao, Y. Preclinical and metabolic profiles of (RR)-formoterol. Ann. Allergy Asthma Immunol. 78 (abstr), **1997**, 94.

96. Boulton, D.W.; Fawcett, J.P. Pharmacokinetics and pharmacodynamics of single oral doses of albuterol and its enantiomers in humans. Clin. Pharmacol. Ther. **1997**, *62*, 138–144.

97. Ramsay, C.M.; Cowan, J.; Flannery, E.; McLachlan, C.; Taylor, D.R. Bronchoprotective and bronchodilator effects of single doses of (S)-salbutamol, (R)-salbutamol and racemic salbutamol in patients with bronchial asthma. Eur. J. Clin. Pharmacol. **1999**, *55*, 353–359.

98. Cockcroft, D.W.; Swystun, V.A. Effect of single doses of S-salbutamol, R-salbutamol, racemic salbutamol, and placebo on the airway response to methacholine. Thorax **1997**, *52*, 845–848.

99. Perrin-Fayolle, M.; Blum, P.S.; Morley, J.; Grosclaude, M.; Chambe, M.T. Differential responses of asthmatic airways to enantiomers of albuterol. Implications for clinical treatment of asthma. Clin. Rev. Allergy Immunol. **1996**, *14*, 139–147.

100. Waldeck, B. Enantiomers of bronchodilating beta2-adrenoceptor agonists: is there a cause for concern? J. Allergy Clin. Immunol. **1999**, *103*, 742–748.

101. Handley, D. The asthma-like pharmacology and toxicology of (S)-isomers of beta agonists. J. Allergy Clin. Immunol. **1999**, *104*, S69–S76.

102. Handley, D.A.; McCullough, J.R.; Crowther, S.D.; Morley, J. Sympathomimetic enantiomers and asthma. Chirality **1998**, *10*, 262–272.

103. Hansel, T.T.; Schwarz, F.; Villiger, B.; Naef, R.; Richardson, B.P.; Morley, J. Anomalous bronchospasm following inhalation of (+) isoprenaline by asthmatics. Agents Actions Suppl. **1993**, *43*, 281–288.

104. Spitzer, W.O.; Suissa, S.; Ernst, P.; Horwitz, R.I.; Habbick, B.; Cockcroft, D.; Boivin, J.F.; McNutt, M.; Buist, A.S.; Rebuck, A.S. The use of beta-agonists and the risk of death and near death from asthma. N. Engl. J. Med. **1992**, *326*, 501–506.

105. Sears, M.R.; Taylor, D.R. The beta 2-agonist controversy. Observations, explanations and relationship to asthma epidemiology. Drug Saf. **1994**, *11*, 259–283.

106. Leff, A.R.; Herrnreiter, A.; Naclerio, R.M.; Baroody, F.M.; Handley, D.A.; Munoz, N.M. Effect of enantiomeric forms of albuterol on stimulated secretion of granular protein from human eosinophils. Pulm. Pharmacol. Ther. **1997**, *10*, 97–104.

107. Mitra, S.; Ugur, M.; Ugur, O.; Goodman, H.M.; McCullough, J.R.; Yamaguchi, H. (S)-Albuterol increases intracellular free calcium by muscarinic receptor activation and a phospholipase C-dependent mechanism in airway smooth muscle. Mol. Pharmacol. **1998**, *53*, 347–354.

108. Nelson, H.S.; Bensch, G.; Pleskow, W.W.; DiSantostefano, R.; DeGraw, S.; Reasner, D.S.; Rollins, T.E.; Rubin, P.D. Improved bronchodilation with levalbuterol compared with racemic albuterol in patients with asthma. J. Allergy Clin. Immunol. **1998**, *102*, 943–952.

109. Handley, D.A.; Tinkelman, D.; Noonan, M.; Rollins, T.E.; Snider, M.E.; Caron, J. Dose-response evaluation of levalbuterol versus racemic albuterol in patients with asthma. J. Asthma **2000**, *37*, 319–327.

110. Handley, D.A. Single-isomer beta-agonists. Pharmacotherapy **2001**, *21*, 21S–27S.

111.   Handley, D.A.; Anderson, A.J.; Koester, J.; Snider, M.E. New millennium bronchodilators for asthma: single-isomer beta agonists. Curr. Opin. Pulm. Med. **2000**, *6*, 43–49.

112.   Boulton, D.W.; Fawcett, J.P. Enantioselective disposition of salbutamol in man following oral and intravenous administration. Br. J. Clin. Pharmacol. **1996**, *41*, 35–40.

113.   Boulton, D.W.; Fawcett, J.P. Enantioselective disposition of albuterol in humans. Clin. Rev. Allergy Immunol. **1996**, *14*, 115–138.

114.   Ward, J.K.; Dow, J.; Dallow, N.; Eynott, P.; Milleri, S.; Ventresca, G.P. Enantiomeric disposition of inhaled, intravenous and oral racemic-salbutamol in man–no evidence of enantioselective lung metabolism. Br. J. Clin. Pharmacol. **2000**, *49*, 15–22.

115.   Mehvar, R.; Jamali, F. Bioequivalence of chiral drugs. Stereospecific versus non-stereospecific methods. Clin. Pharmacokinet. **1997**, *33*, 122–141.

116.   Walle, U.K.; Pesola, G.R.; Walle, T. Stereoselective sulphate conjugation of salbutamol in humans: comparison of hepatic, intestinal and platelet activity. Br. J. Clin. Pharmacol. **1993**, *35*, 413–418.

117.   Borgstrom, L.; Nyberg, L.; Jonsson, S.; Lindberg, C.; Paulson, J. Pharmacokinetic evaluation in man of terbutaline given as separate enantiomers and as the racemate. Br. J. Clin. Pharmacol. **1989**, *27*, 49–56.

118.   Borgstrom, L.; Kennedy, B.M.; Nilsson, B.; Angelin, B. Relative absorption of the two enantiomers of terbutaline after duodenal administration. Eur. J. Clin. Pharmacol. **1990**, *38*, 621–623.

119.   Hartman, A.P.; Wilson, A.A.; Wilson, H.M.; Aberg, G.; Falany, C.N.; Walle, T. Enantioselective sulfation of beta 2-receptor agonists by the human intestine and the recombinant M-form phenolsulfotransferase. Chirality **1998**, *10*, 800–803.

120.   Borgstrom, L.; Liu, C.X.; Walhagen, A. Pharmacokinetics of the enantiomers of terbutaline after repeated oral dosing with racemic terbutaline. Chirality **1989**, *1*, 174–177.

121.   Morgan, D.J. Clinical pharmacokinetics of beta-agonists. Clin. Pharmacokinet. **1990**, *18*, 270–294.

122.   Gumbhir-Shah, K.; Kellerman, D.J.; DeGraw, S.; Koch, P.; Jusko, W.J. Pharmacokinetics and pharmacodynamics of cumulative single doses of inhaled salbutamol enantiomers in asthmatic subjects. Pulm. Pharmacol. Ther. **1999**, *12*, 353–362.

123.   Koster, A.S.; Frankhuijzen-Sierevogel, A.C.; Mentrup, A. Stereoselective formation of fenoterol-para-glucuronide and fenoterol-meta-glucuronide in rat hepatocytes and enterocytes. Biochem. Pharmacol. **1986**, *35*, 1981–1985.

124.   Wilson, A.A.; Wang, J.; Koch, P.; Walle, T. Stereoselective sulphate conjugation of fenoterol by human phenolsulphotransferases. Xenobiotica **1997**, *27*, 1147–1154.

125.   Zhang, M.; Fawcett, J.P.; Kennedy, J.M.; Shaw, J.P. Stereoselective glucuronidation of formoterol by human liver microsomes. Br. J. Clin. Pharmacol. **2000**, *49*, 152–157.

126. Kamimura, H.; Sasaki, H.; Higuchi, S.; Shiobara, Y. Quantitative determination of the beta-adrenoceptor stimulant formoterol in urine by gas chromatography mass spectrometry. J. Chromatogr. **1982**, *229*, 337–345.
127. Arvidsson, P.; Larsson, S.; Lofdahl, C.G. Objective and subjective bronchodilation over 12 hours after inhaled formoterol: individual responses. J. Asthma **1993**, *30*, 459–465.
128. Berges, R.; Segura, J.; Ventura, R.; Fitch, K.D.; Morton, A.R.; Farre, M.; Mas, M.; de La Torre, X. Discrimination of prohibited oral use of salbutamol from authorized inhaled asthma treatment. Clin. Chem. **2000**, *46*, 1365–1375.
129. Sweeny, D.J.; Nellans, H.N. Enantiomeric activation of glucuronidation in dog hepatic microsomes. J. Biol. Chem. **1992**, *267*, 13171–13174.
130. Wong, S.L.; Awni, W.M.; Cavanaugh, J.H.; el-Shourbagy, T.; Locke, C.S.; Dube, L.M. The pharmacokinetics of single oral doses of zileuton 200 to 800 mg, its enantiomers, and its metabolites, in normal healthy volunteers. Clin. Pharmacokinet. **1995**, *29* (suppl. 2), 9–21.
131. Sweeny, D.J.; Nellans, H.N. Stereoselective glucuronidation of zileuton isomers by human hepatic microsomes. Drug Metab. Dispos. **1995**, *23*, 149–153.
132. Machinist, J.M.; Kukulka, M.J.; Bopp, B.A. In vitro plasma protein binding of zileuton and its N-dehydroxylated metabolite. Clin. Pharmacokinet. **1995**, *29* (suppl. 2), 34–41.
133. Eap, C.B.; Guentert, T.W.; Schaublin-Loidl, M.; Stabl, M.; Koeb, L.; Powell, K.; Baumann, P. Plasma levels of the enantiomers of thioridazine, thioridazine 2-sulfoxide, thioridazine 2-sulfone, and thioridazine 5-sulfoxide in poor and extensive metabolizers of dextromethorphan and mephenytoin. Clin. Pharmacol. Ther. **1996**, *59*, 322–331.
134. Jortani, S.A.; Valentour, J.C.; Poklis, A. Thioridazine enantiomers in human tissues. Forensic Sci. Int. **1994**, *64*, 165–170.
135. D'Souza, D.L.; Dimmitt, D.C.; Robbins, D.K.; Nezamis, J.; Simms, L.; Koch, K.M. Effect of alosetron on the pharmacokinetics of fluoxetine. J. Clin. Pharmacol. **2001**, *41*, 455–458.
136. Fuller, R.W.; Snoddy, H.D.; Krushinski, J.H.; Robertson, D.W. Comparison of norfluoxetine enantiomers as serotonin uptake inhibitors in vivo. Neuropharmacology **1992**, *31*, 997–1000.
137. Wong, D.T.; Bymaster, F.P.; Reid, L.R.; Mayle, D.A.; Krushinski, J.H.; Robertson, D.W. Norfluoxetine enantiomers as inhibitors of serotonin uptake in rat brain. Neuropsychopharmacology **1993**, *8*, 337–344.
138. Fjordside, L.; Jeppesen, U.; Eap, C.B.; Powell, K.; Baumann, P.; Brosen, K. The stereoselective metabolism of fluoxetine in poor and extensive metabolizers of sparteine. Pharmacogenetics **1999**, *9*, 55–60.
139. Sidhu, J.; Priskorn, M.; Poulsen, M.; Segonzac, A.; Grollier, G.; Larsen, F. Steady-state pharmacokinetics of the enantiomers of citalopram and its metabolites in humans. Chirality **1997**, *9*, 686–692.
140. Foglia, J.P.; Pollock, B.G.; Kirshner, M.A.; Rosen, J.; Sweet, R.; Mulsant, B, Plasma levels of citalopram enantiomers and metabolites in elderly patients. Psychopharmacol. Bull. **1997**, *33*, 109–112.

141. Eap, C.B.; Yasui, N.; Kaneko, S.; Baumann, P.; Powell, K.; Otani, K. Effects of carbamazepine coadministration on plasma concentrations of the enantiomers of mianserin and of its metabolites. Ther. Drug Monit. 1999, *21*, 166–170.

142. Yasui, N.; Tybring, G.; Otani, K.; Mihara, K.; Suzuki, A.; Svensson, J.O.; Kaneko, S. Effects of thioridazine, an inhibitor of CYP2D6, on the steady-state plasma concentrations of the enantiomers of mianserin and its active metabolite, desmethylmianserin, in depressed Japanese patients. Pharmacogenetics 1997, *7*, 369–374.

143. Denolle, T.; Pellizzoni, C.; Jannuzzo, M.G.; Poggesi, I. Hemodynamic effects of reboxetine in healthy male volunteers. Clin. Pharmacol. Ther. 1999, *66*, 282–287.

144. Rey, E.; Dostert, P.; d'Athis, P.; Jannuzzo, M.G.; Poggesi, I.; Olive, G. Dose proportionality of reboxetine enantiomers in healthy male volunteers. Biopharm. Drug Dispos. 1999, *20*, 177–181.

145. Fleishaker, J.C.; Mucci, M.; Pellizzoni, C.; Poggesi, I. Absolute bioavailability of reboxetine enantiomers and effect of gender on pharmacokinetics. Biopharm. Drug Dispos. 1999, *20*, 53–57.

146. Krause, W.; Kuhne, G.; Sauerbrey, N. Pharmacokinetics of (+)-rolipram and (−)-rolipram in healthy volunteers. Eur. J. Clin. Pharmacol. 1990, *38*, 71–75.

147. Weber-Grandke, H.; Hahn, G.; Mutschler, E.; Mohrke, W.; Langguth, P.; Spahn-Langguth, H. The pharmacokinetics of tranylcypromine enantiomers in healthy subjects after oral administration of racemic drug and the single enantiomers. Br. J. Clin. Pharmacol. 1993, *36*, 363–365.

148. Dahl-Puustinen, M.L.; Perry, T.L.; Jr., Dumont, E.; von Bahr, C.; Nordin, C.; Bertilsson, L. Stereoselective disposition of racemic E-10-hydroxynortriptyline in human beings. Clin. Pharmacol. Ther. 1989, *45*, 650–656.

149. Eap, C.B.; Bender, S.; Gastpar, M.; Fischer, W.; Haarmann, C.; Powell, K.; Jonzier-Perey, M.; Cochard, N.; Baumann, P. Steady state plasma levels of the enantiomers of trimipramine and of its metabolites in CYP2D6-, CYP2C19- and CYP3A4/5-phenotyped patients. Ther. Drug Monit. 2000, *22*, 209–214.

150. Gross, G.; Xin, X.; Gastpar, M. Trimipramine: pharmacological reevaluation and comparison with clozapine. Neuropharmacology 1991, *30*, 1159–1166.

151. Srinivas, N.R.; Hubbard, J.W.; Korchinski, E.D.; Midha, K.K. Enantioselective pharmacokinetics of dl-threo-methylphenidate in humans. Pharm. Res. 1993, *10*, 14–21.

152. Srinivas, N.R.; Hubbard, J.W.; Quinn, D.; Midha, K.K. Enantioselective pharmacokinetics and pharmacodynamics of dl-threo-methylphenidate in children with attention deficit hyperactivity disorder. Clin. Pharmacol. Ther. 1992, *52*, 561–568.

153. Hubbard, J.W.; Srinivas, N.R.; Quinn, D.; Midha, K.K. Enantioselective aspects of the disposition of dl-threo-methylphenidate after the administration of a sustained-release formulation to children with attention deficit-hyperactivity disorder. J. Pharm. Sci. 1989, *78*, 944–947.

154. Fernandez, C.; Martin, C.; Gimenez, F.; Farinotti, R. Clinical pharmacokinetics of zopiclone. Clin. Pharmacokinet. **1995**, *29*, 431–441.

155. Fernandez, C.; Maradeix, V.; Gimenez, F.; Thuillier, A.; Farinotti, R. Pharmacokinetics of zopiclone and its enantiomers in Caucasian young healthy volunteers. Drug Metab. Dispos. **1993**, *21*, 1125–1128.

156. Fernandez, C.; Gimenez, F.; Thuillier, A.; Farinotti, R. Stereoselective binding of zopiclone to human plasma proteins. Chirality **1999**, *11*, 129–132.

157. Lim, W.H.; Hooper, W.D. Stereoselective metabolism and pharmacokinetics of racemic methylphenobarbital in humans. Drug Metab. Dispos. **1989**, *17*, 212–217.

158. Chandler, M.H.; Scott, S.R.; Blouin, R.A. Age-associated stereoselective alterations in hexobarbital metabolism. Clin. Pharmacol. Ther. **1988**, *43*, 436–441.

159. Smith, D.A.; Chandler, M.H.; Shedlofsky, S.I.; Wedlund, P.J.; Blouin, R.A. Age-dependent stereoselective increase in the oral clearance of hexobarbitone isomers caused by rifampicin. Br. J. Clin. Pharmacol. **1991**, *32*, 735–739.

160. Patel, M.; Tang, B.K.; Grant, D.M.; Kalow, W. Interindividual variability in the glucuronidation of (S) oxazepam contrasted with that of (R) oxazepam. Pharmacogenetics **1995**, *5*, 287–297.

161. Brocks, D.R. Anticholinergic drugs used in Parkinson's disease: an overlooked class of drugs from a pharmacokinetic perspective. J. Pharm. Pharm. Sci. **1999**, *2*, 39–46.

162. Eltze, M.; Figala, V. Affinity and selectivity of biperiden enantiomers for muscarinic receptor subtypes. Eur. J. Pharmacol. **1988**, *158*, 11–19.

163. Wedlund, P.J.; Aslanian, W.S.; Jacqz, E.; McAllister, C.B.; Branch, R.A.; Wilkinson, G.R. Phenotypic differences in mephenytoin pharmacokinetics in normal subjects. J. Pharmacol. Exp. Ther. **1985**, *234*, 662–669.

164. Ieiri, I.; Goto, W.; Hirata, K.; Toshitani, A.; Imayama, S.; Ohyama, Y.; Yamada, H.; Ohtsubo, K.; Higuchi, S. Effect of 5-(p-hydroxyphenyl)-5-phenylhydantoin (p-HPPH) enantiomers, major metabolites of phenytoin, on the occurrence of chronic-gingival hyperplasia: in vivo and in vitro study. Eur. J. Clin. Pharmacol. **1995**, *49*, 51–56.

165. Mamiya, K.; Ieiri, I.; Shimamoto, J.; Yukawa, E.; Imai, J.; Ninomiya, H.; Yamada, H.; Otsubo, K.; Higuchi, S.; Tashiro, N. The effects of genetic polymorphisms of CYP2C9 and CYP2C19 on phenytoin metabolism in Japanese adult patients with epilepsy: studies in stereoselective hydroxylation and population pharmacokinetics. Epilepsia **1998**, *39*, 1317–1323.

166. Hooper, W.D.; O'Shea, N.J.; Qing, M.S. Enantioselective pharmacokinetics of ethotoin in humans following single oral doses of the racemate. Chirality **1992**, *4*, 142–147.

167. Villen, T.; Bertilsson, L.; Sjoqvist, F. Nonstereoselective disposition of ethosuximide in humans. Ther. Drug Monit. **1990**, *12*, 514–516.

168. Tomson, T.; Villen, T. Ethosuximide enantiomers in pregnancy and lactation. Ther. Drug Monit. **1994**, *16*, 621–623.

272   **Brocks et al.**

169. Haegele, K.D.; Schechter, P.J. Kinetics of the enantiomers of vigabatrin after an oral dose of the racemate or the active S-enantiomer. Clin. Pharmacol. Ther. **1986**, *40*, 581–586.
170. Rey, E.; Pons, G.; Richard, M.O.; Vauzelle, F.; D'Athis, P.; Chiron, C.; Dulac, O.; Beaumont, D.; Olive, G. Pharmacokinetics of the individual enantiomers of vigabatrin (gamma-vinyl GABA) in epileptic children. Br. J. Clin. Pharmacol. **1990**, *30*, 253–257.
171. Vauzelle-Kervroedan, F.; Rey, E.; Pons, G.; d'Athis, P.; Chiron, C.; Dulac, O.; Dumas, C.; Olive, G. Pharmacokinetics of the individual enantiomers of vigabatrin in neonates with uncontrolled seizures. Br. J. Clin. Pharmacol. **1996**, *42*, 779–781.
172. Tran, A.; O'Mahoney, T.; Rey, E.; Mai, J.; Mumford, J.P.; Olive, G. Vigabatrin: placental transfer in vivo and excretion into breast milk of the enantiomers. Br. J. Clin. Pharmacol. **1998**, *45*, 409–411.
173. Jacqz-Aigrain, E.; Guillonneau, M.; Rey, E.; Macher, M.A.; Montes, C.; Chiron, C.; Loirat, C. Pharmacokinetics of the S(+) and R(−) enantiomers of vigabatrin during chronic dosing in a patient with renal failure. Br. J. Clin. Pharmacol. **1997**, *44*, 183–185.
174. Volosov, A.; Xiaodong, S.; Perucca, E.; Yagen, B.; Sintov, A.; Bialer, M. Enantioselective pharmacokinetics of 10-hydroxycarbazepine after oral administration of oxcarbazepine to healthy Chinese subjects. Clin. Pharmacol. Ther. **1999**, *66*, 547–553.
175. Nau, H.; Hauck, R.S.; Ehlers, K. Valproic acid-induced neural tube defects in mouse and human: aspects of chirality, alternative drug development, pharmacokinetics and possible mechanisms. Pharmacol. Toxicol. **1991**, *69*, 310–321.
176. Barel, S.; Yagen, B.; Schurig, V.; Soback, S.; Pisani, F.; Perucca, E.; Bialer, M. Stereoselective pharmacokinetic analysis of valnoctamide in healthy subjects and in patients with epilepsy. Clin. Pharmacol. Ther. **1997**, *61*, 442–449.
177. Gleiter, C.H.; Schug, B.S.; Hermann, R.; Elze, M.; Blume, H.H.; Gundert-Remy, U. Influence of food intake on the bioavailability of thioctic acid enantiomers. Eur. J. Clin. Pharmacol. **1996**, *50*, 513–514.
178. Eap, C.B.; Finkbeiner, T.; Gastpar, M.; Scherbaum, N.; Powell, K.; Baumann, P. Replacement of (R)-methadone by a double dose of (R,S)-methadone in addicts: interindividual variability of the (R)/(S) ratios and evidence of adaptive changes in methadone pharmacokinetics. Eur. J. Clin. Pharmacol. **1996**, *50*, 385–389.
179. Kristensen, K.; Blemmer, T.; Angelo, H.R.; Christrup, L.L.; Drenck, N.E.; Rasmussen, S.N.; Sjogren, P. Stereoselective pharmacokinetics of methadone in chronic pain patients. Ther. Drug Monit. **1996**, *18*, 221–227.
180. Foster, D.J.; Somogyi, A.A.; Dyer, K.R.; White, J.M.; Bochner, F. Steady-state pharmacokinetics of (R)- and (S)-methadone in methadone maintenance patients. Br. J. Clin. Pharmacol. **2000**, *50*, 427–440.
181. Begg, E.J.; Malpas, T.J.; Hackett, L.P.; Ilett, K.F. Distribution of R- and S-methadone into human milk during multiple, medium to high oral dosing. Br. J. Clin. Pharmacol. **2001**, *52*, 681–685.

182. Grond, S.; Meuser, T.; Zech, D.; Hennig, U.; Lehmann, K.A. Analgesic efficacy and safety of tramadol enantiomers in comparison with the racemate: a randomised, double-blind study with gynaecological patients using intravenous patient-controlled analgesia. Pain **1995**, *62*, 313–320.

183. Grond, S.; Meuser, T.; Uragg, H.; Stahlberg, H.J.; Lehmann, K.A. Serum concentrations of tramadol enantiomers during patient-controlled analgesia. Br. J. Clin. Pharmacol. **1999**, *48*, 254–257.

184. Blouin, R.; Chaudhary, I.; Nishihara, K.; Cox, S. The effects of liver and renal disease on stereoselective serum binding of flurbiprofen. Br. J. Clin. Pharmacol. **1993**, *35*, 62–64.

185. Scott, L.J.; Perry, C.M. Tramadol: a review of its use in perioperative pain. Drugs **2000**, *60*, 139–176.

186. Sindrup, S.H.; Madsen, C.; Brosen, K.; Jensen, T.S. The effect of tramadol in painful polyneuropathy in relation to serum drug and metabolite levels. Clin. Pharmacol. Ther. **1999**, *66*, 636–641.

187. Poulsen, L.; Arendt-Nielsen, L.; Brosen, K.; Sindrup, S.H. The hypoalgesic effect of tramadol in relation to CYP2D6. Clin. Pharmacol. Ther. **1996**, *60*, 636–644.

188. Tinker, J.H. Voices from the past—from ice crystals to fruit flies in the quest for a molecular mechanism of anesthetic action. Anesth. Analg. **1993**, *77*, 1–3.

189. Aboul-Enein, H.Y.; Bojarski, J.; Szymura-Oleksiak, J. The impact of chirality of the fluorinated volatile inhalation anaesthetics on their clinical applications. Biomed. Chromatogr. **2000**, *14*, 213–218.

190. Pauling, L.A. A molecular theory of general anesthesia. Science **1961**, *134*, 15–21.

191. Moody, E.J.; Harris, B.D.; Skolnick, P. The potential for safer anaesthesia using stereoselective anaesthetics. Trends Pharmacol. Sci. **1994**, *15*, 387–391.

192. Dickinson, R.; White, I.; Lieb, W.R.; Franks, N.P. Stereoselective loss of righting reflex in rats by isoflurane. Anesthesiology **2000**, *93*, 837–843.

193. Franks, N.P.; Lieb, W.R. Stereospecific effects of inhalational general anesthetic optical isomers on nerve ion channels. Science **1991**, *254*, 427–430.

194. Jones, M.V.; Harrison, N.L. Effects of volatile anesthetics on the kinetics of inhibitory postsynaptic currents in cultured rat hippocampal neurons. J. Neurophysiol. **1993**, *70*, 1339–1349.

195. Harris, B.D.; Moody, E.J.; Skolnick, P. Stereoselective actions of halothane at GABA(A) receptors. Eur. J. Pharmacol. **1998**, *341*, 349–352.

196. Moody, E.J.; Harris, B.; Hoehner, P.; Skolnick, P. Inhibition of [3H]isradipine binding to L-type calcium channels by the optical isomers of isoflurane. Lack of stereospecificity. Anesthesiology **1994**, *81*, 124–128.

197. Cordato, D.J.; Chebib, M.; Mather, L.E.; Herkes, G.K.; Johnston, G.A. Stereoselective interaction of thiopentone enantiomers with the GABA(A) receptor. Br. J. Pharmacol. **1999**, *128*, 77–82.

198. Nguyen, K.T.; Stephens, D.P.; McLeish, M.J.; Crankshaw, D.P.; Morgan, D.J. Pharmacokinetics of thiopental and pentobarbital enantiomers after intravenous administration of racemic thiopental. Anesth. Analg. **1996**, *83*, 552–558.

199. Mather, L.E.; Edwards, S.R.; Duke, C.C. Electroencephalographic effects of thiopentone and its enantiomers in the rat: correlation with drug tissue distribution. Br. J. Pharmacol. **1999**, *128*, 83–91.

200. Mather, L.E.; Edwards, S.R.; Duke, C.C. Electroencephalographic effects of thiopentone and its enantiomers in the rat. Life Sci. **2000**, *66*, 105–114.

201. Cordato, D.J.; Gross, A.S.; Herkes, G.K.; Mather, L.E. Pharmacokinetics of thiopentone enantiomers following intravenous injection or prolonged infusion of rac-thiopentone. Br. J. Clin. Pharmacol. **1997**, *43*, 355–362.

202. Sasaki, T.; Andoh, T.; Watanabe, I.; Kamiya, Y.; Itoh, H.; Higashi, T.; Matsuura, T. Nonstereoselective inhibition of neuronal nicotinic acetylcholine receptors by ketamine isomers. Anesth. Analg. **2000**, *91*, 741–748.

203. Geisslinger, G.; Hering, W.; Thomann, P.; Knoll, R.; Kamp, H.D.; Brune, K. Pharmacokinetics and pharmacodynamics of ketamine enantiomers in surgical patients using a stereoselective analytical method. Br. J. Anaesth. **1993**, *70*, 666–671.

204. Marhofer, P.; Freitag, H.; Hochtl, A.; Greher, M.; Erlacher, W.; Semsroth, M. S(+)-ketamine for rectal premedication in children. Anesth. Analg. **2001**, *92*, 62–65.

205. Thomas, J.M.; Schug, S.A. Recent advances in the pharmacokinetics of local anaesthetics. Long-acting amide enantiomers and continuous infusions. Clin. Pharmacokinet. **1999**, *36*, 67–83.

206. Cox, C.R.; Faccenda, K.A.; Gilhooly, C.; Bannister, J.; Scott, N.B.; Morrison, L.M. Extradural S(−)-bupivacaine: comparison with racemic RS-bupivacaine. Br. J. Anaesth. **1998**, *80*, 289–293.

207. van der Meer, A.D.; Burm, A.G.; Stienstra, R.; van Kleef, J.W.; Vletter, A.A.; Olieman, W. Pharmacokinetics of prilocaine after intravenous administration in volunteers: enantioselectivity. Anesthesiology **1999**, *90*, 988–992.

208. Burm, A.G.; van der Meer, A.D.; van Kleef, J.W.; Zeijlmans, P.W.; Groen, K. Pharmacokinetics of the enantiomers of bupivacaine following intravenous administration of the racemate. Br. J. Clin. Pharmacol. **1994**, *38*, 125–129.

209. Sharrock, N.E.; Mather, L.E.; Go, G.; Sculco, T.P. Arterial and pulmonary arterial concentrations of the enantiomers of bupivacaine after epidural injection in elderly patients. Anesth. Analg. **1998**, *86*, 812–817.

210. Groen, K.; Mantel, M.; Zeijlmans, P.W.; Zeppenfeldt, B.; Olieman, W.; Stienstra, R.; van Kleef, J.W.; Burm, A.G. Pharmacokinetics of the enantiomers of bupivacaine and mepivacaine after epidural administration of the racemates. Anesth. Analg. **1998**, *86*, 361–366.

211. Burm, A.G.; Cohen, I.M.; van Kleef, J.W.; Vletter, A.A.; Olieman, W.; Groen, K. Pharmacokinetics of the enantiomers of mepivacaine after intravenous administration of the racemate in volunteers. Anesth. Analg. **1997**, *84*, 85–89.

212. Vree, T.B.; Beumer, E.M.; Lagerwerf, A.J.; Simon, M.A.; Gielen, M.J. Clinical pharmacokinetics of R(+)- and S(−)-mepivacaine after high doses of racemic mepivacaine with epinephrine in the combined psoas compartment/ sciatic nerve block. Anesth. Analg. **1992**, *75*, 75–80.

213. Fawcett, J.P.; Kennedy, J.; Kumar, A.; Ledger, R.; Zacharias, M. Stereoselective urinary excretion of bupivacaine and its metabolites during epidural infusion. Chirality **1999**, *11*, 50–55.

214. Palkama, V.J.; Neuvonen, P.J.; Olkkola, K.T. Effect of itraconazole on the pharmacokinetics of bupivacaine enantiomers in healthy volunteers. Br. J. Anaesth. **1999**, *83*, 659–661.

215. Alessi-Severini, S.; Coutts, R.T.; Jamali, F.; Pasutto, F.M. High-performance liquid chromatographic analysis of methocarbamol enantiomers in biological fluids. J. Chromatogr. **1992**, *582*, 173–179.

216. Hartigan-Go, K.; Bateman, D.N.; Daly, A.K.; Thomas, S.H. Stereoselective cardiotoxic effects of terodiline. Clin. Pharmacol. Ther. **1996**, *60*, 89–98.

217. Zobrist, R.H.; Schmid, B.; Feick, A.; Quan, D.; Sanders, S.W. Pharmaco-kinetics of the R- and S-enantiomers of oxybutynin and N-desethyloxybutynin following oral and transdermal administration of the racemate in healthy volunteers. Pharm. Res. **2001**, *18*, 1029–1034.

218. Bertman, L.J.; Advokat, C. Comparison of the antinociceptive and antispastic action of (−)-baclofen after systemic and intrathecal administration in intact, acute and chronic spinal rats. Brain Res. **1995**, *684*, 8–18.

219. Yasuda, S.U.; Zannikos, P.; Young, A.E.; Fried, K.M.; Wainer, I.W.; Woosley, R.L. The roles of CYP2D6 and stereoselectivity in the clinical pharmacokine-tics of chlorpheniramine. Br. J. Clin. Pharmacol. **2002**, *53*, 519–525.

220. Koch, K.M.; O'Connor-Semmes, R.L.; Davis, I.M.; Yin, Y. Stereoselective pharmacokinetics of chlorpheniramine and the effect of ranitidine. J. Pharm. Sci. **1998**, *87*, 1097–1100.

221. Surapaneni, S.; Khalil, S.K. A preliminary pharmacokinetic study of the enantiomers of the terfenadine acid metabolite in humans. Chirality **1994**, *6*, 479–483.

222. Robbins, D.K.; Castles, M.A.; Pack, D.J.; Bhargava, V.O.; Weir, S.J. Dose proportionality and comparison of single and multiple dose pharmacokinetics of fexofenadine (MDL 16455) and its enantiomers in healthy male volunteers. Biopharm. Drug Dispos. **1998**, *19*, 455–463.

223. Wang, C.P.; Howell, S.R.; Scatina, J.; Sisenwine, S.F. The disposition of venlafaxine enantiomers in dogs, rats, and humans receiving venlafaxine. Chirality **1992**, *4*, 84–90.

224. Tanaka, M.; Yamazaki, H.; Hakusui, H.; Nakamichi, N.; Sekino, H. Differential stereoselective pharmacokinetics of pantoprazole, a proton pump inhibitor in extensive and poor metabolizers of pantoprazole—a preliminary study. Chirality **1997**, *9*, 17–21.

225. Tanaka, M.; Ohkubo, T.; Otani, K.; Suzuki, A.; Kaneko, S.; Sugawara, K.; Ryokawa, Y.; Ishizaki, T. Stereoselective pharmacokinetics of pantoprazole, a proton pump inhibitor, in extensive and poor metabolizers of S-mepheny-toin. Clin. Pharmacol. Ther. **2001**, *69*, 108–113.

226. Tybring, G.; Bottiger, Y.; Widen, J.; Bertilsson, L. Enantioselective hydroxylation of omeprazole catalyzed by CYP2C19 in Swedish white subjects. Clin. Pharmacol. Ther. **1997**, *62*, 129–137.

227. Katsuki, H.; Yagi, H.; Arimori, K.; Nakamura, C.; Nakano, M.; Katafuchi, S.; Fujioka, Y.; Fujiyama, S. Determination of R(+)- and S(−)-lansoprazole using chiral stationary-phase liquid chromatography and their enantioselective pharmacokinetics in humans. Pharm. Res. **1996**, *13*, 611–615.

228. Kelly, J.W.; He, L.; Stewart, J.T. High-performance liquid chromatographic separation of ondansetron enantiomers in serum using a cellulose-derivatized stationary phase and solid-phase extraction. J. Chromatogr. **1993**, *622*, 291–295.

229. Liu, J.; Stewart, J.T. High-performance liquid chromatographic analysis of ondansetron enantiomers in human serum using a reversed-phase cellulose-based chiral stationary phase and solid-phase extraction. J. Chromatogr. B Biomed. Sci. Appl. **1997**, *694*, 179–184.

230. Wu, D.F.; Yu, Y.W.; Tang, Z.M.; Wang, M.Z. Pharmacokinetics of (+/−)-, (+)-, and (−)-gossypol in humans and dogs. Clin. Pharmacol. Ther. **1986**, *39*, 613–618.

231. Shelley, M.D.; Hartley, L.; Fish, R.G.; Groundwater, P.; Morgan, J.J.; Mort, D.; Mason, M.; Evans, A. Stereo-specific cytotoxic effects of gossypol enantiomers and gossypolone in tumour cell lines. Cancer Lett. **1999**, *135*, 171–180.

232. Gadd, A.L.; Curtis-Prior, P.B. Comparative effects of (+)-propranolol and nonoxynol-9 on human sperm motility in-vitro. J. Pharm. Pharmacol. **1990**, *42*, 593–594.

233. Fallon, J.K.; Kicman, A.T.; Henry, J.A.; Milligan, P.J.; Cowan, D.A.; Hutt, A.J. Stereospecific analysis and enantiomeric disposition of 3, 4-methylenedioxy-methamphetamine (Ecstasy) in humans. Clin. Chem. **1999**, *45*, 1058–1069.

234. Holm, K.A.; Kindberg, C.G.; Stobaugh, J.F.; Slavik, M.; Riley, C.M. Stereoselective pharmacokinetics and metabolism of the enantiomers of cyclophosphamide. Preliminary results in humans and rabbits. Biochem. Pharmacol. **1990**, *39*, 1375–1384.

235. Williams, M.L.; Wainer, I.W.; Granvil, C.P.; Gehrcke, B.; Bernstein, M.L.; Ducharme, M.P. Pharmacokinetics of (R)- and (S)-cyclophosphamide and their dechloroethylated metabolites in cancer patients. Chirality **1999**, *11*, 301–308.

236. Williams, M.L.; Wainer, I.W.; Embree, L.; Barnett, M.; Granvil, C.L.; Ducharme, M.P. Enantioselective induction of cyclophosphamide metabolism by phenytoin. Chirality **1999**, *11*, 569–574.

237. Roy, P.; Tretyakov, O.; Wright, J.; Waxman, D.J. Stereoselective metabolism of ifosfamide by human P-450s 3A4 and 2B6. Favorable metabolic properties of R-enantiomer. Drug Metab. Dispos. **1999**, *27*, 1309–1318.

238. Granvil, C.P.; Ducharme, J.; Leyland-Jones, B.; Trudeau, M.; Wainer, I.W. Stereoselective pharmacokinetics of ifosfamide and its 2- and 3-N-dechloroethylated metabolites in female cancer patients. Cancer Chemother. Pharmacol. **1996**, *37*, 451–456.

239. Granvil, C.P.; Madan, A.; Sharkawi, M.; Parkinson, A.; Wainer, I.W. Role of CYP2B6 and CYP3A4 in the in vitro N-dechloroethylation of (R)- and

(S)-ifosfamide in human liver microsomes. Drug Metab. Dispos. **1999**, *27*, 533–541.

240. Corlett, S.A.; Parker, D.; Chrystyn, H. Pharmacokinetics of ifosfamide and its enantiomers following a single 1 h intravenous infusion of the racemate in patients with small cell lung carcinoma. Br. J. Clin. Pharmacol. **1995**, *39*, 452–455.

241. Prasad, V.K.; Corlett, S.A.; Abaasi, K.; Heney, D.; Lewis, I.; Chrystyn, H. Ifosfamide enantiomers: pharmacokinetics in children. Cancer Chemother. Pharmacol **1994**, *34*, 447–449.

242. Alshowaier, I.A.; el-Yazigi, A.; Ezzat, A.; Abd el-Warith, A.; Nicholls, P.J. Pharmacokinetics of S- and R-enantiomers of aminoglutethimide following oral administration of racemic drug in breast cancer patients. J. Clin. Pharmacol. **1999**, *39*, 1136–1142.

243. Cockshott, I.D.; Sotaniemi, E.; Cooper, K.J.; Jones, D.C. The pharmacokinetics of Casodex enantiomers in subjects with impaired liver function. Br. J. Clin. Pharmacol. **1993**, *36*, 339–343.

244. Jamali, F. Pharmacokinetics of enantiomers of chiral non-steroidal anti-inflammatory drugs. Eur. J. Drug Metab. Pharmacokinet. **1988**, *13*, 1–9.

245. Mauleon, D.; Artigas, R.; Garcia, M.L.; Carganico, G. Preclinical and clinical development of dexketoprofen. Drugs **1996**, *52* (suppl. 5), 24–45; discussion 46.

246. Jamali, F.; Berry, B.W.; Tehrani, M.R.; Russell, A.S. Stereoselective pharmacokinetics of flurbiprofen in humans and rats. J. Pharm. Sci. **1988**, *77*, 666–669.

247. Jamali, F.; Brocks, D.R. Clinical pharmacokinetics of ketoprofen and its enantiomers. Clin. Pharmacokinet. **1990**, *19*, 197–217.

248. Mehvar, R.; Jamali, F. Stereospecific high-performance liquid chromatographic (HPLC) assay of fenoprofen enantiomers in plasma and urine. Pharm. Res. **1988**, *5*, 53–56.

249. Brocks, D.R.; Jamali, F. Etodolac clinical pharmacokinetics. Clin. Pharmacokinet. **1994**, *26*, 259–274.

250. Hayball, P.J.; Wrobel, J.; Tamblyn, J.G.; Nation, R.L. The pharmacokinetics of ketorolac enantiomers following intramuscular administration of the racemate. Br. J. Clin. Pharmacol. **1994**, *37*, 75–78.

251. Brocks, D.R.; Jamali, F. Clinical pharmacokinetics of ketorolac tromethamine. Clin. Pharmacokinet. **1992**, *23*, 415–427.

252. Jamali, F.; Pasutto, F.M.; Lemko, C. HPLC of ketorolac enantiomers and application to pharmacokinetics in the rat. J. Liq. Chromatogr. **1989**, *12*, 1835–1850.

253. Vakily, M.; Corrigan, B.; Jamali, F. The problem of racemization in the stereospecific assay and pharmacokinetic evaluation of ketorolac in human and rats. Pharm. Res. **1995**, *12*, 1652–1657.

254. Hutt, A.J.; Caldwell, J. The metabolic chiral inversion of 2-arylpropionic acids—a novel route with pharmacological consequences. J. Pharm. Pharmacol. **1983**, *35*, 693–704.

255. Lee, E.J.D.; Williams, K.M.; Day, R.O.; Graham, G.; Champion, D. Stereoselective disposition of ibuprofen enantiomers in man. Br. J. Clin. Pharmacol. **1985**, *19*, 669–674.

256. Wechter, W.J.; Loughhead, D.G.; Reischer, R.J.; VanGiessen, G.J.; Kaiser, D.G. Enzymatic inversion at saturated carbon: nature and mechanism of the inversion of R(−) p-iso-butyl hydratropic acid. Biochem Biophys. Res. Commun. **1974**, *61*, 833–837.

257. Jamali, F.; Singh, N.N.; Pasutto, F.M.; Russell, A.S.; Coutts, R.T. Pharmacokinetics of ibuprofen enantiomers in humans following oral adminis-tration of tablets with different absorption rates. Pharm. Res. **1988**, *5*, 40–43.

258. Cox, S.R.; Brown, M.A.; Squires, D.J.; Murrill, E.A.; Lednicer, D.; Knuth, D.W. Comparative human study of ibuprofen enantiomer plasma concentra-tions produced by two commercially available ibuprofen tablets. Biopharm. Drug Dispos. **1988**, *9*, 539–549.

259. Brocks, D.R.; Jamali, F. The pharmacokinetics of ibuprofen in humans and animals. In *Ibuprofen, A Critical Bibliographic Review*; Rainsford, K.D., Ed.; Taylor and Francis: Philadelphia, 1999; 87–143.

260. Vakily, M.; Jamali, F. Human pharmacokinetics of tiaprofenic acid after regular and sustained release formulations: lack of chiral inversion and stereoselective release. J. Pharm. Sci. **1994**, *83*, 495–498.

261. Evans, A.M.; Nation, R.L.; Sansom, L.N.; Bochner, F.; Somogyi, A.A. Stereoselective plasma protein binding of ibuprofen enantiomers. Eur. J. Clin. Pharmacol. **1989**, *36*, 283–290.

262. Knadler, M.P.; Brater, D.C.; Hall, S.D. Plasma protein binding of flurbiprofen: enantioselectivity and influence of pathophysiological status. J. Pharmacol. Exp. Ther. **1989**, *249*, 378–385.

263. Rendic, S.; Alebic-Kolbah, T.; Kajfez, F.; Sunjic, V. Stereoselective binding of (+)- and (−)-a(benzylphenyl)propionic acid (ketoprofen) to human serum albumin. Farmaco. Edizione. Scientifica. **1980**, *35*, 51–59.

264. Brocks, D.R.; Jamali, F.; Russell, A.S. Stereoselective disposition of etodolac enantiomers in synovial fluid. J. Clin. Pharmacol. **1991**, *31*, 741–746.

265. Muller, N.; Lapicque, F.; Monot, C.; Payan, E.; Dropsy, R.; Netter, P. Stereoselective binding of etodolac to human serum albumin. Chirality **1992**, *4*, 240–246.

266. Lockwood, G.F.; Albert, K.S.; Gillespie, W.R.; Bole, G.G.; Harkcom, T.M.; Szpunar, G.J.; Wagner, J.G. Pharmacokinetics of ibuprofen in man. I. Free and total area/dose relationships. Clin. Pharmacol. Ther. **1983**, *34*, 97–103.

267. Lee, E.J.; Williams, K.M.; Graham, G.G.; Day, R.O.; Champion, G.D. Liquid chromatographic determination and plasma concentration profile of optical isomers of ibuprofen in humans. J. Pharm. Sci. **1984**, *73*, 1542–1544.

268. Berry, B.W.; Jamali, F. Enantiomeric interaction of flurbiprofen in the rat. J. Pharm. Sci. **1989**, *78*, 632–634.

269. Geisslinger, G.; Lotsch, J.; Menzel, S.; Kobal, G.; Brune, K. Stereoselective disposition of flurbiprofen in healthy subjects following administration of the single enantiomers. Br. J. Clin. Pharmacol. **1994**, *37*, 392–394.

270. Day, R.O.; Williams, K.M.; Graham, G.G.; Lee, E.J.; Knihinicki, R.D.; Champion, G.D. Stereoselective disposition of ibuprofen enantiomers in synovial fluid. Clin. Pharmacol. Ther. **1988**, *43*, 480–487.

271. Cox, S.R.; Gall, E.P.; Forbes, K.K.; Gresham, M.; Goris, G. Pharmacokinetics of the R(−) and S(+) enantiomers of ibuprofen in the serum and synovial fluid of arthritis patients. J. Clin. Pharmacol. **1991**, *31*, 88–94.

272. Foster, R.T.; Jamali, F.; Russell, A.S. Ketoprofen enantiomers in synovial fluid. J. Pharm. Sci. **1989**, *78*, 881–882.

273. Davies, N.M. Clinical pharmacokinetics of flurbiprofen and its enantiomers. Clin. Pharmacokinet. **1995**, *28*, 100–114.

274. Nichol, F.E.; Samanta, A.; Rose, C.M. Synovial fluid and plasma kinetics of repeat dose sustained action tiaprofenic acid in patients with rheumatoid arthritis. Drugs **1988**, *35* (suppl. 1), 46–51.

275. Brocks, D.R.; Jamali, F. Enantioselective pharmacokinetics of etodolac in the rat: tissue distribution, tissue binding, and in vitro metabolism. J. Pharm. Sci. **1991**, *80*, 1058–1061.

276. Williams, K.; Day, R.; Knihinicki, R.; Duffield, A. The stereoselective uptake of ibuprofen enantiomers into adipose tissue. Biochem. Pharmacol. **1986**, *35*, 3403–3405.

277. Davies, N.M. Clinical pharmacokinetics of ibuprofen. The first 30 years. Clin Pharmacokinet **1998**, *34*, 101–154.

278. Hall, S.D.; Rudy, A.C.; Knight, P.M.; Brater, D.C. Lack of presystemic inversion of (R)- to (S)-ibuprofen in humans. Clin. Pharmacol. Ther. **1993**, *53*, 393–400.

279. Foster, R.T.; Jamali, F.; Russell, A.S.; Alballa, S.R. Pharmacokinetics of ketoprofen enantiomers in young and elderly arthritic patients following single and multiple doses. J. Pharm. Sci. **1988**, *77*, 191–195.

280. Foster, R.T.; Jamali, F.; Russell, A.S.; Alballa, S.R. Pharmacokinetics of ketoprofen enantiomers in healthy subjects following single and multiple doses. J. Pharm. Sci. **1988**, *77*, 70–73.

281. Skeith, K.J.; Russell, A.S.; Jamali, F. Ketoprofen pharmacokinetics in the elderly: influence of rheumatic disease, renal function, and dose. J. Clin. Pharmacol. **1993**, *33*, 1052–1059.

282. Jamali, F.; Russell, A.S.; Foster, R.T.; Lemko, C. Inversion of R-ketoprofen and lack of enantiomeric interaction in humans. J. Pharm. Sci. **1990**, *79*, 460–461.

283. Jamali, F.; Mehvar, R.; Lemko, C.; Eradiri, O. Application of a stereospecific high-performance liquid chromatography assay to a pharmacokinetic study of etodolac enantiomers in humans. J. Pharm. Sci. **1988**, *77*, 963–966.

284. Brocks, D.R.; Jamali, F.; Russell, A.S.; Skeith, K.J. The stereoselective pharmacokinetics of etodolac in young and elderly subjects, and after cholecystectomy. J. Clin. Pharmacol. **1992**, *32*, 982–989.

285. Boni, J.P.; Korth-Bradley, J.M.; Richards, L.S.; Chiang, S.T.; Hicks, D.R.; Benet, L.Z. Chiral bioequivalence: effect of absorption rate on racemic etodolac. Clin. Pharmacokinet. **2000**, *39*, 459–469.

286. Foster, R.T.; Jamali, F.; Russell, A.S. Pharmacokinetics of ketoprofen enantiomers in cholecystectomy patients: influence of probenecid. Eur. J. Clin. Pharmacol. **1989**, *37*, 589–594.

287. Hamdy, R.C.; Murnane, B.; Perera, N.; Woodcock, K.; Koch, I.M. The pharmacokinetics of benoxaprofen in elderly subjects. Eur. J. Rheumatol. Inflamm. **1982**, *5*, 69–75.

288. Dong, J.Q.; Ni, L.; Scott, C.S.; Retsch-Bogart, G.Z.; Smith, P.C. Pharmacokinetics of ibuprofen enantiomers in children with cystic fibrosis. J. Clin. Pharmacol. **2000**, *40*, 861–868.

289. Rey, E.; Pariente-Khayat, A.; Gouyet, L.; Vauzelle-Kervroedan, F.; Pons, G.; D'Athis, P.; Dubois, M.C.; Murat, I.; Lassale, C.; Goehrs, M., et al. Stereoselective disposition of ibuprofen enantiomers in infants. Br. J. Clin. Pharmacol. **1994**, *38*, 373–375.

290. Choonara, I.A.; Haynes, B.P.; Cholerton, S.; Breckenridge, A.M.; Park, B.K. Enantiomers of warfarin and vitamin K1 metabolism. Br. J. Clin. Pharmacol. **1986**, *22*, 729–732.

291. Schmekel, B.; Rydberg, I.; Norlander, B.; Sjosward, K.N.; Ahlner, J.; Andersson, R.G. Stereoselective pharmacokinetics of S-salbutamol after administration of the racemate in healthy volunteers. Eur. Respir. J. **1999**, *13*, 1230–1235.

292. Knadler, M.P.; Brater, D.C.; Hall, S.D. Stereoselective disposition of flurbiprofen in uraemic patients. Br. J. Clin. Pharmacol. **1992**, *33*, 377–383.

293. Oliary, J.; Tod, M.; Nicolas, P.; Petitjean, O.; Caille, G. Pharmacokinetics of ibuprofen enantiomers after single and repeated doses in man. Biopharm. Drug Dispos. **1992**, *13*, 337–344.

294. Gimenez, F.; Farinotti, R.; Thuillier, A.; Hazebroucq, G.; Wainer, I.W. Determination of the enantiomers of mefloquine in plasma and whole blood using a coupled achiral-chiral high-performance liquid chromatographic system. J. Chromatogr. **1990**, *529*, 339–46.

295. Mather, L.E.; McCall, P.; McNicol, P.L. Bupivacaine enantiomer pharmacokinetics after intercostal neural blockade in liver transplantation patients. Anesth. Analg. **1995**, *80*, 328–335.

7

Stereospecific Pharmacokinetics and Pharmacodynamics: Cardiovascular Drugs

Reza Mehvar
Texas Tech University Health Sciences Center, Amarillo, Texas, U.S.A.

Dion R. Brocks
University of Alberta, Edmonton, Alberta, Canada

1. INTRODUCTION

Cardiovascular diseases are highly prevalent in the population, and many of the drugs used in the treatment of these diseases are chiral. Several of these drugs are used in the treatment of hypertension, and examples include the beta-antagonists, the calcium channel blockers, and the diuretics. In addition, the beta-blockers (Class II antiarrhythmics) and the chiral calcium channel blockers verapamil, gallopamil, and prenylamine (Class IV antiarrhythmics), are also used in the treatment of cardiac arrhythmias along with a number of other chiral agents of the Class I type. A great deal of information is available on drugs used in cardiology, and in this chapter their stereoselective pharmacokinetic and pharmacodynamic properties are presented.

2. BETA-ADRENERGIC ANTAGONISTS

All of the β-adrenergic blockers possess at least one center of asymmetry· that confers optical activity to the drug. Beta-adrenoceptor blocking agents are generally used for the treatment of cardiovascular diseases, such as

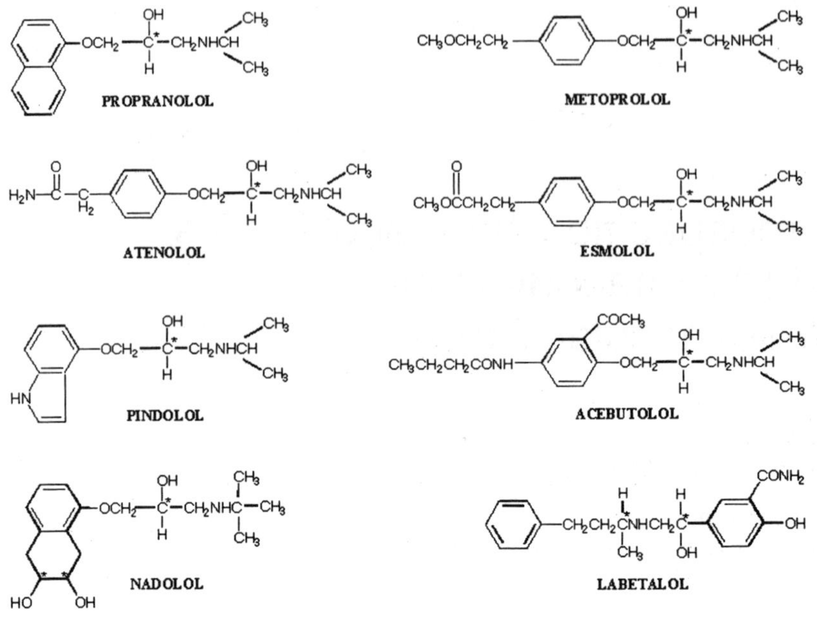

Figure 1 Chemical structure of major beta-adrenoceptor blocking agents. The asterisk denotes the chiral carbon.

hypertension, ischemic heart disease (angina pectoris), and some types of arrhythmias. A common feature of these drugs (Fig. 1) is that there is at least one aromatic ring structure attached to a side alkyl chain possessing a secondary hydroxyl and amine functional group. Each of the available beta blockers has one or more chiral centers in its structure, and in all cases at least one of the chiral carbon atoms residing in the alkyl side chain is directly attached to a hydroxyl group. Timolol is the only β-blocker that is marketed solely as the S-enantiomer for systemic administration; all others are marketed as the racemate or enantiomer combinations. Labetalol, which has two chiral centers (Fig. 1), is marketed as a racemate consisting of four isomers. Nadolol has three chiral centers in its structure (Fig. 1). However, the two ring hydroxyl groups (Fig. 1) are in the cis orientation, allowing for a total of only four isomers.

2.1. Stereoselectivity in Pharmacological Action

Beta blockers display a high degree of stereoselectivity in binding to β-receptors [1–11]. The cardiac beta-blocking activity of the beta blockers usually resides with the S(−) enantiomers [2–8] (S:R activity ratio ranging from 33 to 530) (Table 1). However, the R(+) enantiomer has relatively strong activity in blocking β_2 receptors in the ciliary process (Table 1) [44]. For sotalol, which has R(−) and S(+) conformation, it is the R(−) enantiomer that possesses the majority of the β-blocking activity [8]. Both enantiomers, however, share an equivalent degree of Class III antiarrhythmic potency (Table 1).

Cardevilol is marketed as the racemate for the treatment of hypertension and congestive heart failure [12]. This latter indication is unique among the available beta-blockers, for this condition is normally a contraindication to use of β-blockers. Like other beta-blockers, the S(−) enantiomer of carvedilol is more potent as an antagonist of beta-receptors. Both the R and the S enantiomers, however, are also effective in blocking alpha adrenergic activity [12]. This gives the drug utility in congestive heart failure owing to the combination of decreased vascular resistance (α-adrenergic antagonism) and lack of reflex tachycardia (β-blockade). It should be noted that recent studies involving low doses of selective β-blockers (e.g., metoprolol) display a similar benefit to carvedilol in reducing mortality in patients with heart failure [13].

Labetalol is administered as a combination of four stereoisomers (Table 1) [9–11]. Whereas the RR isomer is mostly responsible for the β blocking activity of the drug, the SR isomer is most potent as an α-adrenoceptor blocker [9]. Both the RS and SS isomers, on the other hand, display weak antagonistic activities against α and β receptors [9]. Stereoselectivity in antagonism of beta adrenoceptors has also been reported [11] for nadolol (Fig. 1), another beta-blocker with four stereoisomers (Table 1).

2.2. General Pharmacokinetic Properties of Beta Blockers

Absorption

Generally, beta-blockers are absorbed from the gastrointestinal tract via passive diffusion, so their absorption is not considered stereoselective. However, some beta-blockers such as talinolol may undergo an intestinal secretion process that may be modestly stereoselective, resulting in an apparent nonlinearity in the kinetics of the drug with increasing oral doses [14]. Nevertheless, despite the suggestion of an active intestinal secretion process, the overall pharmacokinetics of talinolol are not stereoselective [14].

Distribution

Like other basic drugs, beta-blockers bind to both albumin and α_1-acid glycoprotein (AAG) in plasma. For acebutolol [15], pindolol [16], and

Table 1 Stereoselectivity in the Action of Some Beta-Adrenergic Blockers

Drug	Relative activity (ratio)	Experimental model	Biological response	Ref.
Atenolol	$- > +$	Rat	Reduction in heart rate and mean arterial pressure	1
	$- > + (46:1)$	Guinea pig	Beta-blocking activity of heart	2
Betaxolol	$- > + (530:1)$	Rabbit	Beta-blocking activity of heart	3
	$+ > - (190:1)$	Rabbit	Blocking β_2 receptors in ciliary process	
Bucindolol	$- > +$	Dog	Myocardial stimulant, vasodilator	4
Labetalol	RR > SR; RR > RS; RR > SS	Rat	Beta-blocking activity	9
	SR > RR; SR > RS; SR > SS	Rat	Alpha-blocking activity	
	RR-SR > labetalol (2:1)	Rat	Antagonizing pressor effects of phenylephrine and chronotropic effect of isoprenaline	10
	RR-SR > labetalol	Rat	Alpha1-blocking activity	
	RR-SR > labetalol (3:1)	Rat	Beta1-blocking activity	
Metoprolol	$- > + (33:1)$	Rabbit	Beta-blocking activity of heart	3
	$+ > - (10:1)$	Rabbit	Blocking β_2 receptors in ciliary process	
Nadolol	SQ-12151 > SQ-12150 > nadolol > SQ-12148 > SQ-12149	Chinese hamster ovary cells	Binding values to β1, β2, and β3 cloned receptors	11
Pindolol	$- > + (200:1)$	Guinea pig	Blocking β_1 and β_2 receptors	5
Propranolol	$- > + (100:1)$	Rat	Blocking isoprenaline cardiac response	6
	$- = +$	Frog nerve	Local anesthetic effects	
Sotalol	$- > +$	Dog and rabbit	Beta-blocking activity	163
	$- = +$	Dog and rabbit	Class III antiarrhythmic activity	
Timolol	$- > + (44:1)$	Rabbit	Beta-blocking activity of heart	3
	$- = +$	Rabbit	Blocking β_2 receptors in ciliary process	

sotalol [17], which have relatively high free fractions in plasma, the binding appears to be nonstereoselective (Table 2). However, stereoselective binding has been reported [18–23] for propranolol in both whole plasma and individual serum proteins (Table 2). As demonstrated in Table 2, the stereoselectivity in the binding of propranolol to human serum albumin is opposite of that observed for the human AAG. Whereas the free fraction of the (+) enantiomer is higher in AAG, the opposite is true for albumin (Table 2). The overall stereoselectivity in the binding of propranolol to human serum, however, resembles that seen with AAG (Table 2).

A study of propranolol binding in maternal and fetal serum [21] further confirms the importance of AAG in the overall extent of binding and stereoselectivity of propranolol (Table 2). A significantly lower concentration of AAG in fetal blood (14 mg%), compared with that in maternal blood (66 mg%), results in much higher free fractions and a change in the direction of stereoselectivity in binding in the fetal blood (Table 2).

Age and gender do not appear to play a substantial role in the plasma protein binding of propranolol enantiomers (Table 2). A modestly lower unbound fraction of (−)-propranolol in females (10.9%, Table 2) compared with males (12.8%, Table 2) reported in one study [21] was not observed in a subsequent study [22] which reported an unbound fraction of 9.1 and 9.2 in men and women, respectively (Table 2).

The tissue distributions of propranolol [24–26] and pindolol [27] enantiomers have been reported using laboratory animals. These studies collectively suggest that while the concentrations of the enantiomers of beta-blockers in different tissues may be stereoselective, the actual tissue uptake and binding of these drugs in most cases is nonstereoselective. The apparent stereoselectivity in the tissue concentrations of these drugs has been attributed mainly to the stereoselectivity in the plasma protein binding of the drugs. For example, Takahashi et al. [25] demonstrated that, compared with the (+) enantiomer, the (−) enantiomer of propranolol reaches higher concentrations in heart, muscle, gastrointestinal tract, kidney, brain, and lung of rats. However, this apparent stereoselectivity could be explained by higher free fraction of (−)-propranolol in plasma [25]. Similarly, the stereoselective binding of the propranolol enantiomers to plasma proteins is the main reason behind an apparent stereoselectivity in the red blood cell distribution of the drug [28].

It has been demonstrated that both hydrophilic (e.g., atenolol) and lipophilic (e.g., propranolol) beta-blockers are stored in and released from the adrenergic nerve endings [29]. Furthermore, it has been reported [29,30] that the uptake and release of atenolol from the models of adrenergic cells are stereoselective, favoring the more active (−) enantiomer of atenolol by 200 to 500%. Additionally, a study in humans [31] chronically receiving

Table 2 Unbound Fractions of Enantiomers of Chiral Cardiovascular Drugs in Human Specimens. Plasma and Racemate Studied Unless Otherwise Indicated

Drug	Unbound fraction (%)	Ratio	Ref.
Acebutolol	(+), 86 (−), 84	1.0 young	15
	(+), 93 (−), 93	1.0 elderly	
Disopyramide	R(−) > S(+)[a]	~2 R:S	95,96
	R, 57 S, 26	2.2 R:S AAG	102
Gallopamil	S, 5.1 R, 3.5	1.5 S:R	158
	S, 5.6 R, 3.9	1.4 S:R	159
	S alone, 5.9 R alone, 4.0	1.5 S:R	159
Mexiletine	S, 28 R, 20[b]	1.4 S:R in vitro	164
	S, 57 R, 56	1.0 S:R in vivo	92
Nitrendipine	S, 1.6 R, 1.3	1.2 S:R	69
Pimobendan	(+), 2.4 (−), 2.4	1.0 (+):(−)	161
Pindolol	(−), 45 (−), 45	1	16
Pirmenol	(+), 18.5 (−), 14.8	1.3 (+):(−)	157
Propafenone	R, 3.9 S, 2.5	1.6 R:S	97
	R, 7.6 S, 4.8	1.6 R:S	84
Propranolol	(−), 22.0 (+), 25.3	1.2 (+):(−)	18
	(−), 12.7 (+), 16.2	1.3 (+):(−) AAG	
	(−), 64.9 (+), 60.7	1.07 (−):(+) albumin	
	(−), 10.9 (+), 12.2	1.1 (+):(−)	19
	(−), 23.0 (+), 30.2	1.3 (+):(−) AAG	
	(−), 51.0 (+), 48.2	1.06 (−):(+) albumin	
	(−), 20.7 (+), 22.4	1.08 (+):(−) maternal	20
	(−), 40.4 (+), 38.8	1.04 (−):(+) fetal	
	(−), 11.7 (+), 18.0	1.5 (+):(−) young	21
	(−), 12.1 (+), 18.6	1.5 (+):(−) elderly	
	(−), 10.9 (+), 17.8	1.6 (+):(−) female	
	(−), 12.8 (+), 18.8	1.7 (+):(−) male	
	(−), 9.1 (+), 10.8	1.2 (+):(−) male	22
	(−), 9.2 (+), 10.8	1.2 (+):(−) female	
	(−), 17.6 (+), 20.3	1.2 (+):(−)	23
Sotalol	(−), 96 (+), 96	1.0 young	17
	(−), 95 (+), 95	1.0 elderly	
	(−), 65 (+), 62	1.05 (+):(−) supraventricular tachyarrhythmia	44
Tocainide	S, 83-89 R, 86-91	1 S:R	93

(continues)

Table 2 Continued

Drug	Unbound fraction (%)	Ratio	Ref
Verapamil	S, 11 R, 6.4	1.7 S:R	98
	S, 57 R, 40	1.4 S:R albumin	99
	S, 14 R, 7.9	1.8 S:R AAG	
	S, 12 R, 7.0	1.7 S:R serum in vitro	
	S, 12 R, 6	2.0 S:R serum iv	
	S, 23 R, 13	1.8 S:R serum po	
	S, 9.7 R, 5.1	1.9 S:R serum in vitro	100
	S, 9.3 R, 5.0	1.9 S:R serum single oral	
	S, 9.4 R, 5.1	1.8 S:R serum, multiple oral	
	S, 22 R, 17	1.4 S:R maternal	20
	S, 45 R, 35	1.3 S:R fetal	
	S = R	1.0 S:R HDL and LDL	101

[a]Dose-dependent protein binding at unbound serum concentrations of $\geq 0.34\,\mu g/mL$ for both enantiomers [95,96].
[b]The pH of the samples, which substantially affects the mexiletine free fraction [92], was not controlled in this study.
Abbreviations: AAG = human α_1-acid glycoprotein; IV = intravenous; PO = oral.

racemic atenolol, indicated that the (−) enantiomer of atenolol is selectively released into the plasma after exercise. The exercise-induced stereoselective release of atenolol from adrenergic nerve endings significantly changed the (+):(−) concentration ratio of atenolol from 1.18 (at rest) to 0.64 (after exercise) [31]. The effects of exercise on the stereoselective release of atenolol are apparently related to the duration of therapy, because the exercise-induced release of atenolol after a single dose of racemic atenolol was not stereoselective [32]. In contrast to atenolol, the uptake of propranolol, a more lipophilic beta-blocker, into the adrenergic nerve cells or models appears to be via passive diffusion and therefore is not stereoselective [30]. Nonetheless, the stereoselectivity in the storage into and release of beta-blockers from the nerve cells and platelets may have important clinical implications.

Elimination

The elimination of most beta-blockers occurs via hepatic metabolism and/or renal excretion of the unchanged drug. While the lipophilic beta blockers, such as propranolol, are eliminated mostly by metabolism, the more hydrophilic beta blockers, such as atenolol and nadolol, are almost exclusively excreted unchanged in urine. Some aspects of the stereoselective human metabolism of propranolol and metoprolol, two widely studied beta-blockers, will be discussed here.

In humans, propranolol is metabolized by three main pathways; glucuronidation, ring hydroxylation, and side chain oxidation (Fig. 2) [33]. The ring hydroxylation process may occur at either position 4 or 5 (Fig. 2), both of them showing substrate stereoselectivity for R(+)-propranolol [33–35]. Hydroxypropranolol is further conjugated with glucuronic acid, favoring the S(−) enantiomer, or sulfate, favoring the R(+)-propranolol, before excretion into urine [29]. Enantioselectivity in N-dealkylation (Fig. 2) appears to be related to the concentration of the drug; whereas at low substrate concentrations, the R(+) enantiomer is preferentially metabolized, the opposite is true at high propranolol concentrations [33]. On the other hand, based on the urinary excretion of propranolol glucuronides in humans, the formation of propranolol glucuronide appears to favor S(−)-propranolol [34]. Overall, the metabolism of propranolol is faster for the less active R(+) enantiomer, resulting in a higher plasma concentrations of the S(−) enantiomer in humans.

Propranolol metabolism is affected by genetic polymorphism for both CYP1A (mephenytoin hydroxylation) and CYP2D6 (debrisoquine hydroxylation) isozymes in the liver [36,37]. Based on in vivo studies [36] in poor and extensive metabolizers of debrisoquine and mephenytoin and in vitro studies [37] using human liver microsomes and CYP isoforms, it appears that N-dealkylation of propranolol is mainly governed by S-mephenytoin-4-hydroxylase (CYP1A subfamily), whereas ring hydroxylation is predominantly related to debrisoquine isozyme (CYP2D6).

Metoprolol is another beta-blocker that is predominantly eliminated by hepatic metabolism [38]. In humans, metoprolol is eliminated by several oxidation pathways, including benzylic hydroxylation (α-hydroxylation), which results in an active metabolite and accounts for ~10% of the dose [39]. This pathway is stereoselective for S(−)-metoprolol. The major metabolic pathway, however, is O-demethylation and further oxidation to a carboxylic acid metabolite that accounts for 65% of the dose [38]. O-demethylation favors R(+)-metoprolol [39] and facilitates the stereoselectivity observed in the plasma concentrations of metoprolol. A third metabolic pathway (N-dealkylation) accounts for < 10% of the dose in humans [39].

Metoprolol oxidation cosegregates with debrisoquine hydroxylation, and debrisoquine phenotype significantly affects the stereoselective metabolism of the drug [38]. The influence of debrisoquine hydroxylation phenotype on the pharmacokinetics and pharmacodynamics of both propranolol and metoprolol is described later in this chapter.

As mentioned above, renal excretion of the unchanged drug is the primary elimination pathway for hydrophilic beta-blockers such as atenolol, nadolol, and sotalol. In contrast to metabolism, the reported stereoselectivity in the renal clearance of beta-blockers is relatively low, with (−):(+)

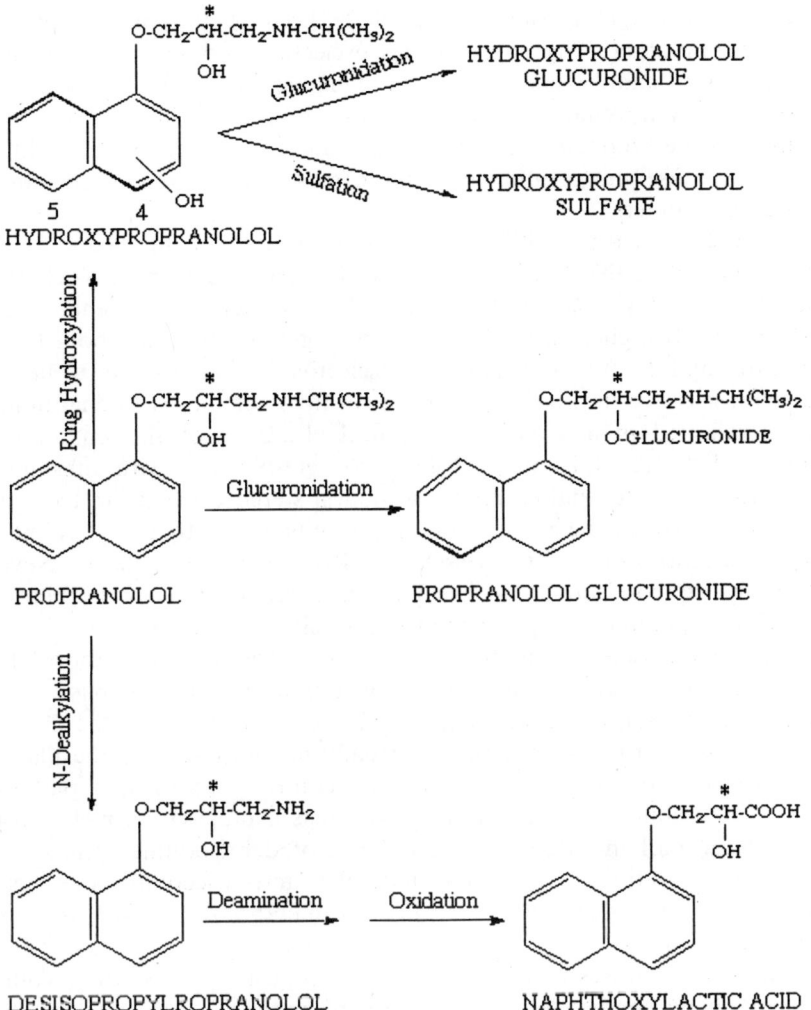

Figure 2 Schematic presentation of main metabolic pathways of propranolol in humans. Asterisk indicates the chiral center.

renal clearance ratios being 0.90, 1.1, 1.2, and 1.05 for metoprolol [40], atenolol [41], pindolol [16], and sotalol [42], respectively. The small degree of stereoselectivity reported for the renal clearance of these drugs is most likely due to an active tubular secretion and/or reabsorption process. A stereoselective plasma protein binding, as a reason for the observed stereoselective renal clearance, is unlikely because the plasma protein binding of pindolol (55%) [16] is not stereoselective, and the protein binding of atenolol and nadolol enantiomers is negligible. Given the low degree of stereoselectivity in renal clearance, for those β-blockers subject to a large fraction of dose excreted in urine, a diminution of renal function might be expected to cause proportionately equal increases in plasma concentrations of both enantiomers.

Except for metoprolol (Table 3) and propranolol (Tables 4 and 5), the stereoselectivity in the pharmacokinetics of other studied beta-blockers is relatively modest (Table 3). For acebutolol, the active S(−) enantiomer attains modestly higher plasma concentrations relative to its antipode (− : + AUC ratio of 1.2 ± 0.1) [43]. This stereoselectivity was suggested to be due to a stereoselective first-pass metabolism in favor of the R(+) enantiomer [43], as reflected in a slightly higher oral clearance of this enantiomer (Table 3). For atenolol and pindolol, two beta-blockers with substantial elimination through renal excretion, a modest stereoselectivity in the renal clearance in favor of the S(−) enantiomer results in slightly (< 20%) higher plasma concentrations of the less active R(+) isomer (Table 3). Nevertheless, the stereoselectivity in the plasma concentrations of acebutolol, atenolol, and pindolol are perhaps of minor clinical significance.

The stereoselectivity in the plasma concentrations of metoprolol is mostly related to the stereoselective metabolism and first-pass effect of the drug. A slight stereoselectivity in the renal clearance of the drug in favor of R(+)-metoprolol (Table 3) virtually has no effect on the plasma stereoselectivity because of the negligible contribution of this pathway to the overall elimination of the drug. Whereas a preferential metabolism of (+)-metoprolol in extensive metabolizers of debrisoquine results in a (−):(+) AUC ratio of 1.37 ± 0.32 [40], the stereoselectivity is reversed [(−):(+) AUC ratio of 0.90 ± 0.06] in poor metabolizers of debrisoquine (Table 3).

After single doses of racemic sotalol administered to healthy volunteers, vitually no stereoseelectivity is observable in the plasma concentrations of the enantiomers. The (−):(+) ratios of C_{max} and AUC were 1.0 and 0.97, respectively, and the differences between enantiomers did not attain the level required for statistical significance [42]. There was some stereoselectivity, however, noted in another study when repeated doses of sotalol were administered to patients with supraventricular tachycardia [44]. After at least 3 days of therapy with 80 or 160 mg of the racemate q 12 h, the (−):(+) of AUCτ were 0.87 and 0.91 for the 80 and 160 mg doses, respectively. For both doses, the differences between enantiomers were significantly different. It is not known why a greater degree of stereoselectivity appears to be present after repeated dose administration.

Table 3 Stereoselective Pharmacokinetics of Some Beta-Blockers in Humans After Single Oral Doses of the Racemates

Drug (Age in yr, n)	Dose, mg	Isomer	C_{max}, ng/mL	T_{max}, h	AUC, ng·h/mL	CL, mL/min/kg	V, L/kg	CL_R, mL/min	$t_{1/2}$, h	Ref.
Acebutolol (19–40, 12)	200	(−)	246±172	2.5±0.9	1230±375	20±5[a]	9.2±2.6[c]	120±38	5.4±1.4	43
		(+)	221±155*		1030±339*	25±7[a],*	11±3.5[c],*	124±40	5.7±1.8	
Acebutolol (60–75, 9)	200	(−)	221±106	2.4±1.5	1380±380	20±5[a]	14±8.9[c]	91±36	7.6±4.0	15
		(+)	209±91		1180±359	24±7[a]	14±7.2[c]	90±36	6.9±3.3	
Atenolol (23–65, 6)	50	(−)	226±136	2.7±1.1	1640±602	1.5±0.4[b]	0.88±0.34	129±32	6.13[d]	41
		(+)	251±138*		1860±652*	1.4±0.3[b]	0.79±0.26	120±29*	6.08[d]	
Carvedilol (53±13, 13)	25	(−)	34.2±22.5	0.67 (median)	125±65.6					67
		(+)	73.5±44.3	1.00 (median)	288±186					
Metoprolol, EM (56±1, 6)	200	(−)			679±388			70±22	2.9±1	40
		(+)			408±409*			75±22*	2.8±1	
Metoprolol, PM (58±11, 6)	200	(−)			3430±623			56±25	7.2±1.5	
		(+)			3800±635*			62±26*	7.7±1.7*	
Pindolol (19–36, 8)	15	(−)	33±7	1.7±0.4	209±73			222±66	2.62[d]	59
		(+)	36±10	1.6±0.5	244±90*			170±55*	2.85[d]	
Sotalol (32±3, 8)	160	(−)	619±164	3.13±0.6	6760±1200			158±38	7.86±1.2	42
		(+)	615±167		6950±850			150±25	8.15±0.7	

[a] Oral clearance or clearance/F.
[b] Systemic clearance.
[c] V/F.
[d] Harmonic half-life.
*Significantly different from the (−)-enantiomer.
Abbreviations: C_{max} = maximum plasma concentration; T_{max} = time to reach C_{max}; AUC = area under the plasma concentration–time curve; CL = clearance; V = volume of distribution; CL_R = renal clearance; $t_{1/2}$ = plasma half-life; EM, extensive metabolizers of debrisoquine; PM = poor metabolizers of debrisoquine; F = oral bioavailability.

Table 4 Stereoselective Pharmacokinetics (Mean $\pm$ SD) of Propranolol in Humans After Intravenous Doses of the Racemate

Dose	Subjects	Isomer	CL, L/min	V, L	$t_{1/2}$, h	Ref.
0.1 mg/kg	4 M, 1 F	$(-)$	1.03 ± 0.27	$286 \pm 52^{a,b}$	3.5 ± 0.5	23
	21–38 yr	$(+)$	$1.21 \pm 0.34*$	$337 \pm 53^{a,b,*}$	3.6 ± 0.6	
$232 \pm 28\,\mu Ci$	12 M, White	$(-)$	0.77 ± 0.14	273 ± 32^c	4.1 ± 0.5	45
long with 16th		$(+)$	0.84 ± 0.16	303 ± 45^c	4.2 ± 0.3	
dose of 80 mg	13 M, Black	$(-)$	$0.95 \pm 0.27**$	$329 \pm 98**$	4.2 ± 0.8	
t.i.d. po		$(+)$	$1.1 \pm 0.3**$	$397 \pm 119**$	4.3 ± 0.9	

[a]Volume of distribution of the beta phase.
[b]Based on a 70-kg subject.
[c]Steady-state volume of distribution.
*Significantly different from the value for the $(-)$-enantiomer.
**Significantly different from the value for the same isomer in the white group.
Abbreviations: CL = systemic clearance; V = volume of distribution; $t_{1/2}$ = plasma half-life.

For propranolol, a higher free fraction of the R(+) enantiomer in blood (Table 2) results in a higher volume of distribution for this enantiomer after the intravenous (IV) administration of the racemate (Table 4). Additionally, the systemic clearance of R(+)-propranolol is slightly, but significantly, higher than that of S(−)-propranolol (Table 4), resulting in a slightly higher plasma concentration of S(−)-propranolol after the IV administration of the drug. However, oral administration magnifies this stereoselectivity (Table 5), presumably owing to a stereoselective first-pass metabolism in favor of R(+)-propranolol [45]. After the oral administration of the racemate, the (−):(+) AUC ratios for propranolol ranged from 1.0 to 1.6 in various studies (Table 5). The possible reasons for this wide range of ratios are explained in the following section.

2.3. Factors Affecting the Pharmacokinetics and Pharmacodynamics of Beta-Blockers

Age and Gender

The effect of age on the stereoselective pharmacokinetics of propranolol has been the subject of several studies [21,46–48], with conflicting results. For instance, while Colangelo et al. [48] and Lalonde et al. [46] reported no significant changes in the oral clearances of the propranolol enantiomers with advancing age, others [21,47] have reported a significant decline in the clearance of both enantiomers in the elderly. Additionally, in contrast to

Table 5 Stereoselective Pharmacokinetics (Mean ± SD) of Propranolol in Humans After Single or Multiple Oral Doses of the Racemate (±) or Pure Enantiomers (− or +)

Drug, Dose (mg)	Subjects	Isomer	C_{max}, ng/mL	AUC, ng·h/mL	CL_O, L/min	$t_{1/2}$, h	Ref.
(±), 80, single	5 M, 3 F, 24–27 yr	(−)	81.7 ± 31	329 ± 118	2.3 ± 0.7	4.5 ± 1.2	51
		(+)	46.5 ± 25*	217 ± 114*	4.0 ± 2.0[a],*	5.2 ± 2.4	
(−), 40, single	10 M, 24 ± 1 yr	(−)	74.6 ± 16	274 ± 83**	2.7 ± 0.7[a],**	4.4 ± 1.0	53
(±), 40, t.i.d.		(−)	27 ± 18	119 ± 80	2.8		
		(+)	20 ± 15*	84 ± 67*	4.0		
(−), 20, t.i.d.		(−)	25 ± 14	98 ± 45	3.4		
(±), 80, single	9 M, 28 ± 9, White	(−)	84.7 ± 37	523 ± 333	1.6 ± 0.8[a]	3.6 ± 0.9	65
		(+)	61.9 ± 33	391 ± 296	2.5 ± 1.8[a]	3.6 ± 1	
	10 M, 27 ± 8, Chinese	(−)	49.8 ± 23***	351 ± 156***	2.6 ± 1.5[a],***	4.0 ± 0.8	
		(+)	34.8 ± 26***	232 ± 111***	4.2 ± 2.5[a],***	3.9 ± 0.8	
(±), 80, t.i.d.	12 M, White	(−)			2.1 ± 0.5	4.1 ± 0.5	45
		(+)			2.9 ± 0.9	4.2 ± 0.3	
	13 M, Black	(−)			3.3 ± 1.7***	4.2 ± 0.8	
		(+)			5.0 ± 4.2***	4.3 ± 0.9	
(±), 80, t.i.d.	6 M, 6 F, 25–33 yr	(−)		387 ± 194	1.7	5.5 ± 2.4	21
		(+)		329 ± 177*	2.0	4.3 ± 1.7	
	6 M, 6 F, 62–79 yr	(−)		475 ± 204	1.4	11.4 ± 5.9****	
		(+)		375 ± 187*	1.8	11.1 ± 4.8****	

(continues)

Table 5 Continued

Drug, Dose (mg)	Subjects	Isomer	C_{max}, ng/mL	AUC, ng·h/mL	CL_O, L/min	$t_{1/2}$, h	Ref.
(±), 80, single	6 M, 24–32 yr	(−)	30±11	152±33	4.56±0.9	3.6±0.7	47
		(+)	18±6	100±20	6.93±1.5	4.3±1.2	
	6 M, 65–80 yr	(−)	42±19****	266±118****	2.76±1.2****	4.8±0.5	
		(+)	27±12****	171±74****	4.55±1.7****	4.8±0.5	
(±), 160, b.i.d.	15 M, 20–35 yr	(−)	290±183	1600±1040	1.5±1.9[a]	3.3±1.3	52
		(+)	275±183	1560±1020	1.5±1.8[a]	3.6±1.2	
(−), 80, b.i.d.		(−)	267±190	1590±1410	1.3±0.9[a]	3.0±1.3	
(+), 80, b.i.d.		(+)	212±140	1120±807	1.9±1.8[a]	3.0±1.2	

[a]Based on a 70-kg subject.
*Significantly different from the value for the (−) enantiomer.
**Significantly different from the value for the (−) enantiomer after the administration of the racemate.
***Significantly different from the value for the same enantiomer in the white group.
****Significantly different from the value for the same enantiomer in the young group.
Abbreviations: C_{max} = maximum plasma concentration; AUC = area under the plasma concentration–time curve; CL_O = oral clearance (clearance/F); $t_{1/2}$ = plasma half-life; F = oral bioavailability.

Zhou et al. [47], who showed no significant differences in the plasma half lives of the propranolol enantiomers between the young and the elderly, Gilmore et al. [21] reported that the half lives of both propranolol enantiomers in the elderly were more than twofold longer than those in the young. These conflicting results are mostly due to different designs of these studies with regard to the dose, the duration of therapy (single versus multiple dosing), the duration of sampling, and the number of subjects used. Overall, these studies suggest that the clearance of both enantiomers of propranolol is reduced in the elderly. Additionally, in the presence of extended duration of sampling (e.g., $\geq 24\,h$), the terminal half-lives of the propranolol enantiomers are apparently longer in the elderly than in the young. However, the reduction in the oral clearance and prolongation of plasma half life appear to be to the same extent for both enantiomers. These data indicate that the known decreased response to beta-blockers in the elderly subjects cannot be explained by stereoselective pharmacokinetic differences between the young and elderly.

For acebutolol, an advance in the age resulted in a decrease in creatinine clearance and an associated decrease in the renal clearance of both enantiomers of the parent drug and its metabolite diacetolol [15]. However, the decrease in the renal clearance of the enantiomers was not stereoselective. On the other hand, a reduction in creatinine clearance with advancing age was associated with a significant decrease in the $(-):(+)$ AUC ratio of the drug (from 1.3 to 1.1 when creatinine clearance declined from 90 to 45 mL/min), suggesting that aging has a stereoselective effect on the other pathways (e.g., metabolism) of acebutolol elimination.

In terms of gender, nonstereoselective studies [49] have clearly shown that the oral clearance, and not the systemic clearance, of propranolol is significantly (63%) higher in men than in women. This difference was attributed to significant increases in the side-chain oxidation and glucuronidation of propranolol during the first pass metabolism in men [49]. Available stereoselective studies [21,22], however, have failed to show a significant difference between males and females in the main pharmacokinetic parameters of propranolol. Again, this discrepancy may be related to methodological differences and the power of statistical tests used in these studies. Future, well-designed studies are needed to test the effects of gender on the stereoselective pharmacokinetics of propranolol.

The importance of delineation of the stereoselective pharmacokinetics in explaining gender-related pharmacodynamics was demonstrated recently for labetalol [50]. Labetalol dose was titrated to a specific antihypertensive effect in 14 men and 5 women with ages ranging from 40 to 63 and 40 to 56 years, respectively. The dose-corrected AUC values for total labetalol in women were 80% higher than those in men. However, the antihypertensive effects were the same for both groups. This discrepancy could be easily explained by stereoselective differences in the pharmacokinetics of labetalol isomers in men and women; whereas the concentrations of the alpha blocking isomer (SR) and two relatively inactive isomers of labetalol (SS and RS) were between 60 to 80% higher in women, the plasma concentrations of the main beta-blocking isomer (RR) were the same in both groups, resulting in similar antihypertensive effects in men and women.

Enantiomer–Enantiomer or Drug–Drug Interactions

The two enantiomers of a racemic drug may interact with each other at different pharmacokinetic or pharmacodynamic levels. This type of interaction has been studied for atenolol [2] and propranolol [51–53]. For atenolol, there was no pharmacokinetic or pharmacodynamic interaction between the two enantiomers; the half-dosed S(−)-atenolol produced the same effect as did the racemic atenolol [2]. Additionally, the plasma concentration–time profiles of S(−)-atenolol were identical after the administration of the racemate and the half-dosed pure enantiomer. On the other hand, both single [51] and multiple [52] dose studies have shown that there is a significant interaction between the enantiomers of propranolol. When administered as pure enantiomer, as opposed to the racemate, R(+)-propranolol tends to show lower plasma concentrations [52]. However, the kinetics of the more active S(−)-enantiomer appear to be the same whether it is administered as a pure enantiomer or racemate [51–53].

In addition to enantiomer–enantiomer interactions, a racemic drug may interact with other drugs stereoselectively. For instance, stereoselective interactions have been reported in man between propranolol and calcium channel blockers [54,55], cimetidine [56], and quinidine [33]. Calcium channel blockers nicardipine [54], diltiazem [55], and verapamil [55] all decreased the first-pass metabolism of both enantiomers of propranolol. However, this inhibitory effect was stereoselective for the R(+) enantiomer in the case of both verapamil [55] and nicardipine [54], resulting in a significant increase in the (+):(−) AUC ratios in plasma. In terms of effects, nicardipine did not increase the blood pressure reduction effect of propranolol [54], a phenomenon that may be explained by a more significant pharmacokinetic effect on the less active R(+) enantiomer. Similarly, cimetidine decreased the oral clearance of R(+)-propranolol to a more significant degree than that of the S(−) enantiomer [56]. As for quinidine, human liver microsome studies [33] indicated that this selective inhibitor of CYPD26 reduced the ring hydroxylation of propranolol in a stereoselective manner in favor of R(+)-propranolol. This was in agreement with in vivo studies [57]. showing 180% and 100% increases in the plasma AUCs of R(+)- and S(−)-propranolol, respectively, as a result of quinidine coadministration. Interestingly, all these studies have shown that the inhibition of the metabolism of propranolol by different drugs is stereoselective for R(+)-propranolol.

In addition to the inhibition of the metabolism of propranolol, verapamil reportedly [58] inhibits the metabolism of metoprolol, another extensively metabolized beta blocker. As mentioned in the metabolism section, the O-demethylation pathway is the main metabolic pathway for the metabolism of metoprolol, accounting for 65% of the dose. Verapamil significantly inhibits this pathway in a stereoselective manner, favoring inhibition of the metabolism of R(+)-metoprolol [58]. The interaction of verapamil and metoprolol results in clinically significant adverse reactions, presumably owing to higher plasma concentrations of metoprolol.

Cimetidine not only reduces the metabolism of beta-blockers such as propranolol, it is also known to act as an inhibitor of tubular secretion of a number of organic cations. Therefore it is not surprising to see that the renal clearance of pindolol is substantially and stereoselectively reduced by coadministration of this drug [59]. The administration of 400 mg cimetidine twice a day for 2 days before and 2 days after pindolol administration resulted in 26% and 34% reductions in the renal clearances of S(−)- and R(+)-pindolol, respectively. Therefore the plasma concentrations of the R(+) enantiomer increased more drastically (47%) than those of the S(−) enantiomer (38%) in the presence of cimetidine [59]. Because renal clearance accounts for only 50% of pindolol elimination, the significant increases in the plasma concentrations of pindolol enantiomers as a result of cimetidine coadministration cannot be explained from the inhibition of its renal clearance only. Apparently, cimetidine also reduces the metabolism of pindolol.

Rate of Drug Input

Theoretically [60,61], the oral rate of input of racemic drugs with stereoselective metabolism may affect the plasma concentration ratio of the enantiomers. This is because the rate of input of the drug into the portal vein may have a different effect on the degree of saturation of the metabolic pathways of the enantiomers, with resultant stereoselectivity in the first-pass metabolism of these drugs. Indeed, as early as 1982, Silber et al. [62] reported that the (−):(+) steady-state plasma concentration ratios of propranolol significantly decreased with an increase in the daily dose of the drug; the ratios were 2.45 ± 1.12, 1.78 ± 0.60, and 1.51 ± 0.05 after the oral multiple doses of 160, 240, and 320 mg/day, respectively. The same trend is observed when results of several studies are combined in Fig. 3. The input rate–dependent change in the ratio may be attributed to a dose-dependent

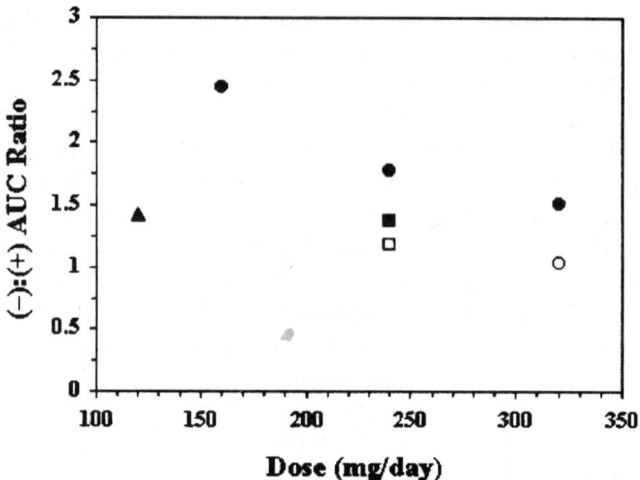

Figure 3 The average steady-state $(-):(+)$ AUC ratio of propranolol as a function of daily dose of the racemic drug in different studies. Key: ●, Ref. 62; ▲, Ref. 53; ■, Ref. 65; □, Ref. 21; ○, Ref. 52.

saturation of the first pass metabolism of propranolol with a greater saturation for the $R(+)$ enantiomer.

In contrast to Silber et al. [62], Bleske et al. [63] could not detect any significant effect of input rate on the stereoselectivity in the propranolol pharmacokinetics. The lack of effect of input rate in the latter study is perhaps because only single and relatively low doses of the drug were used [63]. Simulation studies [60,61] have shown that the effects of input rate on the stereoselective pharmacokinetics are significant in the nonlinear input ranges when the input rate approaches the maximum velocity of the metabolism. Additional pharmacokinetic/dynamic studies are needed to determine the clinical significance of the effects of input rate on the stereoselective metabolism of propranolol.

Genetic Factors

Metabolic Phenotype of the Patient. As mentioned in the Elimination section, the metabolic phenotype of patients may significantly affect the metabolism and consequently the overall pharmacokinetics and pharmacodynamics of propranolol and metoprolol. For propranolol, Ward et al. [36] reported that although the ring hydroxylation and N-dealkylation processes were deficient in poor metabolizers of debrisoquine and mephenytoin,

Table 6 Oral Clearance and Plasma Half-Lives of Propranolol Enantiomers After Oral Administration of the Racemate (80 mg) to Volunteers with Different Phenotypes of Debrisoquine and Mephenytoin Hydroxylation

Group	n	Oral clearance, mL/min		
		(+)-Propranolol	(−)-Propranolol	(+):(−) Ratio
EM[a]	6	2670 ± 697	1910 ± 632	1.42 ± 0.53
PM$_D$	4	1860 ± 1110	1420 ± 788	1.25 ± 0.06
PM$_M$	5	2010 ± 909	1401 ± 595	1.42 ± 0.09
PM$_{D/M}$	1	918	850	1.08

Abbreviations: EM, extensive metabolizers of debrisoquine and mephenytoin; PM$_D$, poor metabolizers of debrisoquine and extensive metabolizers of mephenytoin; PM$_M$, poor metabolizers of mephenytoin and extensive metabolizers of debrisoquine; PM$_{D/M}$, poor metabolizers of both debrisoquine and mephenytoin.
Source: Ref. 36.

respectively, the overall oral clearance of the propranolol enantiomers were modestly reduced in volunteers with one of these deficiencies (Table 6). Additionally, the stereoselectivity in the plasma concentrations in these volunteers were not different from that in extensive metabolizers (Table 6). However, a combined poor metabolism of mephenytoin and debrisoquine (observed only in one volunteer) substantially reduced the oral clearance of both enantiomers and abolished the stereoselectivity in the plasma concentration of propranolol (Table 6). It should be noted that the cosegregation of both deficiencies is expected to be very low (0.4% of the population) [36].

For metoprolol, Lennard et al. [38] demonstrated that after a 200 mg oral dose, the plasma concentrations of racemic metoprolol in poor metabolizers of debrisoquine were six times higher than those in the extensive metabolizers. Additionally, when the beta-blockade effect was plotted against the plasma concentrations of the racemic drug, the relationship was shifted to the right in poor metabolizers [40]. This discrepancy was attributed to the differences in the stereoselectivity in the metabolism and plasma concentrations of the drug in the two groups. Whereas the (−):(+) metoprolol AUC ratio was 1.37 ± 0.32 (mean ± SD) in the extensive metabolizers, the ratio was 0.90 ± 0.06 in poor metabolizers [40]. Therefore, the same concentrations of the racemate would contain less of the active (−) enantiomer in poor metabolizers, shifting the effect–concentration relationship to the right.

Although the metabolism of both propranolol and metoprolol are affected by debrisoquine phenotype, the above data clearly shows that the effect is more substantial for metoprolol. Debrisoquine polymorphism also significantly affects the kinetics of another beta-blocker, bufuralol [64]. In contrast to metoprolol, however, poor metabolism of debrisoquine intensified the stereoselectivity in the plasma concentrations of bufuralol [$(-):(+)$ plasma concentration ratio of 1.8 and 2.6 at 3 h in extensive and poor metabolizers, respectively] [64]. This was due to a significant reduction of the metabolism of $(-)$-bufuralol by ring hydroxylation in poor metabolizers of debrisoquine [64]. Therefore the effect of debrisoquine hydroxylation phenotype on the degree and direction of stereoselectivity in the pharmacokinetics varies with each beta-blocker.

Ethnic Background. It is known that Chinese subjects are more responsive than white subjects to the same dose or plasma concentrations of racemic propranolol. A study [65] investigated whether this could be attributed to higher $(-):(+)$ AUC ratios in Chinese subjects, compared with the ratio in the white population. Although the plasma concentrations of both enantiomers were substantially lower in Chinese volunteers, the ratio was the same in both populations [65], suggesting a pharmacodynamic difference in these two populations with regard to the beta-blockade effect.

In contrast to the Chinese, the black population responds less to the same dose of propranolol, than the white population. Sowinski et al. [45] showed that both the systemic and oral clearances of both enantiomers of propranolol are substantially higher in blacks than in whites. This difference was mostly attributed to a higher intrinsic clearance of propranolol enantiomers, in association with a slightly lower (9%) hepatic blood flow in blacks. The limited available information on the effects of ethnicity on the pharmacokinetics of propranolol suggest that the racial differences in the effects of this drug cannot be attributed to the stereoselectivity in the pharmacokinetics of the drug. Rather, these differences may be due to pharmacodynamic differences among ethnic populations.

Congestive Heart Failure

For highly metabolized drugs, altered cardiac output may influence the drug plasma concentrations in the presence of moderate to high hepatic extraction ratio, conditions that are true for carvedilol enantiomers [66, 67]. Tenero et al. [12] demonstrated that in patients with Class III or IV congestive heart failure, the plasma concentrations of carvedilol enantiomers were substantially higher than those in healthy volunteers. Additionally, patients with Class IV congestive heart failure had consistently higher plasma concentrations (up to twofold) than those patients with Class III congestive heart failure [53]. Like healthy volunteers, patients with congestive heart failure exhibited stereoselectivity in both C_{max} and AUC of the drug [12]. However, the R : S AUC ratios observed in these patients (~ 1.9) were lower than those observed in otherwise healthy young (2.8) and elderly (2.3) volunteers [12], suggesting that the effects of congestive heart failure on the pharmacokinetics of carvedilol are stereoselective.

Although it was not mentioned by the authors, there does appear to be a difference in the level of stereoselectivity in AUC between patients with Class III and Class IV congestive heart failure [12]. The reported mean R(+) : mean S(−) AUC ratios in the Class IV patients were consistently higher (20–43%) than the corresponding ratios in patients with Class III congestive heart failure across a dosage range of 6.25–50 mg administered every 12 h. The ratios of mean R(+) : mean S(−) C_{max} were also consistently higher in the Class IV patients (14–18%). Overall, these data suggest that Class IV congestive heart failure patients attain higher plasma concentrations of both enantiomers, in favor of the R enantiomer, when compared with Class III patients [53].

3. DIURETIC AGENTS

Many of the benzothiazide-type diuretic agents are chiral. As a class, the chiral diuretics include chlorthalidone, indapamide, metolazone, bendroflumethiazide, trichlormethiazide, methyclothiazide, and polythiazide. The diuretics have been largely overlooked from the perspective of stereoselectivity in pharmacodynamics and pharmacokinetics. Stereospecific pharmacokinetic/dynamic data for these drugs are scarce.

The pharmacodynamic properties of the natriuretic/uricosuric agent indacrinone have been studied in subjects given varying ratios of the (+) and (−) enantiomers [68]. With respect to stereoselectivity in pharmacodynamic properties, (−)-indacrinone is a more potent natriuretic agent than the (+) enantiomer. On the other hand, both enantiomers are equipotent with respect to uricosuric activity. The complexities of the pharmacodynamics of indacrinone and the merits of using different ratios of the enantiomers of indacrinone (instead of 1 : 1 in the racemate) are discussed in Chap. 5 (Sec. 3.3).

4. THE 1,4-DIHYDROPYRIDINE CALCIUM CHANNEL ANTAGONISTS

A number of the 1,4-dihydropyridine calcium channel antagonists are chiral and are used as the racemate for the treatment of hypertension and angina.

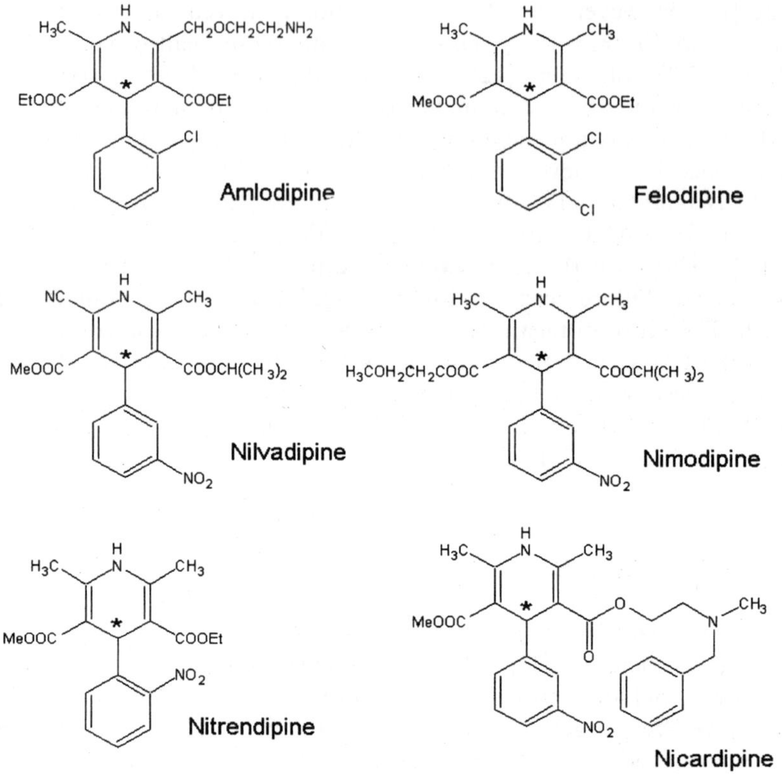

Figure 4 Chemical structures of the 1,4-dihydropyridine calcium channel blockers. Chiral center is denoted by an asterisk.

These drugs include nicardipine, nilvadipine, nitrendipine, felodipine, amlodipine, and nimodipine (Fig. 4). Although verapamil is also a calcium channel blocker, because it is also used as an antiarrhythmic agent, it is discussed later in this chapter along with the antiarrhythmic agents.

4.1. Stereoselectivity in Pharmacological Actions

The pharmacological activities of dihydropyridine calcium channel blockers are complex (see Chap. 5, Sec. 3.1). Generally, the S(−) enantiomers of nitrendipine and felodipine possess the majority of these racemic drugs' ability to antagonize the calcium channels [69]. In a pharmacodynamic assessment using heart rate as a measure of effect, when given alone, the S(−) enantiomers of felodipine and nicardipine showed similar E_{max} and lower EC_{50} values, in comparison with the adminsitration of the racemate [70]. The enantiomers of nilvadipine and nicardipine differ from those of nitrendipine, felodipine, and amlodipine [71] in that their (+) enantiomers are much more potent than antipode in alleviating arterial contractions [72,73].

4.2. Stereoselectivity in Pharmacokinetics

Stereoselectivity of differing degrees has been reported in the pharmacokinetics of this class of drugs (Table 7). For all the drugs in this class, it appears that plasma concentrations of the more active enantiomer are equal to or higher than those of the less active enantiomer. When these drugs have been given intravenously (Table 7), stereoselectivity is markedly less than when given orally, suggesting enantioselective presystemic metabolism. After oral administration, stereoselectivity in plasma concentrations appears least for for amlodipine (between 1.1 and 1.5). The other agents of this class possess stereoselectivity ranging from 1.7 to 2.7 after oral administration (Table 7).

The stereospecific pharmacokinetics of nitrendipine have been studied after oral and IV dosing in healthy volunteers (Table 7) [74]. After IV dosing, the S:R ratio of plasma concentrations was close to one (CL differing by only 7%), but when given orally, the ratio was much higher, reaching about 1.8. Although absolute bioavailability was low for both enantiomers (< 15%), the authors found that F was significantly (75%) higher for the S than the R enantiomer. The Vd and $t_{1/2}$ of the enantiomers were the same after IV dosing. It seems likely that presystemic intestinal metabolism by CYP3A isoenzymes is involved in the stereoselective differences observed in bioavailability after oral dosing.

Grapefruit juice and cimetidine have been shown to inhibit the clearance of both nitrendipine enantiomers after single oral dosing with 20 mg racemic nitrendipine (Table 7) [75]. Both the nitrendipine C_{max} and AUC increased after coingestion of either repeated doses of grapefruit juice or cimetidine prior to and after administration of nitrendipine. A slightly greater inhibitory effect was seen with cimetidine coadministration. The interactions with nitrendipine were nonstereoselective in nature, as the S:R ratios of nitrendipine did not differ between placebo, cimetidine, and grapefruit juice treatments [75].

A similar profile to nitrendipine of S > R plasma concentrations was observed for another dihydropyridine calcium channel blocker, amlodipine (Table 7). In a study employing 18 subjects, in which two salt forms (besylate and maleate) of amlodipine were administered (20 mg racemate), stereoselectivity was noted (Table 7) in the plasma S:R AUC ratio for both

Table 7 Stereospecific Pharmacokinetic Data (Plasma or Serum Unless Stated Otherwise) of Some Selected Calcium Channel Blockers and Pimobendan, an Inotropic Agent. Data Obtained for Racemate After Oral Administration, Unless Otherwise Specified

Drug n (mean age or range)	Dose*	Enantiomer	C_{max}, ng/mL	T_{max}, h	AUC ng·h/mL	$t_{1/2\lambda n}$, h	CL, L/h/kg	Vd, L/kg	Ref.
Amlodipine									
4 (20–45)	16 mg	R	2.21	8	67.5	24			76
		S	3.41	6	76.7	31			
18 (31)	20 mg base as besylate salt	R	5.5	8.7	240	35.5			71
		S	6.1	8.4	351	50.6			
18 (31)	20 mg base as maleate salt	R	5.1	10.9	236	34.3			71
		S	5.7	10.7	330	48.7			
Felodipine									
12 (25)	20 mg	S	20	1	73	12			69
		R	9.8	1	31	11			
Gallopamil									
	50 mg R/S (^{2}H label)	R	57.1	0.83	95.5	4.7			158
		S	46.7	0.83	91.0	5.8			
6 (24–36)	25 mg R alone	R	17.8	0.88	38.5	6			
	25 mg S alone	S	24.7	0.83	54.2	6.7			
Nicardipine									
8 (31)	20 mg	(+)	25		41	1.8			73
		(−)	11		15	2.1			
8 (31)	40 mg	(+)	63.5		106	1.1			73
		(−)	26.4		46	1.5			
7 volunteers	40 mg	(+)	152	0.8	273	1.7			78,165
		(−)	65	0.7	117	1.6			
6 (27–44)	40 mg	(+)			355				78
		(−)			161				
6 (27–44)	40 mg + grapefruit juice	(+)			510				78
		(−)			297				
6 (27–44)	2 mg IV	(+)			46		0.39		78
		(−)			44		0.40		

Subjects (age)	Dose	Enantiomer							Ref
6 (27–44)	2 mg IV + grapefruit juice	(+)			43		0.41		78
		(−)			42		0.43		
Nilvadipine									
3 (20–28)	4 mg	(+)	4.13	0.83	12.4	3.42			72
		(−)	1.5	0.83	4.7	3.17			
Nitrendipine									
12 (25)	20 mg solution	S	30	0.7	52	6.7			69
		R	16	0.7	24	9.0			
9 (25)	20 mg tabs	S	5.3	1.6	20	5.5			75
		R	2.8	1.5	11	5.3			
9 (25)	20 mg + grapefruit juice	S	11	2.0	45	5.8			75
		R	5.9	1.8	25	7.3			
9 (25)	20 mg + cimetidine	S	11	2.9	52	6.9			75
		R	5.8	2.4	24	6.7			
9 (24)	20 mg	S	4.7		15.5	3.3			74
		R	2.9		8.5	3.2			
9 (24)	40 µg/kg IV	S				4.3	1.2	3.9	74
		R				4.0	1.3	3.7	
Pimobendan									
7 cardiac patients (55)	5 mg	(+)	16	0.86	44	2.6			160
		(−)	17	0.87	48	2.9			
8 healthy (20–42)	7.5 mg	(+)	16	1.2	53	2.6			161
		(−)	17	1.2	54	2.9			
8 healthy (20–42)	5 mg IV					1.9	0.86	2.3	161
						1.8	0.81	1.7	
Prenylamine									
8 (24)	200 mg	R	50.7	2.4	502	8.2			156
		S	9.2	2.6	166	24			

*Racemic PO unless stated otherwise (IV or individual enantiomer).

salt forms (~1.4) [71]. A longer $t_{1/2}$ was noted for the S than for the R enantiomer (49 vs. 34 h). Peak plasma enantiomer concentrations were slightly lower for the besylate salt than for the maleate formulation, presumably due to a lower rate of drug absorption [71]. This was in line with the longer t_{max} observed for the besylate salt (Table 7). In another study employing only four subjects, Luksa et al. [76] similarly reported that pharmacologically active S-(−)-amlodipine attained higher plasma concentrations than inactive antipode after oral administration of 16 mg of racemate (Table 7). However, there was less stereoselectivity noted in S : R AUC ratios (1.13) [76] than that observed by Laufen & Leitold (1.4) [71]. Additionally, based on the pharmacokinetic data obtained after administration of racemate or single enantiomers, Luksa et al. [76] reported a lack of pharmacokinetic interaction between enantiomers and an absence of chiral inversion.

For nilvadipine, after administering 4 mg of racemic drug orally to three healthy volunteers, plasma concentrations of the more active (+) enantiomer were higher than antipode (Table 7) [72]. Stereoselectivity was greater than seen for nitrendipine or amlodipine, as the (+):(−) AUC and C_{max} ratios were each 2.8.

Felodipine also displays stereoselectivity in its pharmacokinetics, with reported S : R ratios of C_{max} and AUC of 2.1 and 2.4, respectively, in young healthy male volunteers given oral 20 mg single doses of racemate (Table 7) [69]. The stereoselectivity in t_{max} and $t_{1/2}$ was lower (S : R ≈ 1.1) than in C_{max} and AUC. These same subjects were administered nitrendipine and the prototype achiral 2,4-dihydropyridine derivative nifedipine. The S : R plasma concentration ratios of felodipine were consistently higher than those for nitrendipine over the blood sampling periods. In the study subjects, the AUC of the enantiomers of nitrendipine and felodipine correlated well with the AUC of nifedipine (r ranging from 0.88 to 0.95). This suggested to the authors that a single cytochrome P-450 enzyme, probably CYP3A4, might be primarily involved in the metabolism of each of these drugs, and that it might differ in amount between the different subjects studied [69].

In patients with subarachnoid hemmorhage, nimodipine reportedly displays little enantioselectivity in its pharmacokinetics after IV doses [77]. After oral doses, however, plasma concentrations of the R(+) enantiomer are higher for the first 3 hours postdose. A possible stereoselectivity in first-pass metabolism was offered to explain the finding.

After oral 20 and 40 mg racemic doses to healthy male volunteers in a randomized crossover study, the more active (+) enantiomer of nicardipine attains higher C_{max} and AUC values than antipode (Table 7) [73]. The (+):(−) ratios of nicardipine C_{max} and AUC ranged between 2.2 and 2.8.

Nonlinear increases in AUC were noted between the 20 and 40 mg doses (Table 7), suggesting saturation of an elimination process over this dose range. Uno et al. [78] have recently shown that in healthy male volunteers, nicardipine metabolism in the gastrointestinal tract can be stereoselectively impaired by grapefruit juice. The oral bioavailabilities of both enantiomers were higher after grapefruit juice, although systemic CL, measured from IV doses, was not affected by grapefruit juice. The presystemic metabolism of (−)-nicardipine was inhibited to a greater extent than the (+) enantiomer. The result was an increase of greater than 91% in oral bioavailability of the less active (−) enantiomer. In comparison, the bioavailability of the eutomer was increased by only 43%. The net effect was a reduction in the (+):(−) ratio of AUC after oral doses. Although mean C_{max} and t_{max} were not reported, it appeared from figures in the paper that both C_{max} and t_{max} were higher after grapefruit juice. Shortly after dosing, there was an associated significant increase in the heart rate, after administration of the drug+grapefruit juice combination, over the drug-only situation [78]. No other cardiovascular changes were noted.

5. ANTIARRHYTHMIC AGENTS

Although the chiral antiarrhythmic agents possess some similarities, they also display a number of features that make them unique. From the perspective of stereoselectivity, a range of differences between the enantiomers of each of these drugs exists in their metabolism, excretion, degree of plasma protein binding and volume of distribution, and pharmacodynamic properties.

5.1. Stereoselectivity in Pharmacologic Action

Except for verapamil, which is a calcium channel blocker, drugs listed in Table 8 exert their effects mainly through the blockade of the sodium channel. All of the chiral antiarrhythmic drugs listed in Table 8 have chiral centers (Fig. 5) and are marketed as racemates consisting of two enantiomers. However, in many cases, the pharmacological effects of these drugs are stereoselective (Table 8) [79–91]. Among the listed antiarrhythmics, the highest degree of stereoselectivity has been reported for disopyramide, whose S(+) enantiomer prolongs the QT interval corrected for the heart rate (Q-Tc interval), while the R(−) enantiomer is virtually devoid of this activity [79]. The next degree of stereoselectivity in pharmacological action is reported for the calcium channel antagonist verapamil, whose S(−) enantiomer is approximately 10 times more potent than its antipode (Table 8)

Table 8 Stereoselectivity in the Pharmacologic Action of Some Chiral Antiarrhythmic Drugs

Drug	Relative activity (ratio)	Biological response (species)	Ref.
Disopyramide	S(+) ≫ R(−)	Prolongation of QT intervals (human)	79
Encainide	(+) = (−)	Action potential parameters of cardiac Purkinje fibers (dog)	80
Flecainide	S(+) = R(−)	Prevention of chloroform-induced ventricular fibrillation (mouse) Prevention of ouabain-induced ventricular tachycardia (dog)	81
Mexiletine	R(−) > S(+)	Prevention of ventricular tachycardia (dog)	82
	R(−) > S(+) (2:1)	Tonic block of skeletal muscle sodium channels (frog)	83
Propafenone	R(−) = S(+)	Sodium channel blocking activity (human)	84
	R(−) = S(+)	Antiarrhythmic effect in cardiac Purkinje fibers (dog)	85
	S(+) > R(−) (100:1)	Beta blocking effect on lymphocytes (human)	
Tocainide	R(−) > S(+) (4:1)	Sodium channel blocking activity (human)	86
	R(−) ≫ S(+)	Analgesic effects (mouse)	87
Verapamil	S(−) > R(+) (10:1)	Calcium channel blocking activity (human)	88,89
	S(−) = R(+)	Inhibition of aortic contractions by α_1-agonists (rabbit)	90
	S(−) = R(+)	Reduction of multidrug resistance to vincristine (human)	91

[88,89]. Tocainide and mexiletine also show stereoselectivity in their action with the R(−) enantiomers being four and two times, respectively, more potent than their antipodes in sodium channel blocking activity (Table 8). In contrast, there is no stereoselectivity in the main antiarrhythmic action of encainide, flecainide, and propafenone (Table 8).

It is worthy of note that the enantiomers of the chiral antiarrhythmics may show different degrees of stereoselectivity when interacting with receptors other than those responsible for their primary antiarrhythmic effects. This has been clearly shown for propafenone and verapamil (Table 8). Whereas the enantiomers of propafenone are equipotent in

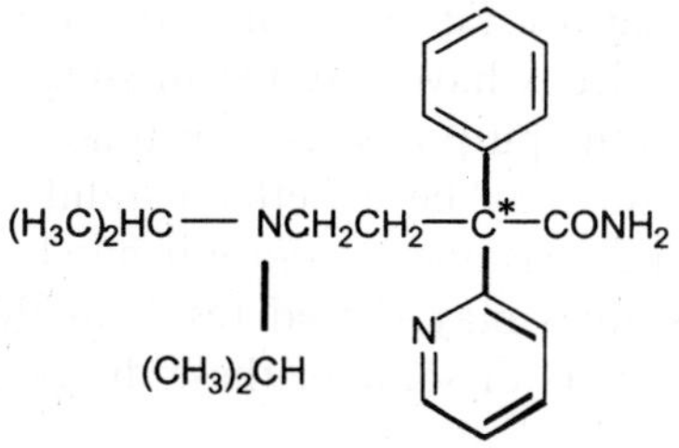

DISOPYRAMIDE

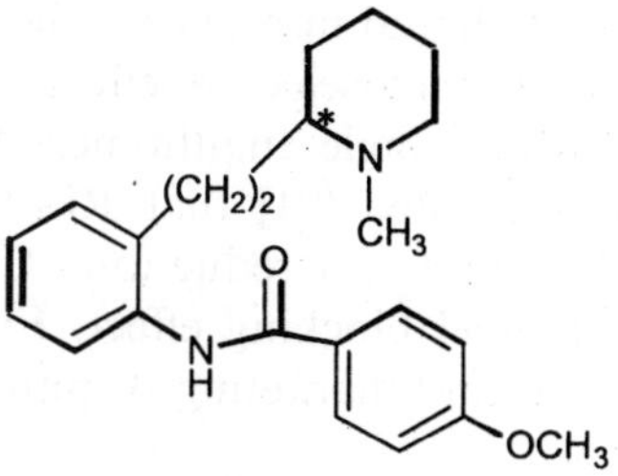

ENCAINIDE

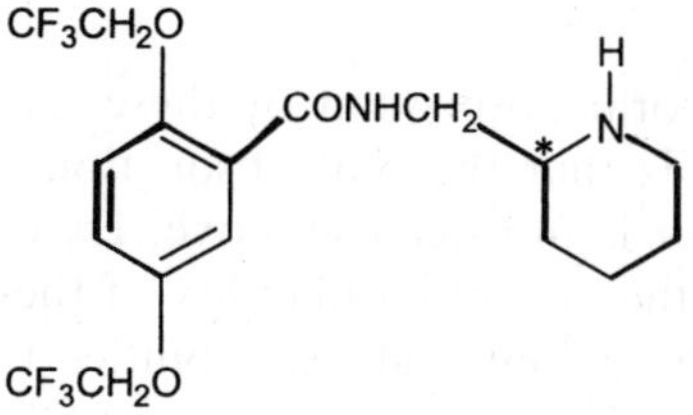

FLECAINIDE

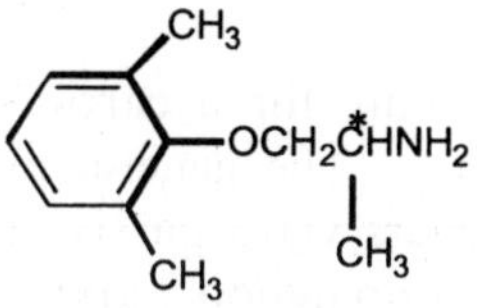

MEXILETINE

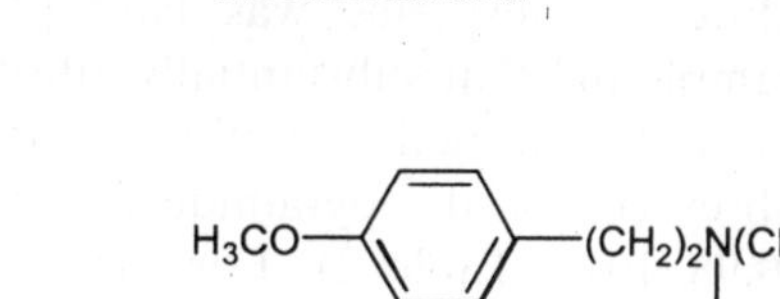

PROPAFENONE

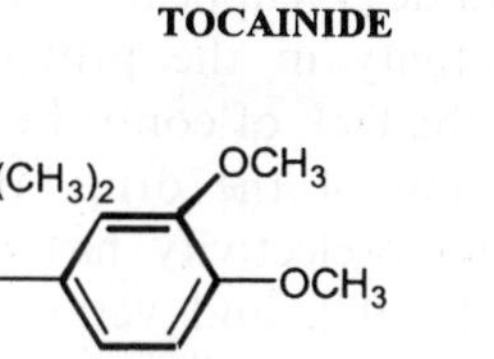

TOCAINIDE

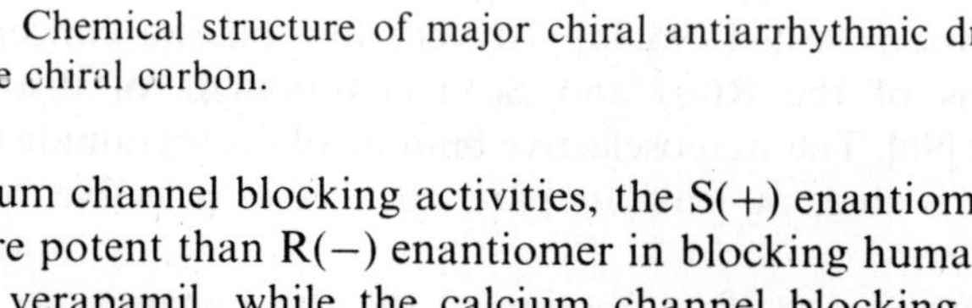

VERAPAMIL

Figure 5 Chemical structure of major chiral antiarrhythmic drugs. The asterisk denotes the chiral carbon.

their sodium channel blocking activities, the S(+) enantiomer is almost 100 times more potent than R(−) enantiomer in blocking human beta receptors [85]. For verapamil, while the calcium channel blocking effect is stereoselective [88,89], both enantiomers are equipotent in opposing the effects of

α_1-agonists [90] and in reducing the multidrug resistance to vincristine [91]. The different stereoselective effects of these drugs have resulted in suggestions for using single enantiomers for different purposes. For instance, it has been suggested [91] that R(+)-verapamil may be a better candidate than the racemate for reduction of multidrug resistance because it has less calcium channel blocking effect. Further studies may indeed result in the development and marketing of pure enantiomers of some of these drugs.

5.2. Stereoselectivity in Pharmacokinetics

Absorption

Evidence is lacking for a carrier-mediated oral absorption of these anti-arrhythmic drugs. The general assumption is that the oral absorption of these drugs occurs via a passive process that is nonstereoselective. Except for encainide, propafenone, and verapamil, the oral bioavailability of these drugs is relatively high (>70%). The reported low oral availabilities for racemic encainide (30–88%), propafenone (5–50%), and verapamil (22%) are mostly due to a significant first-pass metabolism rather than incomplete absorption (see the Metabolism section below).

Distribution

Each of the chiral antiarrhythmics is basic, and all are bound to both albumin and α_1-acid glycoprotein (AAG) in the plasma. Among the anti-arrhythmic drugs listed in Table 2, the plasma protein binding of mexiletine [92] and tocainide [93] appears to be nonstereoselective. An initial report [94] of stereoselectivity in the protein binding of mexiletine was later [92] attributed to the lack of control of the sample pH that substantially affects the free fraction of the drug. In contrast to mexiletine and tocainide, substantial stereoselectivity has been observed for disopyramide [95,96], propafenone [84,97], and verapamil [20,98–101] (Table 2). For disopyr-amide, the protein binding is also concentration dependent within the therapeutic range. The unbound fractions of both enantiomers increase with increasing plasma concentrations of the enantiomers [96]. However, the free fraction of R(−)-disopyramide remains greater than its antipode for all the concentrations tested [95,96]. At very low plasma concentrations, the free fractions of the R(−) and S(+) enantiomers of disopyramide are 0.21 and 0.12 [96]. The stereoselective binding of disopyramide to human plasma proteins is due, at least in part, to a stereoselective binding to AAG [102].

    For propafenone, the free fraction of the R(−) enantiomer is higher than that of the S(+) isomer [84,97] (Table 2). However, no concentration dependency is reported for the protein binding of this drug. Similarly, the protein binding of verapamil [20,98–101] is stereoselective (Table 2) and concentration independent.

Studies using individual plasma proteins [99] have shown that the more active S(−) verapamil binds to a lesser extent to both albumin and AAG (Table 2). However, there is no stereoselectivity in the binding of verapamil to lipoproteins such as HDL and LDL [101]. Gross et al. [99] studied the degree of protein binding of verapamil enantiomers in serum samples obtained from healthy volunteers before dosing (in vitro) and after oral (PO) or IV dosing (Table 2). The degree of binding was the same using the in vitro method and after IV doses. However, after the oral dosing, the free fractions of both enantiomers were twofold higher (Table 2). It was suggested [99] that the higher free fractions after oral dosing may be due to binding competition from verapamil metabolites formed presystemically during the oral absorption. A more recent study [100] using oral verapamil, on the other hand, showed that the free fractions of the enantiomers in predose, after single dose, and at steady state were almost identical (Table 2). Additionally, it was shown that the major metabolites of verapamil do not interfere with the drug binding at clinically relevant concentrations [100]. This discrepancy may be due to methodological differences between the two studies [99,100] including the use of a pseudoracemate [labeled (R) and unlabeled (S) enantiomers] in one of the reports [99].

The stereoselectivity in the plasma protein binding of propafenone and verapamil results in an apparent stereoselectivity in the red blood cell concentrations of these drugs [97,103]. For propafenone, the R(−) enantiomer achieved higher erythrocyte concentrations than its antipode after the in vitro addition of the racemate to the whole blood [97]. This, however, could be simply explained by a higher free fraction of this enantiomer in plasma. The stereoselectivity in the plasma concentrations of propafenone was opposite of that in the red blood cells because the higher penetration of the R(−) enantiomer into the cells resulted in lower concentrations of this enantiomer in plasma [97]. Similarly, a higher free fraction of S(−)-verapamil in plasma than that for the R(+) enantiomer, resulted in an apparent stereoselective erythrocyte uptake in favor of this enantiomer [103]. Interestingly, when the plasma was replaced with buffer (no protein) in the blood, the stereoselectivity in the red blood cell uptakes of both propafenone and verapamil was abolished, confirming the underlying role of plasma proteins for this phenomenon.

In the absence of active transport pathways, the stereoselectivity in the tissue distribution of chiral drugs is affected by both binding to plasma and tissue proteins. Studies [104] in rats with disopyramide, flecainide, and verapamil have shown that the tissue binding of flecainide is not stereoselective. Therefore any apparent stereoselectivity in the distribution of this drug is due to stereoselectivity in its plasma protein binding. However, stereoselective binding to phosphatidylserine, a tissue binding site for basic drugs, was observed for disopyramide and verapamil, with R(+)-verapamil and R(−)-disopyramide being preferentially bound [104]. As for mexiletine, a study [105] in rats showed nonstereoselective distribution of the enantiomers to most tissues except the liver where the concentration of the S(+) enantiomer was more than twice of that of its antipode. This was also associated with a twofold higher tissue:serum concentration ratio for S(+)-mexiletine, as compared with the ratio for the R(−) enantiomer [105]. This apparent stereoselectivity may simply be due to a faster metabolism of R(−)-mexiletine in the liver, which is the focal site for elimination of this drug. Overall, these studies suggest that the stereoselectivity in the tissue distribution of chiral antiarrhythmic drugs is mostly determined by the stereoselectivity in their binding to plasma proteins. However, in the case of verapamil and disopyramide, stereoselective binding to tissue phosphatidylserine may also play a role [104].

Metabolism

Hepatic metabolism plays a significant role in the elimination of each of the antiarrhythmics reviewed here. In fact, for propafenone, verapamil, and mexiletine, metabolism accounts for more than 90% of the elimination. Additionally, the biotransformations of mexiletine, propafenone, encainide, and flecainide cosegregate with polymorphic debrisoquine hydroxylation.

For disopyramide, metabolism accounts for approximately 45% of the dose, the only identified metabolic pathway being mono-N-dealkylation. Earlier in vivo studies [106] reported that the nonrenal unbound clearance of S(+)-disopyramide (6.19 mL/min/kg) was 2.7-fold higher than that of the R(−) isomer (2.32 mL/min/kg). In agreement with in vivo studies, further in vitro studies using human hepatocytes [107] and liver microsomes [108] showed stereoselective N-dealkylation of disopyramide in favor of the S(+) enantiomer. This stereoselectivity was attributed to differences in the affinity of the enantiomers for the binding to the enzyme(s) because the maximum velocities were similar for the two enantiomers [108]. Additionally, very recently [109], it was shown that CYP3A4 is mainly responsible for the N-dealkylation of both enantiomers of disopyramide, raising the possibility of stereoselective interactions between disopyramide and other substrates of CYP3A4.

The metabolism of encainide cosegregates with the debrisoquine hydroxylation pathway, which is subject to genetic polymorphism. In humans, encainide is converted to O-demethyl encainide (ODE), which is further metabolized to 3-methoxy-O-demethyly encainide (MODE). Both of these pathways are reportedly under the same genetic influence [80]. In extensive metabolizers of debrisoquine (EMs), both pathways are stereoselective in favor of the (−) enantiomer. However, in poor metabolizers (PMs), a significantly smaller recovery of the ODE metabolite lacks stereoselectivity. Nevertheless, whereas the metabolism is the major pathway of elimination for encainide in EMs, renal clearance of the unchanged drug is the dominant pathway in PMs [80].

The nonrenal pathways of elimination of flecainide may account for about 60% of the administered dose. The major pathway of metabolism of flecainide in humans results in sequential production of meta-O-dealkylated flecainide (MODF) and meta-O-dealkylated lactam of flecainide (MODLF), both of which may further be conjugated with glucuronide or sulfate (Fig. 6) [110]. It was demonstrated that the metabolism of flecainide (formation of both metabolites) in PMs is impaired when compared with EMs. However, the impairment was stereoselective, in that the metabolism of R(−) enantiomer was inhibited more intensely than that of its antipode [110].

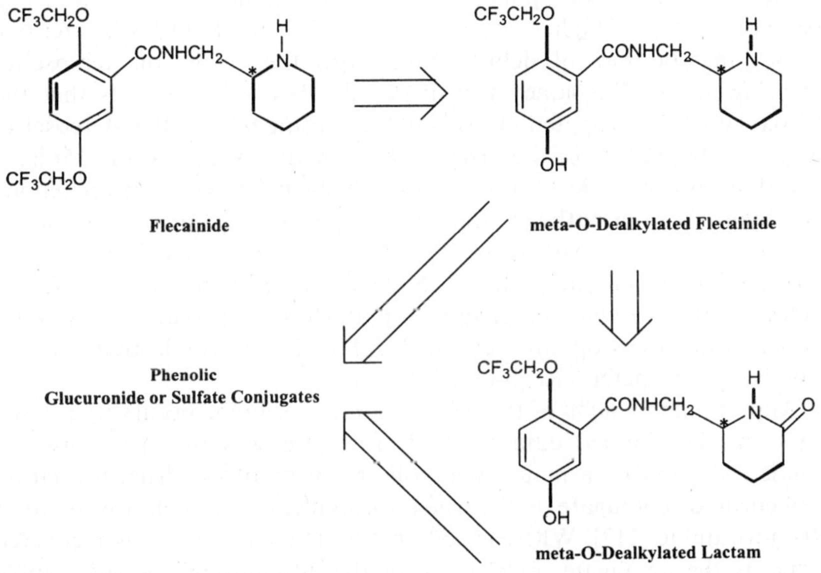

Figure 6 The major metabolic pathways of flecainide. The asterisk denotes the chiral carbon. (From Ref. 110.)

Similar to flecainide, some of the metabolic pathways of mexiletine also cosegregate with debrisoquine hydroxylation phenotype [111,112]. Mexiletine is metabolized extensively ($\sim 90\%$ of the administered dose), major pathways being aliphatic and aromatic hydroxylation, the former leading to hydroxymethyl mexiletine and the latter producing meta- or para-hydroxymexiletine. Both aliphatic and aromatic hydroxylations of mexiletine are reported [111,112] to be under genetic control via debrisoquine hydroxylation phenotype. Abolfathi et al. [112] reported that the stereoselective metabolism of mexiletine observed in humans in favor of the R(−) enantiomer is mostly due to non-CYP2D6-dependent pathways of metabolism of the drug; a high degree of stereoselectivity (R : S ratio of 11) was observed in the formation of N-hydroxymexiletine glucuronide [112]. Additionally, although both aromatic and aliphatic hydroxylation pathways were reduced in PMs, the effect was similar for both enantiomers; therefore the same degree of stereoselectivity was observed in the overall metabolism of mexiletine in poor and extensive metabolizers of debrisoquine.

In humans, propafenone undergoes extensive metabolism ($\sim 100\%$ of the dose) with 5-hydroxy and N-dealkyl propafenone as the major metabolites [113]. Additionally, Kroemer et al. [113] demonstrated that only the 5-hydroxylation pathway of propafenone is associated with the debrisoquine hydroxylation phenotype. In vivo [85], the R(−) enantiomer of propafenone shows a higher clearance than its antipode in both extensive and poor metabolizers of debrisoquine, with the degree of stereoselectivity being almost the same in both populations. This suggests that the 5-hydroxylation of propafenone is not the reason behind the stereoselectivity in the metabolism of the drug. Indeed, in vitro studies [114,115] have shown that the intrinsic clearances of the enantiomers of propafenone for the hydroxylation pathway are very similar after the incubation of the racemate with human liver microsomes. It is also reported [115,116] that the other major metabolic pathway, N-dealkylation, is not stereoselective. Therefore it appears that the metabolic pathways other than 5-hydroxylation and N-dealkylation are responsible for the stereoselectivity in the metabolism of propafenone observed in vivo.

Approximately 60% of the elimination of tocainide occurs by hepatic metabolism. The limited data available for stereoselective metabolism of tocainide indicate that a major metabolic pathway of the drug, formation of a glucuronide conjugate of N-carboxytocainide, is stereoselective in favor of R(−)-tocainide [117]. Whereas 45% of the R enantiomer was recovered in urine as the conjugate, only 1.2% of the administered S(+)-tocainide was found in the urine as this metabolite [117]. The stereoselectivity in the other metabolic pathways of tocainide in humans is not known at this time.

The calcium channel blocker verapamil is almost exclusively eliminated by metabolism in humans. The major metabolic pathways involve N- and O-dealkylation pathways with sequential metabolism of the generated metabolites (Fig. 7). Despite identification of several metabolites (Fig. 7), only half of the administered dose can be accounted for, suggesting the presence of other unidentified metabolites. Overall, the metabolism of verapamil is stereoselective in favor of the more active S(−) enantiomer. However, the direction and degree of stereoselectivity for individual pathways may be different. For example, Nelson and Olsen [118]. determined the amount of various metabolites of verapamil after incubation of the racemate with human liver microsomes. The S:R ratios of different metabolites formed during 60 min of incubation ranged from 0.78 to 1.56 [118]. Additionally, Kroemer et al. [119] used human microsomal preparations to determine the kinetics of formation of the major metabolites of verapamil (Table 9). Although the intrinsic clearance of almost all of the reported pathways showed stereoselectivity in favor of the S enantiomer, the enantiomeric differences were statistically significant only for the formation of norverapamil (Table 9) with an S:R ratio of 1.2. Based on these data (Table 9), the stereoselectivity in the formation of norverapamil is due to a higher affinity of S(−)-verapamil for the responsible enzyme(s). Further studies [120,121] using individual P450 enzymes have shown that P450 3A4, 3A5, 2C8, and 1A2 play vital roles in the stereoselective metabolism of verapamil and sequential metabolism of norverapamil. The identification of these enzymes raises the possibility of interactions between verapamil and other substrates for these enzymes, which is discussed later in this chapter. Nevertheless, the higher in vivo metabolism of S(−)-verapamil, compared with the R(+) enantiomer, is a function of both higher intrinsic metabolic clearance and higher plasma free fraction of this enantiomer.

Renal Excretion

Among the antiarrhythmic drugs reviewed in this chapter (Fig. 5), excretion of the unchanged drug plays an important role for disopyramide, flecainide, and tocainide. Renal clearance of drugs consists of glomerular filtration, with or without additional tubular secretion and/or tubular reabsorption. Whereas glomerular filtration is a passive, nonstereoselective phenomenon, tubular secretion occurs via active processes that may be susceptible to stereoselectivity. Additionally, tubular reabsorption may occur through passive and/or active processes. Therefore if a chiral drug is eliminated only through glomerular filtration, its renal clearance cannot be stereoselective. However, stereoselectivity in plasma protein binding could result in an apparent stereoselectivity in the renal clearance of such drugs. For disopyramide,

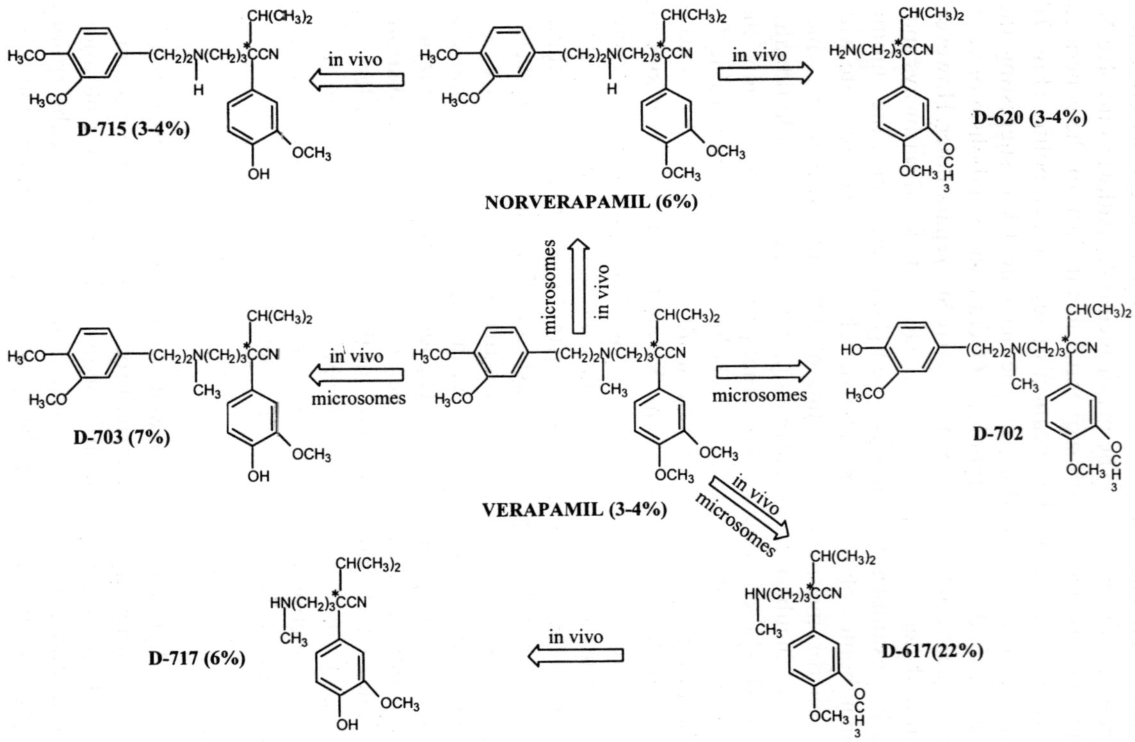

Figure 7 The major metabolic pathways of verapamil observed in vivo and in human liver microsomes. The percentages in the parentheses are the percentages of dose found in human urine as each metabolite. The asterisk denotes the chiral carbon. (From Ref. 119.)

Table 9 Michaelis–Menten Parameters (Mean ± SD) for the Metabolism of Verapamil in Human Liver Microsomes

Metabolite	Isomer	K_M, μM	V_{MAX}, pmol/mg/min	CL_{int}
Norverapamil	S	52.8 ± 41.3*	809 ± 611	18.4 ± 16.1*
	R	63.8 ± 50.4	817 ± 701	14.7 ± 12.4
D-617	S	44.7 ± 18.5	1070 ± 978	27.1 ± 24
	R	92.4 ± 95.3	1160 ± 871	19.6 ± 14
D-702	S	159 ± 142	112 ± 64	1.02 ± 0.75
	R	329 ± 382	146 ± 54	0.84 ± 0.66
D-703	S	59.4 ± 29.3*	309 ± 182*	7.0 ± 4.3
	R	40.9 ± 25.6	174 ± 107	5.4 ± 3.6

*Significantly different from the corresponding value for the R enantiomer.
Abbreviations: K_M = Michaelis–Menten constant; V_{MAX} = maximum velocity of the metabolism; CL_{int} = intrinsic clearance (V_{MAX}/K_M).
Source: Ref. 119.

a short IV infusion of the enantiomers under separate occasions [95] showed unbound renal clearances (mean ± SD) of 338 ± 124 and 182 ± 60 mL/min for the S(+) and R(−) enantiomers, respectively ($p = 0.05$). Each of these values exceeds the average glomerular filtration rate in healthy young humans (∼120 mL/min). Therefore it was suggested that disopyramide undergoes stereoselective tubular secretion, in addition to glomerular filtration, in favor of the S(+) enantiomer [95]. However, other studies [122–124] administering both enantiomers together as a racemate have not been in agreement with this study in which the enantiomers were administered separately. For example, studies administering the racemic drug to healthy subjects via either intravenous [122] or oral [123] routes reported unbound renal clearance values that were higher than the glomerular filtration rate but were similar for both enantiomers. This discrepancy may be due to an interaction [124] between the two enantiomers of disopyramide when administered as a racemate. A stereoselective renal clearance was also reported [125] for disopyramide in pediatric patients after the administration of the racemate. However, the direction of stereoselectivity ($R > S$) [125] was opposite of that reported [95] in adults after the administration of the separate enantiomers. Overall, these data point to the complexities involved in the stereoselective pharmacokinetics of disopyramide which undergoes concentration-dependent protein binding.

For flecainide, the renal clearance of both enantiomers [110,126] appears to exceed the expected value (∼50 mL/min) based on the glomerular filtration rate (120 mL/min) and the free fraction of the drug in plasma (∼0.4). Active secretion also plays a role in the excretion of the enantiomers.

However, it appears that the renal clearances of the enantiomers of flecainide are not stereoselective in both healthy volunteers [110] and patients [126].

In contrast to flecainide, the renal clearance values [117,127] of the enantiomers of tocainide are less than the expected value (~100 mL/min) based on a normal glomerular filtration rate and the drug free fraction (0.85–0.9, Table 2), suggesting involvement of tubular reabsorption for the enantiomers. In agreement with the lack of substantial stereoselectivity in plasma protein binding (Table 2) and the nonstereoselective mechanisms involved in renal excretion of the enantiomers, the renal clearance of S(+)-tocainide (50 mL/min/70 kg) is very close to that of R(−)-tocainide (57 mL/min/70 kg) [127].

In addition to the above drugs, renal clearance plays a major role in the elimination of encainide only in PMs [80]. However, the amount of enantiomers recovered in urine as the parent enantiomer (~60% of the dose) is not stereoselective in these subjects. The role of renal excretion in the elimination of parent enantiomers in EMs is negligible as < 5% of the parent drug was excreted unchanged in urine of these subjects [80].

5.3. Therapeutic Implications for Specific Antiarrhythmic Drugs

Disopyramide

Of the antiarrhythmic drugs reviewed here, the stereoselective pharmacokinetics of disopyramide are the most complex. This is because of the nonlinearity in the plasma protein binding of the drug within the range of therapeutic concentrations as discussed above. Some of the kinetic parameters of disopyramide enantiomers after the IV administration of the racemate or the individual enantiomers are listed in Table 10. Based on these data, the pharmacokinetics of the enantiomers differ if they are administered separately or as the racemate (Table 10), suggesting an enantiomer versus enantiomer interaction for disopyramide. For instance, there were no significant differences between the enantiomers in their total or renal clearance or volume of distribution when they were administered separately. However, both renal and total clearance and volume of distribution of S(+)-disopyramide were significantly less than those of the R enantiomer after administration of the racemate (Table 10). This has been attributed [124] to the displacement of R(−)-disopyramide from the plasma binding sites by the S(+) isomer when the racemate is administered. A higher free fraction of R(−)-disopyramide in plasma significantly increases the clearance of this low hepatic extraction ratio drug. Additionally, a higher

Table 10 Stereoselective Pharmacokinetics (Mean ± SD) of Disopyramide after Intravenous Administration of the Individual Enantiomers (R or S) or the Racemate (RS)

Subjects	Route	Dose, mg	Isomer	CL, mL/min	V, L	CL_R, mL/min	$t_{1/2}$ h	Ref.
5 healthy male	20-min IV infusion	100 R	R(−)	96 ± 12	48 ± 5[a]		6.1[b]	95
		100 S	S(+)	90 ± 24	40 ± 4[a]		5.4[b]	
6 healthy male, 20–30 yr	10-min IV infusion	100/70 kg R	R(−)	111 ± 20	50 ± 9[a]	55 ± 15	5.4 ± 1.2	124
		100/70 kg S	S(+)	111 ± 13	48 ± 6[a]	56 ± 10	5.2 ± 0.7	
		100/70 kg RS	R(−)	130 ± 13	60 ± 12[a]	65 ± 13	4.7 ± 1.0	
			S(+)	83 ± 13*	33 ± 6[a,*]	36 ± 8*	5.6 ± 1.2*	
7 healthy male, 24–34 yr	Bolus	150 RS	R(−)	130 ± 30	89 ± 12[d]	88[c]	7.5[b]	122
			S(+)	77 ± 18*	50 ± 13[d,*]	43[c]	7.4[b]	
1 male & 5 female patients, 5–12 yr	60-min IV infusion	1.7–7.3 mg/kg RS	R(−)	300 ± 110[e]	87 ± 29[e]	110 ± 52[e]	3.4 ± 0.6	125
			S(+)	220 ± 80[e,*]	61 ± 19[e,*]	57 ± 24[e,*]	3.2 ± 0.4	

[a]Steady-state volume of distribution.
[b]Harmonic half-life.
[c]Estimated from the data in Table 2 of Ref. 122.
[d]Volume of distribution of terminal phase.
[e]For comparison with adults, the clearance and volume of distribution values in pediatrics are presented based on a 70 kg subject.
*Significantly different from the R enantiomer.
Abbreviations: CL = systemic clearance; V = volume of distribution; CL_R = renal clearance; $t_{1/2}$ = plasma half-life; IV = intravenous.

Table 11 Stereoselective Pharmacokinetics (Mean ± SD) of Unbound and Total (Bound plus Free) Disopyramide in Humans After Single Oral Doses (100 mg) of the Racemate

Subjects	Isomer	T_{max}, h	C_{max}, ng/mL	AUC, µg·h/mL	CL_O, mL/min	$t_{1/2}$, h[a]
4 M, 4 F,	Free R	2.8 ± 1.2*	0.17 ± 0.03	0.84 ± 0.29	1100 ± 340	3.5
22–33 yr	Free S	1.4 ± 0.7	0.19 ± 0.03	0.90 ± 0.29	1000 ± 360	2.4
	Total R	2.9 ± 1.0	0.58 ± 0.18*	4.3 ± 1.8*	205 ± 102*	3.5
	Total S	2.4 ± 0.2	0.93 ± 0.25	5.7 ± 2.1	156 ± 77	2.8

[a]Harmonic half-life.
*Significantly different from the corresponding value for the S enantiomer.
Abbreviations: T_{max} = time to reach C_{max}; C_{max} = maximum plasma concentration; AUC = area under the plasma concentration–time curve; CL_O = oral clearance; $t_{1/2}$ = plasma half-life.
Source: Ref. 96.

free fraction results in an almost proportional increase in the volume of distribution of this enantiomer. Therefore the half-lives of the enantiomers are similar to each other whether they are administered separately or as the racemate (Table 10).

Takahashi et al. [96] studied the stereoselective pharmacokinetics of free and total (bound plus free) disopyramide in humans after a single 100 mg oral dose of the racemate. It was shown that the plasma concentrations of total S(+)-disopyramide were significantly higher than those of its antipode, thus resulting in a higher area under the plasma concentration–time curve (AUC) for the S enantiomer (Table 11). However, when the free drug concentrations were considered, the plasma concentration–time profiles and AUC values of both enantiomers (Table 11) were very similar. Collectively, these data suggest that a stereoselective and nonlinear plasma protein binding significantly impacts the apparent stereoselective pharmacokinetics of total (bound plus unbound) disopyramide.

Encainide

Probably because of its withrawal from the market in 1991, prompted by its adverse safety profile, only limited information is available for the stereoselective pharmacokinetics or pharmacodynamics of encainide in humans. The antiarrhythmic effects of the drug are not stereoselective [80] (Table 8), and the metabolites are also potent antiarrhythmic agents [128]. Turgeon et al. [80] monitored the steady-state urinary recovery of encainide and its metabolites ODE and MODE in 7 EMs and 3 PMs after the administration of the racemate in the presence and absence of quinidine, a selective inhibitor of CYP2D6. In the absence of quinidine, EMs excreted significant amounts of the dose (~40%) in the from of metabolites with negligible excretion of the parent enantiomers (< 5% of the dose). On the other hand, the metabolites and the parent drug accounted for < 4% and > 60% of the dose, respectively, in PMs. Whereas quinidine did not change the urinary excretion profiles of the enantiomers and metabolites in PMs, it resulted in a substantial reduction in the recovery of MODE and a corresponding increase in the recovery of the parent enantiomers in EMs [80].

Flecainide

As discussed in the metabolism section, the pharmacokinetics of flecainide are influenced by the phenotype of the patient with regard to debrisoquine hydroxylation. Gross et al. [110] studied the kinetics of the racemate in 5 EM and 5 PM volunteers after the administration of a single 50 mg dose of the drug (Table 12). Whereas there was no stereoselectivity in the plasma concentrations of flecainide in EMs (similar AUCs, Table 12), the concentrations of the R enantiomer were generally higher than those of the S enantiomer in PMs (Table 12). The higher concentrations of R-flecainide, relative to those for S-flecainide, are apparently due to a lower nonrenal clearance (metabolism) of this enantiomer in PMs, because there was no stereoselectivity in the renal clearance of the drug (Table 12). Similar data were obtained in a limited study [126] on patients who were under chronic therapy with flecainide (Table 12). The presence and absence of stereoselective plasma concentrations of flecainide in PMs and EMs, respectively, are in contrast to the effect of debrisoquine phenotype on the stereoselectivity of metoprolol where poor metabolism of debrisoquine results in a loss of stereoselectivity [40]. Therefore a particular debrisoquine hydroxylation phenotype should not be associated with a lack or a presence of stereoselectivity for all drugs subjected to this type of polymorphism.

Mexiletine

The fact that the metabolism of mexiletine cosegregates with debrisoquine hydroxylation was not known until recently [112]. Initial studies [129,130] on the stereoselective pharmacokinetics of mexiletine did not distinguish between poor and extensive metabolizers of debrisoquine in their study subjects (Table 13). These studies showed modest or no stereoselectivity in the systemic clearance, the renal clearance, and the terminal plasma half-life of the drug [129,130]. In 1993, Turgeon and colleagues [112] reported the kinetics of the individual enantiomers of mexiletine after a single dose of the racemate in 10 EMs and 4 PMs in the absence and presence of steady-state

Table 12 Stereoselective Pharmacokinetics (Mean ± SD) of Oral Flecainide After the Administration of the Racemate to Extensive and Poor Metabolizers of Debrisoquine

Ref.	Dose	Phenotype (n)	Isomer	AUC, ng·h/mL	CL_O, mL/min	CL_R, mL/min	Urinary Recovery (%)	$t_{1/2}$,[a] h
110	50 mg, single	EM (5)	R(−)	491 ± 98	768 ± 146	274 ± 60	15.8 ± 2.4	7.8
			S(+)	476 ± 91	793 ± 159	290 ± 66	16.2 ± 2.2	8.3
		PM (5)	R(−)	822 ± 205	467 ± 109	322 ± 71	31.2 ± 7.4	13
			S(+)	636 ± 211*	620 ± 172*	319 ± 49	24 ± 7.4*	11
126	150–450 mg/day, multiple	EM (5)	R(−)		395 ± 121	118 ± 45	15.6 ± 5.2	
			S(+)		450 ± 178	111 ± 49	12.9 ± 4.8	
		PM (1)	R(−)		131	60	23.1	
			S(+)		148	64	21.9	

[a]Harmonic half-life
*Significantly different from the R(−) enantiomer.
Abbreviations: AUC = area under the plasma concentration–time curve; CL_O = oral clearance; CL_R = renal clearance; $t_{1/2}$ = plasma half-life; EM, extensive metabolizers of debrisoquine; PM, poor metabolizers of debrisoquine.

Table 13 Stereoselective Pharmacokinetics (Mean ± SD) of Oral Mexiletine After the Administration of the Racemate

Study	Dose	Phenotype (n)	Isomer	AUC, μg·h/mL	CL_O, mL/min	CL_R, mL/min	Urinary Recovery (%)	$t_{1/2}$, h
130	300 mg, single	Not determined (5)	R(−)	3.33 ± 1.26	600 ± 160[a]	43 ± 14[a]	2.5 ± 0.8	9.10 ± 2.9
			S(+)	3.57 ± 1.43	570 ± 180[a]	50 ± 18[a,*]	3.0 ± 1.0*	11.0 ± 3.8*
129	200 mg, single	Not determined (6)	R(−)	2.58 ± 1.77	834 ± 353	37 ± 18[a]	5.14[b]	11.1 ± 1.7
			S(+)	3.38 ± 1.90*	605 ± 254	34 ± 22[a]	6.42[b]	12.5 ± 3.6
112	200 mg, single	EM (10)	R(−)	3.0 ± 1.3	495 ± 245	29 ± 12	5.7 ± 1.9	9.1 ± 3.2
			S(+)	3.1 ± 1.3	475 ± 234	28 ± 13	5.8 ± 2.2	9.6 ± 2.9
		PM (4)	R(−)	4.7 ± 0.1**	233[c]	22[c]	7.6 ± 1.1	14[c]
			S(+)	5.1 ± 0.6**	213[c]	21[c]	7.6 ± 1.5	13[c]
	200 mg, single + Quinidine	EM (10)	R(−)		338 ± 92***	41 ± 17***	11 ± 3***	10 ± 3
			S(+)		333 ± 104***	39 ± 20***	11 ± 4***	11 ± 2
		PM (4)	R(−)		254[c]	28.8[c]	9.8 ± 3.3	15[c]
			S(+)		223[c]	25.2[c]	9.8 ± 4.7	13[c]
92	200 mg, single	Not determined (12)	R(−)	2.8 ± 0.9	553 ± 203[a]	44 ± 60[a]	3.5 ± 3.4	8.1 ± 2.2
			S(+)	2.6 ± 0.9	616 ± 217[a]	50 ± 60[a]	3.7 ± 3.9	8.4 ± 2.6

[a]Adjusted for a 70 kg subject.
[b]Estimated from the data in reference.
[c]Median.
*Significantly different from the R(−) enantiomer.
**Significantly different from the value for the same enantiomer in EMs.
***Significantly different from the value for the same enantiomer in EMs in the absence of quinidine.
Abbreviations: AUC = area under the plasma concentration–time curve; CL_O = oral clearance; CL_R = renal clearance; $t_{1/2}$ = plasma half-life; EM, extensive metabolizers of debrisoquine; PM, poor metabolizers of debrisoquine.

concentrations of quinidine, a selective inhibitor of CYP2D6. In the absence of quinidine therapy, the oral clearance of both mexiletine enantiomers was reduced in PMs (Table 13), resulting in higher AUCs of both enantiomers in this group. However, both enantiomers were affected to the same degree, resulting in no difference in the stereoselectivity in the pharmacokinetics of mexiletine between PMs and EMs (Table 13). Coadministration of quinidine reduced the oral clearance of both enantiomers only in EMs; quinidine had no effect on the pharmacokinetics of the mexiletine enantiomers in PMs (Table 13). These results are in agreement with the fact that quinidine is an inhibitor of CYP2D6 pathway, which is operational only in EMs.

In more recent studies, Turgeon and colleagues have reported the effects of coadministration of caffeine [131] or propafenone [132] on the stereoselective pharmacokinetics of mexiletine in poor and extensive metabolizers of debrisoquine. Caffeine, a substrate for CYP1A2, did not significantly change the plasma concentrations of mexiletine enantiomers in either PMs or EMs [131]. However, it resulted in a slight decrease (~15%) in the urinary recovery of N-hydroxymexiletine generated from the R enantiomer in both EMs and PMs [131]. The metabolism of the S enantiomer was not affected because the N-hydroxymexiletine recovered in urine is mostly from the R enantiomer.

In contrast to caffeine, propafenone coadministration significantly reduced the metabolism of both enantiomers of mexiletine in EMs [132]. However, it did not have a significant effect on the pharmacokinetics of mexiletine enantiomers in PMs. In fact, after coadministration of propafenone to EMs, the plasma concentration–time profiles and pharmacokinetics of mexiletine enantiomers in these subjects were not distinguishable from those in PMs [132]. These results are in agreement with the inhibitory effects of propafenone on the CYP2D6 pathway and are similar to those obtained with quinidine, another CYP2D6 inhibitor (Table 13).

Propafenone

The pharmacokinetics of propafenone enantiomers are summarized in Table 14. The first study on the stereoselective pharmacokinetics of propafenone [84] reported the pharmacokinetics of the individual enantiomers in healthy male volunteers after the oral administrations of the enantiomers separately. Six of the seven volunteers were characterized as EMs and one as PM. The enantiomeric AUCs in the PM subject were ≥ ten times higher than the average value in EMs. Additionally, whereas the oral clearances of the enantiomers were close in the PM subject, the oral clearance of S-propafenone was on average > twice than that of R-propafenone in EMs, resulting in higher plasma concentrations of R-propafenone in EMs

Table 14 Stereoselective Pharmacokinetics (Mean ± SD) of Oral Propafenone After the Administration of the Individual Enantiomers (R or S) or the Racemate (RS)

Study	Dose	Phenotype (n)	Isomer	C_{max}/dose, ng/mL/mg	T_{max}, h	AUC, ng·h/mL	CL_o, L/h	V/F, L	$t_{1/2}$, h
84	250 mg R, single	EM (6)	R	0.57 ± 0.45*	1.0 ± 0.3	522 ± 370	810 ± 663*	4450 ± 3670[a]	4.8 ± 0.6*
		PM (1)	R	2.0	1.5	5110	49.0	591[a]	8.5
	250 mg S, single	EM (6)	S	0.34 ± 0.36	1.0 ± 0.0	290 ± 245	1960 ± 1720	18430 ± 18860	7.8 ± 3.1
		PM (1)	S	2.2	1.5	6490	38.5	587	11
85	150 mg RS q 8 h	EM (5)	R(−)			1700 ± 670	44.1 ± 13.6		
			S(+)			2900 ± 1010	25.4 ± 7.6		
		PM (2)	R(−)			4650, 10000[b]	14.6, 6.78[b]		
			S(+)			6900, 21200[b]	9.84, 3.18[b]		
133	150 mg RS q 6 h	EM (7)	R(−)	3.5 ± 1.4*	2.0 ± 1.0	945 ± 360	87.6 ± 28.8*		3.3 ± 1.9
			S(+)	5.2 ± 1.7**	1.9 ± 0.5	1480 ± 495	55.2 ± 18.0**		4.6 ± 1.9
	150 mg R q 6 h		R(−)	3.4 ± 1.3*	1.6 ± 0.5	2070 ± 630	76.7 ± 20.9*		4.2 ± 1.0*
	150 mg S q 6 h		S(+)	2.6 ± 1.4**	2.0 ± 1.0	1260 ± 720	151 ± 87.0**		2.7 ± 0.9

[a]Steady-state volume of distribution.
[b]Individual values.
*Significantly different from the S enantiomer in the same group.
**Significant difference between the S enantiomer after the administration of the racemate and the enantiomer.
Abbreviations: C_{max} = maximum plasma concentration; T_{max} = time to reach C_{max}; AUC = area under the plasma concentration–time curve; CL_o = oral clearance; V/F = volume of distribution divided by oral bioavailability; $t_{1/2}$ = plasma half-life; EM, extensive metabolizers of debrisoquine; PM, poor metabolizers of debrisoquine.

(Table 14). A subsequent study [85] after the multiple dose administration of the racemate, however, showed an opposite stereoselectivity relative to that found after the administration of the single doses of the enantiomers separately [84]; the plasma concentrations of S-propafenone were higher than those of its antipode when the racemate was administered. Further studies [133] after the administration of the separate enantiomers or the racemate to the same subjects demonstrated that the oral clearance of the R enantiomer is almost the same whether the administered drug is the enantiomer or the racemate. However, the oral clearance of the S enantiomer is significantly reduced in the presence of the other isomer (Table 14). This was attributed to an enantiomer–enantiomer interaction at the level of metabolism where R-propafenone reduces the metabolism of the S enantiomer [85].

A comparison of the oral clearance values for the enantiomers of propafenone in three studies listed in Table 14 reveals a striking difference between the values reported by Brode et al. [84] and those reported by Kroemer and colleagues [85,133]; the oral clearance values in EMs reported by Brode et al. [84] are > 10-fold higher than those in the other two studies [85,133]. One factor contributing to this discrepancy may be that in the study by Brode et al. [84] the enantiomers were administered as a single dose, as opposed to multiple-dose administration of the enantiomers or the racemate in the other two studies (Table 14). It is known that for highly extracted drugs, such as propafenone or verapamil, the drug clearance declines with multiple administration of the drug. An additional contributing factor may be the unusually high (up to 100%) interindividual variability observed in the kinetic parameters reported by Brode et al. [84].

The pharmacokinetic interaction of the two enantiomers of propafenone also has some pharmacodynamic ramifications. Kroemer et al. [85] showed that the beta-blocking effect of 150 mg of the racemate was more intense than that of 75 mg of the S enantiomer administered alone. This apparent discrepancy could be explained by their finding of lower clearance and higher plasma concentrations of S enantiomer in the presence of its antipode (Table 14). These data clearly show that the pharmacological activities of racemates cannot be simply assumed to be a summation of the effects of the individual enantiomers.

The enantiomer–enantiomer interaction observed for propafenone in Caucasians (Table 14) was also found in Chinese subjects [134]. The oral clearance of S-propafenone was reduced to one-half when it was administered as a racemate as opposed to pure enantiomer administration [134]. Additionally, CYP2D6-dependent variability in the pharmacokinetics of propafenone reported in Caucasians (Table 14) was also observed in Chinese subjects [135]. These data suggest that the stereoselective pharmacokinetics and pharmacodynamics of propafenone in Chinese subjects [134,135] are in agreement with those in Caucasians (Table 14).

Another recent study [136] in Chinese subjects measured the concentrations of propafenone enantiomers and propafenone glucuronide diastereomers in plasma after the administration of the racemate. While the S enantiomer of the parent drug was more prevalent in plasma, the opposite was true for the glucuronide conjugate. It was suggested [136] that the stereoselectivity in the plasma concentrations of propafenone may be due, at least in part, to a stereoselective glucuronidation pathway.

Because propafenone undergoes hydroxylation which cosegregates with debrisoquine, interactions with the inhibitors of CYP2D6 are expected. Indeed, a recent study in Chinese subjects [137] showed that fluoxetine therapy results in a partial inhibition of CYP2D6 and a reduction in the oral clearance of both enantiomers of propafenone. Consequently, the plasma concentrations of both enantiomers of propafenone increased as a result of coadministration of fluoxetine. This in vivo study [181] is in agreement with an in vitro report [115] demonstrating the inhibitory effect of fluoxetine on the 5-hydroxylation of propafenone in human liver microsomes. Nevertheless, the fluoxetine–propafenone pharmacokinetic interaction did not result in any changes in the effects of propafenone on PR or QRS intervals [137]. This may be attributed to the relatively modest CYP2D6 inhibitory effects of fluoxetine [137] as opposed to the significant effects of quinidine, which converts EMs to PMs.

Collectively, these data indicate that the stereoselective pharmacokinetics and pharmacodynamics of propafenone are influenced by the debrisoquine phenotype of the patient and the interaction between the two enantiomers. It is not surprising that the pharmacokinetics of the drug are highly variable (Table 14).

Tocainide

The first report [138] on the stereoselective pharmacokinetics of tocainide measured plasma concentrations of tocainide enantiomers obtained occasionally from seven patients with life-threatening ventricular arrhythmia over a three-year period. Although large intra- and inter-individual variability was present, the plasma concentrations of the S enantiomer were always higher than those of R tocainide [138]. Subsequent reports investigated the full pharmacokinetics of tocainide after the IV administration of the racemate to healthy volunteers [127,139] and anephric patients [127] and also after the oral administration of the individual enantiomers in healthy subjects [117] (Table 15). After the IV administration of the racemate, there is no stereoselectivity in the volume of distribution of the

Table 15 Stereoselective Pharmacokinetics (Mean ± SD) of Tocainide After the Administration of the Individual Enantiomers (R or S) or the Racemate (RS)

Subjects	Route	Dose, mg	Isomer	CL, L/h	V, L	CL_R, mL/min	$t_{1/2}$, h	Ref.
7 male & 5 female patients, 45–81 yr	Intermittent IV infusion	250 mg bolus + 500 mg q 8 h (RS)	R(−)	10.0 ± 2.6	131 ± 24[a]		10.3 ± 3.5	139
			S(+)	5.57 ± 1.64*	128 ± 29[a]		17.3 ± 5.6*	
5 healthy male, 24–40 yr	20 min IV infusion	Single 200 mg RS	R(−)	11.0 ± 3.8[b]	176 ± 34[b,c]	57 ± 85[b]	11.7 ± 2.4	127
			S(+)	7.14 ± 2.44[b,*]	173 ± 30[b,c]	50 ± 71[b]	17.1 ± 2.5*	
3 male & 1 female anephric patients, 24–40 yr			R(−)	8.36 ± 0.29[b]	243 ± 77[b,c]	66 ± 16[d]	20.3 ± 6.8	
			S(+)	4.49 ± 0.67[b]	207 ± 78[b,c]	54 ± 26[d]	31.9 ± 10.1*	
6 healthy male, 27 yr (mean)	PO	300 mg R 300 mg S	R(−)	11.7 ± 1.2[e]		70 ± 28	9.7 ± 0.8	117
			S(+)	6.6 ± 0.6[e,*]		58 ± 13	14.5 ± 1.7	

*Significantly different from the R enantiomer.
[a]Steady-state volume of distribution.
[b]Adjusted for a 70 kg subject.
[c]Volume of distribution of terminal phase.
[d]Dialyzer clearance.
[e]Oral clearance (CL/F).
Abbreviations: CL = clearance; V = volume of distribution; CL_R = renal clearance; $t_{1/2}$ = plasma half-life; IV = intravenous; PO = oral.

drug (Table 15). However, the clearance of the S enantiomer is significantly less than that of R-tocainide, resulting in a longer plasma half life for S-tocainide (Table 15). The stereoselectivity in the clearance of tocainide is due, mostly, to a stereoselective nonrenal clearance as the renal clearance of the two enantiomers are very close (Table 15). The pharmacokinetics of tocainide enantiomers after the oral administration of the individual enantiomers [117] appear to be very close to those after the IV injection of the racemate (Table 15), suggesting no significant interactions between the two enantiomers. The similarity of the kinetic parameters after the IV and oral administrations (Table 15) is also in agreement with the high oral bioavailability of tocainide.

In anephric patients, the clearance of both enantiomers of tocainide declined substantially, with the decline being more drastic for the S enantiomer [127] (Table 15). Consequently, there was a stereoselective change in the S:R ratio of clearance from 0.66 in healthy subjects to 0.54 in uremic patients [127]. Because the stereoselectivity in the renal clearance of tocainide is slight (Table 15), the stereoselective reduction of tocainide clearance in uremic patients is most likely due to a stereoselective effect on the metabolism of the drug. Because R-tocainide is a more potent sodium channel blocker than its antipode [86], a change in the stereoselectivity in the plasma concentrations of tocainide in uremic patients may have pharmacodynamic ramifications in these patients.

The excretion of the tocainide enantiomers into saliva was studied by Pillai et al. [140] after the IV injection of the racemic drug. It was reported that both enantiomers are excreted into saliva with saliva:plasma ratios of 2.1 and 3.7 for the S and R enantiomers, respectively. The higher ratio for the R enantiomer was attributed to the involvement of transport mechanisms other than passive diffusion [140]. However, these mechanisms are not known at this time.

Verapamil

The stereoselective pharmacokinetics and pharmacodynamics of verapamil have been studied in detail. In 1984, Eichelabum and colleagues [98] reported the stereoselective pharmacokinetics of verapamil after the IV administration of the individual enantiomers of the drug. Both the clearance and the steady-state volume of distribution of the more active S(−) enantiomer were higher than those for the R enantiomer (Table 16). However, the terminal plasma half-lives were similar for both enantiomers (Table 16). A twofold higher free fraction for the S(−) enantiomer (Table 2) is most likely responsible for the enantiomeric differences in the volume of distribution of the drug (Table 16). However, because the drug has a high

Table 16 Stereoselective Pharmacokinetics (Mean ± SD) of Verapamil After the Administration of the Individual Enantiomers (R or S) or the Racemate (RS)

Subjects	Route	Dose	Isomer	C_{max}, ng/mL	F	CL, mL/min	V, L	$t_{1/2}$, h	Ref.
5, healthy, M, 23–27 yr	5-min IV infusion	5 mg R(−)	R(+)			797±80	192±36[a,b]	4.08[c]	98
		5 mg S(+)	S(−)			1410±155	449±111[a,b]	4.81[c]	
	PO	160 mg RS	R(+)	241±82	50±14	1720±568[d]			141
			S(−)	46.1±16*	20±6*	7460±2160[d,*]			
6, healthy, M, 25–43 yr	PO	160 mg RS alone	R(+)	244±183	46±13	1820±798[a,d]		8.1[c]	144
			S(−)	53.1±50.0*	19±13*	8860±5190[a,d,*]		8.2[c]	
		160 mg RS + Cimetidine	R(+)	325±191	53±10**	1270±518**		8.7	
			S(−)	97.3±75.9**	26±12**	5190±2230[a,d,**]		7.9[c]	
8, healthy, 24–39 yr	15-min infusion	1 mg/kg R	R(+)	1200±547		791±266[a]	266±105[a,b]	6.2±1.0	147
		0.1 mg/kg S	S(−)	125±36		875±322	217±112[a,b]	4.7±3.0	
8, healthy, 63–83 yr	15-min infusion	1 mg/kg R	R(+)	1240±308		665±133[a]	252±56[a,b]	7.4±1.5	
		0.1 mg/kg S	S(−)	117±60		588±224[a,***]	266±126[a,b]	7.1±3.2	
15, healthy, 22 y (mean)	30-min infusion	20 mg RS	R(+)			1010±200	194±46	3.0±1.1	146
			S(−)			1700±387	304±124	2.9±1.9	
	PO, IR	120 mg RS, tid	R(+)	289±97		1010±374[d]		7.3±5.4	
			S(−)	85.8±34.5		4120±1870[d]		6.2±1.9	

Subjects	Dosing	Dose	Enantiomer					
15, healthy, 69 y (mean)	30-min infusion	20 mg RS	R(+)		753 ± 219***	254 ± 77***	6.4 ± 3.4***	
			S(−)		1280 ± 387***	393 ± 221	5.4 ± 4.7	
	PO, IR	120 mg RS, tid	R(+)	328 ± 70	$705 \pm 226^{d,}$***		14 ± 6**	
			S(−)	98.5 ± 27.1	$2680 \pm 1030^{d,}$***		10.8 ± 5.8***	149
14, healthy, F, 20–40 yr	15-min IV infusion	0.1 mg/kg S	S(−)	117 ± 43	1100 ± 245^{a}	$343 \pm 98.0^{a,b}$	5.8 ± 1.6	
11, healthy, F, 60–90 yr			S(−)	121 ± 33	$840 \pm 147^{a,}$***	$364 \pm 147^{a,b}$	7.6 ± 2.6***	
12, healthy, M, 19–37 yr	240 mg RS	PO, IR	R(+)	258 ± 119	1770 ± 886^{a}		6.87 ± 1.51	152
			S(−)	59.0 ± 30.7	8470 ± 4400^{a}		3.74 ± 0.56	

[a] Adjusted for a 70 kg subject.
[b] Steady-state volume of distribution.
[c] Harmonic half-life.
[d] Oral clearance (CL/F).
*Significantly different from the R enantiomer.
**Significantly different from the same enantiomer before cimetidine administration.
***Significantly different from the same enantiomer in young subjects.
Abbreviations: C_{max} = maximum plasma concentration; F = oral bioavailability; CL = clearance; V = volume of distribution; $t_{1/2}$ = plasma half-life; IV = intravenous; PO = oral; IR = immediate release.

hepatic extraction ratio, the protein binding differences cannot account for the significant differences between the enantiomers in their clearance. Consequently, a stereoselective metabolism in favor of S-verapamíl should also be in place. A subsequent work by the same group [141] reported the pharmacokinetics of the individual enantiomers of verapamil after the oral administration of a pseudoracemate. It was shown that the differences between the plasma concentrations of the enantiomers were even more pronounced after oral administration compared with IV injection (Table 16). This was due to more than twofold differences between the first-pass metabolism of the enantiomers in favor of the S(−) enantiomer (Table 16).

The stereoselective first-pass metabolism of verapamil after its oral administration [141] could explain the different total (R plus S) plasma concentration–effect relationships reported [142,143] for verapamil after the oral and IV doses. In 1980, Eichelbaum et al. [142] showed that the slope of the linear regression between the effect (change in P–R interval) and the unresolved verapamil concentrations in plasma was shallower after oral administration compared to IV injection of the same dose. Similarly, when the effect was plotted against the concentrations based on an E_{max} model, the curve for the oral verapamil was substantially shifted to the right [143], indicating that at equal (±)-verapamil concentrations, the IV dose is significantly more effective than the oral dose. This apparent discrepancy is due to the stereoselective first pass-metabolism of verapamil, preferentially removing the more active S(−) enantiomer after the oral administration. Therefore the same concentration of (±)-verapamil after the oral administration contains less of S(−)-verapamil, compared with the IV dosing. These studies clearly illustrate the importance of considering stereochemistry in pharmacodynamics of chiral drugs.

The stereochemical aspects of the interaction of verapamil with cimetidine were reported by Mikus et al. [144] (Table 16) using a pseudoracemate of verapamil. In the absence of cimetidine, the stereoselective pharmacokinetics of verapamil were very close to those reported in a previous study [141] (Table 16). A 6-day pretreatment with 400 mg cimetidine twice daily significantly decreased the oral clearance and increased the oral bioavailability of both enantiomers of verapamil (Table 16) without affecting the degree of the protein binding of the enantiomers. This was due to a reduction in the metabolic clearance of the verapamil enantiomers. The cimetidine-induced increase in the AUC of S-verapamil (150%) was greater than that (118%) for the less active R enantiomer [144]. The increase in plasma AUCs of both enantiomers along with a more pronounced effect for the S enantiomer resulted in an increase in the negative dromotropic effects of the drug [144].

Grapefruit juice, a CYP3A4 inhibitor, was recently shown to increase moderately the steady-state plasma concentrations of the S and R enantiomers of verapamil by 36 and 28%, respectively [145]. The interaction was apparently not stereoselective, as the magnitude of increase in concentration of each enantiomer was similar. There were no significant changes in the concentrations of the enantiomers of norverapamil after grapefruit juice.

The effect of age on the stereoselective pharmacokinetics/dynamics of verapamil has been the subject of several reports [146–150] (Table 16). In an elaborate study, Abernethy and colleagues [146] studied the effects of aging on the pharmacokinetics and pharmacodynamics of verapamil after the single intravenous, steady state IV infusion, and multiple immediate, and controlled release oral doses of racemic verapamil. After the single IV doses, the clearance values of both enantiomers of verapamil were significantly ($\sim$30%) lower in the elderly, compared with young subjects (Table 16). The same conclusion was reached when constant rate infusions were administered to achieve target racemic concentrations of 50, 100, and 200 ng/mL in plasma of young and elderly subjects [146]. Similarly, after the oral administration of both immediate and controlled release verapamil, the oral clearances of both enantiomers were reduced in the elderly to an extent similar to that seen after the IV doses (Table 16), suggesting similar oral bioavailabilities in the young and elderly groups. In terms of pharmacodynamics, when the S enantiomer was modeled for its effect on P–R prolongation, the major difference between the young and elderly populations was that the elderly had substantially lower E_{max} after both IV and oral doses. Additionally, after the IV dose, the elderly had higher EC_{50} values. Based on this study, a previously reported apparent decrease in the sensitivity of older subjects to verapamil cannot be due to an age-dependent stereoselective disposition of verapamil. However, the old age-induced decrease in sensitivity may be explained by a reduction in the E_{max} of S-verapamil in older subjects.

Schwartz and colleagues [147–149] also investigated the effects of aging on the stereoselective pharmacokinetics and pharmacodynamics of verapamil. In their first study [147], 1 and 0.1 mg/kg doses of R(+) and S(−) enantiomers of verapamil, respectively, were administered separately to both young and elderly subjects through 15-min IV infusions. Under these conditions, the clearance of S-verapamil, but not that of R-verapamil, was reduced in the elderly; a reduction in the clearance of R-verapamil in the elderly did not reach statistical significance. Nevertheless, the stereoselectivity in the clearance of verapamil, observed in the young subjects, was lost in the elderly [147]. A subsequent study [149] using S-verapamil in female only subjects (Table 16) confirmed the earlier results [147] using mixed gender volunteers.

The stereoselective effect of aging on the clearance of S-verapamil (and not R-verapamil) reported by Schwartz et al. [147] is in disagreement with the studies of Abernethy et al. [146], who reported nonstereoselective decreases in the clearance of both verapamil enantiomers in the elderly. Schwartz et al. [148] hypothesized that this discrepancy was due to the differences between the two studies regarding the drug administration protocols. Whereas Abernethy et al. [146] used racemic verapamil, Schwartz et al. [147] injected the enantiomers separately. In a follow-up study using the racemate, Schwartz et al. [148] showed that the clearance of R-verapamil after the injection of the racemate was less than that reported by them after the injection of the individual enantiomer [147]. Consequently, it may be concluded that aging results in a reduction in the clearance of both enantiomers of verapamil after the administration of the racemate.

Gupta et al. [150] further confirmed the effects of age on the stereoselective pharmacokinetics of verapamil. Additionally, they showed that the steady-state plasma concentrations of both verapamil and nor-verapamil enantiomers in women were higher than those in men after daily oral doses of 180 mg racemic verapamil [150]. However, the gender difference was lost when the data were corrected for the lean body mass and age [150]. In the absence of additional data, the effects of gender on the stereoselective pharmacokinetics of verapamil do not appear to be substantial.

It has been suggested [60,61] that the rate of input of chiral drugs after the oral administration may affect the stereoselectivity in their pharmacokinetics. Simulations [60] have shown that drugs with Michaelis-Menten type first-pass metabolism whose input rate approaches the maximum rate of metabolism (V_{max}) of the enantiomers are most susceptible to this phenomenon. A study [151] in isolated perfused rat livers demonstrated that when the input rate of racemic verapamil was doubled, the stereoselectivity in the outlet concentration of the drug was lost. In humans, Karim and Piergies [152] showed that the R : S plasma concentration ratios of verapamil at maximum plasma concentration (C_{max}) were formulation dependent; a sustained release formulation resulted in lower total concentrations at time to reach C_{max} (t_{max}), associated with higher R : S ratios, when compared with an immediate release formulation (Fig. 8). Similar input rate-dependent stereoselective pharmacokinetics of verapamil have also been suggested by others [146,153].

With regard to food, it has been shown that the administration of verapamil with food only prolongs the t_{max} of the enantiomers [154] without any significant effect on the major kinetic and dynamic parameters [154,155]. Additionally, the body position of the subjects does not appear to affect the pharmacokinetics/dynamics of the verapamil enantiomers [155].

Collectively, the available data on verapamil show that the stereo-selectivity in the pharmacokinetics of verapamil is due to stereoselective

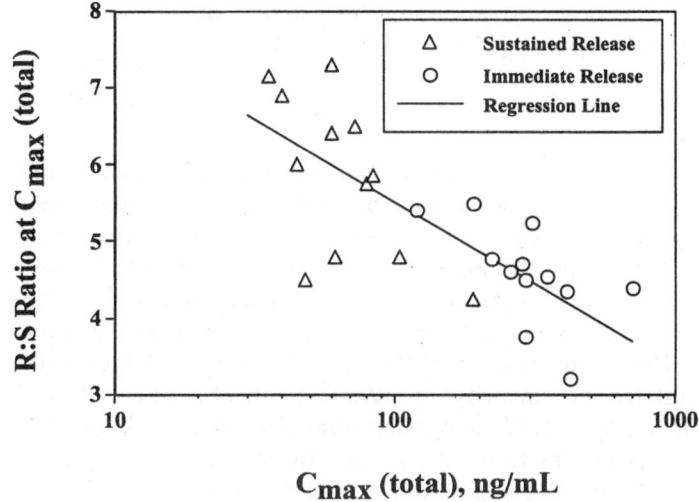

Figure 8 The relationship between the R:S ratio of verapamil at maximum plasma concentration (C_{max}) and the total verapamil (R+S) C_{max} for an immediate- and a sustained-release formulation. (From Ref. 152.)

protein binding and metabolism (including first-pass metabolism) of the drug. Additionally, various factors such as age, input rate, and interacting drugs may affect the stereoselectivity in the pharmacokinetics and perhaps the pharmacodynamics of the drug. Therefore interpretation of the pharmacodynamic data in the absence of kinetic information about the individual enantiomers should be avoided.

6. OTHER ANTIARRHYTHMIC DRUGS AND INOTROPIC AGENTS

6.1. Prenylamine

Prenylamine is a chiral calcium blocker with coronary vasodilating properties; it is used in the treatment of angina pectoris and tachyarrhythmias [156]. The S-(+) enantiomer possesses positive inotropic activity, whereas its antipode exhibits negative inotropic activity. In healthy volunteers [156], R(−)-prenylamine attains much higher plasma concentrations than antipode after oral single and repeated dose administration (Table 7). The R:S ratio of mean AUC was 3 after single doses and 4.3 after multiple doses. The R:S ratio of C_{max} was even higher, with values after single and repeated doses being approximately 5. Stereoselective renal clearance of unchanged drug was observed, although this pathway of elimination was negligible for both enantiomers [156].

6.2. Pirmenol

(±)-Pirmenol is another Class I antiarrhythmic agent that displays stereo-selectivity in its pharmacokinetics after oral dosing. The (+) enantiomer of total (bound plus unbound) species attains a higher steady-state AUC (between 12 and 18%) than (−) enantiomer [157]. The unbound fraction of (+) enantiomer was also significantly (25%) higher than that of the (−) enantiomer (Table 2). When AUC was calculated based on the unbound drug, the (+):(−) ratio increased to between 1.43 to 1.52. Additionally, there was dose-related nonlinearity in plasma protein binding of both enantiomers. The authors [157] suggested that for effective control of premature ventricular contractions in patients, the plasma concentrations of racemic pirmenol should be maintained at 1.5 µg/mL.

6.3. Gallopamil

Gallopamil is a calcium channel blocker that is a structural analogue of verapamil. The S enantiomer of gallopamil has demonstrated higher potency than antipode in negative inotropic and vasodilating effects [158]. The plasma protein binding of the drug is high (> 95%), and it is stereo-selective (unbound S > unbound R) when given as a racemate or when enantiomers are administered separately (Table 2) [158,159].

In healthy human subjects given single 25 mg doses of either S- or R-gallopamil separately, the plasma concentrations displayed similar stereoselectivity as when administered as a psuedoracemate consisting of equal proportions of deuterolabeled enantiomer (Table 7) [159]. However, when given as the pseudo-racemate, the mean AUC of each enantiomer was significantly increased by between 1.7 and 2.4-fold compared to single enantiomer administration (Table 7). For each enantiomer, unbound fraction in plasma remained the same whether given alone or in the presence of antipode (Table 2). The increase in AUC after racemate was attributed to a decrease in first-pass hepatic extraction of each enantiomer secondary to an enantiomeric interaction that affected each enantiomer's clearance to an approximately equivalent degree. A similar result was found in comparing the AUC of 25 and 100 mg doses of enantiopure R enantiomer, as with a four-fold increase in dose the AUC increased by seven-fold. Lack of chiral inversion was also confirmed in the direction R to S enantiomer, as only small levels of S enantiomer, attributed to impurity, were observed in plasma after 100 mg of R enantiomer administration to healthy subjects [159]. Consistent with prior in vitro studies, with respect to pharmacodynamic effect, S-gallopamil, but not R-gallopamil (at doses ranging from 25 to 100 mg of R enantiomer), was found to elicit a significant increase in the electrocardiographic PR interval [159].

6.4. Inotropic Agents

Pimobendan is a chiral, racemic positive type III phosphodiesterase inhibitor that has positive inotropic properties in the heart. The (−) enantiomer has 150% more effect on calcium sensitization of guinea pig cardiac myofilaments than antipode [160]. One of its major metabolites, desmethyl pimobendan, is also chiral. The pharmacokinetics of pimobendan enantiomers were studied in seven patients with dilated cardiomyopathy after single and repeated oral doses of 5 mg daily (Table 7). Substantial uptake of drug into erythrocytes was observed. Little stereoselectivity was noted in the plasma concentrations and pharmacokinetic estimates. However, in the red blood cell fraction of these blood samples [160], concentrations of the (−) enantiomer were significantly higher than those of the (+) enantiomer after both repeat-dose and single-dose administration. Based on assay of the red blood cell fraction, the relative C_{max} and AUC of the (−) to (+) enantiomers were each about 1.6 after single-dose administration of the racemate. As expected based on its short $t_{1/2}$ of 3 h, there was no significant accumulation noted upon repeated administration of 5 mg daily. In healthy Chinese subjects given the drug (Table 7) [161], a similar trend was observed, in that the plasma concentrations after IV and oral doses were much lower and displayed less stereoselectivity than did the red blood cell concentrations.

In neither of these studies by Chu et al. [160,161] was there mention as to how the red blood cells, if at all, were isolated from white blood cells and platelets. This may be a relevant consideration, as for hydroxychloroquine enantiomers it is known that its enantiomers do not accumulate to an appreciable extent in the red cells, but rather in the white blood cells and platelets [162].

7. CONCLUSION

As demonstrated in this chapter, a significant number of cardiovascular drugs are chiral and marketed as racemates. Additionally, most of the marketed chiral cardiovascular agents exhibit stereoselectivity in both their pharmacodynamics and their pharmacokinetics. Because of the relationships observed between plasma concentration and effect for most of these drugs, the stereospecific pharmacokinetics/dynamics of these drugs should be of special clinical relevance. The degree and direction of stereoselectivity in the kinetics/dynamics of these drugs are dependent on the physicochemical properties of drugs in addition to the genetic factors of the patient and/or disease characteristics. Although there are some principles governing the stereoselective interactions between these drugs and proteins in the body, their steroselective pharamcokinetics/dynamics cannot be easily predicted and should be investigated individually. Nevertheless, this information should be of value for design and development of new chiral cardiovascular drugs with improved efficacy/toxicity profiles.

REFERENCES

1. Pearson, A.A.; Gaffney, T.E.; Walle, T.; Privitera, P.J. A stereoselective central hypotensive action of atenolol. J. Pharmacol. Exp. Ther. **1989**, *250*, 759–763.
2. Stoschitzky, K.; Egginger, G.; Zernig, G.; Klein, W.; Lindner, W. Stereoselective features of (R)- and (S)-atenolol: clinical pharmacological, pharmacokinetic, and radioligand binding studies. Chirality **1993**, *5*, 15–19.
3. Nathanson, J.A. Stereospecificity of beta adrenergic antagonists: R-enantiomers show increased selectivity for beta-2 receptors in ciliary process. J. Pharmacol. Exp. Ther. **1988**, *245*, 94–101.
4. Weyl, J.D.; Snyder, R.W.; Hanson, R.C. Differential cardioprotective properties of the l- and d-enantiomers of bucindolol in a canine model of heart failure. Arch. Int. Pharmacodyn. Ther. **1985**, *275*, 4–12.
5. Jeppsson, A.B.; Johansson, U.; Waldeck, B. Steric aspects of agonism and antagonism at beta-adrenoceptors: experiments with the enantiomers of terbutaline and pindolol. Acta. Pharmacol. Toxicol. (Copenh.) **1984**, *54*, 285–291.
6. Barrett, A.; Cullum, V. The biological properties of the optical isomers of propranolol and their effects on cardiac arrhythmias. Brit. J. Pharmacol. **1968**, *34*, 43–55.
7. Kato, Y.; Hirate, J.; Sakaguchi, K.; Ueno, M.; Horikoshi, I. Factors causing age-dependent changes in phenytoin tissue and serum binding in rats. J. Pharmacobiodyn. **1988**, *11*, 402–410.
8. Brittain, R.T.; Drew, G.M.; Levy, G.P. The alpha- and beta-adrenoceptor blocking potencies of labetalol and its individual stereoisomers in anaesthetized dogs and in isolated tissues. Br. J. Pharmacol. **1982**, *77*, 105–114.
9. Gold, E.H.; Chang, W.; Cohen, M.; Baum, T.; Ehrreich, S.; Johnson, G.; Prioli, N.; Sybertz, E.J. Synthesis and comparison of some cardiovascular properties of the stereoisomers of labetalol. J. Med. Chem. **1982**, *25*, 1363–1370.

10. Riva, E.; Mennini, T.; Latini, R. The alpha- and beta-adrenoceptor blocking activities of labetalol and its RR-SR (50:50) stereoisomers. Br. J. Pharmacol. **1991**, *104*, 823–828.

11. Srinivas, N.R.; Barr, W.H.; Shyu, W.C.; Mohandoss, E.; Chow, S.; Staggers, J.; Balan, G.; Belas, F.J.; Blair, I.A.; Barbhaiya, R.H. Bioequivalence of two tablet formulations of nadolol using single and multiple dose data: assessment using stereospecific and nonstereospecific assays. J. Pharm. Sci. **1996**, *85*, 299–303.

12. Tenero, D.; Boike, S.; Boyle, D.; Ilson, B.; Fesniak, H.F.; Brozena, S.; Jorkasky, D. Steady-state pharmacokinetics of carvedilol and its enantiomers in patients with congestive heart failure. J. Clin. Pharmacol. **2000**, *40*, 844–853.

13. Smith, A.J.; Wehner, J.S.; Manley, H.J.; Richardson, A.D.; Beal, J.; Bryant, P.J. Current role of beta-adrenergic blockers in the treatment of chronic congestive heart failure. Am. J. Health. Syst. Pharm. **2001**, *58*, 140–145.

14. Wetterich, U.; Spahn-Langguth, H.; Mutschler, E.; Terhaag, B.; Rosch, W.; Langguth, P. Evidence for intestinal secretion as an additional clearance pathway of talinolol enantiomers: concentration- and dose-dependent absorption in vitro and in vivo. Pharm. Res. **1996**, *13*, 514–522.

15. Piquette-Miller, M.; Foster, R.T.; Kappagoda, C.T.; Jamali, F. Effect of aging on the pharmacokinetics of acebutolol enantiomers. J. Clin. Pharmacol. **1992**, *32*, 148–156.

16. Hsyu, P.H.; Giacomini, K.M. Stereoselective renal clearance of pindolol in humans. J. Clin. Invest. **1985**, *76*, 1720–1726.

17. Carr, R.A.; Pasutto, F.M.; Lewanczuk, R.Z.; Foster, R.T. Protein binding of sotalol enantiomers in young and elderly human and rat serum using ultrafiltration. Biopharm. Drug Dispos. **1995**, *16*, 705–712.

18. Walle, U.K.; Walle, T.; Bai, S.A.; Olanoff, L.S. Stereoselective binding of propranolol to human plasma, alpha 1-acid glycoprotein, and albumin. Clin. Pharmacol. Ther. **1983**, *34*, 718–723.

19. Albani, F.; Riva, R.; Contin, M.; Baruzzi, A. Stereoselective binding of propranolol enantiomers to human alpha 1-acid glycoprotein and human plasma. Br. J. Clin. Pharmacol. **1984**, *18*, 244–246.

20. Belpaire, F.M.; Wynant, P.; Van Trappen, P.; Dhont, M.; Verstraete, A.; Bogaert, M.G. 1995. Protein binding of propranolol and verapamil enantiomers in maternal and foetal serum. Br. J. Clin. Pharmacol. **1995**, *39*, 190–193.

21. Gilmore, D.A.; Gal, J.; Gerber, J.G.; Nies, A.S. Age and gender influence the stereoselective pharmacokinetics of propranolol. J. Pharmacol. Exp. Ther. **1992**, *261*, 1181–1186.

22. Walle, U.K.; Fagan, T.C.; Topmiller, M.J.; Conradi, E.C.; Walle, T. The influence of gender and sex steroid hormones on the plasma binding of propranolol enantiomers. Br. J. Clin. Pharmacol. **1994**, *37*, 21–25.

23. Olanoff, L.S.; Walle, T.; Walle, U.K.; Cowart, T.D.; Gaffney, T.E. Stereoselective clearance and distribution of intravenous propranolol. Clin. Pharmacol. Ther. **1984**, *35*, 755–761.

24. Walle, U.K.; Thibodeaux, H.; Privitera, P.J.; Walle, T. Stereochemistry of tissue distribution of racemic propranolol in the dog. Chirality 1989, 1, 192–196.

25. Takahashi, H.; Ogata, H.; Kanno, S.; Takeuchi, H. Plasma protein binding of propranolol enantiomers as a major determinant of their stereoselective tissue distribution in rats. J. Pharmacol. Exp. Ther. 1990, 252, 272–278.

26. Tamai, G.; Edani, M.; Imai, H. Chiral separation and determination of propranolol enantiomers in rat or mouse blood and tissue by column switching high performance liquid chromatography with ovomucoid bonded stationary phase. Biomed. Chromatogr. 1990, 4, 157–160.

27. Yan, H.; Lewander, T. Differential tissue distribution of the enantiomers of racemic pindolol in the rat. Eur. Neuropsychopharmacol. 1999, 10, 59–62.

28. Takahashi, H.; Ogata, H. Plasma protein binding and blood cell distribution of propranolol enantiomers in rats. Biochem. Pharmacol. 1990, 39, 1495–1498.

29. Walle, T.; Webb, J.G.; Bagwell, E.E.; Walle, U.K.; Daniell, H.B.; Gaffney, T.E. Stereoselective delivery and actions of beta receptor antagonists. Biochem. Pharmacol. 1988, 37, 115–124.

30. Walle, T.; Webb, J.G.; Walle, U.K.; Bagwell, E.E. Stereoselective accumulation of the beta-receptor blocking drug atenolol by human platelets. Chirality 1991, 3, 451–453.

31. Stoschitzky, K.; Lindner, W.; Klein, W. Stereoselective release of (S)-atenolol from adrenergic nerve endings at exercise. Lancet 1992, 340, 696–697.

32. Stoschitzky, K.; Kahr, S.; Donnerer, J.; Schumacher, M.; Luha, O.; Maier, R.; Klein, W.; Lindner, W. Stereoselective increase of plasma concentrations of the enantiomers of propranolol and atenolol during exercise. Clin. Pharmacol. Ther. 1995, 57, 543–551.

33. Marathe, P.H.; Shen, D.D.; Nelson, W.L. Metabolic kinetics of pseudoracemic propranolol in human liver microsomes. Enantioselectivity and quinidine inhibition. Drug Metab. Dispos. 1994, 22, 237–247.

34. Walle, T. Stereochemistry of the in vivo disposition and metabolism of propranolol in dog and man using deuterium-labeled pseudoracemates. Drug Metab. Dispos. 1985, 13, 279–282.

35. Nelson, W.L.; Shetty, H.U. Stereoselective oxidative metabolism of propranolol in the microsomal fraction from rat and human liver. Use of deuterium labeling and pseudoracemic mixtures. Drug Metab. Dispos. 1986, 14, 506–508.

36. Ward, S.A.; Walle, T.; Walle, U.K.; Wilkinson, G.R.; Branch, R.A. Propranolol's metabolism is determined by both mephenytoin and debrisoquin hydroxylase activities. Clin. Pharmacol. Ther. 1989, 45, 72–79.

37. Yoshimoto, K.; Echizen, H.; Chiba, K.; Tani, M.; Ishizaki, T. Identification of human CYP isoforms involved in the metabolism of propranolol enantiomers—N-desisopropylation is mediated mainly by CYP1A2. Br. J. Clin. Pharmacol. 1995, 39, 421–431.

38. Lennard, M.S.; Silas, J.H.; Freestone, S.; Ramsay, L.E.; Tucker, G.T.; Woods, H.F. Oxidation phenotype—a major determinant of metoprolol metabolism and response. N. Engl. J. Med. 1982, 307, 1558–1560.

39. Murthy, S.S.; Shetty, H.U.; Nelson, W.L.; Jackson, P.R.; Lennard, M.S. Enantioselective and diastereoselective aspects of the oxidative metabolism of metoprolol. Biochem. Pharmacol. **1990**, *40*, 1637–1644.

40. Lennard, M.S.; Tucker, G.T.; Silas, J.H.; Freestone, S.; Ramsay, L.E.; Woods, H.F. Differential stereoselective metabolism of metoprolol in extensive and poor debrisoquin metabolizers. Clin. Pharmacol. Ther. **1983**, *34*, 732–737.

41. Mehvar, R.; Gross, M.E.; Kreamer, R.N. Pharmacokinetics of atenolol enantiomers in humans and rats. J. Pharm. Sci. **1990**, *79*, 881–885.

42. Carr, R.A.; Foster, R.T.; Lewanczuk, R.Z.; Hamilton, P.G. Pharmacokinetics of sotalol enantiomers in humans. J. Clin. Pharmacol. **1992**, *32*, 1105–1109.

43. Piquette-Miller, M.; Foster, R.T.; Kappagoda, C.T.; Jamali, F. Pharmacokinetics of acebutolol enantiomers in humans. J. Pharm. Sci. **1991**, *80*, 313–316.

44. Fiset, C.; Philippon, F.; Gilbert, M.; Turgeon, J. Stereoselective disposition of (+/−)-sotalol at steady-state conditions. Br. J. Clin. Pharmacol. **1993**, *36*, 75–77.

45. Sowinski, K.M.; Lima, J.J.; Burlew, B.S.; Massie, J.D.; Johnson, J.A. Racial differences in propranolol enantiomer kinetics following simultaneous i.v. and oral administration. Br. J. Clin. Pharmacol. **1996**, *42*, 339–346.

46. Lalonde, R.L.; Tenero, D.M.; Burlew, B.S.; Herring, V.L.; Bottorff, M.B. Effects of age on the protein binding and disposition of propranolol stereo-isomers. Clin. Pharmacol. Ther. **1990**, *47*, 447–455.

47. Zhou, H.H.; Whelan, E.; Wood, A.J. Lack of effect of ageing on the stereochemical disposition of propranolol. Br. J. Clin. Pharmacol. **1992**, *33*, 121–123.

48. Colangelo, P.M.; Blouin, R.A.; Steinmetz, J.E.; McNamara, P.J.; DeMaria, A.N.; Wedlund, P.J. Age and propranolol stereoselective disposition in humans. Clin. Pharmacol. Ther. **1992**, *51*, 489–494.

49. Walle, T.; Walle, U.K.; Cowart, T.D.; Conradi, E.C. Pathway-selective sex differences in the metabolic clearance of propranolol in human subjects. Clin. Pharmacol. Ther. **1989**, *46*, 257–263.

50. Johnson, J.A.; Akers, W.S.; Herring, V.L.; Wolfe, M.S.; Sullivan, J.M. Gender differences in labetalol kinetics: importance of determining stereoisomer kinetics for racemic drugs. Pharmacotherapy **2000**, *20*, 622–628.

51. Lindner, W.; Rath, M.; Stoschitzky, K.; Semmelrock, H.J. Pharmacokinetic data of propranolol enantiomers in a comparative human study with (S)- and (R,S)-propranolol Chirality **1989**, *1*, 10–13.

52. Paradiso-Hardy, F.L.; Walker, S.E.; Bowles, S.K. Steady-state pharmacokinetics of propranolol enantiomers in healthy male volunteers. Int. J. Clin. Pharmacol. Ther. **1998**, *36*, 370–375.

53. Stoschitzky, K.; Lindner, W.; Egginger, G.; Brunner, F.; Obermayer-Pietsch, B.; Passath, A.; Klein, W. Racemic (R,S)-propranolol versus half dosed optically pure (S)-propranolol in humans at steady state: hemodynamic effects, plasma concentrations, and influence on thyroid hormone levels. Clin. Pharmacol. Ther. **1992**, *51*, 445–453.

54. Vercruysse, I.; Belpaire, F.; Wynant, P.; Massart, D.L.; Dupont, A.G. Enantioselective inhibitory effect of nicardipine on the hepatic clearance of propranolol in man. Chirality **1994**, *6*, 5–10.

55. Hunt, B.A.; Bottorff, M.B.; Herring, V.L.; Self, T.H.; Lalonde, R.L. Effects of calcium channel blockers on the pharmacokinetics of propranolol stereoisomers. Clin. Pharmacol. Ther. **1990**, *47*, 584–591.

56. Donn, K.; Powell, J.; Wainer, I. Stereoselectivity of cimetidine inhibition of propranolol oral clearance. Clin. Pharmacol. Ther. **1985**, *37*, 191.

57. Zhou, H.H.; Anthony, L.B.; Roden, D.M.; Wood, A.J. Quinidine reduces clearance of (+)-propranolol more than (−)-propranolol through marked reduction in 4-hydroxylation. Clin. Pharmacol. Ther. **1990**, *47*, 686–693.

58. Kim, M.; Shen, D.D.; Eddy, A.C.; Nelson, W.L.; Roskos, L.K. Inhibition of the enantioselective oxidative metabolism of metoprolol by verapamil in human liver microsomes. Drug Metab. Dispos. **1993**, *21*, 309–317.

59. Somogyi, A.A.; Bochner, F.; Sallustio, B.C. Stereoselective inhibition of pindolol renal clearance by cimetidine in humans. Clin. Pharmacol. Ther. **1992**, *51*, 379–387.

60. Mehvar, R. Input rate-dependent stereoselective pharmacokinetics: enantiomeric oral bioavailability and blood concentration ratios after constant oral input. Biopharm. Drug Dispos. **1992**, *13*, 597–615.

61. Mehvar, R. Input rate-dependent stereoselective pharmacokinetics: effect of pulsatile oral input. Chirality **1994**, *6*, 185–195.

62. Silber, B.; Holford, N.H.; Riegelman, S. Stereoselective disposition and glucuronidation of propranolol in humans. J. Pharm. Sci. **1982**, *71*, 699–704.

63. Bleske, B.E.; Welage, L.S.; Rose, S.; Amidon, G.L.; Shea, M.J. The effect of dosage release formulations on the pharmacokinetics of propranolol stereoisomers in humans. J. Clin. Pharmacol. **1995**, *35*, 374–378.

64. Dayer, P.; Leemann, T.; Kupfer, A.; Kronbach, T.; Meyer, U.A. Stereo- and regioselectivity of hepatic oxidation in man—effect of the debrisoquine/sparteine phenotype on bufuralol hydroxylation. Eur. J. Clin. Pharmacol. **1986**, *31*, 313–318.

65. Zhou, H.H.; Wood, A.J. Differences in stereoselective disposition of propranolol do not explain sensitivity differences between white and Chinese subjects: correlation between the clearance of (−)- and (+)-propranolol. Clin. Pharmacol. Ther. **1990**, *47*, 719–723.

66. Benet, L.Z.; Oie, S.; Schwartz, J.B. Design and optimization of dosage regimens, pharmacokinetic data. In *Goodman and Gilman's The Pharmacological Basis of Therapeutics*, 9th Ed.; Hardman, J.G., Limbird, L.E, Molinoff, P.B., Ruddon, R.W., Goodman Gilman, A., Eds.; McGraw-Hill: New York, 1996; 1707–1792.

67. Gehr, T.W.; Tenero, D.M.; Boyle, D.A.; Qian, Y.; Sica, D.A.; Shusterman, N.H. The pharmacokinetics of carvedilol and its metabolites after single and multiple dose oral administration in patients with hypertension and renal insufficiency. Eur. J. Clin. Pharmacol. **1999**, *55*, 269–277.

68. Tobert, J.A.; Cirillo, V.J.; Hitzenberger, G.; James, I.; Pryor, J.; Cook, T.; Buntinx, A.; Holmes, I.B.; Lutterbeck, P.M. Enhancement of uricosuric properties of indacrinone by manipulation of the enantiomer ratio. Clin. Pharmacol. Ther. **1981**, *29*, 344–350.

69. Soons, P.A.; Mulders, T.M.; Uchida, E.; Schoemaker, H.C.; Cohen, A.F.; Breimer, D.D. Stereoselective pharmacokinetics of oral felodipine and nitrendipine in healthy subjects: correlation with nifedipine pharmacokinetics. Eur. J. Clin. Pharmacol. **1993**, *44*, 163–169.

70. Soons, P.A.; Cohen, A.F.; Breimer, D.D. Comparative effects of felodipine, nitrendipine and nifedipine in healthy subjects: concentration-effect relationships of racemic drugs and enantiomers. Eur. J. Clin. Pharmacol. **1993**, *44*, 113–120.

71. Laufen, H.; Leitold, M. Enantioselective disposition of oral amlodipine in healthy volunteers. Chirality **1994**, *6*, 531–536.

72. Tokuma, Y.; Fujiwara, T.; Noguchi, H. Plasma levels of (+)- and (−)-nilvadipine after oral dosing with racemic (+)-nilvadipine in man. Res. Commun. Chem. Pathol. Pharmacol. **1987**, *57*, 229–237.

73. Inotsume, N.; Iwaoka, T.; Honda, M.; Nakano, M.; Okamoto, Y.; Naomi, S.; Tomita, K.; Teramura, T.; Higuchi, S. Pharmacokinetics of nicardipine enantiomers in healthy young volunteers. Eur. J. Clin. Pharmacol. **1997**, *52*, 289–292.

74. Soons, P.A.; Breimer, D.D. Stereoselective pharmacokinetics of oral and intravenous nitrendipine in healthy male subjects. Br. J. Clin. Pharmacol. **1991**, *32*, 11–16.

75. Soons, P.A.; Vogels, B.A.; Roosemalen, M.C.; Schoemaker, H.C.; Uchida, E.; Edgar, B.; Lundahl, J.; Cohen, A.F.; Breimer, D.D. Grapefruit juice and cimetidine inhibit stereoselective metabolism of nitrendipine in humans. Clin. Pharmacol. Ther. **1991**, *50*, 394–403.

76. Luksa, J.; Josic, D.; Kremser, M.; Kopitar, Z.; Milutinovic, S. Pharmacokinetic behaviour of R-(+)- and S-(−)-amlodipine after single enantiomer administration. J. Chromatogr. B. Biomed. Sci. Appl. **1997**, *703*, 185–193.

77. Wanner-Olsen, H.; Gaarskaer, F.B.; Mikkelsen, E.O.; Jakobsen, P.; Voldby, B. Studies on concentration-time profiles of nimodipine enantiomers following intravenous and oral administration of nimodipine in patients with subarachnoid hemorrhage. Chirality **2000**, *12*, 660–664.

78. Uno, T.; Ohkubo, T.; Sugawara, K.; Higashiyama, A.; Motomura, S.; Ishizaki, T. Effects of grapefruit juice on the stereoselective disposition of nicardipine in humans: evidence for dominant presystemic elimination at the gut site. Eur. J. Clin. Pharmacol. **2000**, *56*, 643–649.

79. Lima, J.J.; Boudoulas, H. Stereoselective effects of disopyramide enantiomers in humans. J. Cardiovasc. Pharmacol. **1987**, *9*, 594–600.

80. Turgeon, J.; Funck-Brentano, C.; Gray, H.T.; Pavlou, H.N.; Prakash, C.; Blair, I.A.; Roden, D.M. Genetically determined stereoselective excretion of encainide in humans and electrophysiologic effects of its enantiomers in canine cardiac Purkinje fibers. Clin. Pharmacol. Ther. **1991**, *49*, 488–496.

81. Banitt, E.H.; Schmid, J.R.; Newmark, R.A. Resolution of flecainide acetate, N-(2-piperidylmethyl)-2,5-bis(2,2,2-trifluoroethoxy)benzam ide acetate, and antiarrhythmic properties of the enantiomers. J. Med. Chem. **1986**, *29*, 299–302.

82. Turgeon, J.; Uprichard, A.C.; Belanger, P.M.; Harron, D.W.; Grech-Belanger, O. Resolution and electrophysiological effects of mexiletine enantiomers. J. Pharm. Pharmacol. **1991**, *43*, 630–635.

83. De Luca, A.; Natuzzi, F.; Lentini, G.; Franchini, C.; Tortorella, V.; Conte Camerino, D. Stereoselective effects of mexiletine enantiomers on sodium currents and excitability characteristics of adult skeletal muscle fibers. Naunyn. Schmiedebergs Arch. Pharmacol. **1995**, *352*, 653–661.

84. Brode, E.; Muller-Peltzer, H.; Hollmann, M. Comparative pharmacokinetics and clinical pharmacology of propafenone enantiomers after oral administration to man. Methods Find Exp. Clin. Pharmacol. **1988**, *10*, 717–727.

85. Kroemer, H.K.; Funck-Brentano, C.; Silberstein, D.J.; Wood, A.J.; Eichelbaum, M.; Woosley, R.L.; Roden, D.M. Stereoselective disposition and pharmacologic activity of propafenone enantiomers. Circulation **1989**, *79*, 1068–1076.

86. Tricarico, D.; Fakler, B.; Spittelmeister, W.; Ruppersberg, J.P.; Stutzel, R.; Franchini, C.; Tortorella, V.; Conte-Camerino, D.; Rudel, R. Stereoselective interaction of tocainide and its chiral analogs with the sodium channels in human myoballs. Pflugers Arch. **1991**, *418*, 234–237.

87. Franchini, C.; Noja, F.C.; Corbo, F.; Lentini, G.; Tortorella, V.; Bartolini, A.; Ghelardini, C.; Matucci, R.; Giotti, A. Stereoselectivity in central analgesic action of tocainide and its analogs. Chirality **1993**, *5*, 135–142.

88. Ferry, D.R.; Glossmann, H.; Kaumann, A.J. Relationship between the stereoselective negative inotropic effects of verapamil enantiomers and their binding to putative calcium channels in human heart. Br. J. Pharmacol. **1985**, *84*, 811–824.

89. Echizen, H.; Brecht, T.; Niedergesass, S.; Vogelgesang, B.; Eichelbaum, M. The effect of dextro-, levo-, and racemic verapamil on atrioventricular conduction in humans. Am. Heart J. **1985**, *109*, 210–217.

90. Piascik, M.T.; Collins, R.; Butler, B.T. Stereoselective and nonstereoselective inhibition exhibited by the enantiomers of verapamil. Can. J. Physiol. Pharmacol. **1990**, *68*, 439–446.

91. Haussermann, K.; Benz, B.; Gekeler, V.; Schumacher, K.; Eichelbaum, M. Effects of verapamil enantiomers and major metabolites on the cytotoxicity of vincristine and daunomycin in human lymphoma cell lines. Eur. J. Clin. Pharmacol. **1991**, *40*, 53–59.

92. Kwok, D.W.; Kerr, C.R.; McErlane, K.M. Pharmacokinetics of mexiletine enantiomers in healthy human subjects. A study of the in vivo serum protein binding, salivary excretion and red blood cell distribution of the enantiomers. Xenobiotica **1995**, *25*, 1127–1142.

93. Sedman, A.J.; Bloedow, D.C.; Gal, J. Serum binding of tocainide and its enantiomers in human subjects. Res. Commun. Chem. Pathol. Pharmacol. **1982**, *38*, 165–168.

94. McErlane, K.; Igwemezie, L.; Kerr, C. Stereoselective serum protein binding of mexiletine enantiomers in man. Res. Commun. Chem. Pathol. Pharmacol. **1987**, *56*, 141–144.

95. Lima, J.J.; Boudoulas, H.; Shields, B.J. Stereoselective pharmacokinetics of disopyramide enantiomers in man. Drug Metab. Dispos. **1985**, *13*, 572–577.
96. Takahashi, H.; Ogata, H.; Shimizu, M.; Hashimoto, K.; Masuhara, K.; Kashiwada, K.; Someya, K. Comparative pharmacokinetics of unbound disopyramide enantiomers following oral administration of racemic disopyramide in humans. J. Pharm. Sci. **1991**, *80*, 709–711.
97. Mehvar, R. Apparent stereoselectivity in propafenone uptake by human and rat erythrocytes. Biopharm. Drug Dispos. **1991**, *12*, 299–310.
98. Eichelbaum, M.; Mikus, G.; Vogelgesang, B. Pharmacokinetics of (+)-, (−)- and (+/−)-verapamil after intravenous administration. Br. J. Clin. Pharmacol. **1984**, *17*, 453–458.
99. Gross, A.S.; Heuer, B.; Eichelbaum, M. Stereoselective protein binding of verapamil enantiomers. Biochem. Pharmacol. **1988**, *37*, 4623–4627.
100. Johnson, J.A.; Akers, W.S. Influence of metabolites on protein binding of verapamil enantiomers. Br. J. Clin. Pharmacol. **1995**, *39*, 536–538.
101. Mohamed, N.A.; Kuroda, Y.; Shibukawa, A.; Nakagawa, T.; El Gizawy, S.; Askal, H.F.; El Kommos, M.E. Enantioselective binding analysis of verapamil to plasma lipoproteins by capillary electrophoresis-frontal analysis. J. Chromatogr. A **2000**, *875*, 447–453.
102. Hanada, K.; Ohta, T.; Hirai, M.; Arai, M.; Ogata, H. Enantioselective binding of propranolol, disopyramide, and verapamil to human alpha(1)-acid glycoprotein. J. Pharm. Sci. **2000**, *89*, 751–757.
103. Robinson, M.A.; Mehvar, R. Enantioselective distribution of verapamil and norverapamil into human and rat erythrocytes: the role of plasma protein binding. Biopharm. Drug. Dispos. **1996**, *17*, 577–587.
104. Hanada, K.; Akimoto, S.; Mitsui, K.; Mihara, K.; Ogata, H. Enantioselective tissue distribution of the basic drugs disopyramide, flecainide and verapamil in rats: role of plasma protein and tissue phosphatidylserine binding. Pharm. Res. **1998**, *15*, 1250–1256.
105. Igwemezie, L.; Beatch, G.N.; Walker, M.J.; McErlane, K.M. Tissue distribution of mexiletine enantiomers in rats. Xenobiotica **1991**, *21*, 1153–1158.
106. Le Corre, P.; Gibassier, D.; Sado, P.; Le Verge, R. Stereoselective metabolism and pharmacokinetics of disopyramide enantiomers in humans. Drug Metab. Dispos. **1988**, *16*, 858–864.
107. Le Corre, P.; Ratanasavanh, D.; Chevanne, F.; Gibassier, D.; Sado, P.; Le Verge, R.; Guillouzo, A. In vitro assessment of stereoselective hepatic metabolism of disopyramide in humans: comparison with in vivo data. Chirality **1991**, *3*, 405–411.
108. Echizen, H.; Mochizuki, K.; Tani, M.; Ishizaki, T. Interspecies differences in enantioselective mono-N-dealkylation of disopyramide by human and mouse liver microsomes. J. Pharmacol. Exp. Ther. **1994**, *268*, 1518–1525.
109. Echizen, H.; Tanizaki, M.; Tatsuno, J.; Chiba, K.; Berwick, T.; Tani, M.; Gonzalez, F.J.; Ishizaki, T. Identification of CYP3A4 as the enzyme involved in the mono-N-dealkylation of disopyramide enantiomers in humans. Drug Metab. Dispos. **2000**, *28*, 937–944.

110. Gross, A.S.; Mikus, G.; Fischer, C.; Hertrampf, R.; Gundert-Remy, U.; Eichelbaum, M. Stereoselective disposition of flecainide in relation to the sparteine/debrisoquine metaboliser phenotype. Br. J. Clin. Pharmacol. **1989**, *28*, 555–566.

111. Vandamme, N.; Broly, F.; Libersa, C.; Courseau, C.; Lhermitte, M. Stereoselective hydroxylation of mexiletine in human liver microsomes: implication of P450IID6—a preliminary report. J. Cardiovasc. Pharmacol. **1993**, *21*, 77–83.

112. Abolfathi, Z.; Fiset, C.; Gilbert, M.; Moerike, K.; Belanger, P.M.; Turgeon, J. Role of polymorphic debrisoquin 4-hydroxylase activity in the stereoselective disposition of mexiletine in humans. J. Pharmacol. Exp. Ther. **1993**, *266*, 1196–1201.

113. Kroemer, H.K.; Mikus, G.; Kronbach, T.; Meyer, U.A.; Eichelbaum, M. In vitro characterization of the human cytochrome P-450 involved in polymorphic oxidation of propafenone. Clin. Pharmacol. Ther. **1989**, *45*, 28–33.

114. Kroemer, H.K.; Fischer, C.; Meese, C.O.; Eichelbaum, M. Enantiomer/ enantiomer interaction of (S)- and (R)-propafenone for cytochrome P450IID6-catalyzed 5-hydroxylation: in vitro evaluation of the mechanism. Mol Pharmacol. **1991**, *40*, 135–142.

115. Hemeryck, A.; De Vriendt, C.; Belpaire, F.M. Effect of selective serotonin reuptake inhibitors on the oxidative metabolism of propafenone: in vitro studies using human liver microsomes. J. Clin. Psychopharmacol. **2000**, *20*, 428–434.

116. Botsch, S.; Gautier, J.C.; Beaune, P.; Eichelbaum, M.; Kroemer, H.K. Identification and characterization of the cytochrome P450 enzymes involved in N-dealkylation of propafenone: molecular base for interaction potential and variable disposition of active metabolites. Mol. Pharmacol. **1993**, *43*, 120–126.

117. Hoffmann, K.J.; Renberg, L.; Gyllenhaal, O. Analysis and stereoselective metabolism after separate oral doses of tocainide enantiomers to healthy volunteers. Biopharm. Drug Dispos. **1990**, *11*, 351–363.

118. Nelson, W.L.; Olsen, L.D. Regiochemistry and enantioselectivity in the oxidative N-dealkylation of verapamil. Drug Metab. Dispos. **1988**, *16*, 834–841.

119. Kroemer, H.K.; Echizen, H.; Heidemann, H.; Eichelbaum, M. Predictability of the in vivo metabolism of verapamil from in vitro data: contribution of individual metabolic pathways and stereoselective aspects. J. Pharmacol. Exp. Ther. **1992**, *260*, 1052–1057.

120. Kroemer, H.K.; Gautier, J.C.; Beaune, P.; Henderson, C.; Wolf, C.R.; Eichelbaum, M. Identification of P450 enzymes involved in metabolism of verapamil in humans. Naunyn. Schmiedebergs Arch. Pharmacol. **1993**, *348*, 332–337.

121. Tracy, T.S.; Korzekwa, K.R.; Gonzalez, F.J.; Wainer, I.W. Cytochrome P450 isoforms involved in metabolism of the enantiomers of verapamil and norverapamil. Br. J. Clin. Pharmacol. **1999**, *47*, 545–552.

122. Bonde, J.; Pedersen, L.E.; Nygaard, E.; Ramsing, T.; Angelo, H.R.; Kampmann, J.P. Stereoselective pharmacokinetics of disopyramide and interaction with cimetidine. Br. J. Clin. Pharmacol. **1991**, *31*, 708–710.

123. Hasselstrom, J.; Enquist, M.; Hermansson, J.; Dahlqvist, R. Enantioselective steady-state kinetics of unbound disopyramide and its dealkylated metabolite in man. Eur. J. Clin. Pharmacol. **1991**, *41*, 481–484.

124. Giacomini, K.M.; Nelson, W.L.; Pershe, R.A.; Valdivieso, L.; Turner-Tamiyasu, K.; Blaschke, T.F. In vivo interaction of the enantiomers of disopyramide in human subjects. J. Pharmacokinet. Biopharm. **1986**, *14*, 335–356.

125. Echizen, H.; Takahashi, H.; Nakamura, H.; Ochiai, K.; Chiba, K.; Koike, K.; Ogata, H.; Ishizaki, T. Stereoselective disposition and metabolism of disopyramide in pediatric patients. J. Pharmacol. Exp. Ther. **1991**, *259*, 953–960.

126. Birgersdotter, U.M.; Wong, W.; Turgeon, J.; Roden, D.M. Stereoselective genetically-determined interaction between chronic flecainide and quinidine in patients with arrhythmias. Br. J. Clin. Pharmacol. **1992**, *33*, 275–280.

127. McErlane, K.M.; Axelson, J.; Vaughan, R.; Kerr, C.R.; Price, J.D.; Igwemezie, L.; Pillai, G. Stereoselective pharmacokinetics of tocainide in human uraemic patients and in healthy subjects. Eur. J. Clin. Pharmacol. **1990**, *39*, 373–376.

128. Barbey, J.T.; Thompson, K.A.; Echt, D.S.; Woosley, R.L.; Roden, D.M. Antiarrhythmic activity, electrocardiographic effects and pharmacokinetics of the encainide metabolites O-desmethyl encainide and 3-methoxy-O-desmethyl encainide in man. Circulation **1988**, *77*, 380–391.

129. Grech-Belanger, O.; Turgeon, J.; Gilbert, M. Stereoselective disposition of mexiletine in man. Br. J. Clin. Pharmacol. **1986**, *21*, 481–487.

130. Igwemezie, L.; Kerr, C.R.; McErlane, K.M. The pharmacokinetics of the enantiomers of mexiletine in humans. Xenobiotica **1989**, *19*, 677–682.

131. Labbe, L.; Abolfathi, Z.; Robitaille, N.M.; St-Maurice, F.; Gilbert, M.; Turgeon, J. Stereoselective disposition of the antiarrhythmic agent mexiletine during the concomitant administration of caffeine. Ther. Drug. Monit. **1999**, *21*, 191–199.

132. Labbe, L.; O'Hara, G.; Lefebvre, M.; Lessard, E.; Gilbert, M.; Adedoyin, A.; Champagne, J.; Hamelin, B.; Turgeon, J. Pharmacokinetic and pharmacodynamic interaction between mexiletine and propafenone in human beings. Clin. Pharmacol. Ther. **2000**, *68*, 44–57.

133. Kroemer, H.K.; Fromm, M.F.; Buhl, K.; Terefe, H.; Blaschke, G.; Eichelbaum, M. An enantiomer-enantiomer interaction of (S)- and (R)-propafenone modifies the effect of racemic drug therapy. Circulation **1994**, *89*, 2396–2400.

134. Li, G.; Gong, P.L.; Qiu, J.; Zeng, F.D.; Klotz, U. Stereoselective steady state disposition and action of propafenone in Chinese subjects. Br. J. Clin. Pharmacol. **1998**, *46*, 441–445.

135. Cai, W.M.; Chen, B.; Cai, M.H.; Chen, Y.; Zhang, Y.D. The influence of CYP2D6 activity on the kinetics of propafenone enantiomers in Chinese subjects. Br. J. Clin. Pharmacol. **1999**, *47*, 553–556.

136. Chen, X.; Zhong, D.; Blume, H. Stereoselective pharmacokinetics of propafenone and its major metabolites in healthy Chinese volunteers. Eur. J. Pharm. Sci. **2000**, *10*, 11–16.

137. Cai, W.M.; Chen, B.; Zhou, Y.; Zhang, Y.D. Fluoxetine impairs the CYP2D6-mediated metabolism of propafenone enantiomers in healthy Chinese volunteers. Clin. Pharmacol. Ther. **1999**, *66*, 516–521.

138. Sedman, A.J.; Gal, J.; Mastropaolo, W.; Johnson, P.; Maloney, J.D.; Moyer, T.P. Serum tocainide enantiomer concentrations in human subjects. Br. J. Clin. Pharmacol. **1984**, *17*, 113–115.

139. Thomson, A.H.; Murdoch, G.; Pottage, A.; Kelman, A.W.; Whiting, B.; Hillis, W.S. The pharmacokinetics of R- and S-tocainide in patients with acute ventricular arrhythmias. Br. J. Clin. Pharmacol. **1986**, *21*, 149–154.

140. Pillai, G.K.; Axelson, J.E.; Kerr, C.R.; McErlane, K.M. Stereospecific salivary excretion of tocainide enantiomers in man. Res. Commun. Chem. Pathol. Pharmacol. **1984**, *43*, 209–221.

141. Vogelgesang, B.; Echizen, H.; Schmidt, E.; Eichelbaum, M. Stereoselective first-pass metabolism of highly cleared drugs: studies of the bioavailability of L- and D-verapamil examined with a stable isotope technique. Br. J. Clin. Pharmacol. **1984**, *18*, 733–740.

142. Eichelbaum, M.; Birkel, P.; Grube, E.; Gutgemann, U.; Somogyi, A. Effects of verapamil on P-R-intervals in relation to verapamil plasma levels following single I.V. and oral administration and during chronic treatment. Klin. Wochenschr. **1980**, *58*, 919–925.

143. Echizen, H.; Vogelgesang, B.; Eichelbaum, M. Effects of d,l-verapamil on atrioventricular conduction in relation to its stereoselective first-pass metabolism. Clin. Pharmacol. Ther. **1985**, *38*, 71–76.

144. Mikus, G.; Eichelbaum, M.; Fischer, C.; Gumulka, S.; Klotz, U.; Kroemer, H.K. Interaction of verapamil and cimetidine: stereochemical aspects of drug metabolism, drug disposition and drug action. J. Pharmacol. Exp. Ther. **1990**, *253*, 1042–1048.

145. Ho, P.C.; Ghose, K.; Saville, D.; Wanwimolruk, S. Effect of grapefruit juice on pharmacokinetics and pharmacodynamics of verapamil enantiomers in healthy volunteers. Eur. J. Clin. Pharmacol. **2000**, *56*, 693–698.

146. Abernethy, D.R.; Wainer, I.W.; Longstreth, J.A.; Andrawis, N.S. Stereoselective verapamil disposition and dynamics in aging during racemic verapamil administration. J. Pharmacol. Exp. Ther. **1993**, *266*, 904–911.

147. Schwartz, J.B.; Troconiz, I.F.; Verotta, D.; Liu, S.; Capili, H. Aging effects on stereoselective pharmacokinetics and pharmacodynamics of verapamil. J. Pharmacol. Exp. Ther. **1993**, *265*, 690–698.

148. Schwartz, J.B.; Capili, H.; Wainer, I.W. Verapamil stereoisomers during racemic verapamil administration: effects of aging and comparisons to administration of individual stereoisomers. Clin. Pharmacol. Ther. **1994**, *56*, 368–376.

149. Schwartz, J.B.; Capili, H.; Daugherty, J. Aging of women alters S-verapamil pharmacokinetics and pharmacodynamics. Clin. Pharmacol. Ther. **1994**, *55*, 509–517.

150. Gupta, S.K.; Atkinson, L.; Tu, T.; Longstreth, J.A. Age and gender related changes in stereoselective pharmacokinetics and pharmacodynamics of verapamil and norverapamil. Br. J. Clin. Pharmacol. **1995**, *40*, 325–331.

151. Mehvar, R.; Reynolds, J. Input rate–dependent stereoselective pharmacokinetics. Experimental evidence in verapamil-infused isolated rat livers. Drug Metab. Dispos. **1995**, *23*, 637–641.

152. Karim, A.; Piergies, A. Verapamil stereoisomerism: enantiomeric ratios in plasma dependent on peak concentrations, oral input rate, or both. Clin. Pharmacol. Ther. **1995**, *58*, 174–184.

153. Bhatti, M.M.; Lewanczuk, R.Z.; Pasutto, F.M.; Foster, R.T. Pharmacokinetics of verapamil and norverapamil enantiomers after administration of immediate and controlled-release formulations to humans: evidence suggesting input rate–determined stereoselectivity. J. Clin. Pharmacol. **1995**, *35*, 1076–1082.

154. Hashiguchi, M.; Ogata, H.; Maeda, A.; Hirashima, Y.; Ishii, S.; Mori, Y.; Amamoto, T.; Handa, T.; Otsuka, N.; Irie, S.; Urae, A.; Urae, R.; Kimura, R. No effect of high-protein food on the stereoselective bioavailability and pharmacokinetics of verapamil. J. Clin. Pharmacol. **1996**, *36*, 1022–1028.

155. Gupta, S.K.; Yih, B.M.; Atkinson, L.; Longstreth, J. The effect of food, time of dosing, and body position on the pharmacokinetics and pharmacodynamics of verapamil and norverapamil. J. Clin. Pharmacol. **1995**, *35*, 1083–1093.

156. Gietl, Y.; Spahn, H.; Knauf, H.; Mutschler, E. Single- and multiple-dose pharmacokinetics of R-(−)- and S-(+)-prenylamine in man. Eur. J. Clin. Pharmacol. **1990**, *38*, 587–593.

157. Janiczek, N.; Smith, D.E.; Chang, T.; Sedman, A.J.; Stringer, K.A. Pharmacokinetics of pirmenol enantiomers and pharmacodynamics of pirmenol racemate in patients with premature ventricular contractions. J. Clin. Pharmacol. **1997**, *37*, 502–513.

158. Gross, A.S.; Eichelbaum, M.; Morike, K.; Mikus, G. Pharmacokinetics and pharmacodynamics of R- and S-gallopamil during multiple dosing. Br. J. Clin. Pharmacol. **2000**, *49*, 132–138.

159. Gross, A.S.; Mikus, G.; Ratge, D.; Wisser, H.; Eichelbaum, M. Pharmacokinetics and pharmacodynamics of the enantiomers of gallopamil. J. Pharmacol. Exp. Ther. **1997**, *281*, 1102–1112.

160. Chu, K.M.; Shieh, S.M.; Hu, O.Y. Pharmacokinetics and pharmacodynamics of enantiomers of pimobendan in patients with dilated cardiomyopathy and congestive heart failure after single and repeated oral dosing. Clin. Pharmacol. Ther. **1995**, *57*, 610–621.

161. Chu, K.M.; Shieh, S.M.; Hu, O.Y. Plasma and red blood cell pharmacokinetics of pimobendan enantiomers in healthy Chinese. Eur. J. Clin. Pharmacol. **1995**, *47*, 537–542.

162. Brocks, D.R.; Skeith, K.J.; Johnston, C.; Emamibafrani, J.; Davis, P.; Russell, A.S.; Jamali, F. Hematologic disposition of hydroxychloroquine enantiomers. J. Clin. Pharmacol. **1994**, *34*, 1088–1097.

163.   Kato, R.; Ikeda, N.; Yabek, S.M.; Kannan, R.; Singh, B.N. Electrophysio-
       logic effects of the levo- and dextrorotatory isomers of sotalol in isolated
       cardiac muscle and their in vivo pharmacokinetics. J. Am. Coll. Cardiol. **1986**,
       7, 116–125.
164.   McErlane, K.M.; Igwemezie, L.; Kerr, C.R. Stereoselective serum protein
       binding of mexiletine enantiomers in man. Res. Commun. Chem. Pathol.
       Pharmacol. **1987**, *56*, 141–144.
165.   Uno, T.; Ohkubo, T.; Sugawara, K. Enantioselective high-performance
       liquid chromatographic determination of nicardipine in human plasma.
       J. Chromatogr. B. Biomed. Sci. Appl. **1997**, *698*, 181–186.

8

Chiral Inversion

Neal M. Davies
Washington State University, Pullman, Washington, U.S.A.

1. INTRODUCTION

A thorough understanding of the pharmacokinetics of drugs is essential in the determination of safe and effective dosing regimens. In the case of a racemic drug or stereochemically pure enantiomer this implies knowledge of the in vivo behavior of the stereoisomers.

One feature that may exist with chiral xenobiotics is a lack of configurational stability. Some chiral drugs may undergo enzymatic or nonenzymatic interconversion of the enantiomers. Isomerization or enantiomerization is the conversion of one stereoisomeric form into another. When isomerization occurs by the change of configuration at a single chiral center, the process is called epimerization, and when it leads to the formation of a racemate it is termed racemization. Nonenzymatic inversion of drugs is important in the pharmaceutical manufacturing process and has implications for the shelf-life of a drug and the economic feasibility of the resolution. Nonenzymatic inversion can also occur during stereospecific chromatographic procedures that employ precolumn derivatization. Racemization may also occur in physiological fluids such as the acidic environment of the stomach. Enzymatic inversion is concerned with inversion under physiological conditions. Under these conditions, enantiomers may be inverted, as a racemate is enriched in one of the antipodes.

Demonstration of metabolic chiral inversion may have profound consequences for the development of a new pharmaceutical entity. A better understanding of the factors facilitating such interconversions may greatly aid drug development by identifying this feature at an early stage and thereby reducing bioanalytical and toxicology workload. Given the possibility of

351

chiral inversion of racemates and stereochemically pure enantiomers, regulatory agencies are increasingly asking for evidence regarding this phenomenon following administration of racemates or single enantiomers.

The intent of this chapter is to provide a comprehensive, rather than exhaustive, appraisal of chiral bioinversion. Illustrative examples have been selected from drugs currently in use, and from drugs undergoing development that have encountered this metabolic phenomenon. This chapter will discuss enzymatic chiral inversion of drugs and its importance in drug development.

2. INVERSION BY CONJUGATION

2-Arylpropionic acid (2-APA) nonsteroidal anti-inflammatory drugs (NSAIDs) provide one of the most studied illustrated pharmaceutical examples of inversion by conjugation through a coenzyme A (CoA) thioester intermediate. It is also possible that bioinversion of xenobiotics proceed through other conjugation mechanisms such as glutathione [1–4].

2.1. NSAIDs: 2-Arylpropionic Acids

The 2-arylpropionic acid class (2-APA) of nonsteroidal anti-inflammatory drugs (NSAIDs) (Table 1) is characterized by each member having an asymmetric carbon α to the carboxylic acid moiety. The R-enantiomer of this chiral center of some 2-APAs may undergo an in vivo inversion to the S-enantiomer. This inversion process varies substantially between the different members of this class and also varies between species of animal studied. The members of this class that are currently in clinical use include ibuprofen, ketoprofen, tiaprofenic acid, fenoprofen, and flurbiprofen. The majority are marketed as racemates. Naproxen and its sodium salt are internationally marketed as the pure S(+)-enantiomer, while ibuprofen and ketoprofen are now marketed in several European countries as the stereochemically pure S(+)-enantiomer.

Table 1   Chiral 2-Arylpropionic Acid Nonsteroidal Anti-inflammatory Drugs

Alminoprofen	Fenoprofen	Ketoprofen	Pranoprofen
Benoxaprofen	Flunoxaprofen	Loxoprofen	Suprofen
Bermoprofen	Flurbiprofen	Miroprofen	Tiaprofenic acid
Carprofen	Ibuprofen	Naproxen	Thioxaprofen
Cicloprofen	Indoprofen	Piroprofen	Ximoprofen

In vitro studies demonstrated that the anti-inflammatory activity of these 2-APA NSAIDs is largely stereospecific for the S-enantiomer [5,6]. However, some 2-APA NSAIDs undergo a usually unidirectional inversion of the R-enantiomer to the S-enantiomer. Hence the R-enantiomer serves as a prodrug for the S-enantiomer and contributes to the pharmacodynamics of the racemate. Interestingly, when in vivo animal studies were conducted, the two enantiomers of several NSAIDs had similar anti-inflammatory potencies [7–9]. Without stereospecific assay methods, the time-course of the individual enantiomers of these NSAIDs cannot be interpreted, and this may be misleading in relating either toxicity or efficacy to the racemic drug concentration. In view of this interesting metabolic phenomenon, it is not surprising that there has been difficulty in exposing concentration effect relationships with these NSAIDs.

2.2. Mechanisms

The origins of the metabolic inversion phenomenon of the 2-APA NSAIDS stem from the observation that that the urinary metabolites of racemic ibuprofen were dextrorotatory even when the R(−)-enantiomer was administered [8,10,11]. Subsequently, numerous investigations have focused on determining the mechanism involved in this bioinversion process. Initial studies focused on deuterium labeling of ibuprofen analogs in the α-methyl and α-methine protons and suggested a dehydrogenation process as well as an intermediate methylene derivative [12]. An important molecular mechanism of this chiral inversion was determined by Nakamura et al. [13], who reported that the 2α-methinine proton in the chemical structure of ibuprofen was exchanged during chiral inversion. Other work using α-deutro and α-trideutromethyl ibuprofen and deuterated R(−)-ibuprofen indicated that coenzyme A (CoA) and ATP are involved in the process, and a revised postulated mechanism was established after administration of CoA esters of ibuprofen enantiomers [13–18] (Fig. 1). It appears that three defined metabolic steps are involved in the inversion of R-ibuprofen:

1. Thioesterification of R-enantiomer to R-ibuprofen CoA [13] via an adenosine monophosphate intermediate [19] catalyzed by the microsomal long-chain acyl-CoA synthetase [20,21], an enzyme expressed in several isoforms and showing broad distribution.
2. Epimerization of R-ibuprofenoyl-CoA thioester catalyzed by the cytosolic and mitochondrial 2-aryl-CoA-epimerase to S-ibuprofenoyl-CoA via an enolate intermediate [22–24].
3. Nonstereoselective hydrolysis of the acyl-CoA thioesters by one or more hydrolases to yield free S-ibuprofen [17].

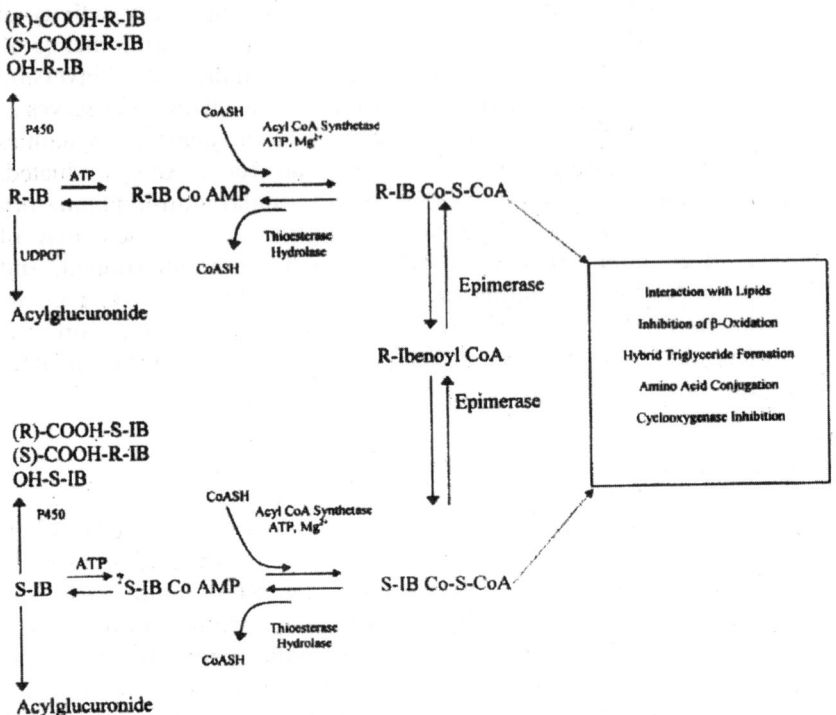

Figure 1 Proposed metabolic scheme for the chiral inversion of ibuprofen extending the mechanism suggested by Nakamura et al. [13] and the possible interactions with lipid biochemistry, amino acid conjugation, and cyclooxygenase.

It also appears that tiaprofenic acid, an NSAID that also undergoes inversion in rats, is not a substrate for purified microsomal rat liver long-chain acyl-CoA synthetase for which R-ibuprofen is a substrate [25]. This data may suggest that metabolic pathways involved in the inversion of tiaprofenic acid and possibly other 2-APA NSAIDs are different from those known for R-ibuprofen. It has been recently reported that in both an in vitro cell-free system and in rat liver homogenates the chiral inversion of ibuprofen was apparent when both CoA and ATP were present; however, the NSAID KE-748 was not inverted [26]. To induce hepatic microsomal and outer mitochondrial long-chain fatty acid CoA ligase, rats were treated with clofibric acid [27]. Whereas chiral inversion of ibuprofen was enhanced significantly compared to controls, this was not the case for R(−)-KE-748.

Using isolated hepatocytes, there was no influence on chiral inversion of .
R-ibuprofen using benzoic acid, a typical substrate of medium-chain fatty
acid CoA ligase in the mitochondrial matrix. However, the chiral inversion
of KE-748 was strongly inhibited [21,26]. KE-748, the active metabolite
of the anti-inflammatory drug KE-298, appears to be catalyzed by a short-
or medium-chain acyl-CoA synthetase which differs from the long-chain
CoA ligase involved in the inversion of R(−)-ibuprofen [26]. Flobufen is
an additional phenyl butyrate aryloxoalkonoic acid that undergoes R to S
inversion experimentally, although information on the enzymes responsible
for its inversion are not known [28].

It is known that there are several long-chain acyl-Coenzyme A synthetases (ACS). ACS1 shares some similar catalytic and antigenic properties to
palmitoyl-CoA synthetase [29]. Nevertheless, there is evidence of multiple
forms of microsomal coenzyme A ligase catalyzing the formation of 2-APA
CoA thioesters with several binding sites as well as a catalytic site. For
instance, R-ketoprofen CoA appears to be catalyzed by an alternative
isoform of a microsomal CoA ligase that is not shared by fenoprofen and
ibuprofen [30,31].

Clinically this bioinversion process does not occur appreciably with
tiaprofenic acid, indoprofen, carprofen, or flurbiprofen, but it does occur
substantially for fenoprofen and ibuprofen, for which 100% and 60% of
the R-enantiomer is inverted, respectively [32–34]. Bioinversion also occurs
for benoxaprofen, an NSAID withdrawn from the market, and it may also
occurs for loxoprofen, suprofen, and CS-670, all 2-APA NSAIDs primarily
utilized clinically throughout Asia [34–37].

The bioinversion process has been shown to be enzymatically mediated. It is a reaction of conjugation with only the R-enantiomer of susceptible
2-APAs being capable of forming an R-acyl-CoA-thioester, with subsequent
epimerization to a mixture of S- and R-acyl-CoA followed by hydrolysis to
yield a product enriched in the S antipode [13,16]. Inversion also appears
to be possible for structurally related phenylbutyrates that have an α-ethyl
moiety in place of the α-methyl group of 2-APAs including the NSAID
flobufen [28] and a platelet aggregation inhibitor indobufen [38,39].

3. LIPIDS

Extramitochondrial fatty acid CoA ligases are destined for lipid synthesis
[40]. Ibuprofen is a substrate for this enzyme, and it has therefore been
demonstrated that a number of 2-APAs, including ketoprofen, ibuprofen,
and fenoprofen, can be stereoselectively incorporated into lipid tissues

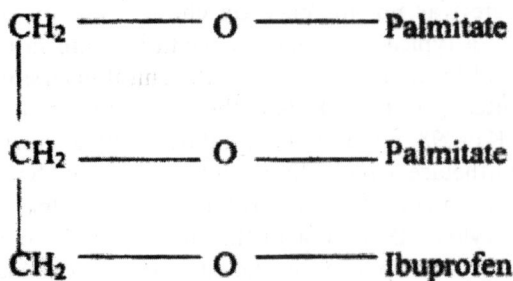

Figure 2   An example of a hybrid-triglyceride in which ibuprofen has replaced the neutral fatty acid palmitic acid.

as triacylglycerols (Fig. 2). However, flurbiprofen fails to be assimilated into glycerol [41–48]. The prior formation of the acyl-CoA thioester appears to be a necessary prerequisite, and this reactive acyl-CoA may undergo inversion or formation of glycerolipid. Following the initial findings with ibuprofen there has been confirmation of the existence of acyl-CoA thioesters and details of the incorporation of several 2-APA's into liver and adipose tissue [42,45]. These later studies have demonstrated that the R-enantiomers are enantioselectively capable of forming hybrid triglycerides, and they stereoselectively incorporated into tissue with the S-enantiomer almost excluded from this process.

Incorporation into adipose tissue is not a general phenomenon of all NSAIDs that undergo bioinversion. NSAIDs such as KE-748 that are substrates for short-chain fatty acid CoA ligase in the mitochondrial matrix are not incorporated into adipose tissue [49].

It has been suggested that the ability of a 2-APA to substitute into glycerolipids is correlated with the ability to inhibit synthesis of fatty acids and sterols. This phenomenon may have potential toxicological significance, since it interferes with normal lipid biochemistry [50]. The hypolipidemic effect by these NSAIDs is the result of depletion of CoA pools via CoA thioester formation and subsequent incorporation into glycerolipids [44,45,51]. Pirprofen has been found to exhibit a concentration-dependent inhibition of mitochondrial β-oxidation of fatty acids, which is postulated to be linked to microvesicular steatosis. Other NSAIDs such as ibuprofen, flurbiprofen, and tiaprofenic acid also inhibit β-oxidation by sequestering cytosolic CoA and inhibiting acetyl-CoA carboxylase [30,31,52–55]. The perturbations observed in hepatocyte intermediary metabolism and mitochondrial function may be attributable to direct effects of R-ibuprofen and R-ibuprofenyl CoA during the process of inversion [30,31]. In addition, there appears to be significant homology of the 2-APA epimerase with enzymes involved in carnitine metabolism. The role in this aspect of fatty acid metabolism is not known [24].

It is known that both the metabolism and the biological effects of 2-APA drugs can be modulated by other xenobiotics that influence the pool of free CoA or the activity of the CoA-dependent enzymes. Clofibrate and possibly other clofibric acid derivatives can increase the expression of long-chain acyl-CoA synthase and increase ibuprofen incorporation and tissue distribution into hybrid lipids in rats and humans [27,56–59]. The addition of valproic acid and pivalic acid, a metabolite of pivampicillin, to suspensions of isolated rat hepatocytes can also interfere with CoA pools and inhibit the formation of ibuprofenoyl-CoA reaction [60,61].

Interestingly, CoA thioesters appear to be common intermediates of amino acid conjugation and conjugation of 2-APA NSAIDs. After administration of racemic ibuprofen, ibuprofen taurine conjugates have been identified in human urine as minor biotransformation products [56]. This metabolite requires prior formation of coenzyme A thioester of ibuprofen. Further experiments using pseudoracemate ibuprofen have determined that the majority of ibuprofen taurine conjugate is derived from S-ibuprofen rather than by way of chiral inversion. This suggests that S-ibuprofen also participates in CoA dependent reactions including metabolic inversion. Similar results also appear with the taurine metabolite of CS-670 in dogs, where both the 2S- and 2R-enantiomers of the *trans*-OH metabolite appear as substrates for the acyl-CoA ligase, but only the CoA thioester with a 2S configuration is a substrate for taurine N-acyltransferase (Fig. 3). These findings are not consistent with the chiral inversion of 2-APA NSAIDs being exclusively mediated via the R-enantiomer [62,63]. A new NSAID, R,S-2-[4-(3-methyl-2-thienyl)phenyl]-propionic acid, undergoing drug development appears to undergo R to S inversion in rats and dogs with a taurine conjugate detected as a metabolite after either R or S administration. This is further evidence that the S-enantiomer, in addition to the R-enantiomer, may form a S-CoA intermediate [64].

4. BIDIRECTIONAL R TO S AND S TO R INVERSION

Members of the 2-APA class of NSAIDs have the potential for in vivo enantiomeric inversion, whereby the R-enantiomer (distomer) may be inverted to the active antipode, the S-enantiomer (eutomer). The S-enantiomer

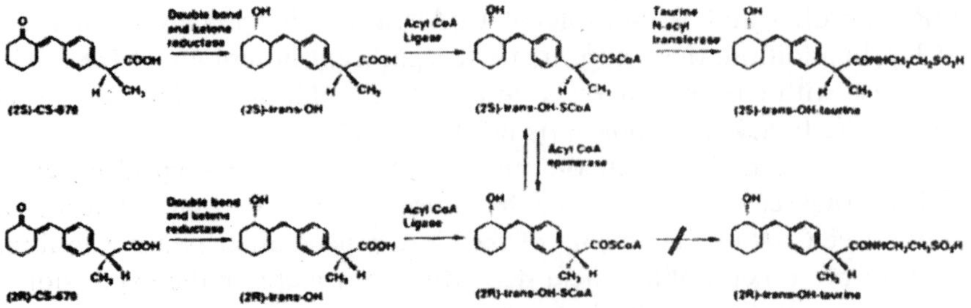

Figure 3 Possible mechanism of stereospecific formation of (2S)-*trans*-OH-taurine from (2S)- and (2R)-CS-670 in dogs. (From Ref. 62.)

appears to be a poor substrate of acyl-CoA ligase, and the chiral inversion of 2-APAs has been thought to be mostly unidirectional.

The reverse reaction, whereby the S-enantiomer inverts to the R-enantiomer, has been reported with some of these 2-APAs in various species, but scientific evidence is rare [65]. The ability of the enzymes that invert various 2-APAs to carry out bidirectional inversion is not surprising and would be consistent with other mammalian and microbial racemases (i.e., alanine racemase, methylmalonyl CoA racemase, and mandelate racemase). S to R inversion of ibuprofen occurred in rats and humans but at a much slower rate than R to S inversion [22,30,56,66–68]. S to R inversion is substantial for ibuprofen in guinea pigs [69]. Significant bidirectional chiral inversion has also been found for 2-phenylpropionic acid in dogs [18]. In mice and perhaps other species there is a rapid and reversible chiral inversion for ketoprofen enantiomers [70,71]. A small amount (∼7%) of S to R inversion of ketorolac is evident in the rat and in humans and is more extensive in the mouse (> 50% in both directions) [72]. In the mold *Verticillum lecanii*, ketoprofen inversion also occurred from S to R [73,74]. Recently, tiaprofenic acid bidirectional inversion has been demonstrated in the rat and in liver homogenates [25]. Oxindanac enantiomers undergo bidirectional inversion in beagle dogs in vivo and in vitro in blood and plasma. It is presumed that this inversion process is enzyme mediated, as it was temperature dependent and was not evident in aqueous solutions [75]. In addition, an analogous phenomenon of bidirectional inversion has also been described for the structurally related 2-aryloxypropionate herbicides such as haloxyfop, diclofop, and fluazifop [76,77].

These observations may not have clinical relevance. However, during preclinical stages in choosing an animal model to study stereochemical aspects of disposition, this information may prove useful and suggest that a priori assumptions should not be made regarding the extent or direction of inversion of chiral drugs. For other drugs, such as mandelic acid, the direction of chiral inversion proceeds from S to R [78]. Recently, XK469, (±)-2{4[(7-chloro-2-quinoxalinyl)oxy]phenoxy} propionic acid, the first topoisomerase IIβ poison with antitumor activity, which is structurally similar to 2-APA NSAIDs, has also been shown to undergo S(−) to R(+) inversion in mice, dogs, and rabbits [79].

5. SPECIES DIFFERENCES IN INVERSION

The metabolic chiral inversion of NSAIDs demonstrates notable instances of the failure of a species to undergo this reaction. For example, clindanac does not undergo bioinversion in rats and mice but does so to an appreciable extent in guinea pigs, where the R(−)-enantiomer is converted to the S-enantiomer [80]. R-flurbiprofen undergoes some quantifiable inversion in the dog and guinea pig, but this occurs to a negligible extent in rat, humans, or gerbils [81]. Ketoprofen is inverted with considerable variability in various animal species. Marked inversion of ketoprofen occurs in rats, dogs, and horses (74–92%) [70,82,83] with the smallest amount of inversion being 27% in gerbils and ∼10% in humans [84,85]. Tiaprofenic acid does not appear to undergo inversion in man [86] but does undergo appreciable inversion in rats [25,87].

Microorganisms are also able to invert several 2-APA NSAIDs. The mold *Verticillum lecanii* inverts R-ibuprofen, fenoprofen, and suprofen as well as the metabolites of racemic ibuprofen [88–90]. For microorganisms, the inversion process also appears to be substrate dependent, as flurbiprofen underwent no inversion using *V. Lecanii* while suprofen underwent extensive bioinversion [74].

Toxicity testing and understanding of metabolism of drugs is carried out using animal and microbial models, and there are a number of examples where the disposition of enantiomers of racemic drugs differs markedly between species. Examination of the pharmacokinetics and metabolism of enantiomers of racemic drugs in various species is necessary for the extrapolation of preclinical safety data to the human situation. Species differences may cause considerable difficulty in interpreting interspecies variation in response to a chiral NSAIDs that undergo significant species-dependent inversion. On the other hand, it may also be possible to select fungi and bacteria for the production of stereochemically pure enantiomers based on their metabolic stereospecificity.

6. SITES OF BIOINVERSION

As the physicochemical characteristics of enantiomers (pKa and partition coefficient) are identical, the passive absorption of individual enantiomers would be expected to be the same.

Several 2-APAs have shown similar time to maximum concentration values (T_{max}) but differences in C_{max}. The observation that the C_{max} S/R ratio of ibuprofen increased proportionally to the duration of residency in the gastrointestinal (GI) tract led to the speculation that the GI tract is a major site of chiral inversion by Jamali et al. (1988) [91], particularly for slow release formulations. This process has been characterized as stereoselective presystemic (chiral inversion) rather than stereoselective absorption.

Similar findings suggesting presystemic GI inversion with ibuprofen have been described by Avgerinos and Hutt [92]. However, a C_{max} S/R ratio alone does not unequivocally demonstrate inversion; this is only demonstrated through an increase in ratio with an increase in T_{max}. Other laboratories have also shown a similar trend of increasing C_{max} S/R ratio with prolongation of T_{max} [93]. After administration of racemic ibuprofen to humans, the T_{max} values for the enantiomers were found to be identical, but the mean C_{max} for the S-ibuprofen exceeded that of the R-isomer [94]. Absorption rate dependency of ibuprofen inversion has also been demonstrated in the rat [95]. Theoretical proof of this presystemic inversion was put forward employing newly developed pharmacokinetic equations [96]. Furthermore, food-induced reduction of the rate of ibuprofen delivery may mimic the effect of a sustained release preparation and also supports the concept of presystemic GI inversion [97].

Using isolated human and rat intestinal tissue, a significant degree of inversion was observed, confirming the GI tract as a site of inversion of 2-APA NSAIDs [98,99]. Other investigators have not replicated this inversion with rat small intestinal homogenates, possibly because of disruption of cellular boundaries by homogenization [16]. The presystemic inversion in the gut wall of rats administered several 2-APAs has been demonstrated [100–102]. There are contradictory findings of inversion in low-protein or protein-free rat gut preparations [101,102] and lack of inversion in an in situ rat intestinal preparation [103].

Data generated following i.v. doses of R-ibuprofen to humans provides strong evidence suggesting systemic inversion of ibuprofen from R to S, as no significant differences were noticed between oral and intravenous routes of administration in healthy subjects [104]. However, solutions of R-ibuprofen were administered in this study, which may be absorbed so rapidly that a negligible amount of the dose may contact the presystemic inversion sites of the distal GI tract. Additionally, a comparison of rapidly absorbed racemic tablet formulations with i.v. doses does not demonstrate different S : R ratios, as this formulation may not provide a sufficiently long residence time in the GI tract for higher S : R plasma concentrations to occur [102,105].

Direct evidence also exists for hepatic bioinversion of 2-APA NSAIDs in perfused intact organ studies [102,103,106–108]. Other experimental evidence for the contribution of the liver to inversion of the 2-APAs in hepatic cell culture and homogenates [13,16,105,109–112] clearly shows that hepatic inversion is dependent on species, substrate, and the method of tissue preparation and isolation. There is often high interassay variability between investigators.

Nevertheless, the epimerase enzyme has been determined to exist in various tissues of rat and guinea pig including the ileum, heart, lung, liver, kidney, and brain [23]. Both the kidney [113] and the brain [114] as well as pancreatic and glial cells [115] are capable of bioinversion of 2-APAs. It has also been shown that ibuprofen and fenoprofen are inverted in isolated perfused rabbit lungs. However, in the presence of protein, these 2-APAs were not metabolized, suggesting that this bioinversion in the lung may be of minimal importance [116]. Additionally, percutaneous penetration studies of ibuprofen through porcine skin indicate that no metabolic inversion occurs through this site [117]. Overall, the epimerase activity appears to be greatest in liver and kidney [69].

7. KETOROLAC: BIOINVERSION / RACEMIZATION

Ketorolac is not a 2-APA NSAID but rather a chiral arylalkanoic NSAID. A small amount ($\sim 7\%$) of S to R inversion of ketorolac is evident in both rats and humans, and a more extensive amount of inversion (> 50%) occurs bidirectionally in the mouse [72,118]. The inversion of ketorolac appears to be enzymatically mediated, as ketorolac adenylate conjugates have been detected in rat liver homogenates and mitochondria. Interestingly, this inversion process was demonstrated to occur without formation of a CoA thioester conjugate as an intermediate, and no lipid incorporation occurred in mice [118–120].

It should also be kept in mind that ketorolac is an example of a chiral NSAID that demonstrates complete spontaneous chemical racemization during chromatographic procedures. The racemization process is dependent on an alkaline pH and ionic strength [121,122]. Further studies are required to clarify the mechanisms of ketorolac bioinversion.

8. INVERSION BY OPPOSING METABOLIC PROCESSES: OXIDOREDUCTION

Reactions of oxidoreduction are an example of chiral inversion that takes place by the intermediacy of two opposing metabolic processes. The alcohol/ketone equilibrium mediated by alcohol dehydrogenase enzyme is an abundant reaction. The dehydrogenation of secondary alcohols to a ketone proceeds with substrate stereoselectivity in oxidation, while the hydrogenation of the ketone metabolite is product selective to one face of the carbonyl group. The consequence of the metabolism of the secondary alcohols may involve chiral inversion of this center, which can result in an altered proportion of the two enantiomers or epimers.

RS-8359 is a selective and reversible inhibitor of A-type monoamine oxidase in clinical development as an antidepressant (Fig. 4). This drug has a secondary hydroxy group at the 7 position of the cyclopentane ring and is a racemate. The R-enantiomer is observed in plasma after administration of the S-enantiomer, and the S to R chiral inversion has been demonstrated to be 45.8% in Sprague-Dawley rats, 15.8% in Wistar rats, 3.8% in mice, 0.8% in dogs, and 4.2% in monkeys. Furthermore, the S-enantiomer is detected in plasma of rats after dosing with the R-enantiomer, suggesting that R to S chiral inversion also occurs (Fig. 5). This drug shows marked species and strain differences in chiral inversion of its cyclopentanol group. The main metabolite produced by rat hepatocytes is the 7-keto form, suggesting that an oxidation/reduction mechanism is involved in the biotransformation of the drug. Further supporting this hypothesis is the stereospecific reduction using the 7-keto metabolite to the alcohol in the R configuration. In addition to dehydrogenase, the complex combination of aldehyde oxidase, glucuronosyl transferase, and cytochrome P-450

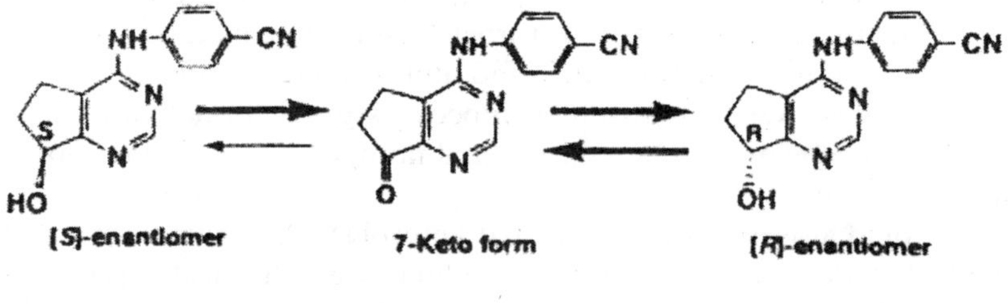

Figure 4   Proposed mechanism for chiral inversion from the [S]- to [R]-enantiomer of RS-8359. (From Ref. 123.)

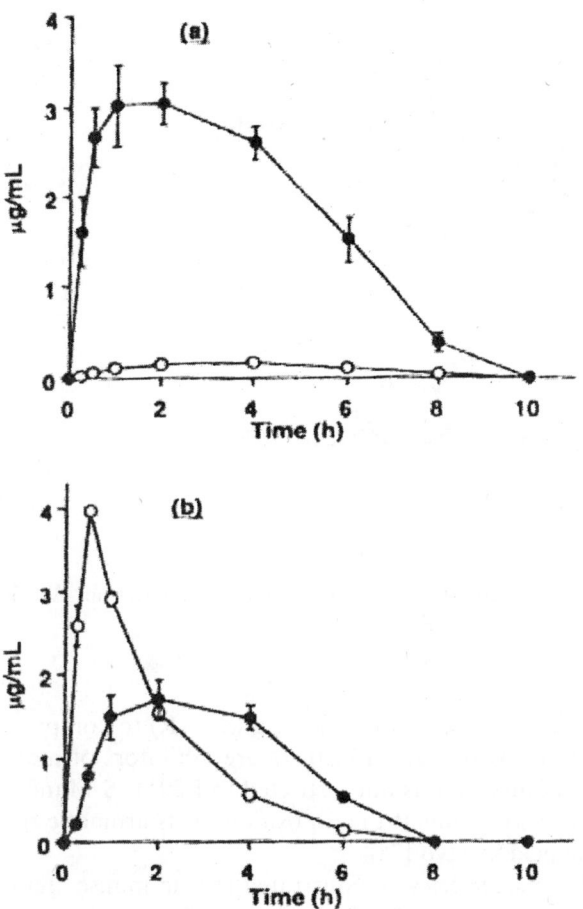

Figure 5 Plasma concentrations of each enantiomer after oral administration of the [R]-enantiomer (a) and [S]-enantiomer (b) of RS-8359 to SD rats (5 mg/kg, $n = 4$). ○, [S]-enantiomer; ●, [R]-enantiomer. (From Ref. 123.)

responsible for the metabolism of RS-8359 may contribute to the species differences in its chiral inversion [123].

E2011 is a new antidepressant, which is a selective monoamine oxidase-A inhibitor with two chiral centers and a hydroxy group in its structure (Fig. 6). One chiral center is in the oxazolidinone ring, and the other is at the α-position of the benzothiazole ring. E2011 is a pure diastereomer with a configuration of R at the oxazolidione ring and S at the α position of the benzothiazole ring (Fig. 6). In rats, E2011 undergoes chiral

(A)

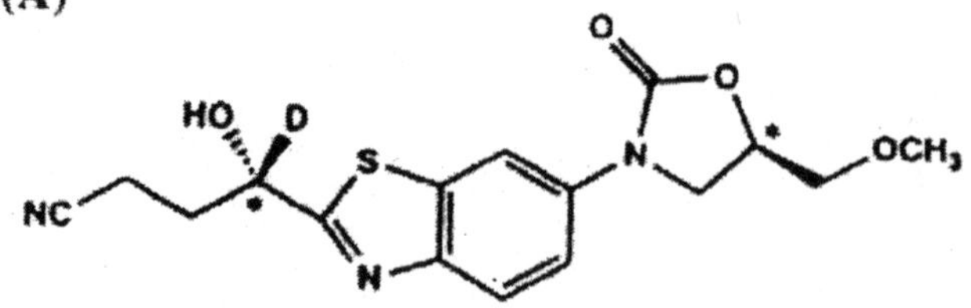

(B)

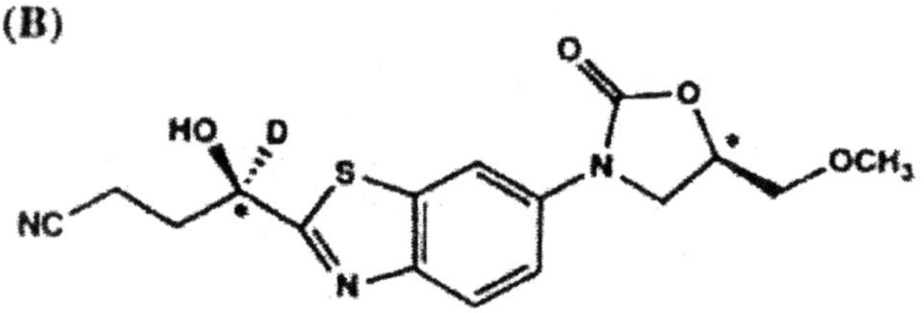

Figure 6 Structure of deuterium-labeled E2011 (A) and deuterium labeled ER-20593 (B). *, chiral carbons. (From Ref. 125.)

inversion at the α position of the benzothiazole ring (S to R) to compound ER20593 (Fig. 6) [124]. Both E2011 and ER20593 are inhibitors of mono-amine oxidase. This type of inversion is not restricted to E2011. S-Mandelic acid, which also has a hydroxy group at the α position of its aromatic ring, can invert to R-mandelic acid in vivo [78].

A subsequent study in six species demonstrated that in mouse, guinea pig, and squirrel monkey, inversion of ER20593 to E2011 occurred, but the reverse reaction did not. In rat, pig, and rhesus monkey, inversion of E2011 to ER20593 occurred but with ratios that were smaller than those in the opposite direction. Using deuterated compounds as well as liver cytosol and microsome studies, it has been demonstrated that inversion occurs owing to oxidation to the carbonyl form, followed by reduction to the original isomer or the other isomer. The documented species differences in this inversion may result from product stereoselectivity of the reductase with NADH [125] (Fig. 7).

BRL35135A is a novel antidiabetic and antiobesity agent and a selective agonist of B_3 adrenoceptors. BRL35135A is a racemate with two asymmetric chiral centers and is the HBr salt of the methyl ester of BRL37344, which is also chiral (Fig. 8). Both BRL35135A and BRL37344 are racemates consisting of RR and SS isomers. When the $[^{14}C](RR)$ isomer

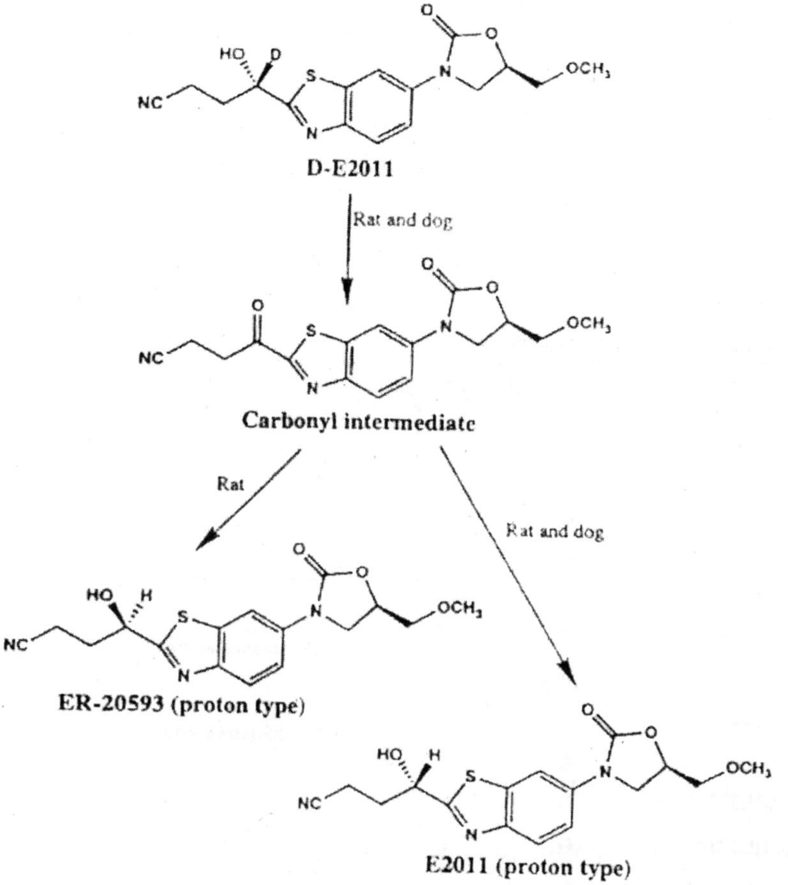

Figure 7 Mechanism of the epimerization of E2011. (From Ref. 125.)

of BRL35135A was administered to rats, both the RR and the SR isomers of BRL37344 were detected in plasma, while only the SS isomer was detected after SS BRL35135A administration. These findings suggest stereoselective chiral inversion in rats, which may occur via enzymatic dehydration of the hydroxy compound to a carbonyl compound, and by rehydrogenation of the carbonyl compound [126]. This is in accordance with the Bauman–Prelog rule, which states that carbonyls are preferentially reduced to the corresponding alcohols with S-configuration in tissues [127].

The stereoselective reduction of labile carbonyl has been observed for many drugs (e.g., loxoprofen and warfarin). E2001, a monoamine oxidase

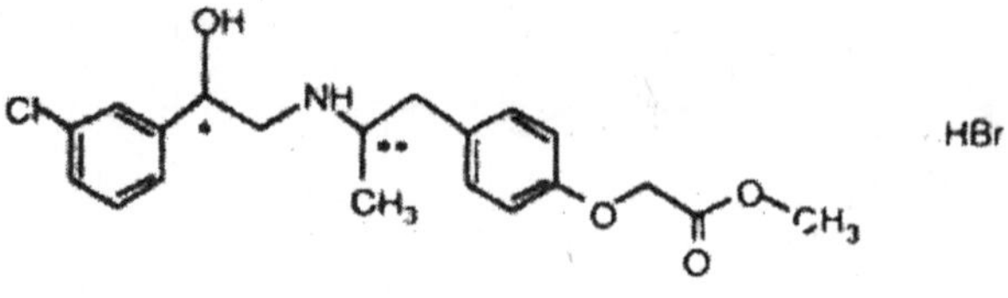

BRL35135A

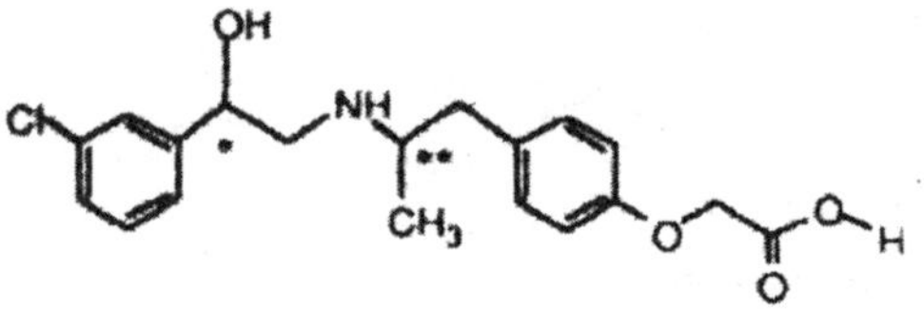

BRL37344

	•	••	
BRL35135A	R	R	} (RR,SS)diastereomer
	S	S	
BRL37344	R	R	} (RR,SS)diastereomer
	S	S	
(RR)BRL37344	R	R	
(RS)BRL37344	R	S	
(SR)BRL37344	S	R	
(SS)BRL37344	S	S	

Figure 8 Chemical structure and stereochemical aspects of BRL37344 and BRL35135A. (From Ref. 126.)

inhibitor, which possesses a secondary alcohol group like that of BRL37344, exhibits chiral inversion through a carbonyl intermediate in a number of species [128].

Tolperisone is an arylalkyl β-aminoketone having an asymmetric carbon atom α to the carbonyl group. The R(−)-enantiomer is a centrally acting muscle relaxant used in the treatment of spastic paralysis, whereas the S(+)-enantiomer shows broncho and peripheral dilating properties.

In rat studies, there was mutual chiral inversion both in vivo and in vitro in plasma and blood [129]. The mechanisms for this inversion are not clear. Lanperisone and eperisone are newly developed members of this group of drugs with lanpersione being developed as a pure R(−)-enantiomer. Currently no information regarding bioinversion or conformation stability of these drugs is available.

Lifibrol is a new drug developed for hypercholesteremia containing a secondary alcohol [130]. In dogs, chiral inversion of the S-enantiomer but not the R-enantiomer was observed. The mechanism is unidirectional, but the magnitude of the process following oral administration compared to intravenous administration suggests possible presystemic metabolism [130]. Synephrine (N-methyl-4,[beta]-dihydroxyphenylethylamine) is a synthetic sympathomimetic drug containing a secondary alcohol and is a component of herbal drugs, Chinese medicines, and citrus plants. It is a chiral compound but is present in nature as the *l*-synephrine. Following ingestion of *l*-synephrine, it is· postulated that chiral inversion occurs, resulting in inversion to *d*-synephrine in urine [131,132].

CS-670 is a racemic NSAID which is readily metabolized to active metabolites trans and unsaturated mono-ols. When *cis*-OH is administered to rats, approximately 9% of the trans diols were excreted in urine. However, the production of *cis*-diols from *trans*-OH was only 0.51% of the dose. The chiral inversion from *cis*-OH to *trans*-OH has been suggested to occur through the saturated ketone intermediate [63,133] (Fig. 9). Nevertheless, as the trans and the unsaturated alcohol (Fig. 9) are active anti-inflammatory analgesics and the cis alcohol has immuonomodulating activity, it is very important to evaluate the disposition of each enantiomer in order to understand pharmacological and toxicological properties of CS-670 [62,133].

Other oxidoreduction examples of inversion are the sulfoxide/sulfide and N-oxide/tertiary amine equilibria. Chiral sulfoxides (e.g., sulindac, flosequinan, pantoprazole) that contain tricoordinated sulfur atoms or tertiary amine N-oxides are first reduced by reductases to the corresponding achiral sulfides or tertiary amines, respectively. The reduced metabolites are nthen reoxygenated by monooxygenases back to the parent compound, which can result in an altered ratio of the two enantiomers or epimers.

For sulfoxides/sulfides, oxidation is catalyzed by cytochrome P-450 and flavin monooxygenases, whereas the reductive metabolism is catalyzed by aldehyde oxidase and/or thiotedocin-linked enzymes. The liver as well as the gut and bacterial flora are potential sites for the formation of sulfide metabolites.

Sulindac, also known as sulindac sulfoxide, is a geometric isomer marketed as the cis configuration. It is marketed as a racemate chiral

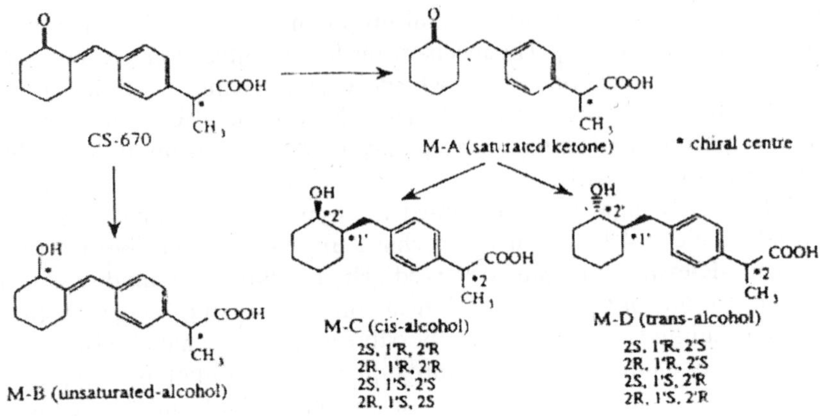

Figure 9 The main metabolic pathway of CS-670 in humans. (From Ref. 133.)

sulfoxide metabolized to the sulfide and sulfone metabolites (Fig. 10). When the metabolite sulindac sulfide was administered to rabbits, there was stereoselective enrichment to 66% R(+)-sulfoxide sulindac [134] (Fig. 10). In rat liver microsomes, the flavin adenine dinucleotide containing monooxygenase also oxygenates the divalent sulfur of sulindac sulfide to yield a single R(+) isomer of the sulfoxide [135]. In humans, acutely and chronically dosed volunteers show an enrichment of R-sulindac sulphoxide in serum and urine after administration of the racemic chiral sulfoxide [136]. Further studies in human liver and kidney microsomes demonstrated that the biotransformation of the sulfide to the sulfoxide is predominantly by flavin containing monooxygenases [137]. Racemic sulindac sulphoxide has activity against 78 kDa gastrin binding protein, which might contribute to the reduction of colorectal carcinoma in vivo [138]. There is currently no information available on the activity of the enantiomers in colorectal carcinoma.

Pantoprazole is a substituted benzimidazole sulfoxide proton pump inhibitor (Fig. 11). Like other proton pump inhibitors such as lansoprazole, all chiral benzimidazoles are administered as racemic mixtures. Pantoprazole is metabolized to pantoprazole sulphone as a major metabolite, and pantoprazole sulfide as a minor metabolite (Fig. 11). The reoxidation of pantoprazole sulfide to pantoprazole occurs in vivo, resulting in chiral inversion of pantoprazole enantiomers. Significant chiral inversion occurred after intravenous and oral administration of (+)-pantoprazole [139]. The mechanism of this inversion has not yet been identified.

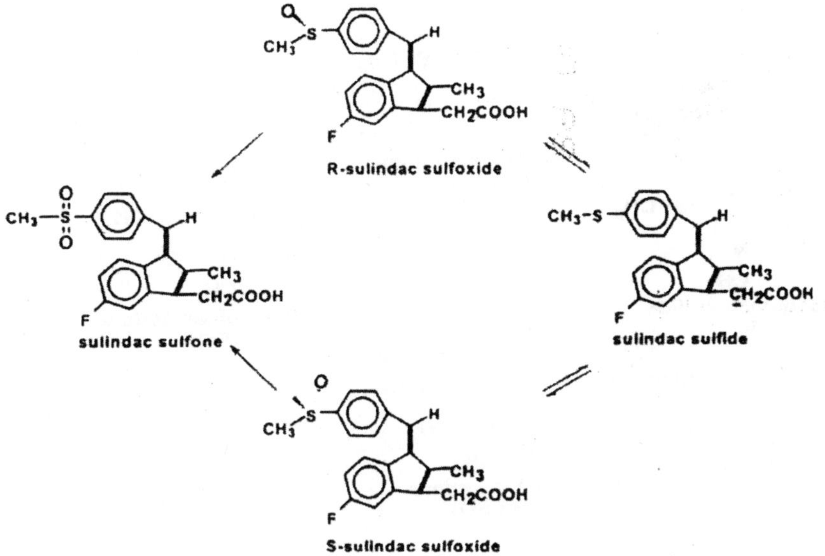

Figure 10 Oxidative and reductive human biotransformations of sulindac in vivo. Racemic sulindac sulfoxide, a prodrug, is reversibly reduced to the active sulfide metabolite, which is reoxidized to either the R- or the S-sulfoxide enantiomer. The sulfoxides are also irreversibly oxidized to a sulfone metabolite. The reduction of the sulfoxides to sulfide is a nonmicrosomal process and therefore is not a confounding parameter in determining microsomal oxidation. (From Ref. 137.)

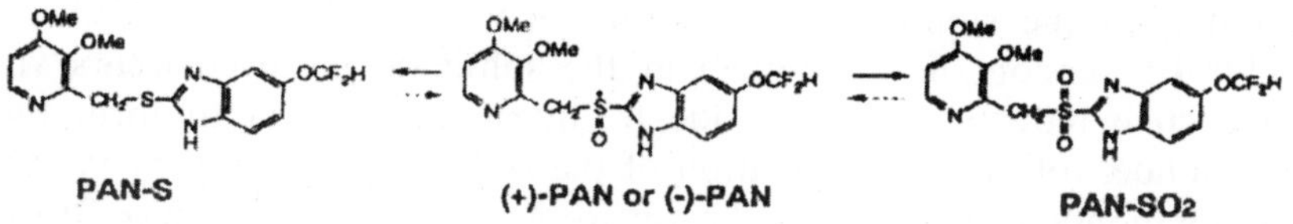

Figure 11 Possible metabolic pathways of pantoprazole. * Position of chiral center. A solid arrow means an identified pathway, and a broken arrow means an unidentified one. (From Ref. 139.)

Flosequinan is also a chiral sulfoxide that is a peripheral vasodilator. Following reduction of the sulfoxide group, the sulfide metabolite is formed, and it is subsequently oxidized to both flosequinan enantiomers (Fig. 12) [140,141]. This chiral inversion occurs in normal rats but not in either antibiotic-treated or germ-free rats after oral administration of each enantiomer. Inversion also appears to be greater when administered orally

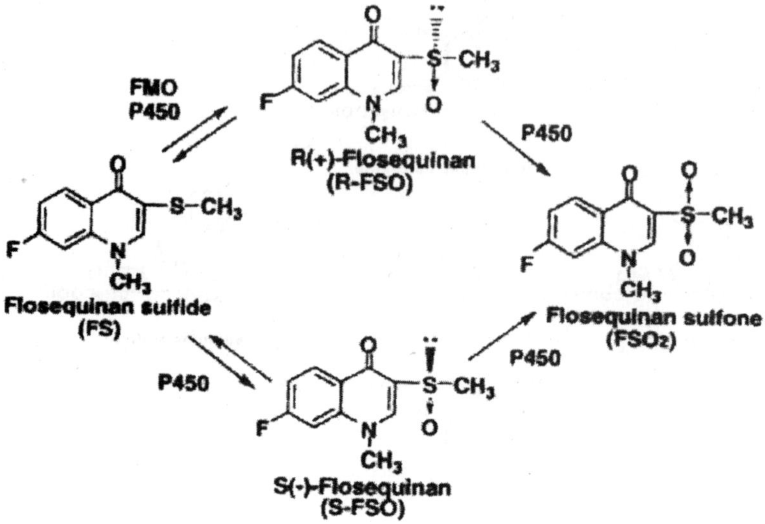

Figure 12 Proposed metabolic pathways and enzymes involved in the metabolism of flosequinan in rats. (From Ref. 142.)

rather than intravenously. Chiral inversion mediated by intestinal bacteria has been postulated, and it has been demonstrated that several strains of facultative anaerobes can reduce flosequinan to the sulfide stereoselectively. Also, it has been shown that some bacteria exclusively reduce S(−)-flosequinan to the sulfide. Based on this experimental data, it appears that chiral inversion of flosequinan at the sulfoxide position occurs via stereoselective reduction of sulfoxide by intestinal bacteria to form the achiral sulfide, followed by oxidation of the sulfide in the body by flavin-containing monooxygenase A1 and cytochrome P450 3A2 to produce R(+)- and S(−)-flosequinan [141,142]. Furthermore, kidney and lung microsomes catalyzed flosequinan to R(+)-flosequinan stereoselectively [143].

9. α-AMINO ACIDS

It was originally thought that only L-enantiomers of amino acids were present in mammalian tissues. There is now considerable evidence for D-enantiomers of amino acids in physiological fluids and tissues of man, and they have for the most part unknown physiological functions [144] (Table 2).

Table 2 Essential and Nonessential Chiral Amino Acids

Essential			Nonessential	
Isoleucine*	Phenylalanine	Alanine	Cysteine	Histidine
Leucine	Threonine*	Arginine	Glutamine	Proline
Lysine	Tryptophan	Asparagine	Glutamic acid	Serine
Methionine	Valine	Aspartic acid	Glycine	Tyrosine

*Have two asymmetric carbon atoms rather than one.

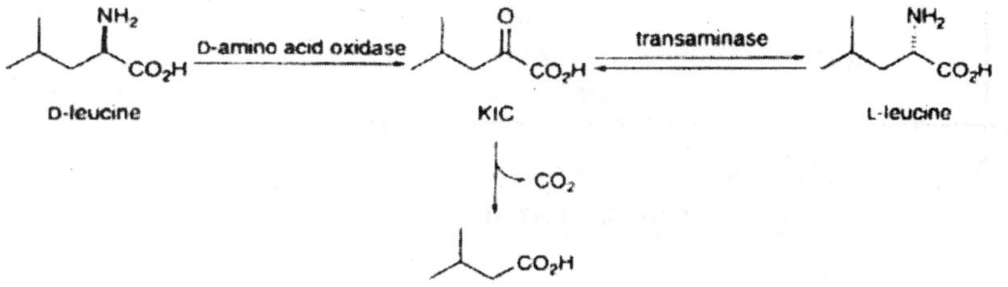

Figure 13 Proposed metabolic pathway of *d*-leucine. (From Ref. 146.)

The branched-chain amino acids L-leucine, L-isoleucine, and L-valine are not synthesized de novo in mammalian species, and thus diet and metabolism of proteins are the only source of these amino acids. The nutritional aspects of D-amino acids are not known, but they may impair digestibility and nutritional quality. Use of D-branched-chain amino acids for growth or maintenance of nitrogen equilibrium depends on the ability of transformation to the L-isomer.

There are several examples of D to L inversion of amino acids in the literature. D-Phenylalanine may have therapeutic properties in endogenous depression and is converted to L-phenylalanine in humans [145]. D-Leucine is inverted to the L-enantiomer in rats. When D-enantiomer is administered, about 30% of the enantiomer is converted to the L-enantiomer with a measurable inversion from L to D-enantiomer. As indicated in Fig. 13, D-leucine is inverted to the L-enantiomer by two steps. It is first oxidized to α-ketoisocarproate (KIC) by D-amino acid oxidase. This α-keto acid is then asymmetrically reaminated by transaminase to form L-leucine. In addition, KIC may be decarboxylated by branched-chain α-keto acid dehydrogenase, resulting in an irreversible loss of leucine (Fig. 13) [146]. D-Valine undergoes a similar two-step inversion process, and this can be antagonized by other amino acids such as D-leucine. The primary factor appears to be interference with the deamination process [147].

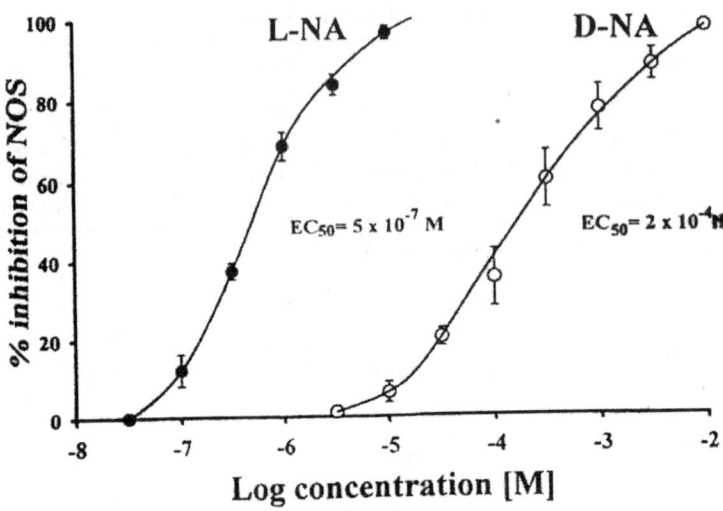

Figure 14 In vitro dose-dependent inhibition of brain NOS activity by nitro-D-arginine (D-NA) or nitro-L-arginine (L-NA). D-NA was 400 times less potent than L-NA (calculated from EC_{50}). (From Ref. 148.)

Nitric oxide synthase (NOS) is an enzyme widely expressed in mammals; it produces nitric oxide by the oxidation of L-arginine to NO and citrulline. Nitro-L-arginine is one of many inhibitors of NOS used experimentally to examine the physiological roles of NO. It is an assumption that D-arginine analogs do not affect NOS activity, and they are thus often used as control agents in experiments using L-arginine analogs to inhibit NOS. However, this assumption does not always appear to be correct. Nitro-D-arginine was 400 times less potent than nitro-L-arginine in in vitro experiments (Fig. 14). However, when nitro-D-arginine was injected in vivo to rats, the L-enantiomer was found to appear rapidly in plasma samples [148] (Fig. 15). There was no bioinversion from L to D evident [148]. This example illustrates that caution must be exercised when interpreting pharmacodynamic data related to systemic administration of D-arginine analogs and perhaps other D-amino acids analogs.

Other amino acids such as L-aspartate and L-serine, can spontaneously racemize with aging in humans [149]. It has been reported that microwaving milk racemizes L-proline to D-proline, which could reduce the milk's nutritional value [150]. For L-serine, a serine racemase has been characterized that catalyzes the direct racemization of L-serine to D-serine. This enzyme can also catalyse D-serine to L-serine but with lesser affinity. As a number

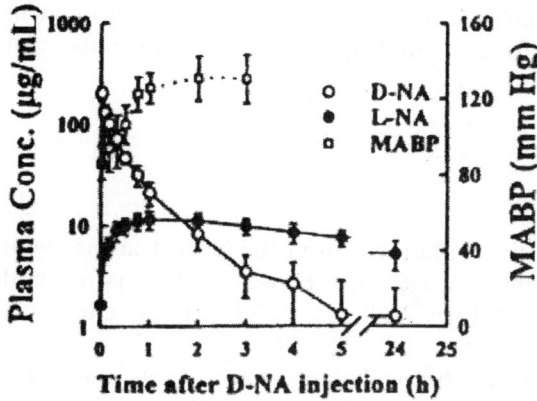

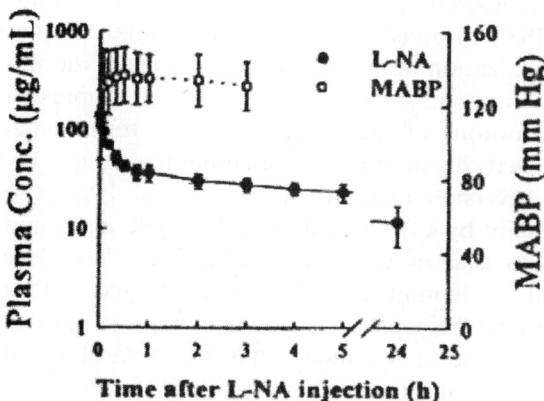

Figure 15 The in vivo conversion of nitro-D-arginine (D-NA) to nitro-L-arginine (L-NA) and the corresponding mean arterial blood pressure (MABP, dotted curve) following acute injection of D-NA (zero time point) (upper panel). The plasma levels of L-NA and blood pressure change after acute L-NA injection (lower panel). There was no conversion of L-NA to D-NA (lower panel). (From Ref. 148.)

of amino acid racemases have been purified in bacteria, this type of racemization may also be evident in mammalian species [151].

The inversion between D- and L- amino acids may be therapeutically important, as some D-amino acids have pronounced immunosuppressive effects [152]. With the advent of biologically active peptides and proteins, the study of racemization of amino acids and consideration of conformational transitions and racemization during process development and in drug development and formulation may be therapeutically important and should not be overlooked [153].

10. THALIDOMIDE

Thalidomide is a former sedative withdrawn from the market in the 1960s due to the formation of severe teratogenicity. However, there is renewed interest in the immonomodulatory and anti-inflammatory effects of thalidomide.

Thalidomide undergoes facile chemical racemization in aqueous media, and it appears that this is a base-catalyzed process [154]. Basic amino acids show a high catalytic potency for thalidomide [155]. A greater extent of racemization has also been observed in human citric plasma compared to buffer [154,155]. Further investigations using physiological concentrations of human serum albumin (HSA) demonstrated that it is the albumin that drastically accelerates the racemization of thalidomide. The rates of racemization differ between rabbit and human plasma and appear to be suppressed by adding increasing concentrations of fatty acid ions. The microsomal enzymes do not have the capacity to racemize the enantiomers as quickly as HSA [154]. The rate of inversion increased with pH and HSA and was impaired to various extents by capric acid, acetylsalicylic acid, and physostigmine. It also appears that in vivo chiral inversion takes place mainly in the circulation and in albumin-rich extravascular spaces with a high concentration of albumin, while inversion is slower in more peripheral sites [156,157]. A similar racemization phenomenon has also been observed for a thalidomide analog, EM12, both in vitro and in vivo [158,159].

Chiral inversion of thalidomide is dependent on the medium employed, and it would be misleading to draw conclusions about in vivo selectivity from in vitro studies. Putative differences in therapeutic or adverse effects of the enantiomers of thalidomide would be abolished by rapid interconversion in vivo and render the development of a streochemically pure enantiomer ineffective.

11. STIRIPENTOL: OXIODOREDUCTION/CONJUGATION/ RACEMIZATION?

Stiripentol is a chiral allylic alcohol that is a novel antiepileptic agent currently undergoing clinical evaluation (Fig. 16). When administered

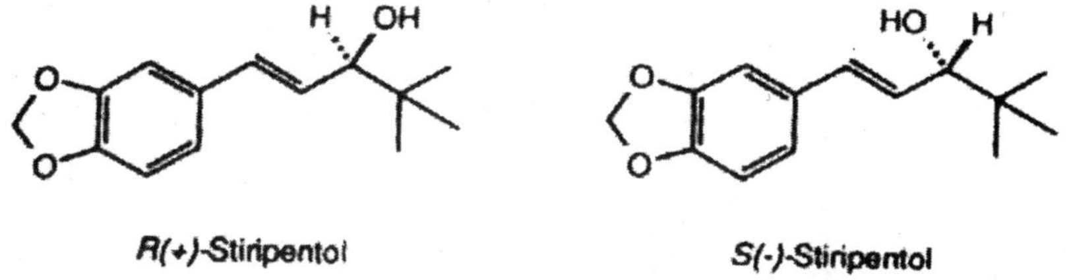

R(+)-Stiripentol   S(-)-Stiripentol

Figure 16 Structures of (R)- and (S)-stiripentol. (From Ref. 161.) http://www.tandf.co.uk/journals

orally, the R-stiripentol enantiomer is extensively transformed into to its antipode unidirectionally (Fig. 17).

The side-chain secondary alcohol of stiripentol undergoes partial oxidation to the corresponding α,β-unsaturated ketone with enantioselective oxidation or reduction of the ketone. Studies have revealed that this inversion process is dependent on the presence of the side-chain carbon–carbon double bond. In the pharmaceutical development of chiral drugs with side-chain secondary alcohols, inversion is apparent, unlike in chiral drugs with saturated alcohols, where the inversion process does not appear to occur [160].

Pretreatment of rats with pentachlorophenol, an inhibitor of sulfation and perhaps of other conjugation reactions, led to a marked decrease in the inversion, suggesting that this chiral inversion of stiripentol is at least in part mediated by an enantioselective conjugation mechanism [160]. Further studies examining the influence of the route of administration demonstrated that this inversion phenomenon was not observed when stiripentol was administered intravenously or intraperitoneally (Fig. 17). Finally, in vitro experiments suggest that stiripentol may undergo partial acid-catalyzed racemization in the rat stomach before absorption from the GI tract [161].

12. THERAPEUTIC IMPLICATIONS OF BIOINVERSION

Bioinversion can lead to variability in both the pharmacokinetics and the pharmacodynamics of chiral drugs. Bioinversion also has implications for preclinical screening and for safety evaluation, and it is a source of variability in response. As chiral inversion may occur for some steroisomeric compounds, an examination of pharmacokinetics and pharmacodynamics both in vitro and in vivo, after the administration of the racemate and the enantiomers, is necessary.

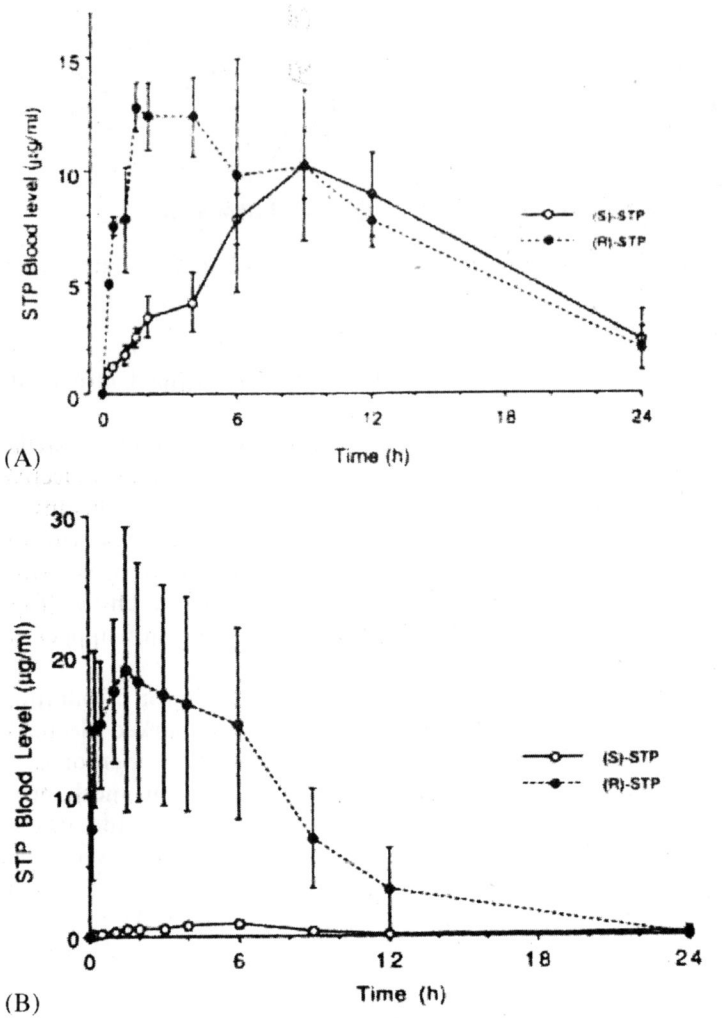

Figure 17 Concentration versus time profiles of the enantiomers of stiripentol (STP) in rat blood following oral administration of (A) (R)-stiripentol (300 mg/kg) and (B) intraperitoneal (R)-stiripentol. Data represent mean ± SD ($N = 3$). (From Ref. 161.) http://www.tandf.co.uk/journals

As some enantiomers are able to undergo a unidirectional inversion, it could be argued that one enantiomer could serve as a prodrug for the other enantiomer. Justification of this argument is only plausible if the extent of inversion is substantial and predictable and the time course to completion is reasonably rapid. Otherwise, unknown variables are introduced that will make dosage regimen predictions difficult.

With respect to 2-APA NSAIDs, diseases such as diabetes and obesity [162] have also been shown to alter the inversion process. In addition, other disease states such as renal failure may interfere with the inversion. For example, S-ketoprofen and its acylglucuronide accumulate in end-stage renal failure, which is consistent with amplification of chiral inversion subsequent to futile cycling [120].

More recent studies have demonstrated that the pattern of stereo-selectivity in serum ibuprofen concentration can be reversed by surgery so that S-ibuprofen is predominant in the control group phase of the study, but not in the postsurgery phase, which suggests the possibility of reduced metabolic chiral inversion after surgery (Fig. 18). This reduced inversion may lead to delayed response and possible treatment failure in patients taking ibuprofen for pain relief after surgery and may have implications for other disease treatments [163].

Differences in absorption phase of ibuprofen dosage forms of sus-pensions and tablets have also been shown to affect the degree of R to S inversion. It is suggested that after administration of tablet, the residence

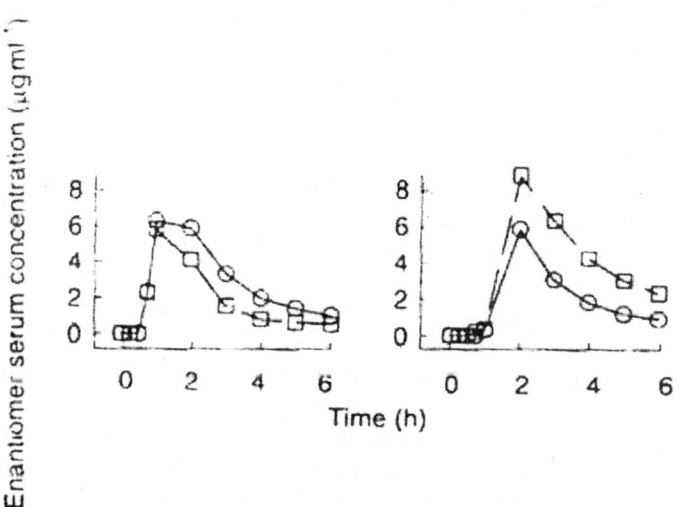

Figure 18 Individual serum ibuprofen enantiomer concentration time curves following oral administration of 200 mg racemic ibuprofen 1 week before (presurgery, left panel) and at the time of first experienced moderate to severe pain (postsurgery, right panel). S-ibuprofen (O), R-ibuprofen (□). (From Ref. 163.)

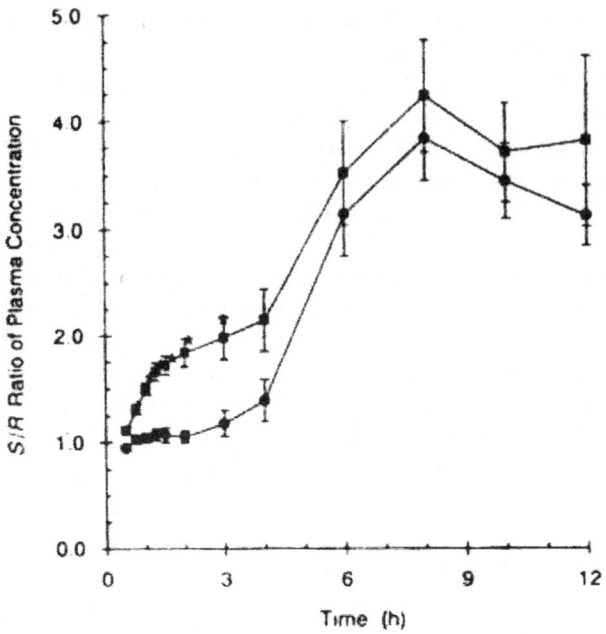

Figure 19   Time course of plasma concentration ratio between S-ibuprofen and R-ibuprofen in the suspension study (■) and tablet study (●). (From Ref. 164.)

time of the dosage form in the stomach is increased, leading to an increased absorption in the stomach and a reduction in the amount delivered to and absorbed or inverted in the intestine [164] (Fig. 19). In this situation, administration of the stereochemically pure S-enantiomer may be justifiable, as it would reduce the formulation-dependent variability in the concentration of the other enantiomer in the body.

Furthermore, drugs or diseases that elevate or reduce CoA thioester formation may also affect metabolic inversion of 2-APA NSAIDs that undergo inversion by conjugation through CoA. These interactions may be due to other drugs competing for the formation of a CoA thioester, sequestering CoA and consequently reducing the 2-APA CoA formation, or sequestering CoA and stimulating 2-APA CoA hydrolysis.

The data presented in this chapter shows that chiral inversion may be modulated by various disease states and pathological conditions as well as by drug formulations. Due to these and other factors, there has been much interest on the part of pharmaceutical companies to develop stereochemically pure enantiomers. For drugs that undergo chiral inversion, the advantages of marketing stereochemically pure drugs are (1) reducing the total dose given, (2) simplifying pharmacokinetic and pharmacodynamic data analysis, (3) removing a source of intersubject variability, and (4) minimizing toxicity of the discarded enantiomer.

However, the issues of chiral drug development must also be analyzed from a pharmacological perspective. With regards to chiral NSAIDs, it must be recognized that the powerful anti-inflammatory properties of NSAIDs may in part be mediated by processes *independent* of cyclooxygenase inhibition. For instance, equipotent inhibition of the synthesis of 5-lipoxygenase by release of leukotriene B_4 and blocking the release from polymorphonuclear leucocytes of β-glucuronidase occurs with similar IC_{50} values for S- and R-ibuprofen [165]. Recent investigations have indicated that the R-enantiomers do contribute to pharmacological activity of these agents. R-flurbiprofen has been shown to be effective in models of pain and nociception in the rat [166–170]. R-flurbiprofen has been shown to have antineoplastic properties in both experimental colorectal and prostate cancer [171,172]. R-ketoprofen is antinociceptive in the guinea pig [173]. A contribution of the R-enantiomer of ibuprofen to therapeutic effects is not only by bioinversion to the S-enantiomer but also via inhibition of cyclooxygenase-2 isoform by R-ibuprofenoyl-CoA thioester [174]. The findings that the R-enantiomers of NSAIDs may have pharmacological effects independent of chiral inversion suggests that an a priori assumption should not be made that the stereochemically pure enantiomer will be necessarily safer or more effective than the racemate. Therefore a decision to employ either the racemate or the stereochemically pure enantiomer in the drug development process should be based on an examination of the relevant and possibly diverse mechanisms of action rather than on the single most obvious effect of chiral xenobiotics that undergo inversion.

It should also be kept in mind that certain analytical and pharmacokinetic considerations may also influence the evaluation of bioinversion data. In order to assess bioinversion, researchers have administered the stereochemically pure enantiomer alone; but pharmacokinetic data demonstrating an absence of the antipode in the bloodstream does not necessarily imply absence of inversion. Undetectable plasma concentrations of enantiomers may be due to insensitivity of the assay, and inversion in the urinary system may also occur, which necessitates urinary excretion data to determine the extent of the inversion. For example, in humans there is little stereoselectivity in the plasma time course for ketoprofen. However, urinary excretion favors the S-enantiomer, which may be the result of inversion [120].

13. CONCLUSION

Awareness and appreciation in the drug development process of conformational stability of chiral drugs may have significant bearing on the interpretation of pharmaceutical, pharmacokinetic, and pharmacodynamic data. This brief overview has highlighted several examples of chiral inversion that can occur in the metabolism of racemic xenobiotics. The decision as to the relative value of marketing a racemate or a pure enantiomer is multifactorial and depends on the magnitude and significance of the pharmacokinetic and pharmacodynamic advantages, the clinical advantages, and the marketing and patenting advantages and must be judged for each drug separately. It is evident that for some racemic compounds chiral bioinversion studies may be central to decision making in this area. In the development of stereochemically pure drugs and racemates, chiral inversion must be taken into account ab initio in the development process, and there is ample justification for detailed consideration of inversion at every stage though drug development and safety evaluation.

REFERENCES

1. Polhuijs, M.; te Koppele, J.M.; Fockens, E.; Mulder, G.J. Glutathione conjugation of the alpha-bromoisovaleric acid enantiomers in the rat in vivo and its stereoselectivity. Pharmacokinetics of biliary and urinary excretion of the glutathione conjugate and the mercapturate. Biochem. Pharmacol. **1989**, *38*, 3957–3962.
2. Polhuijs, M.; Mulder, G.J.; Meyer, D.J.; Ketterer, B. Stereoselective conjugation of 2-bromocarboxylic acids and their urea derivatives by rat liver glutathione transferase 12-12 and some other isoforms. Biochem. Pharmacol. **1992**, *44*, 1249–1253.
3. Polhuijs, M.; Lankhaar, G.; Mulder, G.J. Relationship between glutathione content in liver and glutathione conjugation rate in the rat in vivo. Effect of buthionine sulphoximine pretreatment on conjugation of the two 2-bromoisovalerylurea enantiomers during intravenous infusion. Biochem. J. **1992**, *285* (2), 401–404.
4. Polhuijs, M.; Gasinska, I.; Cherry, W.F.; Mulder, G.J.; Pang, K.S. Stereoselectivity in glutathione conjugation and amidase-catalyzed hydrolysis of the 2-bromoisovalerylurea enantiomers in the single-pass perfused rat liver. J. Pharmacol. Exp. Ther. **1993**, *265*, 1406–1412.
5. Hutt, A.J.; Caldwell, J. The importance of stereochemistry in the clinical pharmacokinetics of the 2-arylpropionic acid non-steroidal anti-inflammatory drugs. Clin. Pharmacokinet. **1984**, *9*, 371–373.

6. Adams, S.S.; Bresloff, P.; Mason, C.G. Pharmacological differences between the optical isomers of ibuprofen: evidence for metabolic inversion of the (−)-isomer. J. Pharm. Pharmac. **1976**, *28*, 256–257.

7. Rubin, A.; Kandler, M.P.; Ho, P.P.; Bechtol, L.D.; Wolen, L.R. Stereoselective inversion of (R)-fenoprofen to (S)-fenoprofen in humans J. Pharm. Sci. **1985**, *74*, 82–84.

8. Adams, S.S.; Cliffe, E.E.; Lessel, B.; Nicholson, J.S. Some biological properties of 2-(4-isoburylphenyl)-propionic acid. J. Pharm. Sci. **1967**, *56*, 1686–1688.

9. Caldwell, J.; Hutt, A.J.; Fournel-Gigleux, S. The metabolic chiral inversion and dispositional enantioselectivity of the 2-arylpropionic acids and their biological consequences. Biochem. Pharmacol. **1988**, *37*, 105–114.

10. Mills, R.F.N.; Adams, S.S.; Cliffe, E.E.; Dickinson, A.; Nicholson, J.S. The metabolism of ibuprofen. Xenobiotica **1973**, *2*, 589–598.

11. Kaiser, D.G.; Vangeissen, G.J.; Reisher, R.J.; Wechter, W.J. Isomeric inversion of ibuprofen (R)-enantiomer in humans. J. Pharm. Sci. **1976**, *65*, 269–273.

12. Wechter, W.J.; Loughead, D.G.; Reischer, R.J.; Van Geissen, G.J.; Kaiser, D.G. Enzymatic inversion at saturated carbon: nature and mechanism of the inversion of R(−) p-iso-butyl hydratropic acid. Biochem. Biophy. Res. Commun. **1974**, *61*, 833–887.

13. Nakamura, Y.; Yamaguchi, T.; Takahashi, S.; Hashimoto, S.; Iwatani, K.; Nakagawa, Y. Optical isomerization mechanism of R(−)-hydrotropic acid derivatives. J. Pharmacobiol. Dyn. **1981**, *4*, S-1.

14. Baillie, T.A.; Adams, W.J.; Kaiser, D.G.; Olanoff, L.S.; Halstead, G.W.; Harpootlian, H.; Van Geissen, G.J. Mechanistic studies of the metabolic chiral inversion of (R)-ibuprofen in humans. J. Pharmacol. Exp. Ther. **1989**, *249*, 517–523.

15. Mayer, J.M. Stereoselective metabolism of anti-inflammatory 2-arylpropionates. Acta Pharm. Nord. **1990**, *2*, 197–216.

16. Knihinicki, R.D.; Williams, K.M.; Day, R.O. Chiral inversion of 2-arylpropionic acid non-steroidal anti-inflammatory drugs—1. In vitro studies of ibuprofen and flurbiprofen. Biochem. Pharmacol. **1989**, *38*, 4389–4395.

17. Knihinicki, R.D.; Day, R.O.; Williams, K.M. Chiral inversion of 2-arylpropionic acid non-steroidal anti-inflammatory drugs—II. Racemization and hydrolysis of (R)- and (S)-ibuprofen-CoA thioesters. Biochem. Pharmacol. **1991**, *42*, 1905–1911.

18. Tanaka, Y.; Shimomura, Y.; Hirota, T.; Nozaki, A.; Ebata, M.; Takasaki, W.; Shigehara, E.; Hayashi, R.; Caldwell, J. Formation of glycine conjugate and (−)-(R)-enantiomer from (+)-(S)-2-phenylpropionic acid suggesting the formation of the CoA thioester intermediate of (+)-(S)-enantiomer in dogs. Chirality **1992**, *4*, 342–348.

19. Menzel, S.; Waibel, R.; Brune, K.; Geisslinger, G. Is the formation of R-ibuprofenyl-adenylate the first stereoselective step of chiral inversion? Biochem. Pharmacol. **1994**, *48*, 1056–1058.

20. Brugger, R.; Garcia Alia, B.; Reichel, C.; Waibel, R.; Menzel, S.; Brune, K.; Geisslinger, G. Isolation and characterization of rat liver microsomal R-ibuprofenoyl-CoA synthetase. Biochem. Pharmacol. **1996**, *52*, 1007–10013.

21. Müller, S.; Mayer, J.M.; Etter, J.C.; Testa, B. Influence of palmitate and benzoate on the unidirectional chiral inversion of ibuprofen in isolated rat hepatocytes. Biochem. Pharmacol. **1992**, *44*, 1468–1470.

22. Sanins, S.M.; Adams, W.J.; Kaiser, D.G.; Halstead, G.W.; Hosley, J. Barnes, H.; Baillie, T.A. Mechanistic studies on the metabolic chiral inversion of R-ibuprofen in the rat. Drug Metab. Dispos. **1991**, *19*, 405–410.

23. Reichel, C.; Bang, H.; Brune, K.; Geisslinger, H.; Menzel, S. 2-Arylpropionyl-CoA epimerase: partial peptide sequences and tissue localization. Biochem. Pharmacol. **1995**, *50*, 1803–1806.

24. Reichel, C.; Brugger, R.; Bang, H.; Geisslinger, G.; Brune, K. Molecular cloning and expression of a 2-arylpropionyl-coenzyme A epimerase: a key enzyme in the inversion metabolism of ibuprofen. Mol. Pharmacol. **1997**, *51*, 576–582.

25. Erb, K.; Brugger, R.; Williams, K.; Geisslinger, G. Stereoselective disposition of tiaprofenic acid enantiomers in rats. Chirality **1999**, *11*, 103–108.

26. Yoshida, H.; Kohno, Y.; Endo, H.; Yamaguchi, J.; Fukushima, K.; Suwa, T.; Hayashi, M. Mechanistic studies on metabolic chiral inversion of 4-(4-methylphenyl)-2-methylthiomethyl-4-oxobutanoic acid (KE-748), an active metabolite of the new anti-rheumatic agent 2-acetylthiomethyl-4-(4-methylphenyl)-4-oxobutanoic acid (KE-298), in rats. Biochem. Pharmacol. **1997**, *53*, 179–187.

27. Knights, K.M.; Addinall, T.F.; Roberts, B.J. Enhanced chiral inversion of R-ibuprofen in liver from rats treated with clofibric acid. Biochem. Pharmacol. **1991**, *41*, 1775–1777.

28. Trejtnar, F.; Wsol, V.; Szotakova, B.; Skalova, L.; Pavek, P.; Kuchar, M. Stereoselective pharmacokinetics of flobufen in rats. Chirality **1999**, *11*, 781–786.

29. Sevoz, C.; Rousselle, C.; Benoit, E.; Buronfosse, T. In vitro study of fenoprofen chiral inversion in rat: comparison of brain versus liver. Xenobiotica **1991**, *29*, 1007–1016.

30. Knights, K.M.; Talbot, U.M.; Baillie, T.A. Evidence of multiple forms of rat liver microsomal coenzyme A ligase catalysing the formation of 2-arylpropionyl-coenzyme A thioesters. Biochem. Pharmacol. **1992**, *44*, 1465–1471.

31. Knights, K.M.; Jones, M.E. Inhibition kinetics of hepatic microsomal long chain fatty acid-CoA ligase by 2-arylpropionic acid non-steroidal anti-inflammatory drugs. Biochem. Pharmacol. **1992**, *44*, 2415–2417.

32. Singh, N.N.; Jamali, F.; Pasutto, F.M.; Russell, A.S.; Coutts, R.T.; Drader, K.S. Pharmacokinetics of the enantiomers of tiaprofenic acid in humans. J. Pharm. Sci. **1986**, *75*, 439–442.

33. Vakily, M.; Jamali, F. Human pharmacokinetics of tiaprofenic acid after regular and sustained release formulations: lack of chiral inversion and stereoselective release. J. Pharm. Sci. **1994**, *83*, 495–498.

34. Jamali, F. Pharmacokinetics of enantiomers of chiral non-steroidal anti-inflammatory drugs. Eur. J. Drug Metab. Pharmacokin. **1988**, *13*, 1–9.

35. Shinohara, Y.; Magara, H.; Baba, S. Stereoselective pharmacokinetics and inversion of suprofen enantiomers in humans. J. Pharm. Sci. **1991**, *80*, 1075–1078.

36. Takasaki, W.; Tanaka, Y. Application of antibody-mediated extraction for the stereoselective determination of the active metabolite of loxoprofen in human and rat plasma. Chirality **1992**, *4*, 308–315.

37. Takasaki, W.; Asami, M.; Muramatsu, S.; Hayashi, R.; Tanaka, Y.; Kawabata, K.; Hoshiyama, K. Stereoselective determination of the active metabolites of a new anti-inflammatory agent (CS-670) in human and rat plasma using antibody-mediated extraction and high-performance liquid chromatography. J. Chromatogr. **1993**, *613*, 67–77.

38. Strolm Benedetti M, Frigerio, E.; Tamassia, V.; Noseda, G.; Caldwell, J. The dispositional enantioselectivity of indobufen in man. Biochem. Pharmacol. **1992**, *43*, 2032–2034.

39. Glowka, F.K. Stereoselective pharmacokinetics of indobufen from tablets and intramuscular injections in man. Chirality **2000**, *12*, 38–42.

40. Yao, K.W.; Mao, L.F.; Lou, M.J.; Schulz, H. The relationship between mitochondrial activation and toxicity of some substituted carboxylic acids. Chem. Biol. Interact. **1994**, *90*, 225–234.

41. Hutt, A.J.; Caldwell, J. The metabolic chiral inversion of 2-arylpropionic acids—a novel route with pharmacological consequences. J. Pharm. Pharmacol. **1983**, *35*, 693–704.

42. Williams, K.; Day, R.; Knihinicki, R.; Duffield, A. The stereoselective uptake of ibuprofen enantiomers into adipose tissue. Biochem. Pharmacol. **1986**, *25*, 3403–3405.

43. Caldwell, J.; Marsh, M.V. Interrelationships between xenobiotic metabolism and lipid biosynthesis. Biochem. Pharmacol. **1983**, *32*, 1667–1672.

44. Fears, R.; Baggaley, K.H.; Alexander, R.; Morgan, B.; Hindley, R.M. The participation of ethyl 4-benzyloxybenzoate (BRL 10894) and other aryl-substituted acids in glycerolipid metabolism. J. Lipid. Res. **1978**, *19*, 3–11.

45. Sallustio, B.C.; Knights, K.M.; Meffin, P.J. The stereospecific inhibition of endogenous triacylglycerol synthesis by fenoprofen in rat isolated adipocytes and hepatocytes. Biochem. Pharmacol. **1990**, *40*, 1414–1417.

46. Carabaza, A.; Suesa, N.; Tost, D.; Pascual, J.; Gomez, M.; Gutierrez, M.; Ortega, E.; Montserrat, X.; Garcia, A.M.; Mis, R.; Cabre, F.; Mauleon, D.; Carcanico, G. Stereoselective metabolic pathways of ketoprofen in the rat: incorporation into triacylglycerols and enantiomeric inversion. Chirality **1996**, *8*, 163–172.

47. Soraci, A.; Benoit, E.; Delatour, P. Comparative metabolism of R(−)-fenoprofen in rats and sheep. J. Vet. Pharmacol. Therap. **1995**, *18*, 167–171.

48. Soraci, A.; Jassaud, P.; Benoit, E.; Delatour, P. Chiral inversion of fenoprofen in horses and dogs: an in vivo–in vitro study. Vet. Res. **1996**, *27*, 13–22.

49. Yoshida, H.; Kohno, Y.; Endo, H.; Hasegawa, M.; Yamaguchi, J.; Tomisawa, K.; Suwa, T. Identification of metabolites of KE-298, a new antirheumatic

drug, and its physiological properties in rats. Biol. Pharm. Bull. **1996**, *19*, 424–429.

50. Mayer, J.M.; Testa, B.; Roy-de Vos, M.; Audergon, C.; Etter, JC. Interactions between the in vitro metabolism of xenobiotics and fatty acids. The case of ibuprofen and other chiral profens. Arch. Toxicol. Suppl. **1995**, *17*, 499–513.

51. Foxworthy, P.S.; Perry, D.N.; Eacho, P.I. Induction of peroxisomal beta-oxidation by nonsteroidal anti-inflammatory drugs. Toxicol. Appl. Pharmacol. **1993**, *1118*, 272–274.

52. Geneve, J.; Hayat-Bonan, B.; Labbe, G.; Degott, C.; Letteron, P.; Freneaux, E.; Dinh, T.L.; Larrey, D.; Pessayre, D. Inhibition of mitochondrial beta-oxidation of fatty acids by pirprofen. Role in microvesicular steatosis due to this nonsteroidal anti-inflammatory drug. J. Pharmacol. Exp. Ther. **1987**, *242*, 1133–1137.

53. Freneaux, E.; Fromenty, B.; Berson, A.; Labbe, G.; Degott, C.; Letterton, P.; Larrey, D.; Pessayre, D. Stereoselective and nonstereoselective effects of ibuprofen enantiomers on mitochondrial beta-oxidation of fatty acids. J. Pharmacol. Exp. Ther. **1990**, *255*, 529–535.

54. Zhao, B.; Geisslinger, G.; Hall, I.; Day, R.O.; Williams, K.M. The effect of the enantiomers of ibuprofen and flurbiprofen on the beta-oxidation of palmitate in the rat. Chirality **1992**, *4*, 137–141.

55. Browne, G.S.; Nelson, C.; Nguyen, T.; Ellis, B.A.; Day, R.O.; Williams, K.M. Stereoselective and substrate-dependent inhibition of hepatic mitochondria beta-oxidation and oxidative phosphorylation by the non-steroidal anti-inflammatory drugs ibuprofen, flurbiprofen, and ketorolac. Biochem. Pharmacol. **1999**, *57*, 837–844.

56. Shirley, M.A.; Guan, X.; Kaiser, D.G.; Halstead, G.W.; Baillie, T.A. Taurine conjugation of ibuprofen in humans and in rat liver In vitro. Relationship to metabolic chiral inversion. J. Pharmacol. Exp. Ther. **1994**, *269*, 1165–1175.

57. Scheuerer, S.; Williams, K.M.; Brugger, R.; McLachlan, A.J.; Brune, K.; Day, R.O.; Geisslinger, G. Effect of clofibrate on the chiral disposition of ibuprofen in rats. J. Pharmacol. Exp. Ther. **1998**, *284*, 1132–1137.

58. Scheuerer, S.; Hall, S.D.; Williams, K.M.; Geisslinger, G. Effect of clofibrate on the chiral inversion of ibuprofen in healthy volunteers. Clin. Pharmacol. Ther. **1998**, *64*, 168–176.

59. Roy-De Vos, M.; Mayer, J.M.; Etter, J.C.; Testa, B. Clofibric acid increases the undirectional chiral inversion of ibuprofen in rat liver preparations. Xenobiotica **1996**, *26*, 571–582.

60. Hall, S.D.; Xiaotao, Q. The role of coenzyme A in the biotransformation of 2-arylpropionic acids. Chem. Biol. Interact. **1994**, *90*, 235–251.

61. Xiaotao, Q.; Hall, S.D. Modulation of enantioselective metabolism and inversion of ibuprofen by xenobiotics in isolated rat hepatocytes. J. Pharmacol. Exp. Ther. **1993**, *266*, 845–851.

62. Asami, M.; Takasaki, W.; Iwabuchi, H.; Haruyama, H.; Wachi, K.; Terada, A.; Tanaka, Y. Stereospecific taurine conjugation of the trans-OH metabolite

(active metabolite) of CS-670, a new 2-arylpropionic acid nonsteroidal anti-inflammatory drug, in dogs. Biol. Pharm. Bull. **1995**, *18*,1584–1589.

63. Asami, M.; Yamamura, M.; Takasaki, W.; Tanaka, Y. Quantitative determination of diol metabolites of CS-670, a new antiinflammatory agent, by capillary column gas chromatography–mass spectrometry. J. Chromatogr. **1995**, *665*, 107–116.

64. Konishi, T.; Nishikawa, H.; Kitamura, S.; Tatsumi, K. In vivo studies on chiral inversion and amino acid conjugation of 2-[4-(3'-methyl-2-thienyl)phenyl]-propionic acid in rats and dogs. Drug Metab. Disp. **1999**, *27*, 158–160.

65. Fournel, S.; Caldwell, J. The metabolic chiral inversion of 2-phenylpropionic acid in rat, mouse and rabbit. Biochem. Pharmacol. **1986**, *35*, 4153–4159.

66. Lee, E.J.D.; Williams, K.; Day, R.; Graham, G.; Champion, M. Stereoselective disposition of ibuprofen enantiomers in man. Br. J. Clin. Pharamcol. **1985**, *19*, 669–674.

67. Chen, C.S.; Chen, T.; Shieh, W.R. Metabolic stereoisomeric inversion of 2-arylpropionic acids. On the mechanism of ibuprofen epimerization in rats. Biochemica Biophysica Acta **1990**, *1033*, 1–6.

68. Chen, C.N.; Wang, P.C.; Song, H.F.; Liu, Y.C.; Chen, C.S. Non-invasive detection of ibuprofen in vivo 13C-NMR signals in rats. Chem. Pharm. Bull. 1006. **1996**, *44*, 204–207.

69. Chen, S.H.; Shieh, W.R.; Lu, P.H.; Harriman, S.; Chen, C.Y. Metabolic stereoisomeric inversion of ibuprofen in mammals. Biochemica Biophysica Acta **1991**, *1078*, 411–417.

70. Aberg, G.; Ciofalo, V.; Pendleton, R.; Ray, G.; Weddle, D. Inversion of (R)- to (S)-ketoprofen in eight animal species. Chirality **1995**, *7*, 383–387.

71. Jamali, F.; Lovlin, R.; Aberg, G. Bidirectional chiral inversion of ketoprofen in CD-1 mice. Chirality **1997**, *9*, 29–31.

72. Jamali, F.; Lovlin, R.; Corrigan, B.W.; Davies, N.M.; Aberg, G. Stereospecific pharmacokinetics and toxicodynamics of ketorolac after oral administration of the racemate and optically pure enantiomers to the rat. Chirality **1999**, *11*, 201–205.

73. Thomason, M.J.; Rhys-Williams, W.; Lloyd, A.W.; Hanlon, G.W. The stereo inversion of 2-arylpropionic acid non-steroidal anti-inflammatory drugs and structurally related compounds by Verticillium lecanii. J. Appl. Microbiol. **1998**, *85*, 155–163.

74. Thomason, M.J.; Rhys-Williams, W.; Hanlon, G.W.; Lloyd, A.W. Bidirectional chiral inversion of 2-phenylpropionic acid in non-growing cultures of Verticillium lecanii. J. Pharm. Pharmacol. **1995**, *47*, 1059P.

75. King, J.N.; Mauron, C.; LeGoff, C.; Hauffe, S. Bidirectional chiral inversion of the enantiomers of the nonsteroidal antiinflammatory drug oxindanac in dogs. Chirality **1994**, *6*, 460–466.

76. Bartels, M.J.; Smith, F.A. Stereochemical inversion of haloxyfop in the Fischer 344 rat. Drug Metab. Dispos. **1989**, *17*, 286–291.

77. Kemal, C.; Casida, J.E. Coenzyme A esters of 2-aryloxyphenoxypropio-nate herbicides and 2-arylpropionate antiinflammatory drugs are potent and

stereoselective inhibitors of rat liver acetyl-CoA carboxylase. Life Sci. **1992**, *50*, 533–540.

78. Drummond, L.; Caldwell, J.; Wilson, H.K. The stereoselectivity of 1,2-phenylethanediol and mandelic acid metabolism and disposition in the rat. Xenobiotica **1990**, *20*, 159–168.

79. Zheng, H.H.; Chan, K.K.; Covey, J.M. Enantioselective determination of R(+) (NSC698215) and S(−) XK469(NSC698216), antitumor agents, in rat plasma vy chiral HPLC. Pharm. Res. Suppl. 11, **2000**, A1014.

80. Tamura, S.; Kununa, S.; Kawai, K.; Kishimopto, S. Optical isomerization of R(−)-clidanac to the biologically active S(+)-isomer in guinea-pigs. J. Pharm. Pharmacol. **1981**, *33*, 701–706.

81. Menzel-Soglowek, S.; Geisslinger, G.; Beck, W.S.; Brune, K. Variability of inversion of (R)-flurbiprofen in different species. J. Pharm. Sci. **1992**, *81*, 888–891.

82. Foster, R.T.; Jamali, F. Stereoselective pharmacokinetics of ketoprofen in the rat. Influence of route of administration. Drug. Metab. Dispos. **1988**, *16*, 623–626.

83. Aboul-Enein, H.Y.; Van Overbeke, A.; Vander Weken, G.; Baeyens, W.; Oda, H.; Deprez, P.; Kruif, A. HPLC on Chiralcel OJ-R for enantiomer separation and analysis of ketoprofen, from horse plasma, as the 9-aminophenanthrene derivative. J. Pharm. Pharmacol. **1998**, *50*, 291–296.

84. Rudy, A.C.; Liu, Y.; Brater, D.C.; Hall, S.D. Stereoselective pharmacokinetics and inversion of (R)- ketoprofen in healthy volunteers. J. Clin. Pharmacol. **1998**, *38*, 3S–10S.

85. Jamali, F.; Russell, A.S.; Foster, R.T.; Lemko, C. Ketoprofen pharmacokinetics in humans: evidence of enantiomeric inversion and lack of interaction. J. Pharm. Sci. **1990**, *79*, 460–461.

86. Vakily, M.; Jamali, F. Pharmacokinetics of tiaprofenic acid in humans: lack of stereoselectivity in plasma using both direct and precolumn derivatization methods. J. Pharm. Pharm. Sci. **1996**, *85*, 638–642.

87. Vakily, M.; Khorasheh, F.; Jamali, F. Dependency of gastrointestinal toxicity on release rate of tiaprofenic acid: a novel pharmacokinetic–pharmacodynamic model. Pharm. Res. **1999**, *16*, 123–129.

88. Hutt, A.J.; Kooloobandi, A.; Hanlon, G.W. Microbial metabolism of 2-arylpropionic acids: chiral inversion of ibuprofen and 2-phenylpropionic acid. Chirality **1993**, *5*, 196–200.

89. Hanlon, G.W.; Kooloobandi, A.; Hutt, A.J. Microbial metabolism of 2-arylpropionic acids: effect of environmental parameters on the growth of Verticillium lecanii and its metabolism of ibuprofen. J. Appl. Bact. **1994**, *76*, 442–447.

90. Thomason, M.J.; Rhys-Williams, W.; Lloyd, A.W.; Hanlon, G.W. Demonstration of the chiral inversion of 2-phenylpropionic acid Verticillium lecanii. Lett. Appl. Microbiol. **1996**, *23*, 417–420.

91. Jamali, F.; Singh, N.N.; Pasutto, F.M.; Russell, A.S.; Coutts, R.T. Pharmacokinetics of ibuprofen enantiomers in humans following oral administration of tablets with different absorption rates. Pharm. Res. **1988**, *5*, 40–43.

92. Avgerinos, A.; Hutt, A.J. Interindividual variability in the enantiomeric disposition of ibuprofen following the oral administration of the racemic drug to healthy volunteers. Chirality **1990**, *2*, 249–256.

93. Borin, M.T.; Khare, S.; Beihn, R.M.; Jay, M. The effect of food on gastro-intestinal (GI) transit of sustained-release ibuprofen tablets as evaluated by gamma scintigraphy. Pharm. Res. **1990**, *7*, 304–307.

94. Geisslinger, G.; Schuster, O.; Stock, K.P.; Loew, D.; Bach, G.L.; Brune, K. Pharmacokinetics of S(+)- and R(−)-ibuprofen in volunteers and first clinical experience of S(+)-ibuprofen in rheumatoid arthritis. Eur. J. Clin. Pharmacol. **1990**, *38*,493–497.

95. Sattari, S.; Jamali, F. Evidence of absorption rate dependency of ibuprofen inversion in the rat. Chirality **1994**, *6*, 435–439.

96. Mehvar, R.; Jamali, F. Pharmacokinetic analysis of the enantiomeric inversion of chiral nonsteroidal antiinflammatory drugs. Pharm. Res. **1988**, *5*, 76–79.

97. Siemon, D.; de Vreies, J.X.; Stozer, F.; Walter-Sack, I.; Dietl, R. Fasting and postprandial disposition of R(−)- and S(+)-ibuprofen following oral administration of racemic drug in healthy individuals. Eur. J. Med. Res. **1997**, *2*, 215–219.

98. Sattari, S.; Jamali, F. Involvement of the rat gut epithelial and muscular layer, and microflora in chiral inversion and acyl-glucuronidation of R-fenoprofen. Eur. J. Drug. Metab. Pharmacokin. **1997**, *22*, 97–101.

99. Jamali, F.; Mehvar, R.; Russell, A.S.; Sattari, S.; Yakimets, W.W.; Koo, J. Human pharmacokinetics of ibuprofen enantiomers following different doses and formulations: intestinal chiral inversion. J. Pharm. Sci. **1992**, *81*, 221–225.

100. Bopp, R.J.; Nash, J.F.; Ridolfo, A.S.; Sheppard, E.R. Stereoselective inversion of (R)-(−)-benoxaprofen to the (S)-(+)-enantiomer in humans. Drug Metab. Disp. **1979**, *7*, 356–359.

101. Berry, B.W.; Jamali, F. Presystemic and systemic chiral inversion of R-(−)-fenoprofen in the rat. J. Pharmacol. Exp. Ther. **1991**, *258*, 695–701.

102. Simmonds, R.G.; Woodage, T.J.; Duff, S.M.; Green, J.N. Stereospecific inversion of (R)-(−)-benoxaprofen in rat and man. Eur. J. Drug Metab. Pharmacokinet. **1980**, *5*, 169–172.

103. Jeffrey, P.; Tucker, G.T.; Bye, A.; Crewe, H.K.; Wright, P.A. The site of inversion of R(−)-ibuprofen: studies using rat in-situ isolated perfused intestine/liver preparations. J. Pharm. Pharmacol. **1991**, *43*, 715–720.

104. Cox, S.R. Effect of route of administration on the chiral inversion of R(−)-ibuprofen. Clin. Pharmacol. Ther. **1988**, *43*, A 146.

105. Tracy, T.S.; Hall, S.D. Metabolic inversion of (R)-ibuprofen. Epimerization and hydrolysis of ibuprofenyl-coenzyme A. Drug Metab. Dispos. **1992**, *20*, 322–327.

106. Cox, J.W.; Cox, S.R.; Van Giessen, G.; Ruwart, M.J. Ibuprofen stereoisomer hepatic clearance and distribution in normal and fatty in situ perfused rat liver. J. Pharmacol. Exp. Ther. **1985**, *232*, 636–643.

107. Yasui, H.; Yamaoka, K.; Nakagawa, T. Moment analysis of stereoselective enterohepatic circulation and unidirectional chiral inversion of ketoprofen enantiomers in rat. J. Pharm. Sci. **1996**, *85*, 580–584.

108. Yasui, H.; Yamaoka, K.; Dote, N.; Nakagawa, T. Moment analysis of stereoselective biliary excretion and chiral inversion of ketoprofen enantiomers in perfused rat liver. J. Pharm. Sci. **1995**, *84*, 1327–1331.

109. Mayer, J.M.; Bartalucci, C.; Maitre, J.; Testa, B. Metabolic chiral inversion of anti-inflammatory 2-arylpropionates: lack of reaction in liver homogenates, and study of methine proton acidity. Xenobiotica **1988**, *18*, 533–543.

110. Sanins, S.M.; Adams, W.J.; Kaiser, D.G.; Halstead, G.W.; Baillie, T.A. Studies on the metabolism and chiral inversion of ibuprofen in isolated rat hepatocytes. Drug Metab. Dispos. **1990**, *18*, 527–533.

111. Müller, S.; Mayer, J.M.; Etter, J.C.; Testa, B. Metabolic chiral inversion of ibuprofen in isolated rat hepatocytes. Chirality **1990**, *2*, 740–748.

112. Tracy, T.S.; Hall, S.D. Determination of the epimeric composition of ibuprofenyl-CoA. Analyt. Biochem. **1991**, *195*, 24–29.

113. Yamaguchi, T.; Nakamura, Y. Stereoselective metabolism of 2-phenylpropionic acid in rat. II. Studies on the organs responsible for the optical isomerization of 2-phenylpropionic acid in rat in vivo. Drug. Metab. Disp. **1987**, *15*, 535–539.

114. Sevoz, C.; Benoit, E.; Buronfosse, T. Thioesterification of 2-arylpropionic acids by recombinant acyl-coenzyme A synthetases (ACS1 and ACS2). Drug Metab. Disp. **2000**, *28*, 398–402.

115. Menzel-Soglowek, S.; Geisslinger, G.; Brune, K. Metabolic chiral inversion of 2-arylpropionates in different tumor cell lines. Agents Actions Suppl. **1993**, *44*, 23–29.

116. Hall, S.D.; Hassanzadeh-Khayyat, M.; Knadler, M.P.; Mayer, P.R. Pulmonary inversion of 2-arylpropionic acids: influence of protein binding. Chirality **1992**, *4*, 349–352.

117. Millership, J.S.; Collier, P.S. Topical administration of racemic ibuprofen. Chirality **1997**, *9*, 313–316.

118. Mroszczak, E.; Combs, D.; Chaplin, M.; Tsina, I.; Tarnowski, T.; Rocha, C.; Tam, Y.; Boyd, A.; Young, J.; Depass, L. Chiral kinetics and dynamics of ketorolac. J. Clin. Pharmacol. **1996**, *36*, 521–539.

119. Grubb, N.; Hall, S.; Brater, D. The role of ketorolac adenylate in chiral inversion of ketrolac. J. Clin. Pharmacol. **1995**, *35*, A939.

120. Grubb, N.G.; Rudy, D.W.; Brater, D.C.; Hall, S.D. Stereoselective pharmacokinetics of ketoprofen and ketoprofen glucuronide in end-stage renal disease: evidence for a 'futile cycle' of elimination. Br. J. Clin. Pharmacol. **1999**, *48*, 494–500.

121. Vakily, M.; Corrigan, B.; Jamali, F. The problem of racemization in the stereospecific assay and pharmacokinetic evaluation of ketorolac in human and rats. Pharm. Res. **1995**, *12*, 1652–1671.

122. Brandl, M.; Conley, D.; Johnson, D.; Johnson, D. Racemization of ketorolac in aqueous solution. J. Pharm. Sci. **1995**, *84*, 1045–1048.

123. Takasaki, W.; Yamamura, M.; Shigehara, E.; Suzuki, Y.; Tonohiro, T.; Hara, T.; Tanaka, Y. Stereoselective pharmacokinetics of RS-8359, a selective and reversible inhibitor of A-type monoamine oxidase, in rats, mice, dogs, and monkeys. Biol. Pharm. Bull. **1999**, *22*, 498–503.

124. Naitoh, T.; Kakiki, M.; Kawaguchi, S.; Kagei, Y.; Horie, T. Stereoselective determination of a new antidepressant, E2011, and its diastereoisomer as a metabolite by high-performance liquid chromatography. J. Chromatogr. **1997**, *694*, 153–161.

125. Naitoh, T.; Kawaguchi, S.; Kakiki, M.; Ohe, H.; Kajiwara, A.; Horie, T. Species differences and mechanism of the epimerization of a new MAO-A inhibitor. Xenobiotica **1998**, *28*, 269–280.

126. Ida, K.; Hashimoto, K.; Kamiya, M.; Muto, S.; Nakamura, Y.; Kato, K.; Mizota, M. Stereoselective action of (R*,R*)-(+/−)-methyl-4-[2-[2-hydroxy-2-(3-chlorophenyl)ethylamino]propyl]-phenoxyacetic acid (BRL37344) on beta-adrenoceptors and metabolic chiral inversion. Biochem. Pharmacol. **1996**, *52*, 1521–1527.

127. Baumann, P.; Prelog, V. Reaktion mit mikroorganismen. Die stereospezifische reduction von stereoisomeren dekalindionen. Helv. Chim. Acta **1958**, *41*, 2362–2379.

128. Toshihiko, N.; Tohol, H.; Shouzi, K.; Hiroshi, O. Mechanism and species difference of enymatic chiral inversion of E2011. Xenobio. Metab. Dispos. **1993**, *8* (suppl), A290.

129. Yokoyama, T.; Fukuda, K.; Mori, S.; Ogawa, M.; Nagasawa, K. Determination of tolperisone enantiomers in plasma and their disposition in rats. Chem. Pharm. Bull. **1992**, *40*, 272–274.

130. Walters, R.R.; Hsu, C.Y.L. Chiral assay methods for lifibrol and metabolites in plasma and the observation of unidirectional chiral inversion following administration of the enantiomers to dogs. Chirality **1994**, *6*, 105–115.

131. Kusu, F.; Matsumoto, K.; Arai, K.; Takamura, K. Determination of synephrine enantiomers in food and conjugated synephrine in urine by high-performance liquid chromatography with electrochemical detection. Anal. Biochem. **1996**, *235*, 191–194.

132. Arai, K.; Jin, D.; Kusu, F.; Takamura, K. Determination of p-hydroxymandelic acid enantiomers in urine by high-performance liquid chromatography with electrochemical detection. J. Pharm. Biomed. Anal. **1997**, *15* (9–10), 1509–1514.

133. Asami, M.; Shigeta, A.; Tanaka, Y. Disposition of CS-670, a novel non-steroidal anti-inflammatory drug, and its metabolites in healthy human volunteers. Chirality **1996**, *8*, 207–213.

134. Duggan, D.E.; Hooke, K.F.; Noll, R.M.; Hucker, H.B.; Van Arman, C.G. Comparative disposition of sulindac and metabolites in five species. Biochem. Pharmacol. **1978**, *27*, 2311–2320.

135. Light, D.R.; Waxman, D.J.; Walsh, C. Studies on the chirality of sulfoxidation catalyzed by bacterial flavoenzyme cyclohexanone monooxygenase

and hog liver flavin adenine dinucleotide containing monooxygenase. Biochemistry **1982**, *21*, 2490–2498.

136. Hamman, M.A.; Hall, S.D. Sulindac sulfoxide (SOX) enantiomer disposition. FASEB J. **1993**, *7*, A482.

137. Hamman, M.A.; Haehner-Daniels, B.D.; Wrighton, S.A.; Rettie, A.E.; Hall, S.D. Stereoselective sulfoxidation of sulindac sulfide by flavin-containing monooxygenases. Comparison of human liver and kidney microsomes and mammalian enzymes. Biochem. Pharmacol. **2000**, *60*, 7–17.

138. Baldwin, G.S.; Rorison, K.A. Structural requirements for the binding of non-steroidal anti-inflammatory drugs to the 78 kDa gastrin binding protein. Biochem. Biophys. Acta **1999**, *1428*, 68–76.

139. Masubuchi, N.; Yamazaki, H.; Tanaka, M. Stereoselective chiral inversion of pantoprazole enantiomers after separate doses to rats. Chirality **1998**, *10*, 747–753.

140. Kashiyama, E.; Yokoi, T.; Todaka, T.; Odomi, M.; Kamataki, T. Chiral inversion of drug: role of intestinal bacteria in the stereoselective sulphoxide reduction of flosequinan. Biochem. Pharmacol. **1994**, *48*, 237–243.

141. Kashiyama, E.; Todaka, T.; Odomi, M.; Tanokura, Y.; Johnson, D.B.; Yokoi, T.; Kamataki, T.; Shimizu, T. Stereoselective pharmacokinetics and interconversions of flosequinan enantiomers containing chiral sulphoxide in rat. Xenobiotica **1994**, *24*, 369–377.

142. Kashiyama, E.; Yokoi, T.; Odomi, M.; Funae, Y.; Inoue, K.; Kamataki, T. Cytochrome P450 responsible for the stereoselective S-oxidation of flosequinan in hepatic microsomes from rats and humans. Drug Metab. Dispos. **1997**, *25*, 716–724.

143. Kashiyama, E.; Yokoi, T.; Odomi, M.; Kamataki, T. Stereoselective S-oxidation and reduction of flosequinan in rat. Xenobiotica **1999**, *29*, 815–826.

144. Friedman, M. Formation, nutritional value, and safety of D-amino acids. Adv. Exp. Med. Biol. **1991**, *289*, 447–481.

145. Lehmann, W.D.; Theobald, N.; Fischer, R.; Heinrich, H.C. Stereospecificity of phenylalanine plasma kinetics and hydroxylation in man following oral application of a stable isotope-labelled pseudo-racemic mixture of L- and D-phenylalanine. Clinica. Chimica. Acta **1983**, *128*, 181–198.

146. Hasegawa, H.; Matsukawa, T.; Shinohara, Y.; Hashimoto, T. Assessment of the metabolic chiral inversion of D-leucine in rat by gas chromatography–mass spectrometry combined with a stable isotope dilution analysis. Drug Metab. Disp. **2000**, *28*, 920–924.

147. Cruz, L.J.; Benson, L.; Berg, C.P. Mutual antagonism in the metabolism of D-valine and D-leucine and antagonism by their analogs. Arch. Biochem. Biophys. **1969**, *135*, 341–349.

148. Wang, Q.; Cwik, M.; Wright, C.J.; Cunningham, F.; Pelligrino, D.A. The in vivo unidirectional conversion of nitro-D-arginine to nitro-L-arginine. J. Pharmacol. Exp. Ther. **1999**, *288*, 270–273.

149. Shapira, R.; Chou, C.H. Differential racemization of aspartate and serine in human myelin basic protein. Biochem. Biophs. Res. Commun. **1987**, *146*, 1342–1349.

150. Lubec, G.; Wolf, C.; Bartosch, B. Aminoacid isomerisation and microwave exposure. Lancet **1989**, *2*, 1392–1393.

151. Wolosker, H.; Blackshaw, S.; Snyder, S.H. Serine racemase: a glial enzyme synthesizing D-serine to regulate glutamate-N-methyl-D-aspartate neurotransmission. Proc. Natl. Acad. Sci. USA **1999**, *96*, 13409–13414.

152. Inoue, Y.; Zama, Y.; Suzuki, M. "D-amino acids" as immunosuppressive agents. Jpn. J. Exp. Med. **1981**, *51*, 363–366.

153. Senderoff, R.I.; Kontor, K.M.; Kreilgaard, L.; Chang, J.J.; Patel, S.; Krakover, J.; Heffernan, J.K.; Snell, L.B.; Rosenberg, G.B. Consideration of conformational transitions and racemization during process development of recombinant glucagon-like peptide-1. J. Pharm. Sci. **1998**, *87*, 183–189.

154. Knoche, B.; Blaschke, G. Investigations on the In vitro racemization of thalidomide by HPLC. J. Chromatogr. A. **1994**, *666*, 235–240.

155. Reist, M.; Carrupt, P.A.; Francotte, E.; Testa, B. Chiral inversion and hydrolysis of thalidomide: mechanisms and catalysis by bases and serum albumin, and chiral stability of teratogenic metabolites. Chem. Res. Toxicol. **1998**, *11*, 1521–1528.

156. Eriksson, T.; Bjorkman, S.; Roth, B.; Fyge, A.; Hoglund, P. Enantiomers of thalidomide: blood distribution and the influence of serum albumin on chiral inversion and hydrolysis. Chirality **1998**, *10*, 223–238.

157. Eriksson, T.; Bjorkman, S.; Roth, B.; Fyge, A.; Hoglund, P. Stereospecific determination, chiral inversion in vitro and pharmacokinetics in humans of the enantiomers of thalidomide. Chirality **1995**, *7*, 44–52.

158. Schmahl, H.J.; Nau, H.; Neubert, D. The enantiomers of the teratogenic thalidomide analogue EM 12: 1. Chiral inversion and plasma pharmacokinetics in the marmoset monkey. Arch. Toxicol. **1988**, *62*, 200–204.

159. Schmahl, H.-J.; Heger, W.; Nau, H. The enantiomers of the teratogenic thalidomide analogue EM 12. 2. Chemical stability, stereoselectivity of metabolism and renal excretion in the marmoset monkey. Toxicol. Lett. **1989**, *45*, 23–33.

160. Zhang, K.; Tang, C.; Rashed, M.; Cui, D., Tombret, F.; Botte, H.; Lepange, F.; Levy, R.H.; Baillie, T.A. Metabolic chiral inversion of stiripentol in the rat. I. Mechanistic studies. Drug Metab. Disp. **1994**, *22*, 544–553.

161. Tang, C.; Zhang, K.; Lepage, F.; Levy, R.H.; Baillie, T.A. Metabolic chiral inversion of stiripentol in the rat. II. Influence of route of administration. Drug Metab. Disp. **1994**, *22*, 554–560.

162. Oian, X.; Hall, S.D. Enantioselective effects of experimental diabetes mellitus on the metabolism of ibuprofen. J. Pharmacol. Exp. Ther. **1995**, *274*, 1192–1198.

163. Jamali, F.; Kunz-Dober, C.M. Pain-mediated altered absorption and metabolism of ibuprofen: an explanation for decreased serum enantiomer concentration after dental surgery. Br. J. Clin. Pharmacol. **1999**, *47*, 391–396.

164. Aiba, T.; Tse, M.M.; Lin, E.T.; Koizumi, T. Effect of dosage form on stereoisomeric inversion of ibuprofen in volunteers. Biol. Pharm. Bull. **1999**, *22*, 616–622.

165. Villanueva, M.; Heckenberger, R.; Strobach, H.; Palmer, M.; Schror, K. Equipotent inhibition by R(−)-, S(+)- and racemic ibuprofen of human polymorphonuclear cell function in vitro. Br. J. Clin. Pharmacol. **1993**, *35*, 235–242.

166. Brune, K.; Beck, W.S.; Geisslinger, G.; Menzel-Soglowek, S.; Peskar, B.M.; Peskar, B.A. Aspirin-like drugs may block pain independently of prostaglandin synthesis inhibition. Experentia **1991**, *47*, 257–261.

167. Geisslinger, G.; Menzel-Soglowek, S.; Beck, W.S.; Brune, K.A. R-flurbiprofen: isomeric ballast or active entity of the racemic compound? Agents Actions Suppl. **1993**, *44*, 31–36.

168. Geisslinger, G.; Ferreira, S.H.; Menzel, S.; Schlott, D.; Brune, K. Antinociceptive actions of R(−)-flurbiprofen—a non-cyclooxygenase inhibiting 2-arylpropionic acid—in rats. Life Sci. **1994**, *54* (10), PL173–177.

169. Lotsch, J.; Geisslinger, G.; Mohammadian, P.; Brune, K.; Kobal, G. Effects of flurbiprofen enantiomers on pain-related chemo-somatosensory evoked potentials in human subjects. Br. J. Clin. Pharmacol. **1995**, *40*, 339–346.

170. Malmberg, A.B.; Yaksh, T.L. Antinociception produced by spinal delivery of the S and R enantiomers of flurbiprofen in the formalin test. Eur. J. Pharmacol. **1994**, *256*, 205–209.

171. McCracken, J.D.; Wechter, W.J.; Liu, Y.; Chase, R.L.; Kantoci, D.; Murray, E.D.; Quiggle, D.D.; Mineyama, Y. Antiproliferative effects of the enantiomers of flurbiprofen. J. Clin. Pharmacol. **1996**, *36*, 540–545.

172. Wechter, W.J.; Murray, E.D. Jr; Kantoci, D.; Quiggle, D.D.; Leipold, D.D.; Gibson, K.M.; McCracken, J.D. Treatment and survival study in the C57BL/6J-APC(Min)/+(Min) mouse with R-flurbiprofen. Life Sci. **2000**, *66*, 745–753.

173. Ghezzi, P.; Melillo, G.; Silvano, S.; Pellegrini, L.; Asti, C.; Porzio, S.; Marullo, A.; Sabbatini, V.; Caseli, G.; Bertini, R. Differential contribution of R and S isomers in ketoprofen anti-inflammatory activity: role of cytokine modulation. J. Pharmacol. Ther. **1998**, *287*, 3, 969–974.

174. Neupert, W.; Brugger, R.; Euchenhofer, C.; Brune, K.; Geisslinger, G. Effects of ibuprofen enantiomers and its coenzyme A thioesters on human prostaglandin endoperoxide synthases. Br. J. Pharmacol. **1997**, *122*, 487–492.

9
Bioequivalency Determination of Racemic Drug Formulations: Is Stereospecific Assay Essential?

Aziz Karim
Takeda Pharmaceuticals North America, Inc., Lincolnshire, Illinois, U.S.A.

Cherukury Madhu
Pfizer, Inc., Ann Arbor, Michigan, U.S.A.

Chyung Cook
Baxter Healthcare, Round Lake, Illinois, U.S.A.

1. INTRODUCTION

Chemical synthesis results in the formation of an optically inactive racemate containing an equal amount of two stereoisomers with identical physico-chemical properties but differing in their property to rotate the plane of polarized light: one enantiomer will rotate the plane in the right direction, (dextrorotatory, d or +) while its antipode will rotate in the opposite direction with the same magnitude (levorotatory, l or −). This optical characteristic, however, does not reflect the three-dimensional configuration of the molecule. A preferred nomenclature system incorporating the three-dimensional configuration of the chiral molecule is based on assigning a priority to atoms attached to the racemic carbon according to the Cahn–Ingold–Prelog rules [1]. Prefixes R- (rectus) and S- (sinister) are assigned to the enantiomers on the basis of their absolute configuration. No relation-·ships exist between the d and l versus R and S nomenclatures. For example, the active S-enantiomer of ibuprofen is dextrorotatory (d or +) while that of verapamil, also an S-enantiomer, is levorotatory (l or $-$). The more active enantiomer is termed the "eutomer," and the less active the "distomer." The distomer is often incorrectly viewed as a passive component of the racemate with little pharmacological or pharmacokinetic (PK) significance. However, in some cases, the distomer may act as an agonist or antagonist at the receptor site or compete for drug metabolizing enzymes and binding sites. In other cases, the two stereoisomers may have different pharmacological effects, and administration of a racemate would represent a combination drug product (e.g., indacrine [2]).

When two or more chiral centers are present in a molecule, then a mixture of diastereoisomers with different physicochemical and maybe different pharmacological properties is obtained. While there are limited examples of stereoselective bioavailability and pharmacokinetic studies of diastereoisomers, this group of stereoisomers has the greatest potential for displaying substantial stereospecificity in their pharmacological responses, and therefore a stereoselective assay for diastereoisomers should be considered in bioavailability and PK studies.

Bioavailability assessment of new drug products and formulations requires assessing the time course of the active moiety in the systemic circulation [3,4]. Large numbers of marketed and investigational drugs contain one chiral center and are administered as racemates. The importance of stereoselective assays in pharmacokinetics (PK), pharmacodynamics (PD), and drug–drug interaction studies is widely recognized, and several recent reviews [5–14] have highlighted how stereoselective disposition and/or dynamics were important in explaining anomalous results obtained on the concentration–effect relationship and drug–drug interaction with racemic drugs. The use of stereoselective assays in comparative bioavailability studies, however, remains controversial [15–29]. Those arguing in favor of the stereoselective assays reason that with advances in analytical methodologies, enantiospecific assays are now widely available and should be used if the predominant pharmacological activity (or toxicity) of the racemate drug is associated with one enantiomer. Those against the use of stereoselective assays reason that such assays are not necessary because the administered racemate contains the same proportion of each enantiomer, and in bioequivalency studies one compares the rate and extent of the drug availability under identical conditions in the same subject. Establishment of bioequivalency by nonspecific (referred to in this chapter as total or R + S) assay therefore assures the bioequivalency of the active enantiomer, and enantiospecific assays would simply add to the cost of the study and drug development.

Both of the above arguments have some merit. It is therefore necessary to consider each racemic drug individually to decide whether some unique characteristics of the individual enantiomers warrant the use of an enantiospecific assay. This chapter will critique various recommendations and proposals that have been made to decide if stereospecific assay is needed in bioequivalency determinations of oral dosage forms containing racemic drugs. Before presenting these proposals, a brief discussion of some important concepts in stereoselective PK such as enantiomer inversion, stereospecific first-pass metabolism, enantiospecific absorption, and chiral excipients will be presented, since these concepts are critical in decision making on the need for a stereoselective assay.

2. ENANTIOMER INTERCONVERSION

Enantiomer interconversion is found for majority of 2-arylpropionate nonsteroidal anti-inflammatory drugs [30–36] (NSAIDs) and is discussed in Chap. 8 in detail. While in vitro prostaglandin synthetase inhibition with some NSAIDs is related to the S-isomer, administration of separate isomers or racemates produce similar in vivo activities. This anomalous finding is due to the thioester-mediated conversion of the inactive R-isomer to the active S-isomer [31]. The conversion of the active S- to inactive R-isomer is not possible owing to the inability of the S-isomer to form the thioester. Ibuprofen, fenoprofen, and benoxaprofen undergo extensive enantiomer conversion, while flurbiprofen, indoprofen, flunoxaprofen, and tiaprofenic acid undergo limited enantiomer conversion [30,31,33]. Since the R-isomer of NSAIDs may be considered inactive, it is possible that with those NSAIDs exhibiting little enantiomer interconversion, equal therapeutic effects can be obtained by administering only the S-isomer at lower doses than that required for the racemic drug, leading to possible reduction in NSAIDs-related gastrointestinal (GI) adverse effects. However, as demonstrated in Chap. 8, the inactivity of the R-enantiomers of NSAIDs may not be entirely true.

There is some evidence that enantiomer interconversion with some NSAIDs occur predominantly in the GI tract. For example with benoxaprofen, the R to S conversion, based on enantiomer AUC measurement, occurs more after oral administration than after intravenous (i.v.) or intraperitoneal (i.p.) administration in rats [34], indicating the importance of the GI tract in this conversion. With ketoprofen in rats, the R/S AUC ratio is similar after i.v. and i.p. administration and is about three times lower compared to this ratio obtained after oral administration, suggesting the importance of the GI tract in the enantiomer interconversion [35,36].

3. STEREOSELECTIVE FIRST-PASS METABOLISM

Many beta-blockers and calcium channel blockers are administered orally as racemates, and they have high oral absorption but low systemic availability of the active moiety owing to high hepatic first-pass metabolism (see Chap. 7 for more detail). If there is high enantioselective first-pass metabolism, and if the enantiomers have different pharmacological characteristics, then the PK/PD study to evaluate the relationship between plasma concentration and response, when a nonspecific assay is used, will depend on the route of drug administration. This phenomenon occurs with racemic propranolol [37–42] and verapamil [43–54].

When both enantiomers have very high hepatic extraction ratios following i.v. administration, they will have similar total body clearance, which will approach hepatic plasma or blood flow. Owing to this flow-limited clearance, differences in intrinsic hepatic clearances of enantiomers may not be evident from their clearance values obtained after i.v. dosing. After oral administration, however, intrinsic clearance differences may lead to large variations in the enantiomer systemic availabilities. This phenomenon is illustrated by propranolol, because the plasma concentration–time curves of the two enantiomers are similar after i.v. administration of racemic drug, systemic clearance of each enantiomer being about 1 L/min approaching hepatic blood flow [38]. However, after oral administration of racemic propranolol, much higher plasma concentrations of the pharmacologically active S-enantiomer are obtained owing to the higher intrinsic clearance of the R-enantiomer. This enantioselective first-pass metabolism results in interesting clinical findings. Coltart and Shand [37] found that when a nonstereoselective assay was used, plasma concentrations of propranolol required to reach 50% of the maximum beta-blockade effect (EC50) were lower after the oral dose than after the i.v. dose (Fig. 1, upper panel). This was attributed to a higher percentage of the active S-enantiomer in the total plasma concentration after the oral dose compared to after the i.v. dose of the racemate. Bleske et al. [42] compared enantiomeric ratios of propranolol after racemic propranolol was administered i.v. (10 mg) and orally as IR (80 mg and 160 mg) and SR (160 mg) formulations. They not only confirmed the enantioselective first-pass metabolism from the oral and i.v. data but also demonstrated significant input-rate-related differences in the enantiomeric ratios at C_{max} (Fig. 1, lower panel).

With Verapamil [44–46] (Fig. 2) and nitrendipine [55], the situation is reversed. Here it is the more active S-enantiomer that has a higher intrinsic clearance and therefore gives substantially lower plasma concentrations after an oral dose than the less active R-enantiomer. Anomalous concentration versus effect (PR prolongation) relationships have also been observed

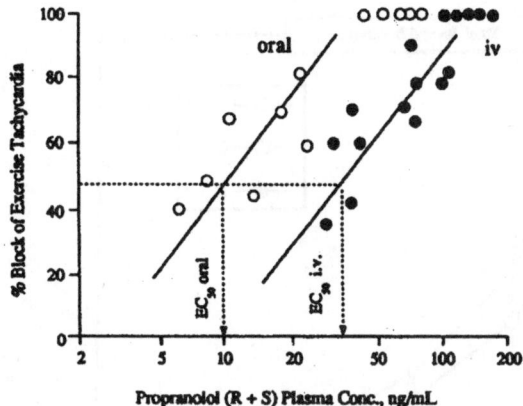

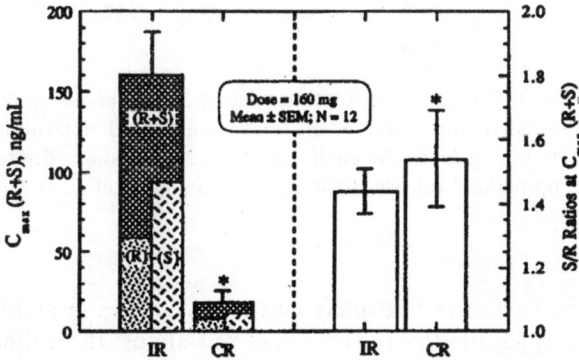

Figure 1 Pharmacokinetic/pharmacodynamic relationship and enantiospecific first-pass metabolism of propranolol, a prototype category II racemic drug. With propranolol, the EC50 value for the beta-blockade, determined on the basis of the nonenantiospecific assay, is lower after the oral dose than after the intravenous dose (upper panel). This is attributed to the greater first-pass metabolism of the less active R-enantiomer (lower panel). Slowing the oral input rate by switching from an immediate-release (IR) to a controlled-release (CR) formulation results in a significant ($* = P < 0.05$ by analysis of variance) increase in the proportion of the more active S-enantiomer at maximum concentration (C_{max}; R + S). (From Ref. 26, with permission.)

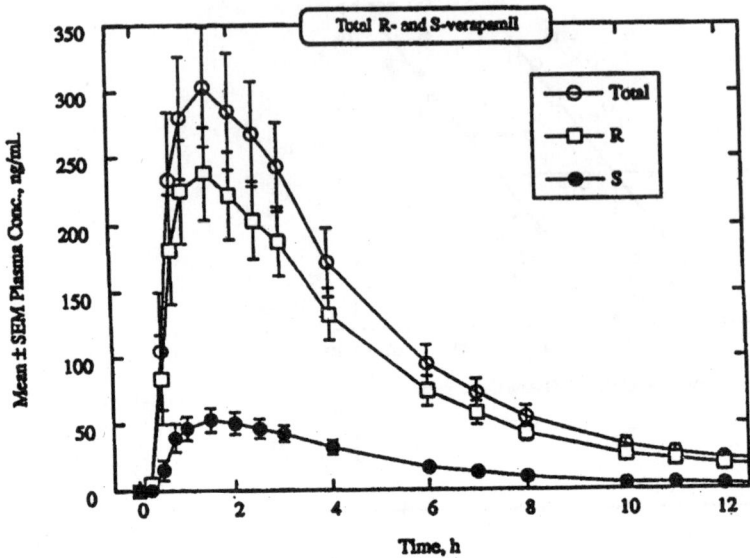

Figure 2 Mean ± SEM plasma concentration–time curves of total verapamil determined by nonstereospecific method and of R- and S-verapamil determined stereospecifically in 12 healthy young men who each received a 240 mg single dose of racemic verapamil as an immediate-release formulation. (From Ref. 53, with permission.)

with verapamil when a nonstereospecific assay was used. Echizen et al. [46] found that the EC50 of verapamil after the oral dose was almost three times higher than the EC50 after the i.v. dose (Fig. 3, upper panel). Different EC50 values were also obtained by Harder et al. [47], depending on whether oral verapamil was given as an IR or a controlled-release (CR) formulation. The latter findings could be attributed to the oral input-rate-related changes in the enantioselective first-pass metabolism (Fig. 3, lower panel).

A classic method of demonstrating enantioselective first-pass metabolism is to administer the racemic drug orally and i.v. in a crossover study and then evaluate the PK of each enantiomer using a stereospecific assay. Stereoselective first-pass metabolism is indicated by significantly different absolute bioavailabilities of the enantiomers. This approach was used to establish low enantioselective first-pass metabolism of ketoprofen [56] and high enantioselectivity of propranolol [38,42] and verapamil [44,45,53]. Enantioselective metabolism of propranolol and verapamil has been studied extensively and found to be influenced by age and gender [40,50]. With verapamil, changes in plasma enantiomeric ratios after administration

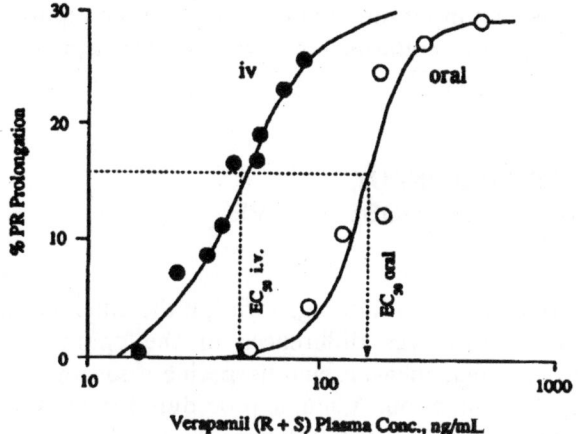

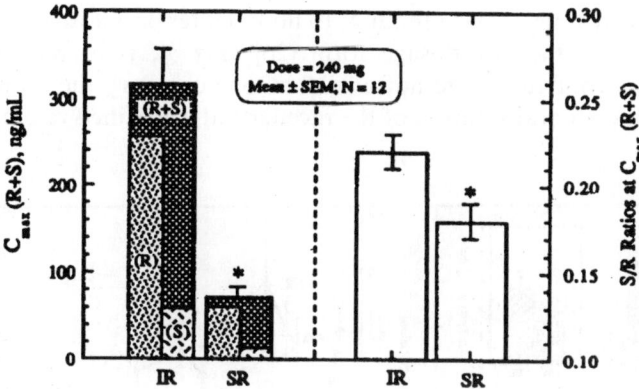

Figure 3 Pharmacokinetic/pharmacodynamic relationship and enantiospecific first-pass metabolism of verapamil. A prototype category III racemic drug. Unlike propranolol, with verapamil the EC50 value for the PR prolongation, determined on the basis of the nonenantiospecific assay, is higher after the oral dose than after the intravenous dose (upper panel). This is attributed to the greater first-pass metabolism of the more active S-enantiomer (lower panel). Slowing the oral input rate by switching from an immediate- (IR) to a sustained-release (SR) formulation results in a significant ($* = P < 0.05$ by paired t test) decrease in the proportion of the more active S-enantiomer at maximum concentration (C_{max}; R + S). (From Ref. 26, with permission.)

of oral formulations with different input rates in humans have been reported [47,53], and these findings have been confirmed in in vitro studies using liver homogenate [54] or liver perfusion [52].

4. DEMONSTRATION OF ORAL INPUT RELATED SATURABLE ENANTIOSELECTIVE FIRST-PASS METABOLISM

As will be discussed in detail later, for oral dosage forms, if the difference in input rate of racemic drug produces differences in the enantiomer concentration ratios in the plasma, then an enantiospecific assay may be required in bioequivalency determination. A simple procedure for assessing input-rate-related differences in the enantioselective metabolism is to determine enantiomeric plasma concentration or AUC ratios after administration of equivalent doses of IR and CR formulations. This approach was used to demonstrate that the percentage of S-verapamil in the plasma was much higher at an early time period (about 4–10 hours) after oral dose of IR verapamil compared to the CR dosage form [26] (Fig. 4, right panel). In cases where CR formulations are not available, one can vary the input rates by administering an oral solution of the racemate at a specified rate by

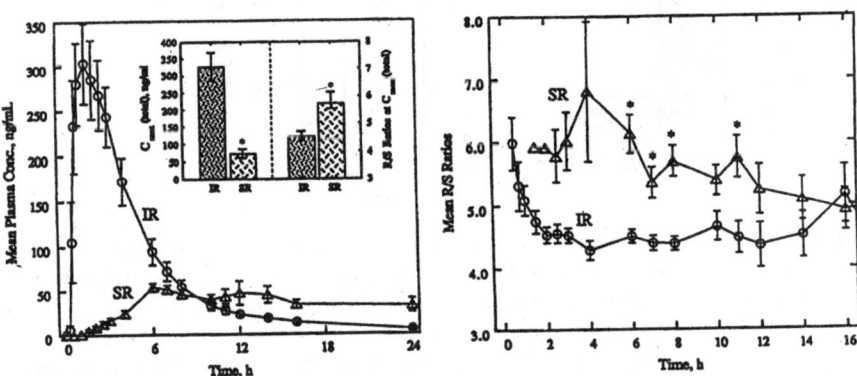

Figure 4 Mean ± SEM plasma concentration–time curves of total verapamil (left panel) and enantiomeric (R/S) ratios (right panel) in 12 healthy young men who each received a 240 mg single dose of racemic verapamil as immediate-release (IR; fasting) and sustained-release (SR; fed) formulations in a randomized crossover study. Insets show mean ± SEM for the individual peak concentration (C_{max}) (total) and R/S ratios associated with these C_{max} (total) values. *$p < 0.05$. (From Ref. 26, with permission.)

means of nasal intubation. Alternatively, the enantiomeric AUC ratios may be determined after oral administration of the doses that represent the extremes of therapeutic range. For a racemic etodolac, an NSAID, Boni et al. [57] studied the effect of altering oral input rate on the enantiomeric composition in the plasma of humans by administering a solution of 400 mg of the racemic drug rapidly over a one-minute period or given slowly as divided doses over 30 min and 90 min in a single-dose three-period crossover study. They noted that the changes in input rate resulted in altered etodolac enantiomeric ratios.

5. ENANTIOSPECIFIC ABSORPTION AND CHIRAL EXCIPIENTS

Some racemic drugs, such as methotrexate [58], leucovorin [59], L-dopa [60], cephalexin [61], and terbutaline [62], exhibit enantioselectivity in absorption. Also, many of the excipients used in the oral dosage forms, such as sugars, cellulose, alginate, and cyclodextrins, are chiral themselves: some of these materials are even used in racemic chromatographic separations. If there is evidence of enantioselective absorption or enantiomer-chiral excipient complexation resulting in enantiospecific dissolution [63,64], then an enantiospecific assay is necessary for clinically meaningful bioequivalency assessment.

6. APPRAISALS OF VARIOUS PROPOSALS ON THE USE OF STEREOSELECTIVE ASSAYS IN BIOEQUIVALENCY STUDIES

The first recommendations on the use of stereoselective assay in bioequivalency studies of racemic drugs were proposed by a Swedish regulatory agency in 1991 [16]. Their recommendations were

> If there is no information on the contribution of each enantiomer to the therapeutic effect, bioequivalence should be shown for both enantiomers.
>
> In cases where the activity resides in only one of the enantiomers, bioequivalence should be based on the concentrations of the eutomer.
>
> If the enantiomers have similar therapeutic activity, bioequivalence should be based on the sum of the enantiomer concentrations.
>
> In a situation where one enantiomer increases or decreases the activity of the other, bioequivalence should be shown for both enantiomers.

In a 1992 conference [17], Analytical Methods Validation: Bioavailability, Bioequivalence and Pharmacokinetic Studies, held in Crystal City, Washington, DC, the following recommendations were made:

All methods used for measurement of stereoisomers should be validated (with emphasis on stereospecificity).

For bioequivalence studies of an existing racemic product, a stereospecific assay is not required if the rate and extent profiles are superimposable (within the usual statistical boundaries).

For new chemical entities, the pharmacokinetic profiles for the stereoisomer should be characterized in normal subjects.

In an article by Nerukar et al. [18] of the U.S. Office of Generic Drugs, published in 1992, it was mentioned that current FDA policy does not require the measurement of individual enantiomers to demonstrate bioequivalence. They indicated that they agreed in principle that "because the pharmacodynamics of the enantiomers of a racemic drug may differ, generic applicants should measure each enantiomer to satisfy the bioequivalence requirements." They also stated, "the office does not wish to impose a requirement for enantiospecific assays in the absence of compelling reasons to do so." Criteria for the compelling reasons were not specified in detail.

In 1993, the European Community [22] issued a Note of Guidance stating that bioequivalence studies supporting generic applications of chiral medicinal products should be based on enantiospecific assays unless the enantiomers show linear kinetics.

With the Swedish recommendations [16], considerable clinical data for each enantiomer are needed before a rational decision on a need for a stereoselective assay can be made. Such clinical data are rarely available for racemic drugs. In addition, a "distomer" for a particular pharmacological effect could be a "eutomer" for another effect, and in some instances it may even be a specific enantiomer ratio that is associated with the optimal therapeutic effect.

The Crystal City 1992 conference [17] recommendation was that if the plasma concentrations of two enantiomers are superimposable, then an enantiospecific assay is not necessary. This recommendation does not distinguish racemic drugs that undergo negligible stereoselective first-pass metabolism from those that undergo extensive and saturable enantiospecific first-pass metabolism. As discussed in detail later, an enantiospecific assay is not necessary in the former case even when the plasma concentration–time curves for each enantiomer are not superimposable. The AUC of individual enantiomers for such drugs is a fixed ratio of the AUC of total $(R + S)$. Moreover, superimposability of the rate and extent of absorption entails statistical comparison of the entire shape of the plasma

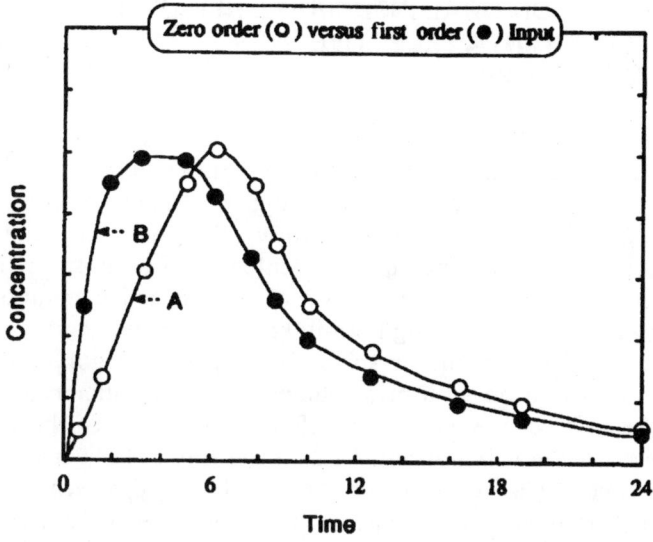

Figure 5 Two hypothetical formulations of a racemic drug with different input rates meeting the current bioequivalency criteria for equivalent peak plasma concentration (C_{max}) and area under the concentration–time curve (AUC). (From Ref. 25, with permission.)

concentration–time curves. Such shape analyses, although desirable, are not the current criteria for judging bioequivalency [29]. Figure 5 illustrates that it is possible for two formulations to have nonsuperimposable concentration–time curves and yet meet the bioequivalency criteria for C_{max} and the AUC, the two metrics currently used for establishing bioequivalency. Hypothetical formulation A, with a pseudo zero-order input rate in Fig. 5, would be considered bioequivalent to formulation B, which has a first-order input rate on the basis of equivalent C_{max} and AUC values. Times to attain maximum concentration (t_{max}) do differ between the two formulations, but this metric, appropriately, is not emphasized in bioequivalency assessment. The commonly used metric C_{max} is confounded by the extent of absorption, and it is a particularly inaccurate index of drug absorption/input rate from a CR dosage form. The need for an accurate metric for input rate in bioequivalency assessment has been widely discussed [65], but to date none has been found to be satisfactory. Recently the concept of partial [66] AUC (AUC from time 0 to t_{max}) has been proposed for input rate assessment. Partial AUC determination has the advantage that it would be able to distinguish formulation A from B in Fig. 5.

7. PROPOSAL BASED ON ORAL-INPUT-RATE-RELATED DIFFERENCES IN ENANTIOSELECTIVE FIRST-PASS METABOLISM

In view of above limitations, a new proposal for stereospecific assay requirements in bioequivalency studies was made by Karim [26] in 1996 and modified slightly by Midha et al. [28] in 1998 (Fig. 6). In this proposal, racemic drugs are divided into three major categories, those with negligible or no stereoselective first-pass metabolism (category I), those with significant enantioselective first-pass metabolism of the less active enantiomer (category II), and those with significant enantioselective first-pass metabolism of the more active and/or toxic enantiomer (category III). Significance should be based on an a priori judgment that a difference of $> 20\%$ ($p < 0.05$) in AUC of the enantiomer after an oral dose of the IR formulation of the racemate is important. Critical PK information needed to judge the importance of a stereospecific assay include the presence of enantioselective versus nonselective first-pass metabolism and saturability in the metabolic (or other elimination/distribution) processes as a result of changes in the dose level (within the therapeutic ranges) and/or oral input rates.

Because of the availability of PK information on each enantiomer, the illustrative examples below have been taken from racemic NSAIDs such as ibuprofen, ketoprofen, flurbiprofen, and etodolac, and from racemic cardiovascular drugs such as propranolol, propafenone, nicardipine, and verapamil. However, the principles and logic presented below are applicable to wide ranges of racemic drugs.

7.1. Category I Drugs

The NSAIDs ketoprofen [36,67] and flurbiprofen [56,68] are good examples of category I racemic drugs, in which $< 20\%$ differences exist in the enantiomeric AUCs after oral doses of racemates (Table 1). Individual enantiomeric concentrations of these racemates are approximately a constant fraction of the total $(R + S)$, and therefore enantiospecific assay is not necessary in bioequivalency assessment. This situation is analogous to the concentration-independent drug plasma protein binding, where separate determinations of the unbound drug concentrations are unnecessary, because unbound concentrations can be derived from the total (bound + unbound) as long as the degree of protein binding is known and remains constant within the therapeutic drug concentration ranges. Similarly, with ketoprofen and flurbiprofen, plasma concentrations of the active S-enantiomer can be determined from the total $(R + S)$ concentration, since

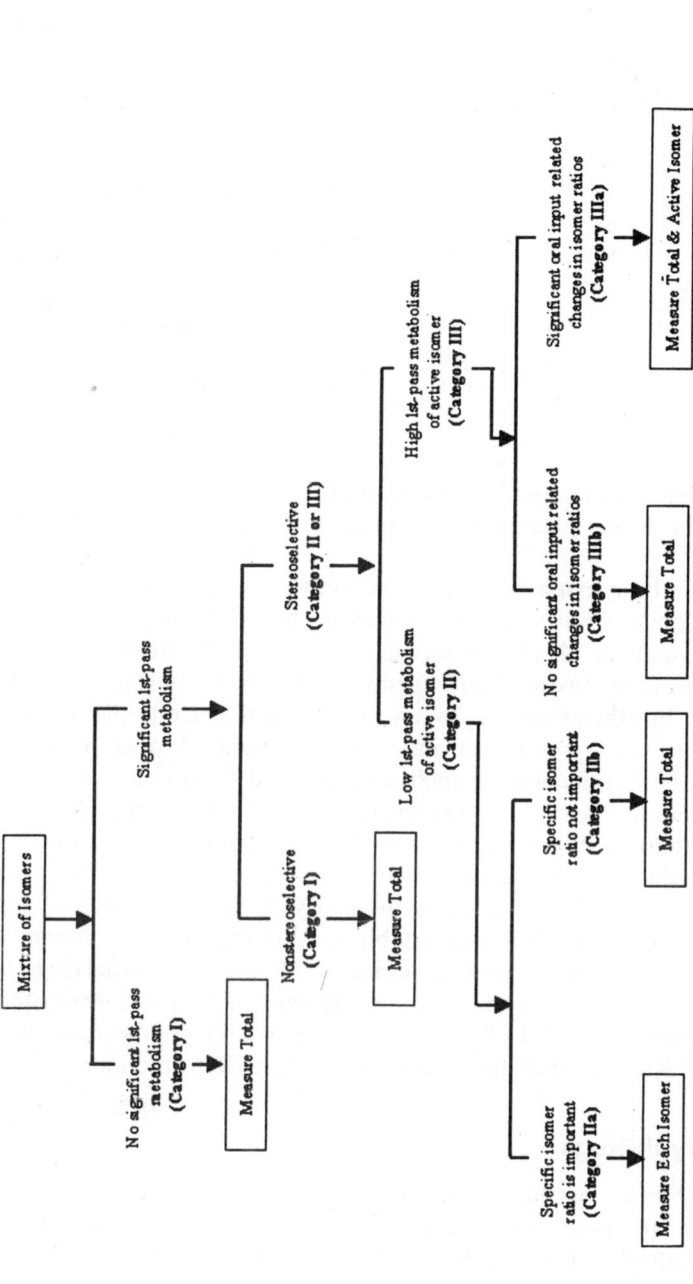

Figure 6 Algorithm for deciding whether an enantiospecific assay should be used in assessing bioavailability/bioequivalency of a racemic drug formulation. Emphasis in this algorithm is on the enantiospecific first-pass metabolism, because this is a more commonly occurring phenomenon. This algorithm is equally applicable, however, to other enantiospecific saturable processes, such as enantiospecific active transport in absorption, enantiospecific plasma protein and/or tissue binding, and enantiospecific biliary and/or renal excretion. (From Refs. 26 and 28, with permission.)

Table 1 Enantiomeric AUC Ratios After Oral Administration (Immediate Release) of Representative Racemic Nonsteroidal Anti-inflammatory (NSAID) and Cardiovascular Drugs

Drugs	Active enantiomer	S/R AUC ratio (oral)	Drug category	Ref.
NSAIDs				
Flurbiprofen	S	1.13	I	Jamali et al. [56]
Etodolac	S	0.09	III	Brocks and Jamali [88]
Ibuprofen	S	1.69	II	Geisslinger et al. [78]
Ketoprofen	S	0.94	I	Geisslinger et al. [36]
Cardiovascular				
Propafenone	S	1.73	II	Kroemer et al. [71]
Propranolol	S	1.43	II	Bleske et al. [42]
Verapamil	S	0.21	III	Karim and Piergies [53]

the S-enantiomer concentration proportion in the total drug concentration is not dependent on the plasma concentrations of the total, and it remains constant.

There have been extensive discussions on the need for enantiospecific versus nonselective assays for comparative bioavailability studies of flurbiprofen. Jamali et al. [69] reported no significant differences in the AUC of flurbiprofen between two IR formulations when a nonstereospecific assay was used, but then found a 7.8% ($p < 0.05$) difference in the AUC of the S-enantiomer when a stereospecific assay was used. Hooper et al. [70] commented that bioequivalency conclusions derived from either enantiospecific or nonselective assays of flurbiprofen in the Jamali et al. [69] study were not meaningfully different. Midha et al. [28] determined the geometric mean and 90% confidence intervals for the log (test/reference) AUC ratios of each enantiomer of flurbiprofen in the study by Jamali et al. [69] and concluded that bioequivalency was established between the two formulations whether enantiospecific or nonspecific assay was used. Applying the new proposal outlined in Fig. 6, an enantiospecific assay for flurbiprofen, a category I drug [26] (AUC S/R ratio = 1.13), would not be necessary, especially when its bioavailability is compared from IR formulations.

7.2. Category II Drugs

The IR formulations of these racemic drugs exhibit significant (> 20%) enantioselective first-pass metabolism of the pharmacologically less active enantiomer, resulting in the enrichment of the active enantiomer in the total.

Propranolol is a classic example of a category II drug, and, as mentioned before, the less active R-enantiomer undergoes greater first-pass metabolism to 4-hydroxy-propranolol [38]. Nonstereospecifically determined propranolol predominantly contains the active S-enantiomer, so an enantiospecific assay for such category II racemic drugs is not essential in the comparative bioavailability studies when the input rates of the dosage forms do not differ substantially. An exception would be when a specific enantiomeric ratio is critical for optimal pharmacological effect, and this ratio changes with the oral dose level and/or input rates.

Enantiomers of the antiarrhythmic drug propafenone [71–74] have differing antiarrhythmic activity profiles. In addition, S-propafenone is reported to exert greater potency of effect (> 40-fold) over its optical antipode in blocking beta-adrenergic activity [73,74]. PK studies using stereospecific assays have demonstrated that the S/R AUC ratios in the plasma are significantly greater than one (Table 1) in extensive metabolizers of propafenone [71]. Therefore there is enrichment of the active S-enantiomer in the plasma as a result of stereoselective first-pass metabolism. These findings would make propafenone a category II drug for which an enantiospecific assay would not be required in bioequivalency assessment. This conclusion is supported by data from Midha et al. [28], who applied current FDA recommended statistical tests for assessing bioequivalency and found that the test and reference IR formulations of propafenone were bioequivalent whether a stereospecific or a nonspecific assay was used.

7.3. Category II Drug with Racemic Inversion

An important example of a category II drug undergoing racemic inversion is ibuprofen, an NSAID containing a racemic center α to the carboxyl group. Cyclooxygenase inhibitory activity of ibuprofen resides mainly with the S-enantiomer [75–79]. Minor differences between the enantiomer potencies are obscured in vivo, however, because of the inversion of the less active R-enantiomer to its more active S-antipode. Jamali et al. [79] reported that this inversion was influenced by the residence time of the dosage form in the GI tract, and they were able to demonstrate a relationship between the t_{max} of S-ibuprofen and the S/R enantiomeric ratios. Results of a more recent study by Hall et al. [80], however, demonstrated the systemic nature of the racemic inversion. They obtained similar AUCs for R- and S-ibuprofen after oral and i.v. administration of the racemic drug. Because oral ibuprofen in their study was given only as an IR formulation, the impact of greater residence time of the drug formulation in the GI tract on the presystemic inversion of ibuprofen remains unresolved.

Formulation-dependent enantiomeric ratio differences of ibuprofen were found by Avgerinos et al. [81]. Plasma enantiomeric ratio–time profiles with the CR formulation (enantiomeric S/R ratio range, 1.2–2.6) showed decreased fluctuations during a dose interval compared with the IR formulation (S/R range, 1.3–4.2). Such chiral aspects of bioavailability assessment (Fig. 7) can only be elucidated with the stereoselective assay. Its clinical relevance, however, remains unknown.

Cox et al. [82] and Walker and Hardy [83] compared bioavailability of ibuprofen from IR formulations using enantiospecific and nonspecific assays. The former investigators derived identical conclusions for C_{max} and AUC, although formulation-dependent differences in the elimination rate constants (ke) were only detected with the enantiospecific assay. Walker and Hardy [83] concluded that formulations deemed bioequivalent on the basis of nonstereospecific assays were not bioequivalent with respect to the S-enantiomer.

From the above reports, it would appear prudent to use an enantiospecific assay when comparing bioavailability of ibuprofen formulations, especially when the formulations are designed to have modified drug release characteristics.

7.4. Category III Drugs

With category III racemic drugs, it is the pharmacologically more active enantiomer that undergoes greater first-pass metabolism. The principal argument here for measuring plasma concentrations of the active enantiomer is that total concentrations determined by nonstereospecific assay mostly reflect the concentrations of the less active enantiomer. Examples of category III drugs include verapamil [46] and etodolac [54] (Table 1).

With verapamil, the pharmacologically more active S-enantiomer is metabolized substantially more than its less active antipode (Table 1). Sahajwalla et al. [49] calculated that a 7% difference in the total drug may result in as much as 30% change in the concentration of the active S-verapamil. Here again one can draw an analogy with the protein binding phenomenon. Small changes in the extent of protein binding of extensively bound drug can result in large changes in the concentrations of the pharmacologically active unbound drug. For example, with an initial total concentration of 100 ng/mL of a hypothetical drug that is only 40% bound, a 10% decrease in the protein binding will produce only a 6.6% increase (from 60 to 64 ng/mL) in the unbound concentration. The same 10% change will produce an almost 1,000% increase (from 1 to 10.9 ng/mL) of a drug that is 99% protein bound. In this respect, category III drugs may be perceived as extensively bound drugs that undergo concentration-dependent

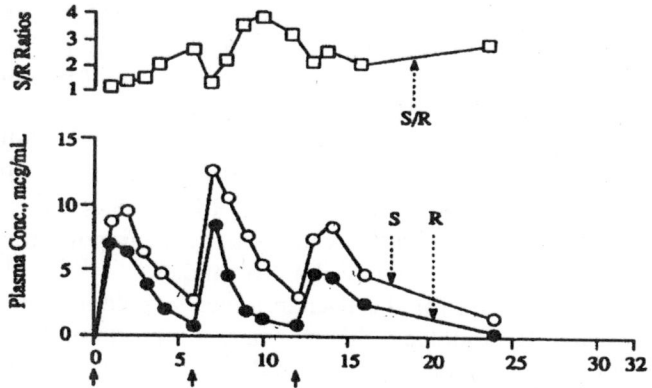

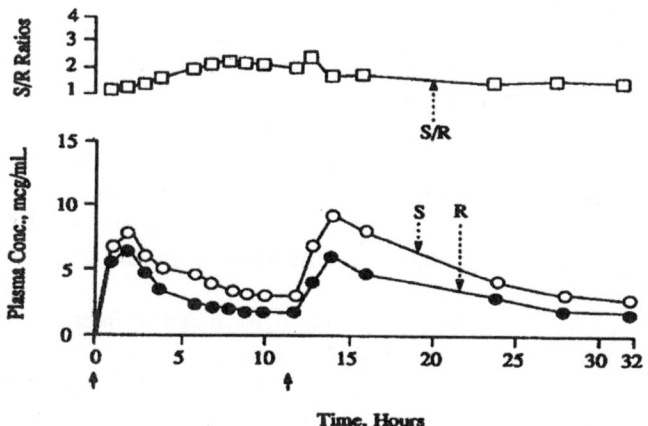

Figure 7 Ibuprofen, a prototype category II racemic drug with chiral inversion. Mean ($n=4$) plasma concentration–time curves of R-ibuprofen, S-ibuprofen, and enantiomeric S/R ratios after oral administration of racemic ibuprofen given as an immediate-release (upper panel; three 200 mg doses 6 hours apart) and controlled-release formulation (lower panel; two 300 mg doses 12 hours apart). The arrows on the time axes indicate the time of drug administration. (From Ref. 26, with permission.)

(nonlinear) protein binding; and just as determination of the unbound drug is important, here so is the determination of active enantiomer necessary for a category III drug with nonlinear-oral-input-related enantioselective first-pass metabolism.

With another calcium channel blocker, nicardipine [85], it is the pharmacologically more active (+)-enantiomer that undergoes less first-pass metabolism than its less active antipode. Here the concentration of the total drug predominantly reflects that of the active enantiomer, and therefore stereospecific assay would not be essential in a comparative bioavailability study of this racemic drug. The two calcium channel blockers verapamil and nicardipine are excellent examples of why the PK/PD profiles of each racemic drug should be looked at individually in assessing the importance of enantiospecific assays in comparative bioavailability studies.

Although it is less common for the R-enantiomer to predominate in plasma for racemic NSAIDs, etodolac is different in this respect and has an S/R ratio of about 0.10 (Table 1), and thereby the PK of racemic etodolac is dominated mostly by the less active R-enantiomer. The S/R concentration ratio in plasma also varies with time and formulation, with highest S/R ratio occurring prior to t_{max} of S-etodolac [84,86–88]. The effect of altered oral input rate on the enantiomeric composition of etodolac was studied by Boni et al. [57]. They found that the C_{max} values for the total drug determined by nonstereospecific assay contained higher percentages of the pharmacologically active S-etodolac when 400 mg of racemic etodolac was given rapidly over 1 min when compared to the same dose given more slowly over 30 to 90 min. In view of these input-rate-related differences in the enantiomeric ratios in the plasma, they recommended that an enantiospecific assay should be used in the meaningful bioequivalency studies of racemic etodolac formulations. These recommendations are consistent with the algorithm presented in Fig. 6.

The above-described oral-input-related changes in enantiomeric ratios in the plasma for ibuprofen, propranolol, etodolac, and verapamil have been documented only after administration of the IR and CR formulations or after oral input rates that differ substantially. An important issue that needs to be addressed is: what if the two formulations compared are CR formulations whose input rate differences are not that substantial? Can a relatively small difference between the rate of release/dissolution produce significant changes in the concentrations of the active enantiomer that cannot be detected by measurement of the total drug concentration?

Mehvar has addressed this question by simulation [89] and by liver perfusion studies [54], but such studies are known to have limitations. A more reliable method would be to study a correlation between the input rate of racemic drug formulation and the plasma enantiomeric ratio changes in humans. The difficulties, both statistical and practical, of implementing and interpreting the results of such clinical studies would be formidable. A simpler, more practical approach would be to compare the bioavailability of such category III drug formulations with enantiospecific assays and compare AUC and C_{max} values after administration of equivalent doses of IR and CR. If the enantiomeric ratios of AUC and C_{max} values are not significantly different between the two formulations under these widely divergent input rates, then an enantiospecific assay will not be required in evaluating the bioequivalency of formulations whose in vitro dissolution rate differences are not great.

8. FOOD AND DRUG ADMINISTRATION (FDA) CURRENT GUIDANCE ON ENANTIOSPECIFIC ASSAYS IN BIOEQUIVALENCY STUDIES

This issue has been addressed in detail in the next chapter. However, a brief overview is presented here. Recent guidance [29] by the FDA issued in October 2000 discusses when an enantiospecific assay may be used in bioequivalency studies. The FDA recommendations are that nonstereospecific assay may be used for bioequivalence assessment of most racemic drug formulations. Measurement of individual enantiomers in bioequivalency studies is recommended only when all of the following conditions are met: (1) the enantiomers exhibit different pharmacodynamic characteristics, (2) the enantiomers exhibit different PK characteristics; (3) primary efficacy/ safety activity resides with the minor enantiomer; and (4) nonlinear absorption is present (as expressed by a change in the enantiomeric concentration ratio with change in the input rate of the drug) for at least one of the enantiomers. In such case, bioequivalency criteria should be applied to the enantiomers separately. The above FDA criteria for requiring an enantiospecific assay are very similar to those for the racemates belonging to category IIIa in Fig. 6.

9. CONCLUSIONS

Overall, various factors are known to affect the systemic absorption of orally administered drugs; so the PK/PD characteristics of each racemic drug should be considered individually when deciding whether an enantiospecific assay is necessary in comparative bioavailability studies. Drugs that have insignificant or nonstereoselective first-pass metabolism (category I) usually do not require stereoselective assays. Those drugs for which the less active enantiomer undergoes greater first-pass metabolism (category II) also generally do not require stereoselective assays. It is with racemic drugs, for which the pharmacologically more active and/or toxic enantiomer undergoes greater first-pass metabolism (category III), that an enantiospecific assay becomes important. Even then, it is necessary to establish that the metabolism (or distribution/elimination) of the active and/ or toxic enantiomer is saturable and that changes in the amount administered or in the oral input rates can significantly affect enantiomeric ratios in plasma. The need for stereoselective assays in comparative bioavailability studies should be the exception rather than the rule; however, stereoselective determinations are important for certain racemic drugs, particularly when administered in modified-release dosage forms.

REFERENCES

1. Cahn, R.S.; Ingold, C.K.; Prelog, V. The specification of asymmetric configuration in organic chemistry. Experientia 1956, 12, 81–124.
2. Vlasses, P.H.; Irvin, J.D.; Huber, P.B.; Lee, R.B.; Ferguson, R.K.; Schrogie, J.J.; Zacchei, A.G.; Davies, R.O.; Abrams, W.B. Pharmacology of enatiomers and (−) p-OH metabolite of indacrinone. Clin. Pharmacol. Ther. 1981, 29, 798–807.
3. Balant, L.P.; Benet, L.Z.; Blume, H.; Bozler, G.; Breimar, D.D.; Echelbaum, M., et al. Is there a need for more precise definitions of bioavailability? Eur J Clin Pharmacol 1991, 40, 123–126.
4. Ravis, W.R.; Owen, J.S. Sterochemical considerations in bioavailability studies. In: Generics and Bioequivalence; Jackson, A.J., Ed.; CRC Press: Ann Arbor (MI), 1994; 113–137.
5. Ariens, E.J. Stereochemistry as a basis for sophisticated nonsense in pharmacokinetics and clinical pharmacology. Eur. J. Clin. Pharmacol. 1984, 26, 663–668.
6. Toon, S.; Low, L.K.; Gibaldi, M.; Trager, W.F.; O'Reilly, R.A.; Motley, C.H.; Goulart, M.A. The warfarin-sulfinpyrazone interaction: stereochemical considerations. Clin. Pharmacol. Ther. 1986, 39, 15–24.
7. O'Reilly, R.A. Ticrynafen-racemic warfarin interaction, hepatotoxic or stereoselective? Clin. Pharmacol. Ther. 1982, 32, 356–361.
8. Giacomini, K.M.; Nelson, W.L.; Pershe, R.A.; Valdivieso, L., Turner-Tamiyasu, K.; Blaschke, T.F. In vivo interaction of the enatiomers of disopyramide in human subjects. J. Pharmacokinet. Biopharm. 1986, 14, 335–356.
9. Evans, A.M.; Nation, R.L.; Sansom, L.N.; Bochner, F., Somogyi, A.A. Stereoselective drug disposition: potential for misinterpretation of drug disposition data. Br. J. Clin. Pharmacol. 1988, 26, 771–780.
10. Lee, E.J.D.; Williams, K.M. Chirality, clinical pharmacokinetic and pharmacodynamic considerations. Clin. Pharmacokinet. 1990, 18, 339–345.
11. Tucker, G.T.; Lennard, M.S. Enantiomer specific pharmacokinetics. Pharmacol. Ther. 1990, 45, 309–329.

12. Campbell, D.B. Stereoselectivity in clinical pharmacokinetics and drug development. Eur. J. Drug Metab Pharmacokinet **1990**, *15*, 109–125.
13. Caldwell, J. The importance of stereochemistry in drug action and disposition. J. Clin. Pharmacol. **1992**, *32*, 925–929.
14. Gibaldi, M. Stereoselective and isozyme-selective drug interactions. Chirality **1993**, *5*, 407–413.
15. Birkett, D.J. Racemates or enantiomers: regulatory approaches. Clin. Exp. Pharmacol. Physiol. **1989**, *16*, 479–483.
16. Some views from the Medical Product Agency of the documentation for chiral drugs. Uppsala: Registration Division, Medical Product Agency, 1991.
17. Shah, V.P.; Midha, K.K.; Dighe, S., McGilveray, I.J.; Skelly, J.P.; Yacobi, A., et al. Analytical methods validation: bioavailability, bioequivalence and pharmacokinetic studies (conference report). Pharm Res **1992**, *9*, 588–592.
18. Nerukar, S.G.; Dighe, S.V.; Williams, R.L. Bioequivalence of racemic drugs. J. Clin. Pharmacol. **1992**, *32*, 935–943.
19. Wechter, W.J. From controversy to resolution: bioequivalency of racemic drugs. A symposium on the dynamics, kinetics, bioequivalency, and analytical aspects of stereochemistry. J. Clin. Pharmacol. **1992**, *32*, 915–916.
20. Nitchuk, W.M. Regulatory requirements for generic chiral drugs. J. Clin. Pharmacol. **1992**, *32*, 953–954.
21. Jamali, F. Stereochemistry and bioequivalence. J. Clin. Pharmacol. **1992**, *32*, 930–934.
22. EC-Note for Guidance: Investigation of Chiral Active Substances, CPMP Working Party on Quality of Medicinal Products. CPMP Working Party on Safety of Medicinal Products. CPMP Working Party on Efficacy of Medicinal Products. 1993 Oct, III/3501/91EN.
23. Marzo, A. How incoming guidelines on chiral drugs could impact on the international scenary of drug development. Boll Chim Farmaceutico **1993**, *132*, 267–271.
24. Nation, R.L.; Sansom, L.N. Bioequivalence requirements for generic products. Pharmac. Ther. **1994**, *62*, 41–55.
25. Nation, R.L. Chirality in new drug development: clinical pharmacokinetic considerations. Clin Pharmacokinet **1994**, *27*, 249–255.
26. Karim, A. Enantiospecific assays in comparative bioavailability studies of racemic drug formulations: nice to know or need to know? J. Clin. Pharmacol. **1996**, *36*, 490–499.
27. Mehvar, R.; Jamali, F. Bioequivalence of chiral drugs. Clin. Pharmacokinet. **1997**, *33*, 122–141.
28. Midha, K.K.; Gordon, M.; Maureen, J.R.; John, W.H. The impact of stereoisomerism in bioequivalence studies. J. Pharm. Sci. **1998**, *7*, 797–802.
29. Guidance for Industry. Bioavailability and Bioequivalence Studies for Orally Administered Drug Products—General Considerations. U.S. Department of Health and Human Services. Food and Drug Administration. Center for Drug Evaluation and Research (CDER), October 2000. http://www.fda.gov/cder/guidance/index.htm.

30. Hutt, A.J.; Caldwell, J. The importance of stereochemistry in the clinical pharmacokinetics of the 2-arylpropionic acid non-steroidal anti-inflammatory drugs. Clin. Pharmacokinet **1984**, *9*, 371–373.

31. Caldwell, J.; Hutt, A.J.; Fournel-Gigleux, S. The metabolic chiral inversion and dispositional enantioselectivity of the 2-arylpropionic acids and their biological consequences. Biochem. Pharmacol. **1988**, *37*, 105–114.

32. Evans, A.M. Enantioselectivity pharmacodynamics and pharmacokinetics of chiral nonsteroidal antiinflammatory drugs. Eur J. Clin. Pharmacol. **1992**, *42*, 237–256.

33. Jamali, F. Pharmacokinetics of enantiomers of chiral non-steroidal antiinflammatory drugs. Eur. J. Drug Metab Pharmacokinet **1994**, *26*, 259–274.

34. Simmonds, R.G.; Woodage, T.J.; Duff, S.M.; Green, J.N. Stereospecific inversion of R-benoxaprofen in rat and man. Eur J Metab Pharmacokinet **1980**, *5*, 169–172.

35. Foster, R.T.; Jamali, F. Stereoselective pharmacokinetics of ketoprofen in the rat. Influence of route of administration. Drug Metab Dispos. **1988**, *16*, 623–626.

36. Geisslinger, G.; Menzel, S.; Wissel, K.; Brune, K. Pharmacokinetics of ketoprofen enantiomers after different doses of racemate. Br. J. Clin. Pharmacol. **1995**, *40*, 73–75.

37. Coltart, D.J.; Shand, D.G. Plasma propranolol levels in quantitative assessment of beta-adrenergic blockade in man. Br. Med. J. **1970**, *3*, 731–734.

38. Von Bahr, C.; Hermansson, J.; Tawara, K. Plasma levels of (+) and (-)-propranolol and 4-hydroxypropranolol after administration of racemic (+)-propranolol in man. Br J. Clin. Pharmacol. **1982**, *14*, 79–82.

39. Paxton, J.W.; Norris, R.M. Propranolol disposition after acute myocardial infarction. Clin. Pharmacol. Ther. **1984**, *36*, 337–342.

40. Gilmore, D.A.; Gal, J.; Gerber, J.G.; Nies, A.S. Age and gender influence the stereoselective pharmacokinetics of propranolol. J. Pharmacol. Exp. Ther. **1992**, *261*, 1181–1186.

41. Egginger, G.; Lindner, W.; Brunner, G.; Stoschitzky, K. Direct enantiospecific determination or (R)- and (S)-propranolol in human plasma: application to pharmacokinetic studies. J. Pharm. Biomed. Anal. **1994**, *12*, 1537–1545.

42. Bleske, B.E.; Welage, L.S.; Rose, S.; Amidon, G.L.; Shea, M.J. The effect of dosage release formulations on the pharmacokinetics of propranolol stereoisomerism in humans. J. Clin. Pharmacol. **1995**, *35*, 374–378.

43. Eichelbaum, M.; Ende, M.; Remberg, G.; Schomerus, M.; Dengler, H.J. The metabolism of D, L-[14C] verapamil in man. Drug Metab Dispos. **1979**, *7*, 145–148.

44. Vogelgesang, B.; Echizen, H.; Schmidt, E.; Eichelbaum, M. Stereo-selective first-pass metabolism of highly cleared drugs: studies of the bioavailability of L- and D-verapamil examined with a stable isotope technique. Br. J. Clin. Pharmacol. **1984**, *18*, 733–740.

45. Eichelbaum, M.; Mikus, G.; Vogelgesang, B. Pharmacokinetics of (+)-, (−)- and (+/−)-verapamil after intravenous administration. Br. J. Clin. Pharmacol. **1984**, *17*, 453–458.

46. Echizen, H.; Vogelgesang, B.; Eichelbaum, M. Effects of *d*, *l*-verapamil on atrioventricular conduction in relation to its stereoselective first-pass metabolism. Clin. Pharmacol. Ther. **1985**, *38*, 71–76.

47. Harder, S.; Thurmann, P.; Siewert, M.; Blume, H.; Huber, T.; Rietbrock, N. Pharmacodynamic profile of verapamil in relation to absolute bioavailability: investigations with a conventional and a controlled-release formulation. J. Cardiovasc. Pharmacol. **1991**, *17*, 207–212.

48. Kroemer, H.K.; Echizen, H.; Heidemann, H.; Eichelbaum, M. Predictability of the in vivo metabolism of verapamil from in vitro data: contribution of individual metabolic pathways and stereoselective aspects. J. Pharmacol. Exp. Ther. **1992**, *260*, 1052–1057.

49. Sahajwalla, C.G.; Longstreth, J.; Karim, A.; Purich, E.D.; Cabana, B.E. Consequences in pooling R- + S-verapamil in bioequivalence assessment (abstract). J. Clin. Pharmacol. **1992**, *32*, 961.

50. Sasaki, M.; Tateishi, T.; Ebihara, A. The effects of age and gender on the stereoselective pharmacokinetics of verapamil. Clin. Pharmacol. Ther. **1993**, *54*, 278–285.

51. Harder, S.; Siewert, M.; Thurmann, P.; Blume, H.; Rietbrock, N.; Siewert, B. Plasma concentrations of S-verapamil after single doses of two different galenic formulations of racemic verapamil. Arzneimittelforschung **1993**, *43*, 520–523.

52. Mehvar, R.; Reynolds, J.M.; Robinson, M.A.; Longstreth, J.A. Enantiospecific kinetics of verapamil and nor-verapamil in isolated perfused rat livers. Pharm. Res. **1994**, *11*, 1815–1819.

53. Karim, A.; Piergies, A. Verapamil stereoisomerism: enantiomeric ratios in plasma dependent on peak concentrations, oral input rate, or both. Clin. Pharmacol. Ther. **1995**, *58*, 174–184.

54. Mehvar, R.; Reynolds, J. Input rate-dependent stereoselective pharmacokinetics: experimental evidence in verapamil-infused isolated rat livers. Drug Metab Dispos. **1995**, *23*, 637–641.

55. Soon, P.A.; Breimer, D.D. Stereoselective pharmacokinetics of oral and intravenous nitrendipine in healthy male subjects. Br J. Clin. Pharmacol. **1991**, *32*, 11–16.

56. Jamali, F.; Berry, B.W.; Tehrani, M.R.; Russell, A.S. Stereoselective pharmacokinetics of flurbiprofen in humans and rats. J. Pharm. Sci. **1988**, *77*, 666–669.

57. Boni, J.P.; Korth-Bradley, J.M.; Richards, L.S.; Chiang, S.T.; Hicks, D.R.; Benet, L.Z. Chiral bioequivalence: effect of absorption rate on racemic etodolac. Clin. Pharmacokinet. **2000**, *39*, 450–469.

58. Hendel, J.; Brodthagen, H. Enterohepatic cycling of methotrexate estimated by use of the D-isomer as a reference marker. Eur. J. Clin. Pharmacol. **1984**, *26*, 103–107.

59. Schilsky, R.L.; Choi, K.E.; Vokes, E.E.; Guaspari, A.; Guarnieri, C.; Whaling, S.; Liebnet, M.A. Clinical pharmacology of the stereoisomers of leucovorin during repeated oral dosing. Cancer **1989**, *63* (suppl 6), 1018–1021.

60. Wade, D.N.; Mearrick, P.T.; Morris, J.L. Active transport of L-dopa in the intestine. Nature **1973**, *242*, 463–465.

61. Tamai, I.; Ling, H-Y.; Timbul, S-M.; Nishikido, J.; Tsuji, A. Stereo-specific absorption and degradation of cephalexin. J. Pharm. Pharmacol. **1988**, *40*, 320–324.

62. Borgstrom, L.; Nyberg, L.; Jonsson, S.; Lindberg, C.; Paulson, J. Pharmacokinetic evaluation in man of terbutaline given as separate enantiomers and as the racemate. Br. J. Clin. Pharmacol. **1989**, *27*, 49.

63. Duddu, S.P.; Vakilynejad, M.; Jamali, F.; Grant, D.J.W. Stereoselective dissolution of propranolol hydrochloride from hydroxypropyl methylcellulose matrices. Pharm. Res. **1993**, *10*, 1648–1653.

64. Aubry, A-F.; Wainer, I.W. An in vitro study of the stereoselective dissolution of (rac)-verapamil from two sustained release formulations. Chirality **1993**, *5*, 84–90.

65. Chen, M.L. An alternative approach for assessment of rate of absorption in. bioequivalence studies. Pharm. Res. **1992**, *9*, 1380–1385.

66. Chen, M.L.; Lesko, L.; Williams, R.L. Measures of exposure versus measures of rate and extent of absorption. Clin. Pharmacokint. **2001**, *40*, 565–572.

67. Foster, R.T.; Jamali, F.; Russell, A.S.; Alballa, S.R. Pharmacokinetics of ketoprofen enantiomers in healthy subjects following single and multiple doses. J. Pharm. Sci. **1988**, *77*, 70–73.

68. Knadler, M.P.; Hall, S.D. High-performance liquid chromatographic analysis of the enantiomers of flurbiprofen and its metabolites in plasma and urine. J. Chromatogr. **1989**, *494*, 173–182.

69. Jamali, F.; Collins, D.S.; Berry, B.W.; Molder, S.; Cheung, R.; McColl, K.; Cheung, H. Comparative bioavailability of two flurbiprofen products: stereospecific versus conventional approach. Biopharm. Drug Dispos. **1991**, *12*, 435–445. .

70. Hooper, W.D.; Dickinson, R.G.; Gal, J. Enantiospecific versus non-enantiospecific assays in comparative bioavailability studies with racemic drugs. Biopharm. Drug Dispos. **1992**, *13*, 383–387.

71. Kroemer, H.K.; Funk-Brentano, C.; Silberstein, D.J., et al. Stereoselective disposition and pharmacologic activity of propafenone enantiomers. Circulation **1989**, *79*, 1068–1076.

72. Stoschitzky, K.; Klein, W.; Stark, G., et al. Different stereoselective effects of (R)- and (S)-propafenone: clinical, pharmacologic, electrophysiologic, and radioligand binding studies. Clin. Pharmacol. Ther. **1990**, *47*, 740–746.

73. Groschner, K.; Linder, W., et al. The effect of the stereoisomers of propafenone and diprafenone in guinea-pig heart. Br. J. Pharmacol. **1991**, *102*, 669–674.

74. Stark, U.; Stark, G.; Stoschitzky, et al. Stereoselective electro-physiological effects of propafenone in Langendorff perfused guinea pig heart. Basic Res. Cardiol. **1992**, *87*, 87–97.

75. Lee, E.J.; Williams, K.; Day, R.; Graham, G.; Champion, D. Stereoselective disposition of ibuprofen enantiomers in man. Br. J. Clin. Pharmacol. **1985**, *19*, 669–674.

76. Jamali, F.; Singh, N.N.; Pasutto, F.M.; Russell, A.S.; Coutts, R.T. Pharmacokinetics of ibuprofen enantiomers in humans following oral administration of tablets with different absorption rates. Pharm. Res. **1988**, *5*, 40–43.

77. Evans, A.M.; Nation, R.L.; Sansom, L.N.; Bochner, F.; Somogyi, A.A. The relationship between the pharmacokinetics of ibuprofen enantiomers and the dose of racemic ibuprofen in humans. Biopharm. Drug Dispos. **1990**, *11*, 507–518.

78. Geisslinger, G.; Schuster, O.; Stock, K-P.; Loew, D.; Bach, G.L.; Brune, K. Pharmacokinetics of S(+)- and R(-)-ibuprofen in volunteers and first clinical experience of (S)-ibuprofen in rheumatoid arthritis. Eur. Clin. Pharmacol. **1990**, *38*, 493–497.

79. Jamali, F.; Mehvar, R.; Russell, A.S.; Sattari, S.; Yakimets, W.W.; Koo, J. Human pharmacokinetics of ibuprofen enantiomers following different doses and formulations: intestinal racemic inversion. J. Pharm. Sci. **1992**, *81*, 221–225.

80. Hall, S.D.; Rudy, A.C.; Knight, P.M.; Brater, D.C. Lack of presystemic inversion of (R)- to (S)-ibuprofen in humans. Clin. Pharmacol. Ther. **1993**, *53*, 393–400.

81. Avgerinos, A.; Noormohammadi, A.; Hutt, A.J. Disposition of ibuprofen enantiomers following the oral administration of a novel controlled release formulation to healthy volunteers. Int. J. Pharm. **1991**, *68*, 97–103.

82. Cox, S.R.; Brown, M.A.; Squires, D.J.; Murrill, E.A.; Lednicer, D.; Knuth, D.W. Comparative human study of ibuprofen enantiomer plasma concentrations produced by two commercially available ibuprofen tablets. Biopharm. Drug Dispos. **1988**, *9*, 539–549.

83. Walker, S.E.; Hardy, B.G. Alterations in apparent bioequivalency of ibuprofen based on isomer analysis (abstract). J. Clin. Pharmacol. **1992**, *32*, 957.

84. Jamali, F.; Mehvar, R.; Lemko, C.; Eradiri, O. Application of stereospecific high-performance liquid chromatography assay to a pharmacokinetic study of etodolac enantiomers in humans. J. Pharm. Sci. **1988**, *77*, 963–966.

85. Iwaoka, T.; Inotsume, N.; Inoue, J.; Naomi, S.; Okamoto, Y.; Higuchi, S., et al. Determination of (+)- and (−)-nicardipine concentrations in human serum and their correlation with the antihypertensive effect after oral administration of racemic nicardipine. Eur. J. Clin. Pharmacol. **1995**, *48*, 345–349.

86. Brocks, D.R.; Jamali, F. The pharmacokinetics of etodolac enantiomers in the rat: lack of pharmacokinetic interaction between enantiomers. Drug Metab. Dispos. **1990**, *18*, 471–475.

87. Brocks, D.R.; Jamali, F.; Russell, A.S. Stereoselective disposition of etodolac enantiomers in synovial fluid. J. Clin. Pharmacol. **1991**, *131*, 741–746.

88. Brocks, D.R.; Jamali, F. Etodolac clinical pharmacokinetics. Clin. Pharmacokinet. **1994**, *26*, 259–274.

89. Mehvar, R. Stereochemical considerations in pharmacodynamic modeling of chiral drugs. J. Phar. Sci. **1992**, *81*, 199–200.

10
Regulatory Considerations in Drug Development of Stereoisomers

Chandra Sahajwalla
Food and Drug Administration, Rockville, Maryland, U.S.A.

1. INTRODUCTION

This chapter deals with some of the regulatory issues that need to be considered at the time of developing a drug with chiral center(s). Before we discuss regulatory issues in developing chiral drugs, a brief discussion of drug development and its regulatory process is warranted.

1.1. Drug Development and the Regulatory Process

The regulatory drug development process for new drugs has evolved over the years. Development of drug law dates back to 1906 when the Federal Food and Drug act was enacted, but the FDA had no role in premarketing evaluation of drugs. In 1962, for the first time, a drug had to be shown to be effective. Subsequently, guidelines to format and content of the clinical and statistical sections of the drug application were issued in 1988. Only then were there attempts to discern dose response relationships in adverse events, and to examine the rates of adverse events in demographics (age, race, gender) and other subgroups (metabolic status, renal function).

The views expressed in this article are those of the author's and do not reflect the official policy of the FDA. No official support or endorsement by the FDA is intended or should be inferred.

419

Cumulative drug development eras [1] could be identified as the era of safety, of efficacy, and now of the individualization of drug therapy.

In general, once a drug company (sponsor/applicant) has identified a chemical for development as a drug, a significant amount of research has to be carried out before it can be approved for marketing as a drug. The areas of research for a drug can broadly be categorized into chemistry, pharmacology/ toxicology, clinical (determining safety and efficacy), clinical pharmacology, and biopharmaceutics, and for certain drugs also microbiology.

Research data in all these areas are summarized and submitted in a New Drug Application (NDA), which is reviewed by the scientists in the regulatory agencies for appropriateness and acceptability. In assessing acceptability of the data, regulatory agencies focus on assuring that the drug will be safe and effective, and sufficient information will be available to individualize the dose in special populations (e.g., by gender, age, and race), disease state (hepatic and renal insufficiency), and in the presence of other drugs (drug–drug interactions). These data are then summarized into a product label for a drug approved by the regulatory agencies. Therefore the product label is an excellent source of information about the drug product. Readers are encouraged to refer to the product labels and drug reviews available on the FDA Internet site (http://www.fda.gov/cder/ drug/default.htm New Prescription Drug Approvals) for recently approved racemates and enantiomers. This will provide the reader with an overview of information that was considered in the approval of such drugs.

For the development of any drug, general scientific and regulatory principles remain the same and are known to the drug development community. Presence of chiral center(s) adds a different challenge to the known principles, especially for drugs that show enantioselective pharmacokinetics (PK) and/or pharmacodynamics (PD).

1.2. Regulatory Guidance

Traditional bases for regulatory considerations are the code of federal regulations (CFR), the domestic guidances issued by drug regulatory agencies, the International Conference on Harmonization of Technical Requirements for Restoration of Pharmaceuticals for Human Use (ICH) guidance, the reviewer MAPP (manual of policies and procedures), current scientific standards, and precedents.

In order to facilitate drug development and provide transparency in the drug regulatory and approval process, regulatory agencies have issued several topic-specific guidances in the last decade. These guidance documents provide current thinking and general acceptable approaches. It should be noted that these documents are guidance and not regulations. Therefore if the applicant wishes to choose a different path from that outlined in a guidance, it is acceptable as long as the data provided by the alternative approach addresses the questions and concerns about the drug.

Several guidances issued by the regulatory agencies worldwide address many aspects of drug development. As noted in the previous chapters of this book, there have been significant advances in synthetic chemistry for chiral compounds and in the ability to assess therapeutic implications (i.e., analytical chemistry to measure isomers), to effect surrogate and biomarkers, and to perform pharmacogenomic procedures. Since technological advances permit the production of a single enantiomer on a commercial scale, the FDA and other regulatory agencies have issued their policy with respect to new drugs that consist of more than one stereoisomer. Technological advances provide choices to the sponsors as to what path of the development should be taken for a drug candidate with a chiral center. These choices include the development of a racemate versus an enantiomer or an enantiomer of an already marketed racemate.

General guidance for development of chiral drugs [2] has been issued by European, Canadian, and US regulatory agencies and will be discussed in this chapter. The main guidance documents, which contain information on the development of chiral compounds, are as follows:

1. The FDA's policy statement for the development of new stereoisomeric drugs, issued by the FDA in 1992.
2. Investigations of chiral active substances issued by a commission of the European countries in 1994.
3. Stereochemical issues in chiral drug development, issued as the Therapeutic Product Programme, Canada, in 2000.

For any drug development, one has to design the program and experiments to answer the questions that need to be addressed before a drug can be accepted for marketing [3]. While developing a drug with chiral center(s), the main question is the contribution of each isomer to the safety and efficacy when it is administered as a racemate or as an enantiomer (if inter-conversion occurs). The decision to develop a racemate or enantiomer is that of the sponsors. When developing a racemate, issues related to acceptable manufacturing controls of synthesis and impurities, appropriate pharmacological and toxicological assessments, characterization of ADME (absorption, distribution, metabolism, and elimination), and clinical evaluation should be addressed. The following paragraphs discuss each of these areas in detail.

2. CHEMISTRY

For any drug being developed, manufacturing and control procedures to assure the identity, quality, purity, and strength of the drug substance and the drug product are essential. In addition, the following considerations for chiral drug substances and drug products are necessary.

2.1. Drug Substance, Drug Product, and Stability

·For enantiomeric and racemic drug substances, stereochemically specific identity tests and/or stereospecific assay methods should be developed and available. In addition to measuring·optical rotation, the Canadian guidance recommends to include melting point, chiral chromatography, optical rotary dispersion, circular dichroism, and NMR using chiral shift reagents. The choice of the controls should be based upon the product's composition, the method of manufacture, and its stability characteristics. At the time of writing this chapter, a document outlining guidance for setting chemistry specifications for drug substances and drug products for chiral compounds is under development and will be available at the US FDA website when finalized.

The stability protocol for enantiomeric drug substances and drug products should include methods capable of assessing the stereochemical integrity of the drug substance and drug product. If it is demonstrated that stereochemical conversion does not occur, stereospecific assays might not be needed. Complete description including the critical factors of the manufacturing process used to obtain the single enantiomer or racemate should be provided. In regulatory submissions to the authorities in Europe and Canada, the identity and stereoisomeric purity of the starting material, key intermediates, and chiral reagents should be established.

2.2. Labeling

The labeling should include a unique established name and a chemical name with the appropriate stereochemical descriptors.

3. PHARMACOLOGY/TOXICOLOGY

3.1. Pharmacology

For racemates, the pharmacological activity of the individual enantiomers should be characterized for the principal and any other important effects, with respect to potency, specificity, and maximum effect. To monitor in vivo interconversion and disposition, the pharmacokinetic profile of each isomer should be characterized in animals and later compared to the pharmacokinetic profile obtained in Phase 1 clinical studies.

3.2. Toxicology

It is generally sufficient to carry out toxicity studies on the racemate. If toxicity other than that predicted from the pharmacological properties of the drug occurs at relatively low multiples of the exposure planned for clinical trials, the toxicity study should be repeated with the individual isomers to ascertain whether only one enantiomer was responsible for the toxicity. If toxicity of significant concern resides in a single isomer, the development of single isomer with the desired pharmacological effect would be desirable.

Where questions exist regarding the definition of "significant toxicity," one should discuss the issue with the appropriate clinical division within the regulatory agency where the drug will be reviewed.

3.3. Impurity Limits

The concentration of each isomer should be determined and limits defined for all isomeric components, impurities, and contaminants for the compound tested preclinically that is intended for use in clinical trials. The maximum allowable level of impurity in a stereoisomeric product employed in clinical trials should not exceed that present in the material evaluated in nonclinical toxicity studies.

4. DEVELOPING RACEMATES OR SINGLE ENANTIOMERS

The US, European, and Canadian guidance documents state that it is the sponsor's decision to develop a racemate or an enantiomer for a chiral compound. However, regulatory agencies evaluate the rationale provided by the sponsor in reaching that decision. Therefore it is advantageous to discuss the drug development plan with the regulatory agency. It is essential to develop quantitative analytical assays for quantitation of individual enantiomers in samples obtained from in vivo studies early in drug development.

Let us discuss different scenarios that may arise from a racemate consisting of two enantiomers. In this case, there could be four different scenarios with respect to the PK and PD of the two enantiomers: same PK and PD (efficacy and toxicity) profiles (I: + +); different PK but same PD profile (II: − +); same PK but different PD profile (III: + −); or different PK and PD profiles (IV: − −).

Racemate vs. Enantiomer

	PD(+)	PD(−)
PK(+)	+ + (I)	+ − (III)
PK(−)	− + (II)	− − (IV)

For case I and II, one may consider developing a racemate or enantiomer. However, for case II, if PK is significantly enantioselective, one may consider developing the enantiomer that is more bioavailable, so that a lower dose may be needed. For case III, if safety is not an issue, that is, efficacy is greater in one compared to the other enantiomer, but toxicity is not a problem, one may consider developing either a racemate or enantiomer. For case IV, developing the enantiomer with most beneficial PD is more appropriate, especially if safety is the concern.

If little difference in activity and disposition of the enantiomers exists, racemates may be developed. Development of a single enantiomer is particularly desirable when one enantiomer has a toxic or undesirable pharmacological effect that the other enantiomer lacks.

However, to make the above decisions, one needs to know stereoselectivity in ADME, including protein binding, and in PD (efficacy and safety). Based on the general PK/PD knowledge of the drug, one would study special populations and drug interactions accordingly. Generally, during development of chiral drugs, studies would be conducted in the same way as for an achiral drug. The main difference, however, is the decision about measuring the racemate or each of the enantiomers.

Use of chiral versus achiral analytical assays has been the subject of much discussion in the literature. If the pharmacokinetic profile is the same for both isomers, or a fixed ratio between the plasma levels of enantiomers is demonstrated in the target population, an achiral assay or an assay that monitors one of the stereoisomers may be sufficient.

As stated in the 1992 policy statement, the FDA invites discussion with sponsors concerning whether to pursue the development of the racemate or the individual enantiomer. All information developed by the sponsor or available from the literature that is relevant to the chemistry, pharmacology, toxicology, or clinical actions of the stereoisomers should be included in the IND and NDA submissions.

5. CLINICAL AND BIOPHARMACEUTICAL CONSIDERATIONS

It has been recommended that, if observed differences in activity and disposition of the enantiomers are minimal, racemates may be developed. However, the development of a single enantiomer is particularly desirable when only one of the enantiomers has a toxic or undesirable pharmacological effect. Further investigation of the properties of the individual enantiomers and their active metabolites is warranted if unexpected toxicity or pharmacological effect occurs with clinical doses of the racemate. Signals of adverse events may be explored in animals, but human testing may also be essential. It should be recognized that toxicity or unusual pharmacological properties might reside not in the parent isomer, but in an isomer-specific metabolite. In summary, both enantiomers should be evaluated clinically, and the development of only one enantiomer should be considered when both enantiomers are pharmacologically active but differ significantly in potency, specificity, or maximum effect. However, the development of a single enantiomer may not be warranted if one of the isomers is essentially inert. When both enantiomers are found to carry desirable but different properties, the development of a mixture of the two, not necessarily the racemate, as a fixed combination might be reasonable.

If a racemate is being studied, the pharmacokinetics of the two isomers and their potential interconversion should be studied in Phase 1. Based on Phase 1 or 2 pharmacokinetic data in the target population, it should be possible to determine whether an achiral assay or monitoring of just one enantiomer, where a fixed ratio is confirmed, will be sufficient for pharmacokinetic evaluation.

5.1. Developing an Isomer

If a single isomer is being developed and no interconversion occurs, enantiospecific assays need not be used. However, if the antipode is formed in vivo, that should be considered as a metabolite and addressed accordingly. If a racemate has been studied, and a single enantiomer is being developed, an abbreviated pharmacology/toxicology evaluation could be conducted, which would allow for the use of the existing knowledge of the racemate. This would usually include the longest repeat-dose toxicity (up to 3 months), and the reproductive toxicity segment II study in the most sensitive species, using the single enantiomer being developed. These studies should include a positive control group, which consists of treatment with the racemate. If no differences in toxicity profiles are found between the enantiomer and the racemate treatment groups, no further studies would be needed. If single enantiomer is more toxic, the explanation should be provided with supporting data, and implications for human studies should be considered.

In humans, evaluation should include the determination of interconversion and the comparison of the PK profile of the individual isomer when administered alone versus when administered as a racemate. Some examples of recent single isomer approvals for previously marketed racemic drugs include escitalopram (S-citalopram), esomeprazole (S-omeprazole), and focalin (d-isomer of methylphenidate). Readers are encouraged to refer to their product labels and the FDA reviews available on the FDA website (http://www.fda.gov/cder/drug/default.htm New Prescription Drug Approvals) to gain an understanding of what data were submitted for the approval of these drugs. In case of rapid in vivo interconversion, regulatory guidance issued by the Therapeutic Products Program indicate that the antipode should be considered as a metabolite during the drug development process.

5.2. Biopharmaceutics

Bioavailability and bioequivalence have been discussed in detail in Chap. 9. There has been considerable discussion in the literature on the need to use enantiospecific assays to assess bioequivalence [5–16]. A guidance published by the FDA, Bioavailability and Bioequivalence Studies for Orally Administered Drug Products—General Considerations, addresses this issue and is summarized below (see also Fig. 1 for a decision tree).

For bioavailability studies, measurement of individual enantiomers may be important. For bioequivalence studies, this guidance recommends measuring of the racemate using an achiral assay. Measurement of the individual enantiomers in bioequivalence studies is recommended only when all of the following conditions are met (Fig. 1): (1) the enantiomers exhibit different pharmacodynamic characteristics; (2) the enantiomers exhibit different pharmacokinetic characteristics; (3) primary efficacy/safety activity resides with the minor enantiomer; and (4) nonlinear availability is present (as expressed by a change in the enantiomer concentration ratio with a change in the input rate of the drug) for at least one of the enantiomers.

A guidance issued by the Therapeutic Products Programme states that in general, when comparing solid dosage forms of similar type

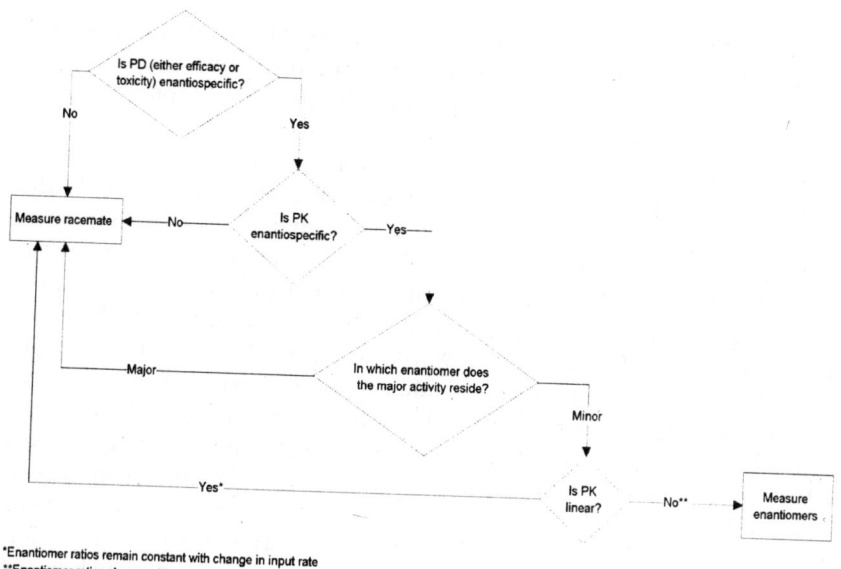

Figure 1 When to use an enantiospecific assay for a racemate.

(e.g., two immediate release formulations), total drug concentrations can be measured. When conducting a comparative bioavailability study of solid dosage forms of different types, such as a modified release dosage form and an immediate release dosage form (for definitions please refer to the guidance in Ref. 2), enantioselective assays should be used. This guidance also calls for meeting the bioequivalence criteria for each enantiomer for drugs whose rate of release and/or absorption affect the in vivo enantiomeric ratio of the drug (e.g., the nonlinear first-pass effect).

5.3. Clinical Pharmacology

The role of clinical pharmacology during drug development is to optimize the dose and dosing regimen with the goal in mind of reducing the undesirable effects and producing beneficial responses in a predictable manner. To achieve individualization of dose, maximum knowledge of the factors contributing to the PK and PD variability has to be acquired [4]. As stated by Temple [4], "Clinical pharmacologists are trained to take a broad view of drugs, recognizing that they not only have the pharmacologic property of primary interest but often other properties as well, that a 'drug' is really many drugs (isomers, active metabolites) with different properties,

and that the properties of drugs should affect how they are dosed and used."
Planning the right type of studies to answer the questions that prescribing
physicians and patients pose will optimize the drug development process.
Proper planning and asking the right questions early in the drug
development process will save costs by minimizing the time and eliminating
the studies that do not contribute to the overall understanding of the drug
being investigated. The Office of Clinical Pharmacology and Biopharmaceutics at the FDA has described its mission as "to assure that an individual
patient receives the right drug, in the right dose at the right time and in the
right dosage form." To achieve this goal for every drug has its unique
challenges, but one has to work toward achieving this goal if patient and
physician are to have utmost confidence in the drug product. Recent
significant advances in science related to pharmacogenomics, the discovery
of the role of transporters, and the emphasis by the regulatory agencies on
product quality issues provide an important role for clinical pharmacologist
in the entire drug development process. It is now accepted that
enantiospecific ADME is more likely than not, and disease states could
have enantiospecific effects on the pharmacokinetics of drugs, which in turn
could have significant impact on pharmacodynamic response. Therefore,
when developing a chiral compound, one should assess the need for
enantiospecific determinations in special populations (e.g., age, gender, and
race) and drug interactions. This decision will be based on knowledge of
enantiospecific assessments during early Phase 1 studies and the animal and
in-vitro data available.

6. EXCLUSIVITY

Combination drug policy described at 21 CFR 300.50 has not been applied
to chiral products. Thus it is not required that applicants demonstrate the
contribution of each enantiomer to the effectiveness of the racemic drugs.
This approach was not adopted, because the combining of the two
enantiomers is not deliberate, the activities of the enantiomers may be
similar, and in the past the separation of the enantiomers was difficult.
In contrast, the mixtures of diastereoisomers are considered combination
drugs subject to the combination drug policy, because they are readily
separated, and their activities are often very different. At present, the
marketing exclusivity period for developing a single isomer of previously
approved racemate is three years.

   The FDA requested comments (62 FR 2167, January 15, 1997) on the
appropriate period of marketing exclusivity for drug products whose active
ingredient is a single enantiomer of a racemate that is an active ingredient

of a previously approved drug product. The following questions were posed in this notice, and 26 firms or individuals commented on them. These responses are summarized here.

1. What period of marketing exclusivity would best effectuate the 1984 amendments' dual policy goals of increasing drug price competition and providing incentives for the development of innovative drug products?

Twenty-three responses addressed this question. In general, innovator firms supported increased market protection for enantiomers, whereas generic pharmaceutical firms submitting comments supported the shorter period of exclusivity, which would permit earlier generic competition.

2. Would granting a five-year period of exclusivity to enantiomers of previously approved racemates encourage medically significant pharmaceutical innovation?

Nineteen comments addressed this question. Granting five-year marketing exclusivity to enantiomers of previously approved racemates will not necessarily encourage medically significant pharmaceutical innovation. The opposite may be true if such a policy encourages innovator firms to pursue racemate-to-enantiomer switches, without any demonstrated gain in efficacy or safety, at the expense of other development efforts. Benefits, to date, of the single enantiomers developed from racemates have been elusive.

3. If the pharmacological action of each enantiomer is described in the approved NDA for the racemate, should a subsequently submitted application for an enantiomer of the racemate receive different treatment for exclusivity purposes than if the pharmacological action of each enantiomer is not described in the approved NDA for the racemate drug product?

Twelve of the 19 comments answered this question in the negative.

4. If the agency were to assess requests for exclusivity for enantiomers of previously approved racemates on a case-by-case basis, what criteria should the agency apply?

Eleven of the 26 responses directly addressed this question and varied criteria were suggested.

5. Compared with other drug products, what are the costs of and technical barriers to obtaining safety and efficacy data for a drug product whose active ingredient is a single enantiomer of a previously approved racemate?

Fifteen responses directly addressed this question. Only two of these stated that they believed that the costs and technical barriers were substantially greater for single enantiomers. The increased costs of and technical barriers to obtaining safety and efficacy data for an enantiomer compared to a racemate are minimal.

      6. How many drug products (whether approved, or subject of pending NDAs, or in development) are likely to be affected by this policy?

Estimates of the number of drug products likely to be affected by a reversal of the FDA position ranged from two to three per year to "over 200" with no period stated. Maybe the higher figures include racemic drugs for which there is a potential for development as single enantiomers rather than single enantiomers actually under development.

7. CONCLUSIONS

Regulatory principles for chiral drugs are the same as for any other drug, that is, to provide safe and effective drugs based on sound scientific principles. Appropriate information to adjust the dose/dosage regimen should be available. For chiral drugs, additional considerations are

      1. The availability of a validated, sensitive, and specific analytical method to determine concentrations of the racemate and each isomer in biological fluids early on during drug development
      2. Decisions based on a scientific rationale for the development of a racemate versus a single enantiomer
      3. Appropriate characterization of drug substance and product
      4. Characterization of pharmacology and toxicology of each enantiomer with respect to potency, specificity, and maximum effect
      5. Basic understanding of stereoselectivity in ADME and PD (efficacy and safety parameters) in animal and human studies
      6. Attention to stereoselectivity in drug interactions, covariates like gender, age, ethnicity, and disease states

In general, for PK assessment of chiral compounds, the main difference from other drugs is the decision whether to use an enantioselective or an achiral assay to characterize the pharmacokinetics of each isomer or racemate, respectively.

    At present, the marketing exclusivity period in the US, for developing a single isomer from a previously approved racemate, is 3 years.

Some examples of chiral drugs recently approved or submitted are citalopram, which is a racemate, and esomeprazole (S-isomer of omeprazole), escitalopram (the S-isomer of citalopram), and focalin (the dextrorotary isomer of methylphenidate), which are the single isomers of already approved racemates. Product label and reviews on the FDA website for these drugs provide excellent references and examples for readers interested in learning more about the data required for development of chiral drugs.

ACKNOWLEDGMENT

The author would like to acknowledge the contribution of the FDA working group members on enantiomer exclusivity, for the information on exclusivity provided in this chapter.

REFERENCES

1. Temple, R. Policy developments in regulatory approval. Stat. Med. **2002**, *21*, 2939–2948.
2. Web site addresses for regulatory agencies:

http://www.hc-sc.gc.ca/hpb/	Canada
http://www.ifpma.org/ich1.html	ICH
http://www.mhw.go.jp/english/index.html	Japan
http://www.health.gov.au/tga/	Australia
http://eudraportal.eudra.org/	EMEA

3. Lesko, L.J.; Williams, R.L. The question based review, a conceptual framework for good review practices. Applied Clinical Trials **1999**, *8*, 52–62.
4. Temple, R.J. The clinical pharmacologist in drug regulation: the US perspective. Br. J. Clin. Pharmacol. **1996**, *42*, 73–79.
5. Srinivas, N.R.; Barbhaiya, R.H.; Midha, K.K. Enantiomeric drug development: issues, considerations, and regulatory requirements. J. Pharm. Sci. **2001**, *90*, 1205–1215.
6. Midha, K.; Gordon McKay.; Rawson, M.; Hubbard, J. The impact of stereoisomerism in bioequivalence studies. J. Pharm. Sci. **1998**, *87*, 797–802.
7. Karim, A. Enantioselective assays in comparative bioavailability studies of racemic drug formulations: nice to know or need to know? J. Clin. Pharmacol. **1996**, *36*, 490–499.
8. Mehvar, R.; Jamali, F. Bioequivalence of chiral drugs. Clin. Pharmacokinetinet. **1997**, *33*, 122–141.
9. Jamali, F. Stereochemistry and bioequivalence. J. Clin. Pharmacol. **1992**, *32*, 930–934.

10. Marzo, A. How incoming guidelines on chiral drugs could impact on the international scenery of drug development: IUPAC. Boll. Chim. Farm. **1993**, *132*, 267–271.
11. Nation, R.L.; Sansom, L.N. Bioequivalence requirements for generic products. Pharmacol. Ther. **1994**, *62*, 41–55.
12. Nerurkar, S.G; Dighe, S.V.; Williams, R.L. Bioequivalence of racemic drugs. J. Clin. Pharmacol. **1992**, *32*, 935–943.
13. Sahajwalla, C.G.; Longstreth, J.; Karim, A.; Purich, E.D.; Cabana, B.E. Consequences in pooling R-+S-erapamil in bioequivalence assessment. J. Clin. Pharmacol. **1992**, *32*, 961.
14. Birkett, D.J. Racemate or enantiomers: regulatory approaches. Clin. Exp. Pharmacol. Physiol. **1989**, *16*, 479–483.
15. Nitchuk, W.M. Regulatory requirements for generic chiral drugs. J. Clin. Pharmacol. **1992**, *32*, 953–954.
16. Wechter, W.J. From controversy to resolution: bioequivalency of racemic drugs: a symposium on the dynamics, kinetics, bioequivalency, and analytical aspects of stereochemistry. J. Clin. Pharmacol. **1992**, *32*, 915–916.

Index